# Atlas
# of Oregon
# Wildlife

*This volume was made possible by the support and sponsorship of the following organizations:*

 **National Gap Analysis Program**
Biological Resources Division
U.S. Geological Survey

 **Oregon Department of Fish and Wildlife**

 The Nature Conservancy

**U.S. Environmental Protection Agency***

Oregon Natural Heritage Advisory Council

U.S. Forest Service*

Oregon Chapter, The Wildlife Society

**Biodiversity Research Consortium**

Strategic Environmental Research and Development Program **SERDP**
Strategic Environmental Research and Development Program
Improving Mission Readiness Through Environmental Research

*This document has not been subject to agency review.

# Atlas of Oregon Wildlife

### DISTRIBUTION, HABITAT, AND NATURAL HISTORY

*Blair Csuti*
*A. Jon Kimerling*
*Thomas A. O'Neil*
*Margaret M. Shaughnessy*
*Eleanor P. Gaines*
*Manuela M. P. Huso*

Oregon State University Press
Corvallis, Oregon

The paper used in this publication meets the guidelines for permanence and durability of the Committee for Production Guidelines for Book Longevity of the Council on Library Resources and the minimum requirements of the American National Standard for Permanence of Paper for Printed Library Materials, ANSI Z39.48-1984

**Library of Congress Cataloging in Publication Data**

Atlas of Oregon Wildlife : distribution, habitat, and natural history / Blair Csuti ... [et al.]

  p.   cm.

Includes bibliographical references (p.   ) and index.

ISBN 0-87071-395-7 (alk. paper)

1. Zoology—Oregon. I. Csuti, Blair A.

QL202.A85 1997

591.9795—dc21                                                    97-8467

                                                                    CIP

 **Oregon State University Press**
**101 Waldo Hall**
**Corvallis, OR 97331-6407**
**Phone 541-737-3166 • Fax 541-737-3170**

# Contents

# Acknowledgments

The *Atlas of Oregon Wildlife* summarizes information dating from the early days of European exploration of the Pacific Northwest to the present. The first records of many species from Oregon were reported in the journals of the Lewis and Clark Expedition (1804-1806). Most specimens collected by Lewis and Clark were deposited with the Peale Museum, Independence Hall, Philadelphia, and were incorporated into the collections of the Academy of Natural Sciences of Philadelphia, established in 1812. We are indebted to those individuals who, over the years, have collected, recorded, and catalogued Oregon wildlife specimens for natural history museums and private collections in the United States and Europe, as well as those who have compiled and evaluated over 100,000 museum records used in the development of this book. The 35 museums that have shared their information with us are listed in Appendix IV.

We appreciate Teri Waldron's dedication and tenacity while compiling these museum data for the Oregon Species Information System over the past six years. Others who helped collect and compile information for this effort include: Charlie Bruce, Dan Edwards, Richard Green, Loree Havel, Anita McMillan, Wanda McKenzie, Kim Mellon, Harry Nehls, Melissa Platt, Mark Stern, Harold Sturgis, and Faye Weekly. Range Bayer, Merlin Eltzroth, Craig Groves, Steve Summers, and Laura Todd contributed data sets that were invaluable to this effort. Jeff Waldon, Fish and Wildlife Information Exchange, provided valuable assistance with computer programming and information management.

The Oregon Actual Vegetation Map, prepared by Jimmy Kagan and Steve Caicco, was used in the development of the species distribution maps. Greg White, C & G White Cartography; Mike Wing, E&S Environmental Chemistry; and Chris Kiilsgaard, Cascade Geodata, assisted the Oregon Department of Fish and Wildlife in developing distribution maps. Charley Barrett, Becky Frasier, Trevor Jones, Kevin Sahr, and Denis White, Oregon State University, also assisted in map compilation and production. Additional draft maps were prepared by Troy Merrill, Eva Strand, and Nancy Wright, University of Idaho. Roger Cole, Rob Solomon, Mark Stern, and countless volunteers helped the Oregon Natural Heritage Program prepare its data base on species distribution. Reviewers of the range maps include Lois F. Alexander, Brad Bales, Joe Beatty, Andy Blaustein, Charlie Bruce, Heidi Brunkel, Bruce Bury, Chris Carey, Alan Contreras, Larry Cooper, Doug Cottam, John Crawford, Dan Edwards, Leonard Erickson, Terry Farrell, Richard Forbes, Jeff Gilligan, Rebecca Goggans, Marc Hayes, Mark Henjum, Richard F. Hoyer, Gary Ivey, Bob Jarvis, David Johnson, George Keister, A. Ross Kiester, Jim Lemos, John Loegering, Harry Nehls, S. Kim Nelson, Eric Pelren, Joe Pesek, Michael Pope, Claire Puchy, Mark Stern, Alan St. John, Robert Storm, Cynthia Tait, Neil Teneyck, Jim Thrailkill, Chuck Trainer, Madeleine Vander Heyden, Dan VanDyke, Walt VanDyke, and Simon Wray. Alan Contreras and Richard Forbes reviewed the text for bird and mammal species accounts, respectively. Jennifer Deaton and Margaret Donsbach proofread the manuscript. We thank all the above for generously agreeing to review parts of this atlas; the final maps and text remain the responsibility of the authors. Bud Adams, Larry Bright, and Bill Haight provided encouragement and support for development of the Oregon Species Information System. Jim Greer, A. Ross Kiester, Donavin Leckenby, Russell Peterson, J. Michael Scott, Steve Williams, and Tom Williams provided financial, logistical, and moral support for this project.

Agencies contributing support were: Oregon Department of Fish and Wildlife (ODFW), the Wildlife Diversity Program); U.S. Fish and Wildlife Service, Federal Aid Program, Wildlife Restoration Project (under Grant W-87-R to ODFW); National Biological Service (now Biological Resources Division, U.S. Geological Survey); Oregon State Office, U.S. Fish and Wildlife Service; and Oregon Natural Heritage Advisory Council. The U.S. Environmental Protection Agency (EPA) provided support to The Nature Conservancy (under Cooperative Assistance Agreement C R 820694-01-0). We are grateful to the Oregon Chapter of The Wildlife Society for funding the printing costs of the map of Oregon Wildlife Habitat Types, included with this atlas as a fold-out map.

Data compilation and map production for this atlas were supported by the Biodiversity Research Consortium (BRC), a group of United States government agencies, academic, and nongovernmental institutions performing coordinated research on biodiversity assessment and management methods. The BRC acknowledges the support of Cooperative Research Agreement PNW 92-0283 between the USDA Forest Service and Oregon State University, Inter-agency Agreement DW12935631 between the EPA and the USDA Forest Service, and the USDA Forest Service, Pacific Northwest Research Station.

We are grateful to Barry Garrison, California Wildlife Habitat Relationships Program, California Department of Fish and Game, and the California Interagency Wildlife Task Group for permission to reproduce line drawings from the three-volume publication, *California's Wildlife* (Zeiner et al. 1988, 1990a, 1990b), and for providing original drawings prepared by Lisa Hall. In addition, we thank Dr. Richard DeGraaf for permission to reproduce line drawings for some breeding bird species from *New England Wildlife* (DeGraaf and Rudis 1983). In addition to these two sources, line drawings on pages 325, 329, 356, 364, 369, 371, 376, 381, 399, 405, 410, 418, 437, and 441 are reprinted from *Mammals of the Pacific States: California, Oregon, and Washington*, by Lloyd G. Ingles (1965), with permission of the publishers, Stanford University Press. Line drawings on pages 68, 69, and 129 are reprinted from *Birds of the Pacific Northwest*, by E.J. Larrison (1971), with permission of the publishers, University of Idaho Press. Line drawings on pages 12, 18, and 32 are reprinted from *Amphibians and Reptiles of Western North America*, by Robert C. Stebbins (1954), McGraw-Hill Inc., with permission of the author. The line drawing of the Trumpeter Swan, page 83, is reprinted from *Waterfowl of North America*, by Paul A. Johnsgard (1975), Indiana University Press, with permission of the author. The line drawing of the Solitary Sandpiper, page 141, is reprinted from *A Popular Handbook of the Ornithology of Eastern North America, Volume 1: the Land Birds*, by T. Nuttall, revised and annotated by M. Chamberlain (1891), Little Brown and Company.

We would like to thank Jo Alexander, Managing Editor, OSU Press, without whose perspective and editing skills this atlas would not have been possible.

# Introduction

Oregon has a rich diversity of wildlife: 426 species of native terrestrial vertebrates breed in Oregon, and 91 additional species visit the state during winter or migration (Puchy and Marshall 1993). Another 98 species show up occasionally, and 27 species are introduced, bringing Oregon's total wildlife diversity to 642 species, or more than 42% of all terrestrial vertebrates found in the United States and Canada.

Despite this wealth of wildlife, the only comprehensive accounts of the distribution, habitat, and natural history for most of Oregon's species are over half a century old. In 1936, Vernon Bailey published *The Mammals and Life Zones of Oregon,* and, in 1940, Ira Gabrielson and Stanley Jewett wrote *Birds of Oregon.* While some more recent reference works deal with particular groups of animals (Gilligan et al. 1994, Leonard et al. 1993, Maser et al. 1981, Nussbaum et al. 1983, Storm and Leonard 1995), we lack a comprehensive guide to the current distribution, habitat, and natural history of Oregon's wildlife.

The present work grew out of efforts by the National Biological Service, the Oregon Department of Fish and Wildlife (ODFW), the Oregon Natural Heritage Program (ONHP), and other cooperating agencies and organizations to determine the distribution of two components of Oregon's biological diversity: vegetation cover and its associated wildlife species. This program, known as "Gap Analysis," is intended to provide wildlife managers with the information they need to anticipate and prevent loss of biodiversity. Details of the methods of Gap Analysis are presented in Scott et al. (1993) and Scott and Csuti (1997); further information, including availability of state map products, is available from the Gap Analysis World Wide Web Home Page: *http://www.gap.uidaho.edu/ gap.* While the information gathered for Gap Analysis is maintained as digital data in a geographic information system, this atlas provides a larger audience with access to current knowledge about Oregon's wildlife. Only limited detail can be shown on state maps small enough to fit on an atlas page, but these should provide the user with a guide to where species can be expected to occur in appropriate habitats within Oregon's varied landscapes.

Regardless of their scale, all range maps are predictions about where species may be found. These predictions are more likely to be correct when applied to larger areas (a whole mountain range or major watershed) over several breeding seasons. Many species have special habitat requirements that are irregularly distributed over the landscape. For example, even though frogs lay their eggs in streams or ponds, it is impossible to map the location of every stream or pond in the state. Because of this, the most important thing to remember when using the maps in this atlas is that a species will be found in an area only if its special habitat requirements are satisfied. In addition, the distribution of some species, especially birds, may vary from year to year. The maps presented here serve as a guide to the habitats and general distribution of each species. They can direct you to areas where field studies can determine if a species has found the right combination of habitat elements that enable it to establish and maintain a population.

# Methods

Animal distribution maps are developed by plotting the locations at which a species has been observed. Observations can be recent or old, well documented or suspect. Prior to the last few decades, many field naturalists collected specimens of the species they encountered and deposited them in natural history museums. These specimens form the basis of much of our knowledge about animal distributions, although recent habitat changes may invalidate older records. Specimens are infrequently collected these days except to answer specific research questions. Instead, reliable observations of many species are recorded through a variety of volunteer programs, such as the U.S. Fish and Wildlife Service's Breeding Bird Survey or the Audubon Society's Christmas Bird Counts.

A very simple type of distribution map places dots on a map where a species has been seen. If the peripheral dots are connected, the result is a continuous range map. Of course, not every area within its range is equally likely to be inhabited by a species. In addition, biological surveys in Oregon, as in other regions, have not sampled the state evenly, so we have better collection data for certain parts of the state. In the past, distribution maps depicted a species present everywhere within the limits of its range, often including large areas of unsuitable habitat (Csuti 1971, Hall 1981). The maps in this atlas represent an attempt to address this problem by linking species distributions with a map of wildlife habitats, thereby excluding areas of inappropriate habitat from a species' predicted distribution.

Our distribution maps synthesize data from information systems developed by ONHP, in cooperation with the Biodiversity Research Consortium, and ODFW. To prepare their maps, ONHP placed 441 equal area hexagons over a map of the state of Oregon (see Appendix III for examples). Each hexagon covers 635 square kilometers, or about 160,000 acres. The hexagons were developed for the Environmental Monitoring and Assessment Program (EMAP) of the U.S. Environmental Protection Agency (White et al. 1992). Here they serve as a spatial accounting unit for individual locality records. The sources (specimen record, expert opinion, literature citation, state reference book) used to predict the presence of a species in a hexagon are recorded in a data base by ONHP. Most species have been recorded in relatively few hexagons in Oregon. Museum specimen records contributed just over 3,000 primary sources used to attribute species to individual hexagons. ONHP asked experts on each class of vertebrate to provide their opinions about the probable presence of species in hexagons for which ONHP had no locality record. The resulting hexagon distribution maps represent an intermediate step in the development of final distribution maps.

The Oregon Department of Fish and Wildlife queried 35 natural history museums across the country for specimen records from the state of Oregon. Over 100,000 museum specimen records document the distribution of Oregon's wildlife species in ODFW's Oregon Species Information System (OSIS). These specimen records constitute the best data source for confirming the presence (or former presence) of a species in a particular part of the state. Specimen records documenting the presence of a species in a county are maintained as digital data in OSIS. The Oregon Department of Fish and Wildlife invites feedback from users concerning the distribution, habitats, or natural history of Oregon's wildlife species. Comments should be sent to: Wildlife Atlas, Oregon Department of Fish and Wildlife, Wildlife Division, P.O. Box 25, Portland, OR 97209; email *wildlife.atlas@state.or.us*.

ODFW drew on the resources of OSIS to prepare distribution maps, keyed to the presence of a species in any of Oregon's 36 counties or 10 physiographic provinces. The physiographic provinces of Oregon were first delineated by Franklin and Dyrness (1973), and have recently been revised by ONHP (1995). While the spatial accounting units created by the intersection of counties and physiographic provinces are usually much larger than EMAP hexagons, knowledge of species distribution by county and physiographic province allows ODFW to predict species presence in areas of the state that have not been systematically surveyed for wildlife. Counties are also the traditional geographic subdivision noted on museum specimen labels. ODFW independently solicited expert review of its distribution maps, providing an additional level of quality control.

Another component needed to generate distribution maps is a map of wildlife habitats. This map began as a map of vegetation cover types produced for the Oregon Gap Analysis Program. It is based on photo-interpretation of 1:250,000 LANDSAT Multispectral Scanner imagery. Most LANDSAT scenes were taken in summer 1988, although a few scenes from 1986 were used for a strip along the Oregon/Idaho border. Information from U.S. Forest Service and Bureau of Land Management land cover maps were used to assist in labeling the state-wide vegetation map. A total of 133 vegetation cover types were identified. These vegetation types have been grouped into 30 habitat types according to their faunal similarity (O'Neil et al. 1995). A 1:750,000 scale map of the resulting wildlife habitats has been produced through the sponsorship of the Oregon Chapter of The Wildlife Society. Copies may be obtained from the Oregon Chapter of The Wildlife Society, P.O. Box 2214, Corvallis, OR 97339-2214 (1997 price $7.00, including shipping and handling). [Note: Due to a computer error, an area of juniper-sagebrush woodland along the Crooked River south of Lake Billy Chinook, Jefferson County, was color-coded "Open Water" on the Oregon Wildlife Habitat Types map.]

The final component needed to produce distribution maps is a table associating each species with the habitat types within which it is likely to occur. OSIS consolidates information about the habitat preferences of species drawn from existing wildlife-habitat relationships documents (e.g., Thomas 1979) and ecological literature. The wildlife habitats in OSIS, described in the next section, are clusters of vegetation cover types inhabited by similar groups of species (O'Neil et al. 1995). Each species is then predicted to occur in stands of appropriate habitat. By combining the habitat information contained in OSIS with geographic distribution by hexagon, county, or physiographic province, we restrict the predicted distribution of a species to stands of appropriate habitat types within its distributional limits. This intersection of tabular and spatial habitat data insures, for example, that forest-dwelling species are not mapped in neighboring unforested landscapes. The major limitation remains our inability to map important but small habitat features. These features are described in the text, and should be used in conjunction with the maps to assess the probability of encountering a species in a particular area. In addition to using habitat associations to predict species' distributions, ODFW also incorporated elevational constraints to help refine the distribution of many amphibians and reptiles and some species of mammals. Details about this computerized approach to distribution mapping can be found in Butterfield et al. (1994) and Csuti (1996).

The computer-generated maps in this atlas therefore represent an intersection of three data sets. First, the limits of distribution of a species are described in terms of predicted presence or absence in a geographic unit. Next, a habitat map, based on vegetation cover types and the species that occupy them, is created. Then each species is assigned to a set of habitat types. The final map represents all habitat types associated with a particular species within its distributional limits in Oregon.

Cartography is the art and science of making maps. On maps, the real world is represented at reduced scale by lines, symbols, shadings, and other graphic conventions. Maps are only a representation of reality and cannot depict all the detail one encounters in nature. The maps in this atlas (scale 1:4,300,000) make unavoidable generalizations about the distributions of species. For example, the state hydrologic network is used to represent the distribution of 16 species closely associated with water (e.g., Belted Kingfisher), but only larger rivers and streams are shown. This representation of rivers and streams does not mean that other suitable streams within the state do not support populations of kingfishers, only that cartographic constraints prevent us from displaying every stream on our maps. Other species (e.g., Mallard) are widespread throughout Oregon, but are found only in localized microhabitats (e.g., farm ponds, lakes). We have presented the distribution of 22 such species using a stippled pattern, to emphasize the discontinuous nature of their distribution.

Certain species are rare and occur only in a few localities within the state, even though apparently suitable habitat occurs elsewhere. We would over-estimate the range of such species if we predicted them to be present in all habitat that was similar to that where they are known to live. For this group of species, we consulted other sources, including state and federal agency surveys, to restrict their distribution to areas in which they are known to occur.

Winter distribution of birds is less predictable than their breeding distribution. Birds tend to move from one area to another during winter, often in response to changes in the weather. Some species that ordinarily migrate south may remain in Oregon during mild winters. Although Christmas Bird Counts, sponsored by the National Audubon Society, provide the best information on winter distribution of birds, they take place over a span of several weeks in late fall and early winter. This variation may introduce some inconsistency to our understanding of winter bird distribution. National maps based on these records have been compiled by Root (1988). Because of the uncertain nature of winter distributions, we have not attempted to map winter ranges of birds using habitat associations. Maps of winter distribution of birds are presented in Appendix III in a simplified, small-scale format. These maps should provide users with sufficient information to determine which birds might be present in a part of the state during the course of a winter. This level of detail is consistent with the dynamic nature of winter bird distributions.

There are 91 additional bird species that neither winter nor breed in Oregon but stop over in the state during their spring and fall migrations to breeding grounds in Canada and Alaska or their winter homes to the south. Some of these birds travel to and from South America, enjoying the austral summer while their breeding habitat is frozen over. Many, but not all, of these species are shorebirds and waterfowl, and they can be seen on nearly any piece of open water sometime during migration. Because they are "just traveling through," and because their appearance is to some extent unpredictable, we do not attempt to map the distribution of these species. Gilligan et al. (1994) give the reader more specific details about migratory and winter distribution of birds in Oregon. More detailed information about the distribution and abundance of birds in winter is presented in *Northwest Birds in Winter* by Alan Contreras, Oregon State University Press, fall 1997.

The Oregon Field Ornithologists is sponsoring a five-year volunteer project to gather observations of breeding birds throughout Oregon. Publication of these results will add significantly to our knowledge of the breeding distribution of Oregon's birds. Volunteers interested in being a part of this effort through the year 2000 should contact the Oregon Breeding Bird Atlas Project, P.O. Box 2189, Corvallis, OR 97339. In addition, the Defenders of Wildlife administers a statewide, public wildlife data collection project called NatureMapping. Information on the distribution of amphibians, reptiles, birds, and mammals is solicited from the public and recorded for use by concerned citizens, natural resource agencies, policy makers, and land use planners. Data may be submitted in hard-copy format or on customized software. A web site with data entry capabilities will be available in mid-1997. Cooperative partners include Oregon Department of Fish and Wildlife, Oregon Breeding Bird Atlas Project, and Wolftree, Inc. Volunteers interested in participating should contact the Defenders of Wildlife, 1637 Laurel St., Lake Oswego, OR 97034; (503) 697-3222; *http://www.naturemapping.org.*

We have described the global range and summarized the taxonomy and natural history of each native and commonly encountered introduced terrestrial vertebrate that breeds in Oregon from OSIS, The Nature Conservancy's Vertebrate Characterization Abstract data base, standard reference works, regional guides and, in some instances, original literature. For each species, selected references providing additional ecological or taxonomic details are listed. Some sources used to compile species notes expressed measurements in English units. We have elected to retain the original measurements rather than introduce rounding errors by converting to metric units. The length of an average adult (including tail, when present) is given in inches and centimeters for each species. However, users should be aware that there is often considerable size variation among adults of a species, and that males and females may differ in size. We have followed species nomenclature that has been compiled by The Nature Conservancy. In most cases, this nomenclature follows that of generally accepted reference works (American Ornithologists' Union 1983 and supplements, Collins 1990, Jones et al. 1992, Wilson and Reeder 1993). Other common or Latin names used for species in the literature are mentioned in *Comments*. We follow the sequence of amphibian, reptile, and mammal species provided by Nussbaum et al. (1983) and Hall (1981), which is more consistent with the American Ornithologists' Union (1983) sequence of bird names than are the alphabetical arrangements of Collins (1990) and Jones et al. (1992). Consistent with the American Ornithologists' Union (1983), we capitalize English names of birds. Except for recent taxonomic revisions, the list of native species that breed in Oregon follows Puchy and Marshall (1993), supplemented by more recent information on breeding status given in Gilligan et al. (1994). We hope that, in combination with species distribution maps, the information in this atlas will provide interested citizens with a basic understanding of and appreciation for the diverse wildlife of our remarkable state and serve as a useful reference for serious naturalists and wildlife professionals.

## Wildlife Habitats of Oregon

### 1. Alpine
Open to closed communities of dwarf shrubs, grasses and forbs, including alpine fell fields and other rocky areas. Some alpine grasslands in northeastern Oregon occur in a matrix with whitebark pine, but on Steens Mountain they occur with stands of aspen. Usually found above 7,000 feet. Idaho fescue is an important grass in these communities and mountain big sage is also frequent.

### 2. Perennial Bunchgrass
A grassland community dominated by tall grasses (such as bluebunch wheatgrass, bottlebrush squirreltail, Sandberg bluegrass or Idaho fescue) often interspersed with shrubs (e.g., big sagebrush) or occasional junipers or other trees.

### 3. Alkali Grassland
These are usually seasonal wetlands in flat, poorly drained floodplains or internally drained basins of historic lake beds. Alkali grasslands are dominated by a fairly continuous cover of tall to medium grasses interspersed with rushes or sedges and occasional forbs. This community is found in northeastern Oregon valley bottoms and in alkaline areas of southeastern Oregon.

### 4. Mountain Big Sagebrush
Medium tall shrub communities found on higher elevation plateaus, slopes, and rocky flats with moderate to good soil development. Mountain big sagebrush (*Artemisia tridentata* ssp. *vaseyana*) is usually the dominant or co-dominant shrub, but sometimes occurs in a mosaic with low sagebrush. There is often an understory of Idaho fescue. Mountain snowberry communities, found between mountain big sagebrush and alpine communities, are also included in this wildlife habitat type.

### 5. Agricultural
This anthropogenic cover type includes all manner of lands under cultivation, including pasture, row crops, dryland and irrigated wheat fields, alfalfa, and orchards. A surprising variety of native wildlife species make use of agricultural lands, especially where there are small remnants of riparian vegetation or other trees (native and introduced) and shrubs in the agricultural matrix.

### 5A. Urban
This anthropogenic cover type includes all areas that are dominated by urban or industrial development, even if there are remnant patches of native vegetation scattered within the urban-industrial mosaic. Many urban areas in western Oregon retain

considerable amounts of natural vegetation and support a diverse fauna. Urban areas are also heavily used for shelter by wintering birds throughout the state during periods of inclement weather.

## 6. Canyon Shrubland

This is a tall to medium shrub community that occurs on steep slopes of foothill or mountain canyons in eastern Oregon. Along the southern part of the east slope of the Cascades, it merges with southwest Oregon chaparral cover types. Idaho fescue is often present in the understory. Typical shrubs include bitterbrush, serviceberry, and bittercherry. Big sagebrush often grows nearby, and there may be occasional junipers or ponderosa pines in these communities.

## 7. Early Shrub-Tree

This habitat type includes early successional stages of forests following clear-cutting or fires. In this community, the regenerating forest has not yet grown sufficiently to provide any significant canopy closure, however a variety of shrub species provide good ground cover. Where stumps and downed logs are abundant, as following a fire, this community type can provide habitat for many species adapted to open areas with good cover. Recent clear-cuts and fires are easily identified in satellite imagery.

## 8. Lava Field

There are several examples of recent lava flows, mostly in eastern Oregon, which are unvegetated or have sparse grass-shrub vegetation in pockets of soil accumulation. These areas with little or no soil provide no habitat for burrowing rodents.

## 9. Big Sagebrush

Big sagebrush (*Artemisia tridentata*) communities dominate vast expanses of eastern Oregon valleys where there is moderate to good soil development. This is a tall shrub community dominated by Wyoming and basin big sagebrush. Low sagebrush may occur in areas with shallow soil. A variety of native bunchgrasses may be found in the understory. Rabbitbrush is often found in areas that are grazed or otherwise disturbed. Scattered western junipers commonly occur in this habitat type.

## 10. Low Sagebrush

Low sagebrush communities are found on ridge tops, plateaus, or gentle slopes in eastern Oregon where, typically, there is little soil development. In addition to low sagebrush, these communities may be dominated by rigid sagebrush, budsage, or black sagebrush. They are generally low to medium shrub mosaics with a variety of bunchgrasses in the understory. In some examples, there are scattered junipers and ponderosa pines.

## 11. Mixed Sagebrush

This is a habitat composed of a collection of plant communities in which various species of sagebrush occur as a mixture or mosaic. Big sagebrush is usually a component plant, but other plants include silver sagebrush, rabbitbrush, bitterbrush, low sagebrush, and mountain big sagebrush. These are low to tall shrub communities that have either native or introduced grass understories. They usually occur across wide valleys with non-alkaline soils. Silver sagebrush communities, however, are typical of moist, semi-alkaline flats or valley bottomlands.

## 12. Quaking Aspen

This is a low woodland or forest community, ranging from wet streamside areas to moist mountain slopes. It is usually found on steep slopes, around seeps in desert mountains, or on north slopes in plateau grasslands. Quaking aspen is the characteristic tree, and snowberry is the most common shrub. The stands often include Engelmann spruce, lodgepole pine, or broad-leaved trees.

## 13. Western Juniper

Western juniper is a widespread tree that is diagnostic of the many types of open woodland habitats with understories of various scattered shrubs and grasses. Big sagebrush, low sagebrush, mountain mahogany, bitterbrush, and mountain big sagebrush all are found within western juniper woodlands. A variety of native grasses are found in the understory. A unique mixture of ponderosa pine and western juniper found at the Lost Forest of northern Lake County is included in this habitat type.

## 14. Idaho Fescue Grassland

This complex of grasslands occurs in eastern Oregon on plateaus, canyons, and flatlands of the Columbia Basin. Native grasslands on canyon slopes and plateaus, characterized by Idaho fescue, and cheatgrass grasslands in basins are included in this habitat type. It is characterized by a general lack of trees and only sparse occurrence of shrubs, mainly rabbitbrush. Ponderosa pine may be common in canyons.

## 15. Ponderosa Pine-White Oak

Arid woodlands and dry forests of ponderosa pine, often mixed with other trees, including Douglas-fir, lodgepole pine and Oregon white oak are represented in this habitat type. In southern Oregon and on the east slope of the northern Cascades, Oregon white oak is commonly associated with ponderosa pine. Most of the original and some of the remaining woodland of the Blue Mountains typify this habitat type. Often there is a savannah-like structure to the woodland, with an understory of Idaho fescue or bluebunch wheatgrass.

### 16. Saltmarsh

This habitat consists of Oregon's coastal saltmarshes, including both vegetated and unvegetated intertidal marshes dominated by sedge, rush, or bulrush, with silty, sandy, or rocky substrates.

### 17. Douglas-fir Mixed Conifer

These are closed-canopy mid-elevation (3,000 to 6,500 feet) mixed coniferous forests found primarily in the Blue Mountains but also in parts of the east slope of the northern Cascades. Common trees include Douglas-fir, true fir, western larch, lodgepole pine, and ponderosa pine. Rarely does a single species dominate. The second-growth mixed forests of the Blue Mountains following logging or disturbance are included in this habitat type.

### 18. Mountain Hemlock

These forests and woodlands occur at mid- to high elevations (mostly above 4,500 feet but sometimes as low as 3,500 feet) on both slopes of the Cascades. Mountain hemlock is the characteristic tree, but stands are often mixed with other trees, including true fir, Douglas-fir, western white pine, subalpine fir, and lodgepole pine. Except at the highest elevations it usually forms a closed canopy forest.

### 19. Montane-Lodgepole Pine

This habitat consists of a number of mid- to high elevation (above 3,500 feet) closed canopy forest types that occur on the east slope of the Cascades and in the Blue Mountains. Lodgepole pine is conspicuous in most but not all of these forests. The tree species that are also present include Engelmann spruce, true fir, Douglas-fir, western larch, and whitebark pine. Many of the forests of the Blue Mountains that are recovering from fires are of this type.

### 20. Siskiyou Mixed Conifer

The Siskiyou Mountains and, to a lesser extent, the west slopes of the south Cascades, have one of the world's most diverse coniferous forest floras. The forest types in this region are grouped together because of their geography and faunal similarity as much as for their common floristics. They occur at elevations from 2,000 to 7,300 feet. At lower elevations, Oregon's redwood and mixed evergreen forests of Douglas-fir, tanoak, and madrone are included. Higher up the slopes are forests of Jeffrey pine (growing on serpentine soil), lodgepole pine, ponderosa pine, incense cedar, and true fir. At elevations above 4,000 feet, some sites support open-canopied forests of red fir, white fir, Douglas-fir, and lodgepole pine.

### 21. Mixed Conifer-Deciduous

This is the habitat type that covers most of the forested lowlands of western Oregon and the lower elevations of the west slope of the Cascades. Along the coastal strip the forest is dominated by Sitka spruce, mixed with western hemlock and western red cedar. Farther south, Sitka spruce becomes mixed with grand fir and Douglas-fir. Most of the second-growth forests of western Oregon are dominated by young Douglas-fir and western hemlock, which form a mosaic with various deciduous hardwood species such as bigleaf maple and red alder. In remaining areas of older forests, Douglas-fir and western hemlock form a closed canopy under which a variety of other trees and shrubs occur to form a multi-storied forest with a rich and diverse assemblage of plants on the forest floor.

### 22. Playa

This habitat denotes the barren, alkali flats of southeastern Oregon. During the winter they are flooded by rains and can be important to migratory birds. They are generally devoid of vegetation but there may be patches of saltgrass and an occasional clump of greasewood. Large playas are found in many Pleistocene lake basins.

### 23. Marsh

Marshes are typically flooded for much of the year. The most common freshwater marshes contain some open water that is surrounded by encroaching growths of hardstem bulrush, cattail, and burreed. These are important habitats for migratory and breeding waterfowl and a variety of other aquatic species. This habitat type also includes remnants of Willamette Valley tufted hairgrass prairie, which is flooded during the winter, and reed canarygrass wetlands of disturbed areas, such as those around reservoirs and farm ponds.

### 24. Wet Meadow

There are two types of wet mountain meadows, one in the Blue and Ochoco Mountains and the other in valleys of the east slope of the Cascades. Both are found on alluvial soils along stream channels or in valley bottoms. One type is dominated by tall sedges (genus *Carex*) and the other by tufted hairgrass or bluegrass. Frequently there will be stringers of willow riparian vegetation along streams that run through these meadows.

### 25. Riparian

This complex of riparian habitat types includes black hawthorn riparian that formerly covered broad, low-elevation floodplains in eastern Oregon (now common only in the Grande Ronde Valley). The two most common riparian types in Oregon are dominated by cottonwoods and willows, which form tall forest or shrub communities along the banks of rivers and streams. These communities are rich in wildlife, especially breeding birds, but are sensitive to grazing. In the valleys of western Oregon, streams are frequently bordered with Oregon ash and black cottonwood, often within a matrix of agricultural and pasture lands. Included in this habitat type, because of proximity, are the forests and woodlands of Oregon white oak and Douglas-fir that mix with farmland and towns around the edge of the Willamette, Rogue, and Umpqua valleys.

## 26. Open Water

While there may be open water in some marshes and wetlands, at least temporarily, true "open water" refers to large bodies of standing water (lakes, reservoirs) and permanent rivers and streams. Many aquatic species are restricted to open water habitats. Digital information on the location of small bodies of open water is not yet available for Oregon, so this microhabitat is underrepresented in predictions of distributions of species associated with open water.

## 27. Chaparral

This is a wildlife habitat type found in foothills and mountain slopes around the valleys of southwestern Oregon. It includes medium to tall shrublands (often dominated by buckbrush and manzanita) in which there are scattered stands of trees (ponderosa pine, Jeffrey pine, lodgepole pine, Oregon white oak) and open forests and woodlands of oak and madrone with a shrub understory.

## 28. Inland Dunes

This habitat includes open sand dunes with scattered medium to tall shrubs (often big sagebrush, bitterbrush, or horse-brush) that are located on flats at the margins of inland playas. Occasional bunchgrasses, rhizomatous grasses, and forbs occur among widely spaced shrubs. The most significant inland sand dunes are the Alvord, Catlow, Guano, and Christmas Valley dunes.

## 29. Salt-Desert Shrub

This group of plant communities is dominated by shrubs of the family Chenopodiaceae (shadscale, black greasewood, spiny hopsage, saltsage, and winterfat). This family has a special photosynthetic pathway that improves water conservation in the hot, arid environments of southeastern Oregon. It is a habitat of low to medium tall scattered shrubs growing on flat desert pavements, low alkaline dunes, around playas, or on gentle slopes above playas. This habitat is often surrounded by sagebrush growing on higher, less alkaline soil. Grasses and annual, often succulent, forbs grow widely spaced in the understory.

## 30. Coastal Dunes

These are open coastal sand dunes with areas partially to totally stabilized by introduced grasses and shrubs, native grasses and shrubs, and tree islands. They are mostly open dunes with scattered islands of pine forests, shrubs, and beachgrass. There are often extensive deflation plain wetlands between the dunes. Shorepine is the most widespread tree species in the dunes. They are entirely coastal, and salt spray and desiccation are major ecological factors.

## Status Codes

*Federal Status* (U. S. Fish & Wildlife Service Endangered Species Act)
FE = Federally listed as Endangered
FT = Federally listed as Threatened
C = Candidate for listing as Threatened or Endangered
SC = Species of Concern

*State Status* (Oregon Department of Fish and Wildlife)
SE = State listed as Endangered
ST = State listed as Threatened
Sensitive = State listed as Sensitive

*Global (and State) Ranks* (Oregon Natural Heritage Program)
State Ranks refer only to the status of a species during its breeding season in Oregon.
G1 (S1) = Critically imperiled globally (within the state); typically 5 or fewer occurrences.
G2 (S2) = Imperiled globally (within the state); typically 6-20 occurrences.
G3 (S3) = Rare or uncommon but not imperiled; typically 21 to 100 occurrences (within the state).
G4 (S4) = Not rare and apparently secure, but with cause for long term concern; usually more than 100 occurrences (within the state).
G5 (S5) = Demonstrably widespread, abundant, and secure.
GU (SU) = Unrankable without further information.
G? (S?) = Unranked at present.
G_ (SE) = State Exotic (introduced into Oregon)

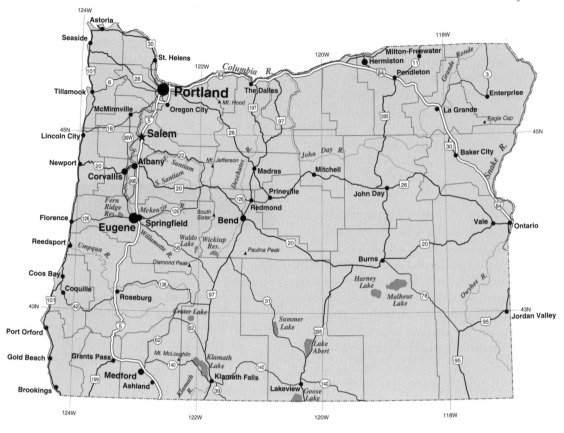

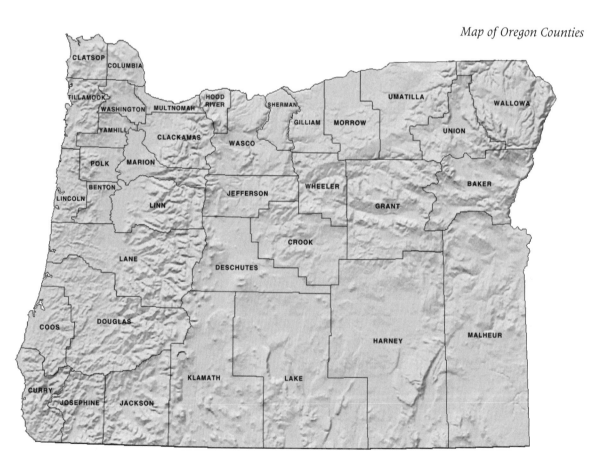

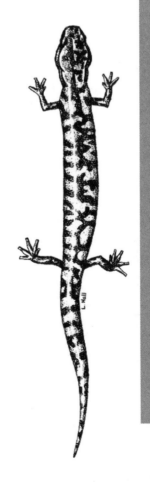

# Amphibians

# Northwestern Salamander
## *(Ambystoma gracile)*

*Order: Caudata*
*Family: Ambystomatidae*
*State Status: None*
*Federal Status: None*
*Global Rank: G5*
*State Rank: S5*
*Length: 9 in (23 cm)*

**Global Range:** Pacific Coast of North America from southern Alaska to the Gualala River, northern California.

**Habitat:** The northwestern salamander resides in meadows, woodlands, and coniferous or deciduous forests, ranging from sea level to 10,000 feet. It spends most of its life underground in rodent burrows and under rocks and logs near water, and breeds in ponds, lakes, and slow streams.

**Reproduction:** Eggs are laid in clusters in quiet water and attached to vegetation 1-3 feet below the surface. Average clutch size is 80 eggs (range 30-270). It normally breeds from January to May, though breeding may start as late as August at high elevations. Eggs are laid from March to May and hatch from May to July in 30-60 days, depending on temperature.

**Food Habits:** Adults feed primarily on insects, earthworms, molluscs, and other terrestrial invertebrates. Larvae feed on zooplankton and other aquatic invertebrates.

**Ecology:** The northwestern salamander is active at night and searches for food on the forest floor under debris. Nonpaedomorphic populations migrate between breeding and nonbreeding habitats, usually on rainy nights.

**Comments:** It is preyed on by introduced trout and may be sensitive to pesticides and herbicides that enter breeding waters.

*References:* Nussbaum et al. 1983, Snyder 1963, Stebbins 1985, Taylor 1977, Titus 1990.

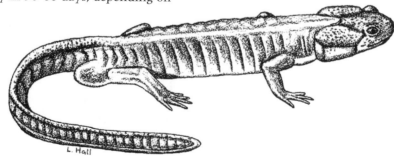

L. Hall

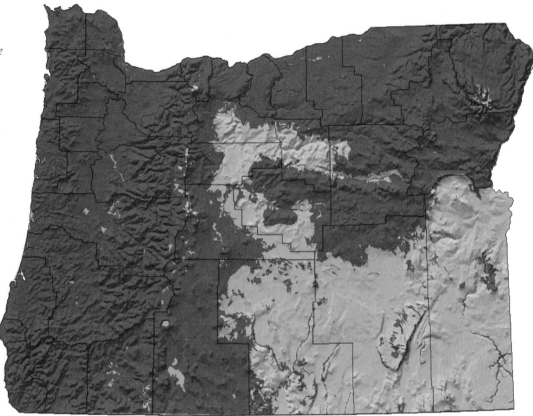

## Long-toed Salamander
### (Ambystoma macrodactylum)

*Order: Caudata*
*Family: Ambystomatidae*
*State Status: None*
*Federal Status: None*
*Global Rank: G5*
*State Rank: S5*
*Length: 6 in (15 cm)*

**Global Range:** Northern British Columbia south along the coast to central Oregon and inland to the Rocky Mountains of Idaho. Also found farther south in the Sierra Nevada of California to Lake Tahoe, with an isolated population in Santa Cruz County, California.

**Habitat:** The long-toed salamander is found in a wide variety of habitats, including semiarid sagebrush desert, dry woodlands, humid forests, and alpine meadows. It lives under bark and rocks near ponds, lakes, and streams. It retreats underground during hot, dry, or cold weather.

**Reproduction:** The breeding season varies according to climate; it can breed as early as January and February at low elevations and from April to July in the mountains. The clutch size ranges from 85 to 411 eggs. This salamander becomes sexually mature at 2-3 years, but larvae may not transform until the second year. Eggs are laid in quiet water, either singly or in clusters of 5-100 eggs. Adults migrate to breeding waters with the beginning of fall rains, usually in October and November.

**Food Habits:** Larvae eat zooplankton, immature insects, and aquatic snails. Adults eat insects, earthworms, spiders, and other invertebrates.

**Ecology:** The long-toed salamander searches for prey under surface objects, usually near water, and is active at night. Predators of larvae include aquatic insects and garter snakes. Garter snakes, ring-necked snakes, and bullfrogs prey upon adults.

**Comments:** This is the most common and widespread species of *Ambystoma* in the Northwest.

*References*: Anderson 1967, 1968, Farner 1947, Ferguson 1961.

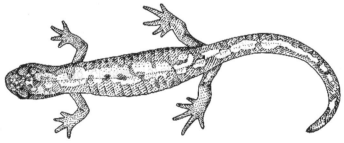

# Tiger Salamander
## *(Ambystoma tigrinum)*

*Order: Caudata*
*Family: Ambystomatidae*
*State Status: Sensitive*
*Federal Status: None*
*Global Rank: G5*
*State Rank: SU*
*Length: 12 in (30 cm)*

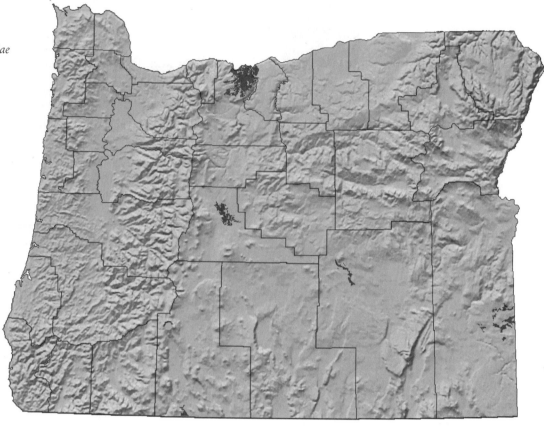

**Global Range:** The tiger salamander is widespread in the United States east of the Rocky Mountains. Occurs in eastern Washington, along the central coast, and in the Central Valley of California. There are isolated specimen records, some possibly introduced, from Oregon.

**Habitat:** This salamander occurs in a wide variety of habitats. Terrestrial adults spend most of their time underground. Though rarely seen, they may be abundant near breeding ponds. They breed in a wide range of environments, from clear mountain ponds to temporary turbid pools in the lowlands.

**Reproduction:** It breeds in April or May, laying up to 1,000 eggs singly or in clusters. The larvae emerge after a few weeks and may transform into adults during the first, second, or third summer. They can also become paedomorphic.

**Food Habits:** Larvae eat aquatic invertebrates and vertebrates (especially amphibian larvae). Adults eat any small animal that can be captured and swallowed, including vertebrates.

**Ecology:** Little is known about the life history of this species in Oregon. Its surface activity is usually associated with rainfall. It usually breeds in ponds free of fishes.

**Comments:** Natural populations are thought to occur along the Columbia River, but other specimens from Oregon may have been introduced. This salamander has been sold as fish bait in eastern Washington, increasing the likelihood of introductions. Marshall et al. (1996) report five records of the blotched tiger salamander (*Ambystoma tigrinum melanostictum*) in eastern Oregon that they believe represent native populations (Mosier, Hood River County; south of Klamath Falls, Klamath County; Moon Reservoir and Hines, Harney County; and Leslie Gulch, Malheur County).

*References:* Harte and Hoffman 1989, Leonard et al. 1993, Nussbaum et al. 1983.

## Cope's Giant Salamander
### (Dicamptodon copei)

*Order: Caudata*
*Family: Dicamptodontidae*
*State Status: Sensitive*
*Federal Status: None*
*Global Rank: G3*
*State Rank: S2*
*Length: 8 in (20 cm)*

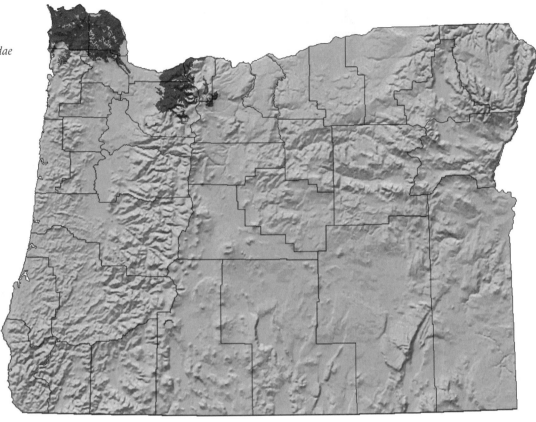

**Global Range:** Coastal mountains and the west slope of the Cascade Range in Washington. The species is known from five sites in Oregon, in mountainous areas just south of the Columbia River, from Clatsop County to Oneonta Gorge, Multnomah County.

**Habitat:** Cope's giant salamander is found in moist forested areas in clear, cold streams (where the water temperature is usually between 8° and 14° C), brooks, and ponds with gravel bottoms and boulders. They may be seen on wet rocks and vegetation on rainy nights, but ordinarily occur under rocks, slabs of bark, or other cover in streams.

**Reproduction:** Most information on this species is inferred from life history of *Dicamptodon ensatus*. It probably breeds from spring to fall, laying a clutch of 20-115 eggs in nest chambers under stones. Females guard the eggs—which may take up to 9 months to hatch—from predators. Most reproduce as neotenic larvae.

**Food Habits:** This salamander takes a wide variety of aquatic organisms, including insect larvae, fish eggs, tadpoles, and young salmonids and sculpins.

**Ecology:** Predators include garter snakes, other *Dicamptodon* larvae, and water shrews. The larval form can travel overland, and may disperse this way. It is seldom found in water warmer than 18° C.

**Comments:** This species was first described in 1970. Terrestrial adults are very rare; only three metamorphosed adults have been found. Its requirement for clear, cold water may make it sensitive to logging activities.

*References:* Antonelli et al. 1972, Good 1989, Jones and Corn 1989, Leonard et al. 1993, Marshall 1992, Nussbaum 1970.

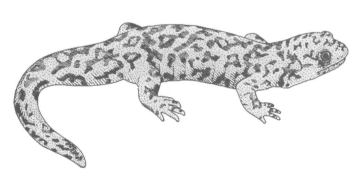

# Pacific Giant Salamander
## *(Dicamptodon tenebrosus)*

*Order: Caudata*
*Family: Dicamptodontidae*
*State Status: None*
*Federal Status: None*
*Global Rank: G5*
*State Rank: S4*
*Length: 12 in (30 cm)*

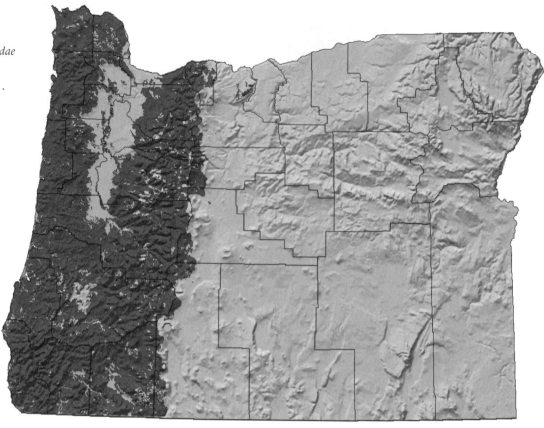

**Global Range:** Coast Ranges and Cascade Mountains from southern British Columbia to San Francisco Bay. It is also found in the Rocky Mountains of central Idaho, although this population is considered a distinct species (*Dicamptodon aterrimus*) by some authorities.

**Habitat:** The Pacific giant salamander is found in humid mixed conifer and deciduous forests and riparian zones, occurring as high as 6,000 feet. Downed logs are required for shelter and foraging. Larvae occupy cold, clear water of streams, rivers, lakes, and ponds.

**Reproduction:** The reproductive biology of this species is poorly known. It probably breeds in the spring. Eggs are usually laid in deep-water nest chambers, where the female remains until the eggs hatch. In coastal populations, eggs take about 200 days to hatch. The clutch size is about 100-200 eggs. The larval period lasts 2 years in small streams, but in large streams and lakes the salamander can be neotenic.

**Food Habits:** This salamander is an opportunistic carnivore. Aquatic larvae feed on insect larvae, tadpoles, other salamanders, and small fish. Adults eat land snails, slugs, insects, small mammals, other amphibians, lizards, and even birds.

**Ecology:** This species requires clear and cold water, making siltation of streams and the removal of riparian vegetation (which helps maintain cool water temperatures) detrimental. Adults search for prey on the forest floor under ground debris, and have been known to climb trees in search of food.

**Comments:** This species is common in suitable habitat. Larvae have been observed in streams that run through oak woodlands and chaparral in the Rogue River Valley. In Good's 1989 revision of the genus, Oregon populations were reclassified from *Dicamptodon ensatus* to *Dicamptodon tenebrosus*.

*References:* Bury 1972, Good 1989, Jones et al. 1990, Nussbaum 1969, Nussbaum and Clothier 1973.

6

# Cascade Torrent Salamander
## *(Rhyacotriton cascadae)*

*Order: Caudata*
*Family: Dicamptodontidae*
*State Status: Sensitive*
*Federal Status: None*
*Global Rank: G2G3*
*State Rank: S3*
*Length: 4 in (10 cm)*

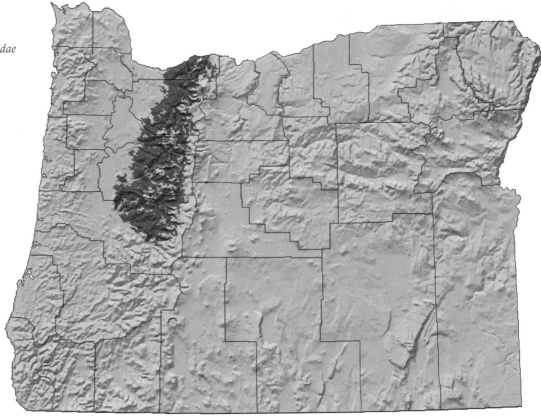

**Global Range:** Coastal mountains from the Olympic Peninsula in Washington south to Mendocino County, California and in the Cascade Mountains of southern Washington and northern Oregon, with a local disjunct population in the southern Oregon Cascades.

**Habitat:** The torrent salamander is most abundant in rocks bathed in a constant flow of cold water, also occurring in cool rocky streams, lakes, and seeps. It tends to remain in the splash zone of streams and spray zones of waterfalls; these microhabitats usually occur within conifer or alder forests.

**Reproduction:** These salamanders breed from March to June, laying a clutch of 7-16 eggs which take 7-10 months to hatch. Larvae metamorphose in 4 to 5 years and reach sexual maturity at 6 to 7 years of age.

**Food Habits:** The diet consists of aquatic and semi-aquatic invertebrates, including amphipods, springtails, fly larvae, worms, snails, and spiders. They search for prey under rocks and other objects in streams.

**Ecology:** These species are very dependent on nearly continuous access to cold water and are therefore not found in areas where timber harvest has resulted in decreases in cover and increases in water temperature or siltation. They can be found moving about in forests during wet weather.

**Comments:** Good and Wake (1992) have recently revised the genus *Rhyacotriton* based on allozyme variation, describing three new species from populations formerly considered *Rhyacotriton olympicus* (Olympic salamander). Populations in Oregon's northern coast ranges are *R. kezeri* (Columbia torrent salamander). Only the southern torrent salamander (*R. variegatus*), which occurs in coastal mountains south of Tillamook County and in the Cascade Mountains of eastern Douglas County, is a federal "species of concern." The Cascade torrent salamander (*R. cascadae*) is found in the Oregon Cascades as far south as northern Lane County. Current ecological information does not differentiate between the three new species, so this account applies to populations of all of Oregon's torrent salamanders.

*See following page for range maps for Columbia torrent salamander and southern torrent salamander.*

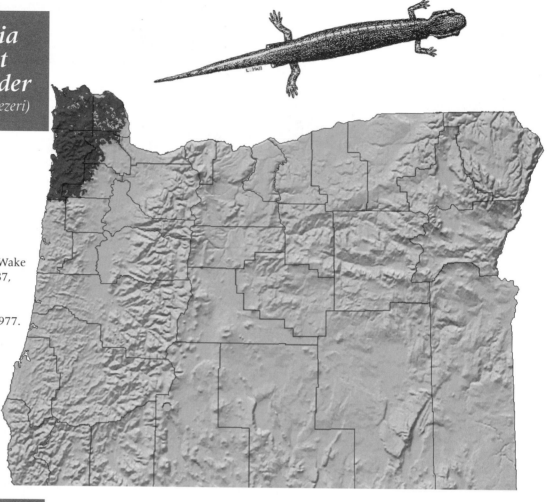

## Columbia Torrent Salamander
### (Rhyacotriton kezeri)

*State Status: Sensitive*
*Federal Status: None*
*Global Rank: G2G3*
*State Rank: S3*

*References:* Good and Wake 1992, Good et al. 1987, Leonard et al. 1993, Marshall et al. 1996, Nussbaum and Tait 1977.

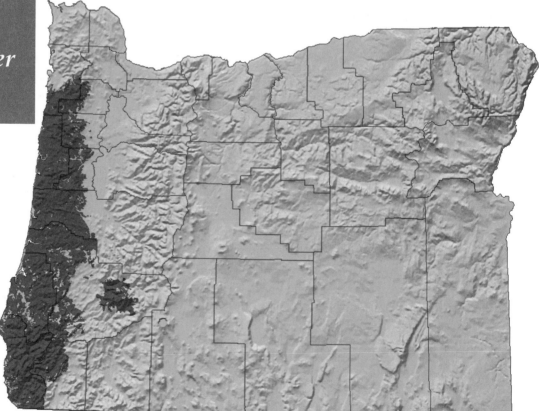

## Southern Torrent Salamander
### (Rhyacotriton variegatus)

*State Status: Sensitive*
*Federal Status: SC*
*Global Rank: G3*
*State Rank: S3*

# Clouded Salamander
## (Aneides ferreus)

*Order: Caudata*
*Family: Plethodontidae*
*State Status: Sensitive*
*Federal Status: None*
*Global Rank: G4*
*State Rank: S4*
*Length: 5 in (13 cm)*

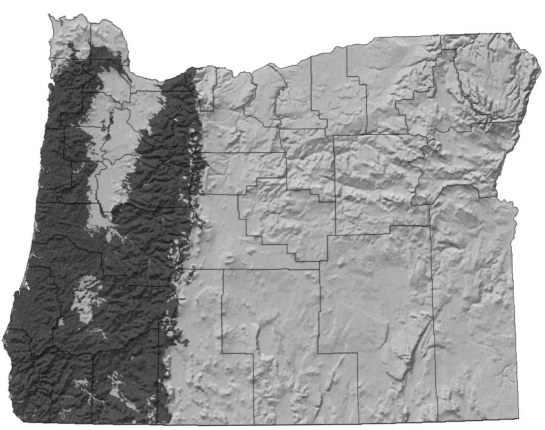

**Global Range:** Coastal mountains and Cascade Range from southwestern Washington to northern California. The population found on Vancouver Island may be introduced.

**Habitat:** The clouded salamander is primarily a forest dweller, found in moist areas under downed logs and other debris. It is often common in clearings caused by fire or timber harvest if large downed logs are present.

**Reproduction:** Eggs are laid from spring to early summer and hatch in September and October. The average clutch size is 14 eggs (range 8-18) and eggs are brooded by the female until they hatch. The incubation period is about 2 months. Females breed in alternate years.

**Food Habits:** Clouded salamanders eat small invertebrates, including ants, beetles, mites, spiders, and springtails. In one Oregon study 57% of the diet comprised ants.

**Ecology:** This salamander can climb trees to heights of at least 20 feet. It is inactive in cold weather. It is often found in groups deep inside downed logs.

**Comments:** Populations of this salamander can adapt to timber harvest practices that leave downed logs scattered in clearings, but not to young reforested areas lacking coarse woody debris.

*References:* Marshall 1992, McKenzie 1970, McKenzie and Storm 1970, Stelmock and Harestad 1979, Storm and Aller 1947.

# Black Salamander
## *(Aneides flavipunctatus)*

*Order: Caudata*
*Family: Plethodontidae*
*State Status: Sensitive*
*Federal Status: None*
*Global Rank: G4*
*State Rank: S2*
*Length: 5 in (13 cm)*

**Global Range:** From the mountains of Jackson and Josephine counties, Oregon, south in coastal mountains to Santa Cruz County, California.

**Habitat:** The black salamander is usually found near streams, in talus slopes, or under rocks and logs. It inhabits open woodlands, and mixed coniferous and mixed coniferous-deciduous forests.

**Reproduction:** Nothing is known of its reproductive biology in Oregon. Eggs are probably laid in the spring and hatch in the fall, emerging with the fall rains. One nest from California had 15 eggs. Like that of other plethodontid salamanders, the nest was in a moist chamber below ground and was tended by the female.

**Food Habits:** This species feeds on a variety of terrestrial invertebrates, such as spiders, snails, slugs, earthworms, beetles, and springtails. The diet varies seasonally, and larger individuals take larger prey items.

**Ecology:** Little is known about its specific habitat requirements. It seems to occur in moist situations in the southern part of its range and can be found among moss-covered rocks in Oregon. It is known to climb trees like its relative, the clouded salamander.

**Comments:** This species has been found on rocky slopes formed by road cuts. The Oregon populations differ in color from California populations, which are greenish rather than black, and may warrant recognition as a separate subspecies.

*References:* Leonard et al. 1993, Lynch 1981, 1985, Marshall 1992.

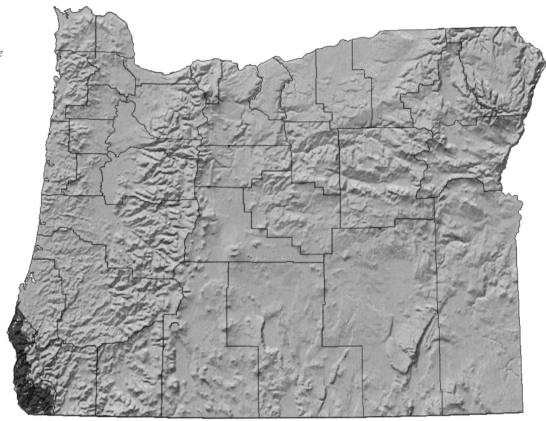

# California Slender Salamander
## *(Batrachoseps attenuatus)*

*Order: Caudata*
*Family: Plethodontidae*
*State Status: Sensitive*
*Federal Status: None*
*Global Rank: G5*
*State Rank: S2*
*Length: 5 in (13 cm)*

**Global Range:** This species ranges south from the southern Oregon coast through the Coast Ranges to Monterey County, California. Scattered populations also exist on the west slope of the central Sierra Nevada of California. The northern limit of this species' range is in extreme south coastal Oregon.

**Habitat:** In Oregon, this species is restricted to lower-elevation forests along the southern coast, including hardwood, redwood, and other coniferous forests. It also occurs in open areas with scattered trees and is generally found under logs, rocks, or other objects in contact with the ground.

**Reproduction:** The reproductive biology of the California slender salamander has not been studied in Oregon. In California, eggs are laid in moist places under litter or logs, or in burrows. The average clutch of 10 eggs (range 4-25) is laid in the fall, and young emerge in the spring, reaching sexual maturity in about 3 years.

**Food Habits:** This salamander eats small invertebrates such as springtails, earthworms, aphids, larval and adult beetles, millipedes, and amphipods.

**Ecology:** This species has a small home range, seldom moving more than a few feet in any direction. It can be dense in good habitat, but it tends to be distributed in patches.

**Comments:** Distribution and biology of the Oregon populations are poorly known. There is much local genetic variation in California populations of this species, and the taxonomic status of the Oregon population may change with further study.

*References:* Hendrickson 1954, Maiorina 1976, Marshall 1992, St. John 1982.

# Oregon Slender Salamander
### (Batrachoseps wrighti)

*Order: Caudata*
*Family: Plethodontidae*
*State Status: Sensitive*
*Federal Status: None*
*Global Rank: G3*
*State Rank: S3*
*Length: 3.5 in (9 cm)*

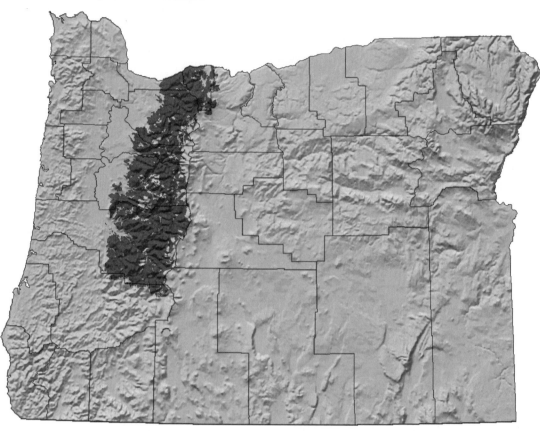

**Global Range:** The only amphibian endemic to Oregon, this species is found on the west slope of the Cascade Range from the Columbia River south to southern Lane County.

**Habitat:** This salamander is found under bark or moss in mature and second-growth Douglas-fir forests, as well as under rocks or logs in stands of moist hardwood forests within coniferous forest landscapes. Usually it is absent from recent clear-cuts, but it is sometimes found in talus and lava fields near the crest of the Cascades.

**Reproduction:** A clutch of 6 eggs (range 3-11) is laid in the spring and hatches after about 4 months. Eggs are guarded by the female.

**Food Habits:** As is typical of small salamanders, the Oregon slender salamander eats small invertebrates like springtails, mites, insect larvae and adults, spiders, snails, and earthworms.

**Ecology:** This species can be found near the surface during fall and spring, but retreats underground in late spring and summer.

**Comments:** The species seems to be common in local areas but is rare or absent throughout large portions of its range. It is known to occur at only a few dozen localities. The impact of timber harvest practices on this species and its general distribution in the Cascade Mountains need to be investigated.

*References:* Marshall 1992, Stebbins 1949c, Tanner 1953.

# Ensatina
## *(Ensatina eschscholtzii)*

*Order: Caudata*
*Family: Plethodontidae*
*State Status: None*
*Federal Status: None*
*Global Rank: G5*
*State Rank: S5*
*Length: 4 in (10 cm)*

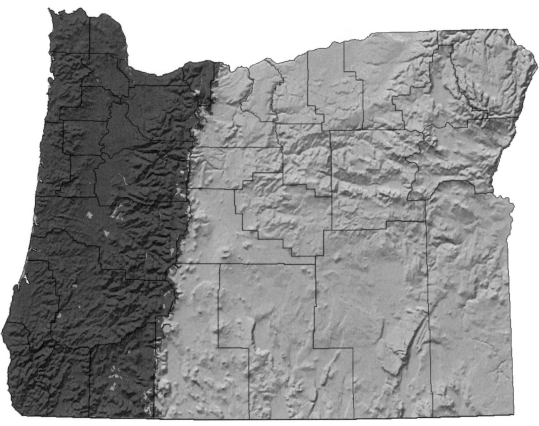

**Global Range:** Southwestern British Columbia south along the coastal, Cascade, and Sierra Nevada mountains to the Transverse and Peninsular ranges of southern California and extreme northern Baja California.

**Habitat:** Ensatinas are found under logs, rocks, or other debris on forest and woodland floors. Although they require moist microhabitats, they are not associated with open water.

**Reproduction:** Courtship and egg-laying can occur from fall through spring. Eggs are laid in leaf litter, logs, or rodent burrows. The female broods the eggs until they hatch and the young emerge with the fall rains. Average clutch is 11 eggs (range 5-16).

**Food Habits:** Food consists of small arthropods such as spiders, springtails, beetles, mites, termites, and ticks, but larger individuals will take earthworms and snails as well.

**Ecology:** This species can be quite abundant in appropriate habitat. Densities of 0.3-2.9 individuals per square meter have been observed near Portland. As a defense against predators, this species secretes a white, sticky, poisonous mucus along the top of the tail when disturbed.

**Comments:** The painted salamander (*E. e. picta*) is found only in coastal mountains of southern Oregon and adjacent northern California, where it intergrades with the more widely distributed Oregon salamander (*E. e. oregonensis*).

*References:* Altig and Brodie 1971, Brown 1974, Stebbins 1949a, 1949b, 1954.

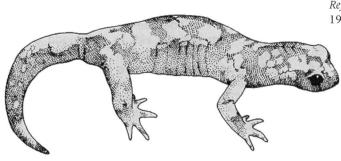

13

# Dunn's Salamander
## (Plethodon dunni)

*Order: Caudata*
*Family: Plethodontidae*
*State Status: None*
*Federal Status: None*
*Global Rank: G4*
*State Rank: S4*
*Length: 6 in (15 cm)*

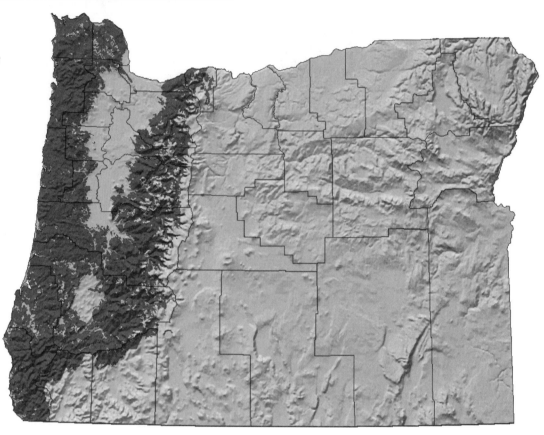

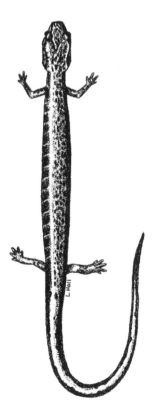

**Global Range:** Coastal and Cascade mountains from southwestern Washington to northwestern California. Most of the range of this species is in Oregon.

**Habitat:** This species is found in and around streams and seeps in coniferous or mixed forests, often in the splash zone or under rocks, logs, moss, or leaf litter. It prefers moss-covered rock rubble and talus along permanent streams and seeps.

**Reproduction:** The majority of eggs are laid in the spring and hatch in late summer or fall. Eggs are brooded by the female and hatch after an incubation period of about 70 days. The average clutch is 13 eggs (range 4-18). Eggs are deposited in rock crevices and moist talus near permanent cold-water streams and seeps.

**Food Habits:** Dunn's salamander feeds on small invertebrates, including springtails, isopods, mites, flies, beetles, and worms and is occasionally cannibalistic.

**Ecology:** They may be active in any month at lower elevations, but have long periods of winter inactivity at higher elevations.

**Comments:** An unstriped form of Dunn's salamander found in the Coast Range around Marys Peak, Benton County, has been considered a separate species, *Plethodon gordoni* (Marys Peak salamander) by some authors.

*References: Altig and Brodie 1971, Brodie 1970, Dumas 1955, 1956.*

# Del Norte Salamander
### *(Plethodon elongatus)*

*Order: Caudata*
*Family: Plethodontidae*
*State Status: Sensitive*
*Federal Status: SC*
*Global Rank: G3*
*State Rank: S2*
*Length: 6 in (15 cm)*

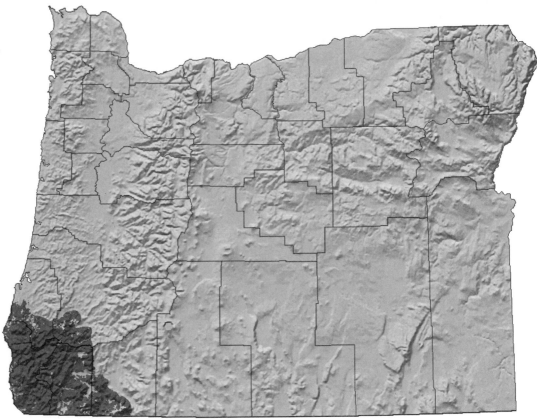

**Global Range:** This species is found only in extreme southwestern Oregon and extreme northwestern California.

**Habitat:** This salamander occurs in moist, rocky areas within forests, such as rock rubble and talus slopes. It can tolerate dryer conditions than Dunn's salamander and is occasionally found in decaying logs and under litter on the forest floor. In one study, 91% of 406 individuals were collected in old-growth Douglas-fir forests.

**Reproduction:** Little is known about reproduction in this species. The average clutch size is 8 eggs. Nests are probably in moist cavities in rock talus or under logs. Like other members of the genus, eggs are probably laid in the spring, hatch in the fall, and are tended by the female.

**Food Habits:** It feeds on small invertebrates such as springtails and larval and adult beetles, as well as some butterfly and moth larvae, leafhoppers, and millipedes.

**Ecology:** This species is closely associated with old-growth coniferous forests. It may be sensitive to activities such as timber harvest and resulting habitat fragmentation.

**Comments:** It is possible that road construction, which increases rock rubble, may increase habitat for this species. During hot, dry weather it retreats deep into rock crevices. Some taxonomists regard the Siskiyou Mountains salamander (*Plethodon stormi*) to be a subspecies of *P. elongatus*.

*References:* Brodie 1970, Bury and Johnson 1965, Livezey 1959, Marshall 1992, Marshall et al. 1996, Welsh 1990.

15

# Larch Mountain Salamander
*(Plethodon larselli)*

*Order: Caudata*
*Family: Plethodontidae*
*State Status: Sensitive*
*Federal Status: SC*
*Global Rank: G2*
*State Rank: S2*
*Length: 4 in (10 cm)*

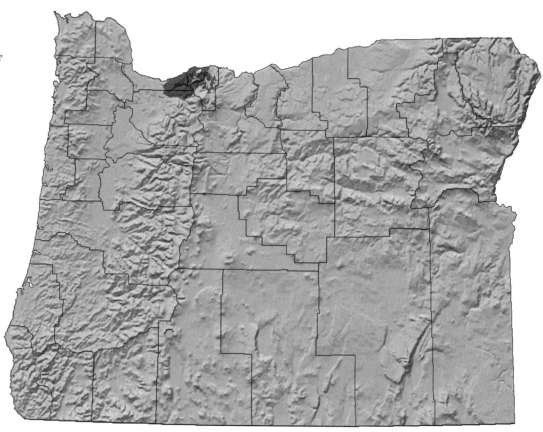

*References:* Altig and Brodie 1971, Herrington and Larsen 1985, 1987; Howard et al. 1983, McAllister 1995.

**Global Range:** Once thought to be restricted to the Columbia River Gorge of Washington and Oregon, it is now known to occur at several locations in the Cascade Mountains of Washington as far north as Snoqualmie Pass.

**Habitat:** It is found in talus slopes within areas of Douglas-fir forests. The talus may have a covering of moss that is kept moist by the forest overstory. In very hot or cold conditions, the salamander retreats deep into the talus.

**Reproduction:** This species breeds in both spring and fall. Although no nests have been found, they probably occur deep in talus slopes. Based on counts of ovarian eggs, average clutch size is thought to be 7.

**Food Habits:** Typical of the genus, this species feeds on smaller invertebrates such as mites and springtails. Larger individuals also take snails and earthworms.

**Ecology:** The Larch Mountain salamander tolerates dry conditions and is rarely found in wet areas. Distribution is more continuous on north-facing slopes.

**Comments:** The distribution and abundance of this species is poorly known. Disturbance of the talus habitat of this species should be avoided because microclimates within the talus can be destroyed by movement of rocks.

*Order: Caudata*
*Family: Plethodontidae*
*State Status: Sensitive*
*Federal Status: SC*
*Global Rank: G2Q*
*State Rank: S2*
*Length: 5 in (13 cm)*

**Global Range:** This salamander is a local endemic found in the Siskiyou Mountains along the Oregon-California border. It seems to be restricted to a 377-km² area in the Applegate River drainage of Oregon and the Seiad Creek drainage of California.

**Habitat:** It is found primarily among loose rock rubble or talus on north-facing slopes or in dense wooded areas. It is sometimes found under bark or logs on the forest floor, but always near talus.

**Reproduction:** Reproduction of this species has been poorly studied but, like other members of the genus, the female probably lays eggs in chambers deep in talus slopes in the spring and broods them until they hatch in the fall. Clutch size averages 9 eggs (range 2-18).

**Food Habits:** It eats small invertebrates such as mites, spiders, ants, springtails, and beetles.

**Ecology:** It has been seen to emerge at night even in dry summer weather to feed on insects at the surface of talus or on the forest floor. Because summers are hot and dry in the range of this species, it is especially sensitive to disturbance of talus microhabitats.

**Comments:** Some taxonomists consider the Siskiyou Mountains salamander to be a subspecies of the Del Norte Salamander (*Plethodon elongatus*). A study conducted in 1974 estimated the total population of the species at just over 3,000,000 individuals.

*References:* Leonard et al. 1993, Marshall 1992, Nussbaum 1974.

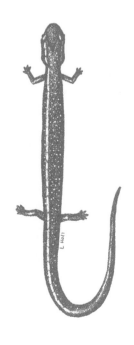

# Western Redback Salamander
*(Plethodon vehiculum)*

*Order: Caudata*
*Family: Plethodontidae*
*State Status: None*
*Federal Status: None*
*Global Rank: G5*
*State Rank: S5*
*Length: 4 in (10 cm)*

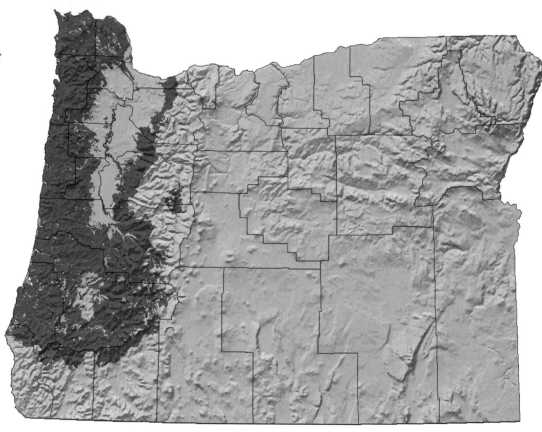

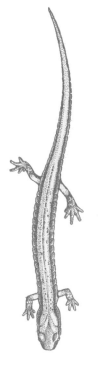

**Global Range:** Vancouver Island and coastal southern British Columbia south in the coastal and Cascade Mountains to southern Oregon.

**Habitat:** A forest dweller, the western redback salamander occurs under logs, rocks, and debris throughout humid coniferous and deciduous forests. It is found in moist microhabitats, but is not closely associated with running water. This species is scarce to absent above snowline.

**Reproduction:** It breeds from November to March, laying a clutch of 10 (range 6-19) eggs, which the female broods until they hatch in the fall. Females probably breed every other year.

**Food Habits:** This salamander feeds on terrestrial invertebrates such as mites, springtails, beetles, sowbugs, ants, and spiders.

**Ecology:** The western redback salamander is inactive in both cold and hot weather. Temperature must be above freezing for courtship. Juveniles remain in the nest until fall rains begin.

**Comments:** This is the most widespread and abundant woodland salamander in the Pacific Northwest.

*References:* Dumas 1956, Hanlin et al. 1979, Peacock and Nussbaum 1973.

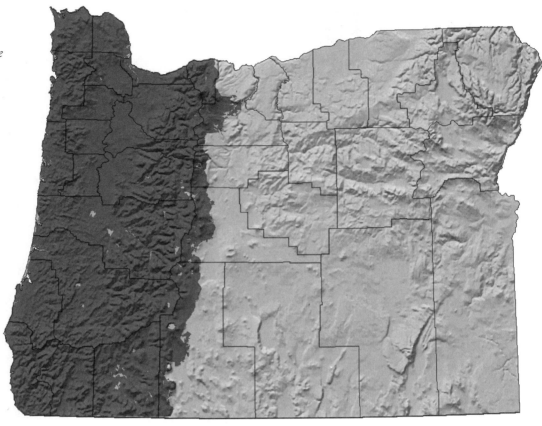

## Roughskin Newt
### *(Taricha granulosa)*

*Order: Caudata*
*Family: Salamandridae*
*State Status: None*
*Federal Status: None*
*Global Rank: G5*
*State Rank: S5*
*Length: 7 in (18 cm)*

**Global Range:** Coastal mountains and lowlands from southern Alaska to Santa Cruz County, California; it is also found in the Cascade Mountains in Washington, Oregon, and northwestern California.

**Habitat:** The roughskin newt occurs in a wide variety of habitats, from grasslands to conifer and mixed forests; it requires ponds, lakes, or streams for breeding and is usually near a source of water. It hides under rocks, bark, or logs, but is often seen moving about in daylight.

**Reproduction:** These salamanders breed from December to July, the onset of breeding being earlier at lower elevations. Up to 30 eggs are laid singly and are attached to vegetation or other objects. The eggs hatch in 20-26 days and metamorphose before winter. Sexual maturity is reached at 3-4 years of age.

**Food Habits:** Larvae eat zooplankton and small aquatic invertebrates. Adults feed on earthworms, slugs, snails, and insect larvae, as well as amphibian eggs and tadpoles.

**Ecology:** The roughskin newt can migrate considerable distances to breeding ponds. It is preyed upon by common garter snakes. The average life span is 12 years, but individuals may live up to 26 years. The species can be very abundant in appropriate habitat.

**Comments:** This is the most commonly encountered northwestern salamander. The skin secretions of this salamander contain a potent toxin (tetrodotoxin) that protects the newt from most common predators and can be fatal to humans. It is chemically identical to the toxin found in puffer fish (such as the fugu fish of Japan), which has been responsible for many human deaths.

*References:* Brodie 1968, Davis and Twitty 1964, Pimentel 1960, White 1977.

# Western Toad
## (Bufo boreas)

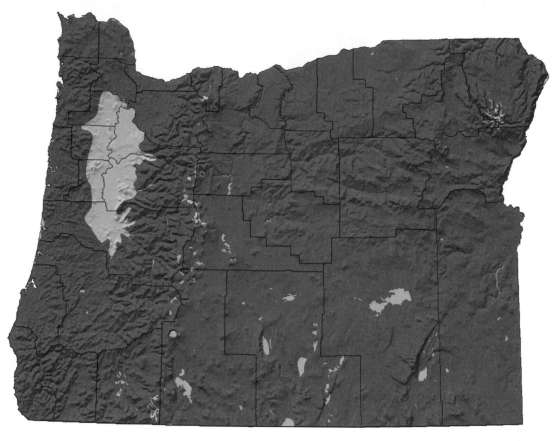

*Order: Anura*
*Family: Bufonidae*
*State Status: Sensitive*
*Federal Status: None*
*Global Rank: G4*
*State Rank: S4*
*Length: 4.5 in (11 cm)*

**Global Range:** Widespread west of the Rocky Mountains from southern Alaska and northern British Columbia south to Baja California, but not found in the Sonoran Desert of Arizona.

**Habitat:** The western toad can be found in a wide variety of habitats—in deserts, chaparral, grasslands, woodlands, and forests—from sea level to above timberline, provided some source of water is available for breeding.

**Reproduction:** These toads breed from February to July, depending on elevation. Females deposit strings of 12,000 jelly-covered eggs (range 30-16,000) in shallow water. Depending on water temperature, eggs hatch in 3 to 10 days. Tadpoles metamorphose in a few months (again, the exact time depends on temperature).

**Food Habits:** Tadpoles feed on algae and bottom detritus. Adults feed on the usual array of small invertebrates, including ants, beetles, spiders, earthworms, and crayfish.

**Ecology:** The western toad has adapted to human-modified environments such as irrigation canals. It can be very abundant—over 2,000 adults have been

observed in a single 100-hectare lake—though mortality between the egg stage and adulthood is over 99%.

**Comments:** This species has disappeared from areas where it was once common. It may be sensitive to human-induced changes in the global environment, such as increased ultra-violet radiation caused by thinning of the ozone layer. The reasons for local declines remain unknown.

*References:* Marshall et al. 1996, Olson 1989, Samallow 1980, Schonberger 1945, Smits 1984.

## Woodhouse's Toad
### *(Bufo woodhousii)*

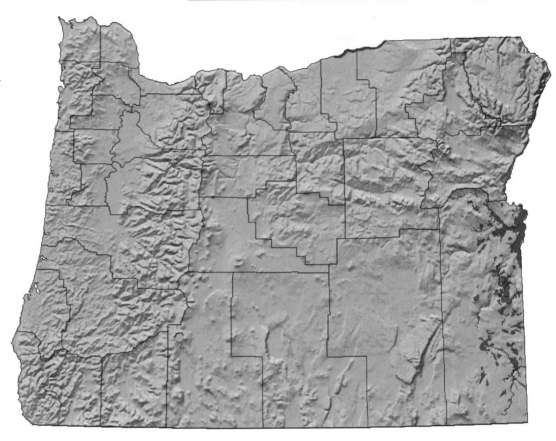

*Order: Anura*
*Family: Bufonidae*
*State Status: Sensitive*
*Federal Status: None*
*Global Rank: G5*
*State Rank: S2*
*Length: 4 in (10 cm)*

**Global Range:** Widespread throughout the eastern, central, and southwestern United States, but present only locally on the eastern edges of the Pacific states.

**Habitat:** Not as adapted to ephemeral waters as the western toad, Woodhouse's toad breeds in permanent waters of ponds, streams, rivers, reservoirs, and irrigation canals, though it has been seen breeding in shallow,

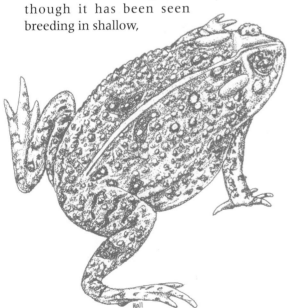

temporary ponds in Morrow County. It appears to have entered Oregon along the Columbia and Snake rivers, being found in agricultural and semi-desert habitat near those rivers.

**Reproduction:** Females lay a clutch of up to 25,000 eggs following spring rains. In Oregon, individuals have been observed breeding in late May to June. The tadpoles metamorphose in 1 to 2 months, and are sexually mature at 2 years of age.

**Food Habits:** Larvae eat a variety of algae and other organic matter found in breeding waters. Adults feed on a wide variety of small invertebrates.

**Ecology:** Like the western toad, this species burrows underground during hot, dry, or cold weather. Congregations of males vocalize around breeding ponds. This species may be subject to predation by introduced species (e.g., bullfrogs, bass).

**Comments:** Woodhouse's toad is peripheral to Oregon, known from near the Columbia River in Morrow County and near Nyssa and the Owyhee Reservoir in Malheur County.

*References:* Leonard et al. 1993, Marshall et al. 1996, Nussbaum et al. 1983, Woodward 1982.

21

# Pacific Chorus Frog
## (Pseudacris regilla)

*Order: Anura*
*Family: Hylidae*
*State Status: None*
*Federal Status: None*
*Global Rank: G5*
*State Rank: S5*
*Length: 1.5 in (4 cm)*

**Global Range:** Southern British Columbia to southern Baja California. The range extends east to the Rocky Mountains in Montana and all but the eastern part of Nevada. There are isolated populations in the California desert.

**Habitat:** This frog occurs in a variety of habitat types, from sagebrush deserts and grasslands to forests, from sea level to 11,000 feet. It is usually found in grass or low shrubs near water in the breeding season, but it can wander far from the nearest water source.

**Reproduction:** The breeding season varies from February to June, depending on temperature. Females lay several hundred eggs in clusters of 9 to 70. Eggs hatch in about 3 weeks, and tadpoles metamorphose 2 months later, enabling this species to exploit ephemeral ponds for breeding.

**Food Habits:** Larvae are thought to feed on a diet of aquatic algae and organic detritus. Adults eat a variety of small invertebrates, including small beetles, ants, spiders, isopods, and leafhoppers.

**Ecology:** Pacific chorus frogs are often very abundant, and are preyed upon by a variety of carnivores. Introduced bullfrogs appear to be so predacious on Pacific chorus frogs that the two are not found in the same ponds. Large insects have been known to catch tadpoles.

**Comments:** This species was formerly known as the Pacific treefrog (*Hyla regilla*).

*References:* Hedges 1986, Perrill and Daniel 1983, Schaub and Larsen 1978, Whitney and Krebs 1975.

*Order: Anura*

*Family: Leiopelmatidae*

*State Status: Sensitive*

*Federal Status: SC*

*Global Rank: G3G4*

*State Rank: S3*

*Length: 2 in (5 cm)*

**Global Range:** Forested coastal and interior mountains of the Northwest, from southern British Columbia to northern California. In the interior, it is found in the Blue Mountains of Oregon and the Rocky Mountains of Idaho and western Montana, although some herpetologists feel these populations are a different species.

**Habitat:** This species is restricted to cold, fast-flowing permanent streams from sea level to near timberline, normally in forests, though they are sometimes found in streams flowing through non-forested areas. Adults emerge at night to forage on the forest floor up to 75 feet away from streams. Larvae attach to smooth rock surfaces in well-oxygenated parts of streams.

**Reproduction:** Fertilization is internal, an adaptation to life in fast-flowing streams. Mating takes place in the fall, and 40 eggs (range 30-98) are laid early the following summer. Larvae hatch 6 weeks later, in August or September. Tadpoles transform from 1 to 4 years later, depending on local temperatures.

**Food Habits:** Larvae feed primarily on diatoms, algae, and pollen, which they scrape off rocks in streams. Adults feed on a wide variety of small invertebrates.

**Ecology:** Tailed frogs are not found in lakes or ponds; their chief requirement is cold, clear stream water and loss of riparian vegetation and sedimentation can render streams unsuitable. Their larvae are taken in large numbers by Pacific giant salamander larvae.

**Comments:** Environmental changes resulting from timber harvest are thought to cause extinction of local populations and fragmentation of habitat for this species.

*References:* Adams 1993, Brown 1975, Corn and Bury 1989, Daugherty and Sheldon 1982, Metter 1964, Welsh 1990, Werntz 1969.

# Great Basin Spadefoot
## (Scaphiopus intermontanus)

*Order: Anura*
*Family: Pleobatidae*
*State Status: None*
*Federal Status: None*
*Global Rank: G5*
*State Rank: S5*
*Length: 2.5 in (6 cm)*

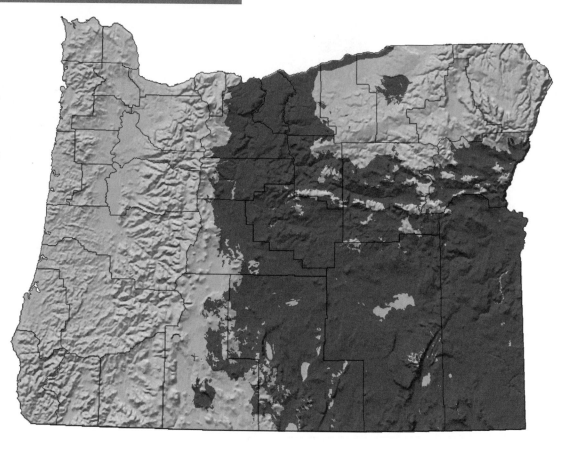

**Global Range:** Along the east side of the Cascade and Sierra Nevada mountains from southern British Columbia south to southern Nevada and east to the Rocky Mountains.

**Habitat:** The Great Basin spadefoot is found from sagebrush deserts through pinyon-juniper woodlands into open coniferous forests, usually in sandy areas near water. It uses a variety of temporary or permanent waters for breeding, including rain pools, temporary streams, roadside or irrigation ditches, and pond or reservoir edges.

**Reproduction:** This species begins breeding after spring rains. Females lay a total of 300-500 eggs in groups of 20-40. Eggs hatch quickly, usually within 2 or 3 days. The tadpole stage is likewise accelerated, lasting only a few weeks, probably as an adaptation to rapidly drying water sources in its arid habitat. These toads are mature by their third summer.

**Food Habits:** Tadpoles generally eat algae and organic debris, but some individuals eat animal material. Adults eat insects, including ants, beetles, grasshoppers, crickets, and flies.

**Ecology:** These toads are active on the surface, especially at night, following spring and summer rains. They are adapted to burrowing in sand or soft soil, allowing them to survive cold, hot, or dry periods. They are prey to a variety of mammalian and avian predators.

**Comments:** The species may have expanded its range in some areas by using irrigation ditches. However, the conversion of sagebrush deserts to pasture or grain crops has eliminated the toad from parts of its former range. Some taxonomists place this toad in the genus *Spea*.

*References:* Bragg 1965, Svihla 1953, Tanner 1939.

## Red-legged Frog
### (Rana aurora)

*Order: Anura*
*Family: Ranidae*
*State Status: Sensitive*
*Federal Status: SC*
*Global Rank: G4*
*State Rank: S3S4*
*Length: 3.5 in (9 cm)*

**Global Range:** Coastal mountains and west slope of the Cascade and Sierra Nevada mountains from southern British Columbia to northern Baja California.

**Habitat:** This frog occurs in meadows, woodlands, and forests, but is usually found near ponds, marshes, and streams, favoring areas with dense ground cover and aquatic or overhanging vegetation. In non-breeding season, it occurs up to 300 yards from standing water.

**Reproduction:** The breeding period is short (1-2 weeks), and may begin as early as February at low elevations. About 2,000 eggs (range 750-4,000) are attached to aquatic vegetation in quiet waters. Eggs hatch in about a month and tadpoles metamorphose about 4 months later. Sexual maturity is attained at 3 or 4 years of age.

**Food Habits:** Larvae eat aquatic algae and other organic debris. Adults eat invertebrates such as beetles, isopods, insect larvae, and other aquatic insects.

**Ecology:** The red-legged frog has short egg and tadpole stages, allowing it to use temporary ponds for breeding. It is preyed upon heavily by introduced bullfrogs.

**Comments:** This species is declining seriously in the Willamette Valley. Several recent surveys have failed to detect this species at sites in the valley where it was once common to abundant.

*References:* Blaustein and Wake 1990, Licht 1971, Marshall 1992, Storm 1960.

# Foothill Yellow-legged Frog
## *(Rana boylii)*

*Order: Anura*
*Family: Ranidae*
*State Status: Sensitive*
*Federal Status: SC*
*Global Rank: G3*
*State Rank: S3?*
*Length: 3 in (7 cm)*

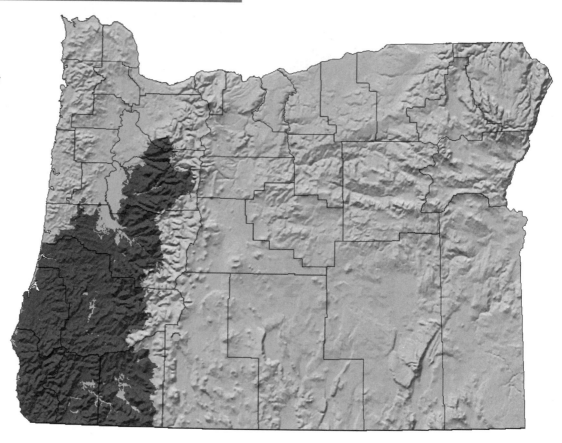

**Global Range:** Coastal and Cascade mountains of Oregon, south through the coastal and Sierra Nevada mountains of California to the Tejon Pass, northern Los Angeles County, and a few scattered but questionable localities as far south as northern Baja California.

**Habitat:** These frogs are generally found in permanent slow-flowing streams in a variety of habitat types, including grassland, chaparral, and coniferous or deciduous forests and woodlands. They prefer streams with rocky bottoms, stream-side vegetation, and sloping banks.

**Reproduction:** This species breeds after high water subsides, usually from late April to early June. Females lay clutches of about 1,000 eggs, which hatch in 5 days. Tadpoles metamorphose in 3 to 4 months.

**Food Habits:** Tadpoles feed on algae, plant tissue, and organic debris. Adults eat a variety of both aquatic and terrestrial invertebrates, including grasshoppers, hornets, ants, flies, beetles, and mosquitoes.

**Ecology:** The streams inhabited by this frog may dry to a series of potholes connected by trickles in summer. Small adults have been found 50 meters from permanent water on moist outcrops. If startled while along streambanks, they will jump into the water and hide under rocks. Garter snakes are a major predator of this species.

**Comments:** This frog was once considered common in its Oregon range, but some populations have declined greatly. However, a survey in 1984 found it still common along streams in Curry County. Present status and reasons for population declines are unknown.

*References:* Dumas 1966, Green 1986, Marshall 1992, Zweifel 1955.

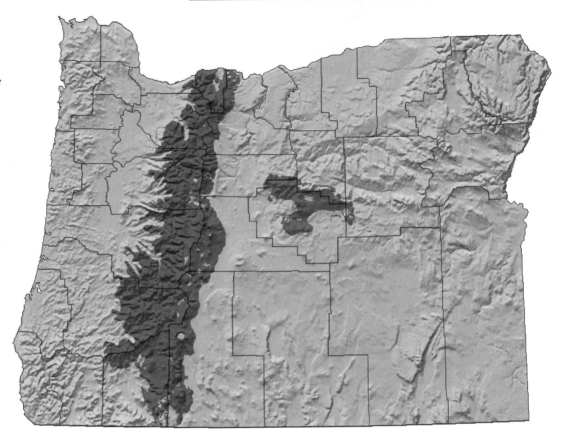

*Order: Anura*
*Family: Ranidae*
*State Status: Sensitive*
*Federal Status: SC*
*Global Rank: G4*
*State Rank: S3*
*Length: 2.5 in (6 cm)*

**Global Range:** Olympic Mountains of Washington and the Cascade Range from the Canadian border south to the vicinity of Mt. Lassen in California.

**Habitat:** Closely associated with water, this species occurs in lakes, ponds, and small streams that run through meadows. It is rarely found below 2,600 feet, but occurs up to timberline.

**Reproduction:** The Cascades frog breeds early, as soon as water is free of ice and snow. Females lay about 300 eggs (range 100-600). Tadpoles form schools during their 2-month larval period, and metamorphose in August or September, although some may overwinter as larvae. They reach sexual maturity after 3 years.

**Food Habits:** Larvae eat algae and other organic matter. Adults feed on a variety of small invertebrates.

**Ecology:** These frogs hibernate in mud or soil saturated by spring-water, sometimes up to 75 meters from standing water. They appear to be bottom feeders, preferring muddy or silty substrate of shallow waters.

**Comments:** Although their distribution was always discontinuous, they were formerly very abundant in some locations. One report of 30 locations studied since the 1970s found 80% had disappeared by 1990.

*References:* Blaustein and Wake 1990, Briggs 1987, Briggs and Storm 1970, Marshall 1992.

# Bullfrog
## (Rana catesbeiana)

*Order: Anura*
*Family: Ranidae*
*State Status: None*
*Federal Status: None*
*Global Rank: G5*
*State Rank: SE*
*Length 7.5 in (19 cm)*

**Global Range:** Native to most of North America east of the Rocky Mountains from southern Canada to the Gulf Coast of Mexico. Introduced and established in western North America from southern British Columbia to northern Mexico. Also introduced to Hawaii, some Caribbean Islands, Italy, and Japan.

**Habitat:** This species is always found in or near water. It is tolerant of a wide variety of aquatic habitats so long as there is permanent water. It occupies rivers, lakes, reservoirs, ponds, sloughs, marshes, or irrigation ditches where emergent vegetation grows at water's edge. It is rare or absent from colder, high-elevation waters.

**Reproduction:** In the Northwest, bullfrogs breed from May to August. The clutch size ranges from about 5,000 to over 15,000 eggs, which are deposited on the surface of the water and sink prior to hatching. Young overwinter as tadpoles and metamorphose the following summer. Bullfrogs do not breed until their second or third year.

**Food Habits:** Tadpoles eat algae and other submerged aquatic vegetation. They may also eat some animal matter. Adults are voracious carnivores that eat a variety of animals, including earthworms, insects, crayfish, frogs and their larvae, snakes, turtles, birds, and small mammals.

**Ecology:** Male bullfrogs are territorial during the breeding season. Bullfrogs overwinter in mud on the bottom of ponds. They may travel overland to ponds at least 1 mile away, usually moving during rainy weather. Bullfrogs in the Willamette Valley have been observed feeding on salamanders traveling across blacktop roads during rainy nights.

**Comments:** Bullfrog predation is thought to be responsible for the decline of a number of native aquatic species in Oregon, including the western pond turtle, spotted frog, and native fish. They are a game species in Oregon, but they are so prolific that control is difficult.

*References:* Hayes and Jennings 1986, Leonard et al. 1993, Nussbaum et al. 1983, Stebbins 1985.

*Order: Anura*
*Family: Ranidae*
*State Status: Sensitive*
*Federal Status: None*
*Global Rank: G5*
*State Rank: S2?*
*Length: 4 in (10 cm)*

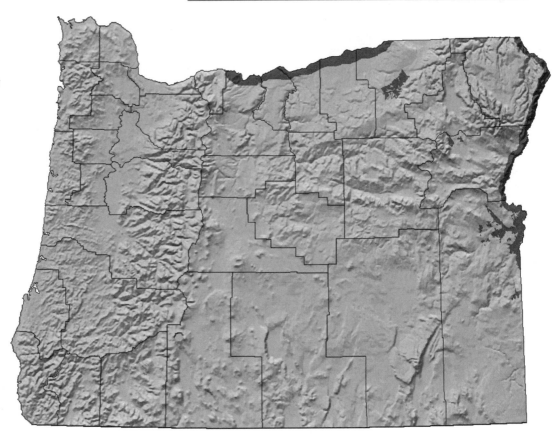

*References:* Hammerson 1982, Leonard et al. 1993, Marshall 1992, Pace 1974.

**Global Range:** Widely distributed from Nevada through the Plains states to New England and in most of Canada east of British Columbia and south of the Northwest Territories. Enters the Northwest in areas adjacent to the Columbia and Snake rivers.

**Habitat:** This frog lives in marshes, wet meadows, vegetated irrigation canals, ponds, and reservoirs. It prefers quiet or slowly flowing waters. It avoids areas without cover, but breeding waters have been found in a variety of habitat types.

**Reproduction:** Populations in the Northwest have not been studied well. However, elsewhere this species is an early breeder. Females lay clutches of several thousand eggs, which hatch in about a month. Tadpoles metamorphose 2 months after hatching and reach sexual maturity in 2 to 3 years.

**Food Habits:** Larvae eat algae, plant tissue, and other organic debris. Carnivorous adults eat both invertebrates (spiders, insects, snails, leeches) and vertebrates (birds, tadpoles, small frogs, small snakes, and fish).

**Ecology:** This species may forage far from water in damp meadows. It hibernates underwater during cold weather.

**Comments:** This frog is known in Oregon mostly from older records, and recent surveys have failed to find it in the state. It has disappeared from one Washington location, apparently due to predation by introduced bullfrogs.

# Spotted Frog
## (Rana pretiosa)

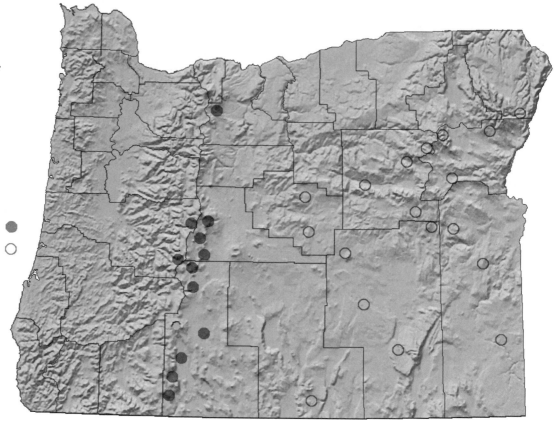

Order: Caudata
Family: Ranidae
State Status: Sensitive
Federal Status: C
Global Rank: G3G4
State Rank: S2
Length: 3.5 in (9 cm)

Rana pretiosa ●

Rana luteiventris ○

**Global Range:** Northern British Columbia and coastal southern Alaska south to the Rocky Mountains of Idaho, Montana, and Utah. Both interior and coastal mountains of the Pacific Northwest. Isolated populations occur in northern Nevada and Utah.

**Habitat:** The spotted frog frequents waters and associated vegetated shorelines of ponds, springs, marshes, and slow-flowing streams and appears to prefer waters with a bottom layer of dead and decaying vegetation. It is found in aquatic sites in a variety of vegetation types, from grasslands to forests.

**Reproduction:** This species breeds from February to June, depending on temperature. Females lay from 700 to 2,400 eggs. Most tadpoles metamorphose their first year, but some overwinter as larvae.

**Food Habits:** Larvae have a diet of algae, plant material, and other organic debris. Adults eat insects (ants, beetles, mosquito larvae, grasshoppers), spiders, mollusks, tadpoles, crayfish, and slugs.

**Ecology:** This is a highly aquatic species, usually found near cool, permanent, quiet water. Females are reported to lay egg masses in communal clusters at locations that may be used in successive years.

**Comments:** Once thought to be common west of the Cascades, spotted frogs have disappeared from the Willamette Valley and are found only at sites that do not support bullfrogs. They are currently known from less than 2 dozen sites in the high Cascades. Spotted frogs occur locally in eastern Oregon, but a recent study (Green et al. 1996) suggests these populations belong to another species. Green et al. (1997) have named the western Oregon populations the Oregon spotted frog (*Rana pretiosa*) and the eastern Oregon populations the Columbia spotted frog (*Rana luteiventris*).

References: Briggs 1987, Dumas 1966, Marshall 1992, Schonberger 1945.

*Reptiles*

# Painted Turtle
*(Chrysemys picta)*

*Order: Testudines*
*Family: Emydidae*
*State Status: Sensitive*
*Federal Status: None*
*Global Rank: G5*
*State Rank: S2*
*Length: 8 in (20 cm)*

**Global Range:** Widespread in the eastern and central United States, reaching the West Coast only in the Northwest. There are isolated and apparently introduced populations at several locations in the interior West.

**Habitat:** The painted turtle is found in shallow, quiet waters with a muddy or sandy substrate. They live in lakes, ponds, marshes, and small streams located in a variety of surrounding vegetation types. Basking sites and aquatic vegetation at water's edge are also important.

**Reproduction:** This species breeds from May to July, with females laying clutches of 5 to 8 eggs (range 4-20) in nests dug in soft ground. The nests may be up to several hundred meters from water. Hatchlings usually winter in the nest and reach maturity 4 to 6 years after hatching, depending on climate.

**Food Habits:** This turtle eats both plants—including algae, duckweed, and bulrush—and animal matter—including spiders, beetles, insect larvae, earthworms, crayfish, fish, frogs, and tadpoles. Young are more carnivorous, and adults are more herbivorous.

**Ecology:** This turtle hibernates buried in bottom mud in lakes, ponds, or streams, and spends a considerable amount of time basking.

**Comments:** The painted turtle appears to be declining in Oregon due to lack of recruitment. Predation on young by introduced bullfrogs may be responsible for the decline.

*References:* Congdon and Gatten 1989, Gibbons 1968, Marshall 1992, Moll 1973.

*Order: Testudines*
*Family: Emydidae*
*State Status: Sensitive*
*Federal Status: SC*
*Global Rank: G3*
*State Rank: S2*
*Length: 7 in (18 cm)*

**Global Range:** From the Columbia River south to northwestern Baja California. Most populations are west of the crests of the Cascade, Sierra Nevada, Transverse, and Peninsular mountains, although interior populations (possibly introductions) have been reported in the past. Apparently extirpated from western Washington.

**Habitat:** The western pond turtle prefers quiet water in small lakes, marshes, and sluggish streams and rivers. It frequents bodies of water with muddy or rocky bottoms, but requires basking sites, such as logs, rocks, mudbanks, or cattail mats. Nests can be several hundred meters from water in a variety of vegetation types.

**Reproduction:** Clutches of 5-13 eggs are laid in terrestrial nests from late May to August. Eggs hatch in about 12 weeks. Hatchlings are thought to overwinter in the nest, and emerge in the spring. They reach sexual maturity in about 10 years and may live 40 years or more.

**Food Habits:** Turtles are scavengers and opportunistic predators, eating both plant and animal food. They take insects, earthworms, molluscs, crayfish, fish, tadpoles, and frogs.

**Ecology:** Western pond turtles usually hibernate in bottom mud, but are sometimes seen basking during warm winter days in the Willamette Valley. The size of their home range varies from a fraction of an acre to over 7 acres, and they may hibernate as far as 1,600 feet from water.

**Comments:** Formerly quite common in the Willamette Valley, they have declined by as much as 96-98% since the beginning of the century. A recent survey found only six areas supporting more than 20 turtles. They are thought to be more common in the Rogue and Umpqua river systems. Introduced predators, such as bullfrogs and bass, may eliminate young turtles from a population.

*References:* Bury 1986, Holland 1993, Marshall 1992, Marshall et al. 1996.

# Northern Alligator Lizard
## (Elgaria coerulea)

*Order: Squamata*
*Family: Anguidae*
*State Status: None*
*Federal Status: None*
*Global Rank: G5*
*State Rank: S5*
*Length: 10 in (25 cm)*

**Global Range:** From southern British Columbia and Vancouver Island south through the Rocky Mountains to northern Idaho and western Montana, and in the Cascade, Coastal, and Sierra Nevada mountains as far south as central California.

**Habitat:** This lizard prefers humid areas, such as the edges of meadows in coniferous forests, and is also found in riparian zones. This is the only lizard found in the cool coastal forests of northern Oregon.

**Reproduction:** This species mates in April and May. The eggs are retained in the oviduct, and young are born about 3 months later. The average litter is 4 (range 2-8). In northern California, females attain sexual maturity when they are 32 to 44 months old.

**Food Habits:** Like the southern alligator lizard, this species eats small invertebrates (termites, beetles, ticks, spiders, millipedes, and snails), and occasionally takes small birds, mammals, and other lizards.

**Ecology:** This species is adapted to lower temperatures and higher elevations than most reptiles. It is active during the day, and hunts for prey on the ground and under objects. It seems to require high humidity in feeding areas.

**Comments:** This lizard hibernates in winter; the duration of the inactive period varies with local climate. It may be at risk where it is sympatric with the cinnabar moth, which accumulates toxins while feeding on tansy ragweed that are fatal to northern alligator lizards. Cinnabar moths were introduced into southwestern Oregon to control this poisonous plant, which is native to Europe. Some herpetologists place this lizard in the genus *Gerrhonotus*.

*References:* Good 1988, Nussbaum et al. 1983, Stewart 1985, Vitt 1973.

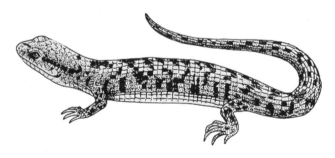

# Southern Alligator Lizard
### *(Elgaria multicarinata)*

*Order: Squamata*
*Family: Anguidae*
*State Status: None*
*Federal Status: None*
*Global Rank: G5*
*State Rank: S5*
*Length: 11 in (28 cm)*

**Global Range:** In the northern part of its range, this lizard is found in southern Washington and northern Oregon east of the Cascades. From the Columbia River south to its southern limits in northern Baja California, it occurs mostly west of the Cascade and Sierra Nevada mountains.

**Habitat:** The southern alligator lizard is found in a variety of habitats from grassland and chaparral to oak woodlands and edges of open coniferous forests, as well as riparian zones and moist canyon bottoms. It requires thickets, brush heaps, downed logs, or rock piles for cover.

**Reproduction:** This lizard mates from April to June, and deposits clutches of 12 (range 5-20) eggs from June to August. Hatchlings emerge 2 months later, in September and October.

**Food Habits:** This carnivorous lizard feeds primarily on small invertebrates (slugs, spiders, centipedes, scorpions, beetles, grasshoppers, and crickets), but also is known to feed on bird eggs, nestlings, other lizards, and small mammals.

**Ecology:** The southern alligator lizard is active during the day in cold weather, but becomes nocturnal in the warmer parts of the year. It searches for prey on the ground and under surface objects. This species sometimes enters water to escape from predators.

**Comments:** Apparently this lizard is sensitive to the skin secretions of amphibians, and does not feed on them. Some herpetologists place this lizard in the genus *Gerrhonotus*.

*References:* Brodie et al. 1969, Goldberg 1972, Good 1988.

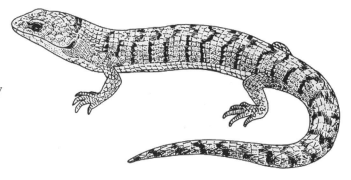

# Mojave Black-collared Lizard
## (*Crotaphytus bicinctores*)

*Order: Squamata*
*Family: Iguanidae*
*State Status: Sensitive*
*Federal Status: None*
*Global Rank: G5*
*State Rank: S2*
*Length: 13 in (33 cm)*

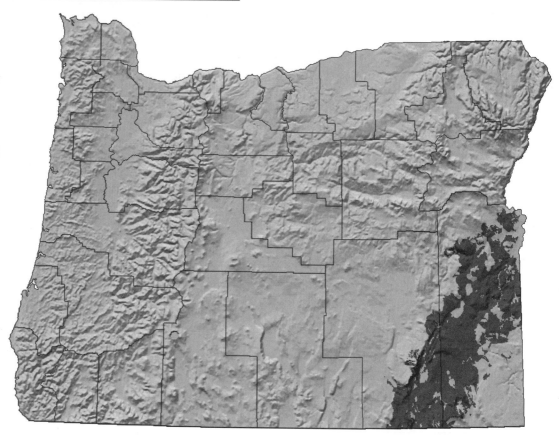

**Global Range:** Southeastern Oregon and adjacent Idaho, south through the western Great Basin and Mojave deserts to southern Baja California.

**Habitat:** This lizard is found in a variety of desert shrub vegetation types, but is most dependent on the presence of rock outcrops, boulders, or talus slopes.

**Reproduction:** Females probably lay clutches of 3-8 eggs in June or July and may lay two clutches a year. Some females reach sexual maturity in 1 year. Reproduction has not been studied in Northwest populations.

**Food Habits:** This lizard is an aggressive carnivore that eats a variety of other reptiles and large insects, such as crickets and grasshoppers, as well as some plant material.

**Ecology:** It can be active on warm winter days, but is rarely seen above ground when the air temperature is below 50° F. Males are territorial in the breeding season.

**Comments:** This species reaches the northern edge of its range in southeastern Oregon, and is uncommon and discontinuously distributed there. A study in 1985 found 28 occupied areas in Oregon, but there are probably more. This large and brightly colored lizard is subject to overcollecting. This species was formerly known as the desert collared lizard (*Crotaphytus insularis*)

*References:* Marshall 1992, Montanucci 1983, Whitaker and Maser 1981.

## Longnose Leopard Lizard
*(Gambelia wislizenii)*

*Order: Squamata*
*Family: Iguanidae*
*State Status: None*
*Federal Status: None*
*Global Rank: G5*
*State Rank: S4*
*Length: 12 in (30 cm)*

**Global Range:** From southeastern Oregon and southern Idaho south through the Great Basin and arid Southwest to California, Baja California, and the mainland of Mexico.

**Habitat:** These lizards are found in open desert shrublands, particularly where islands of sand have accumulated around shrubs, and are absent where a dense grass understory would inhibit their ability to run.

**Reproduction:** A clutch of 4 to 7 eggs is laid in a burrow in sandy soil in late May or June in the Northwest. The eggs hatch in about 5 to 7 weeks. Sexual maturity is reached at 2 years of age.

**Food Habits:** This lizard eats large insects, such as grasshoppers, crickets, and beetles, and also takes small vertebrates, including pocket mice, side-blotched lizards, whiptails, and fence lizards. Some plant material (flowers, berries) is eaten when available.

**Ecology:** Active only during the warmer months, longnose leopard lizards are active hunters, becoming bipedal at top speeds. These lizards use rodent burrows for shelter, and are most abundant where rodent burrows are common.

**Comments:** This species reaches the northern limit of its range in Oregon. It is an uncommon lizard, but is encountered more frequently in southeastern Oregon than is its close relative, the Mojave black-collared lizard. Two isolated populations along the Columbia River are thought to be extinct.

*References:* Mitchell 1984, Tanner and Krogh 1974, Whitaker and Maser 1981.

# Short-horned Lizard
## (Phrynosoma douglassii)

Order: Squamata
Family: Iguanidae
State Status: None
Federal Status: None
Global Rank: G5
State Rank: S4?
Length: 4 in (10 cm)

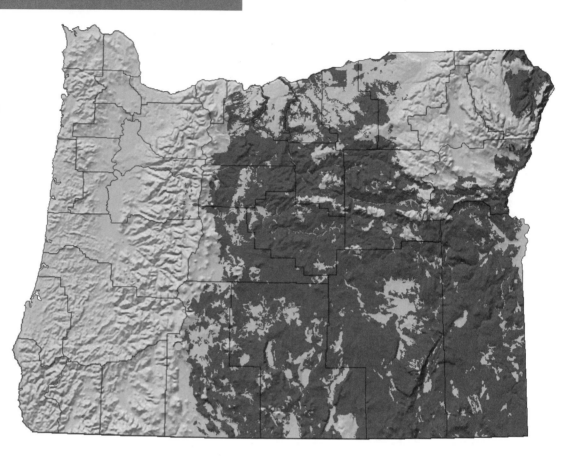

**Global Range:** Most of interior western North America from southern Canada to central Mexico. Nowhere does it reach the Pacific Coast.

**Habitat:** The short-horned lizard occurs in sagebrush deserts, juniper woodlands, and open coniferous forests. It prefers open areas with sandy soils, but is also found on rocky soil.

**Reproduction:** This species mates soon after emerging from winter hibernation, usually just after snowmelt. The female bears 5 to 10 (range 3-15) live young 2 months after fertilization. They reach sexual maturity in 2 years.

**Food Habits:** Ants make up a large part of the diet, but beetles, caterpillars, spiders, and sowbugs are also eaten.

**Ecology:** This lizard burrows into the soil when inactive. It is preyed upon by the usual range of avian, reptilian, and mammalian predators.

**Comments:** More cold-tolerant than other horned lizards, it can be common at elevations as high as the Cascade passes. Early mating and live bearing of young may allow it to exploit colder habitats than its relatives.

*References:* Dumas 1964, Pianka and Parker 1975, Whitaker and Maser 1981.

*Order: Squamata*
*Family: Iguanidae*
*State Status: Sensitive*
*Federal Status: None*
*Global Rank: G5*
*State Rank: S3*
*Length: 5.5 in (14 cm)*

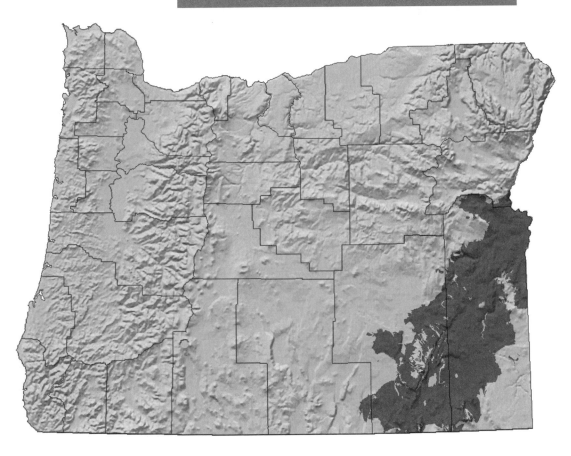

**Global Range:** Southeast Oregon and southwest Idaho south through the Great Basin, Mojave, and Sonoran deserts to extreme northern Mexico.

**Habitat:** The desert horned lizard is found in flat or gently rolling deserts covered with sagebrush or salt-desert shrub. It prefers areas with scattered bushes and loose, sandy soil, but sometimes occurs in rocky areas or on hardpan.

**Reproduction:** Clutches of 7 or 8 (range 2-16) eggs are laid around early June in the Northwest. Hatchlings appear after a 50-60 day incubation period, and mature by their second spring.

**Food Habits:** Primary foods are ants and beetles. It also feeds on insect larvae, spiders, crickets, flies, and small grasshoppers.

**Ecology:** This lizard burrows a few centimeters under the sand at night. Duration of seasonal activity varies with climate, but adults are rarely seen after mid-July. A wide variety of predators take this lizard, including Prairie Falcons, Loggerhead Shrikes, whipsnakes, and leopard lizards.

**Comments:** This is an uncommon species that is collected easily and its populations may be depleted in parts of its Oregon range by over-collecting. Although they are occasionally found together, this species usually occurs at lower elevations than the short-horned lizard.

*References:* Marshall 1992, Pianka and Parker 1975, Tanner and Krogh 1973, Whitaker and Maser 1981.

# Sagebrush Lizard
## (Sceloporus graciosus)

*Order: Squamata*
*Family: Iguanidae*
*State Status: None*
*Federal Status: None*
*Global Rank: G5*
*State Rank: S5*
*Length: 5.5 in (14 cm)*

**Global Range:** Western United States from near the Canadian border in Montana west to the Pacific Ocean and south to Mexico. Not found in southern Arizona or southwestern New Mexico.

**Habitat:** As their common name implies, these lizards are found in sagebrush habitats, but also occur in chaparral, juniper woodlands, and coniferous forests. They require well-illuminated open ground near cover and are primarily ground dwellers.

**Reproduction:** A clutch of 4 eggs (range 2-7) is laid in June or July in a hole dug in loose, well-aerated soil. There may be 2 clutches a year. The hatchlings emerge in mid-August, and reach maturity about 2 years later.

**Food Habits:** They eat a variety of small invertebrates, including crickets, beetles, flies, ants, wasps, bees, mites, ticks, and spiders.

**Ecology:** This lizard is inactive during the winter months. In Oregon, it is seldom found above 1,700 meters elevation. It can be found along river bottoms in the coastal redwood forests of southwestern Oregon.

**Comments:** This is the most common lizard of the sagebrush plains of southeastern Oregon, often occurring at high densities.

*References:* Marcellini and Mackey 1970, Tinkle 1973, Woodbury and Woodbury 1945.

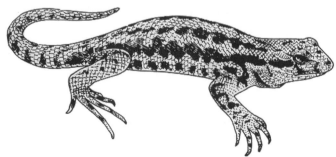

## Western Fence Lizard
*(Sceloporus occidentalis)*

*Order: Squamata*
*Family: Iguanidae*
*State Status: None*
*Federal Status: None*
*Global Rank: G5*
*State Rank: S5*
*Length: 7 in (18 cm)*

**Global Range:** Western United States west of the Rocky Mountains to the Pacific Coast and south to northern Baja California. Absent from the arid deserts of southeastern California and Arizona.

**Habitat:** This lizard occupies a wide range of habitats, from desert canyons and grasslands to coniferous forests. It requires vertical structure in its habitat, such as rock piles or logs. It is absent from dense, humid forests and flat desert valleys.

**Reproduction:** Breeding season begins in April. Clutches of 8 eggs (range 3-17) are laid in a pit dug by the female and covered with loose soil. Eggs hatch in about 2 months, and juveniles reach sexual maturity in the spring of their second year.

**Food Habits:** Western fence lizards are insectivorous, and feed on crickets, grasshoppers, beetles, ants, wasps, leafhoppers, and aphids. Some spiders are taken as well.

**Ecology:** Generally, this species is found only in mountain ranges in arid southeastern Oregon. They are inactive in cold weather, and are active longer in warmer parts of the state.

**Comments:** Adult males are territorial during the breeding season, and often can be observed basking in the morning sun on top of rocks or logs.

*References:* Davis and Ford 1983, Davis and Verbeek 1972, Marcellini and Mackey 1970.

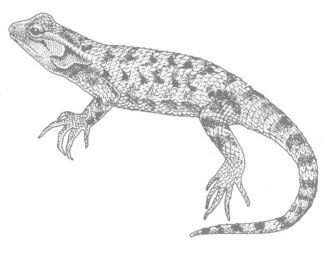

41

# Side-blotched Lizard
## (Uta stansburiana)

*Order: Squamata*
*Family: Iguanidae*
*State Status: None*
*Federal Status: None*
*Global Rank: G5*
*State Rank: S5*
*Length: 5 in (13 cm)*

**Global Range:** From eastern Washington and Oregon south through the Great Basin to the desert Southwest, central and southern California, and northern Mexico.

**Habitat:** The side-blotched lizard occurs in arid and semi-arid regions with scattered bushes and/or trees. In Oregon, it is found in sagebrush, juniper, and shadscale habitats. It is often found on sandy bottoms of washes or canyons, especially in the presence of scattered rocks.

**Reproduction:** Females lay 1 or 2 clutches of 3 (range 2-5) eggs per year, beginning from late April to early June. The eggs are buried in sandy soil, and hatch in 2 months. Hatchlings from early clutches may reach maturity their first year, but those from later clutches may not mature until their second year.

**Food Habits:** This small lizard feeds mostly on smaller invertebrates such as spiders, mites, ticks, sowbugs, beetles, flies, ants, and small grasshoppers.

**Ecology:** This lizard is usually the most abundant reptile in communities where it occurs. There may be significant winter mortality in the Northwest;

over 500 dead side-blotched lizards were found in a winter hibernaculum in central Oregon.

**Comments:** Predators include a variety of other reptiles and predatory birds such as shrikes and American Kestrels. This species has a small home range, perhaps less than a radius of 50 feet.

*References:* Nussbaum and Diller 1976, Parker and Pianka 1975, Tinkle 1967, Whitaker and Maser 1981.

*Order: Squamata*
*Family: Scincidae*
*State Status: None*
*Federal Status: None*
*Global Rank: G5*
*State Rank: S5*
*Length: 8 in (20 cm)*

**Global Range:** North America west of the Rocky Mountains from southern British Columbia south to northern Arizona and, in coastal mountains and valleys, to southern Baja California.

**Habitat:** The western skink is found in moist places, such as under rocks or logs, in a variety of habitats from grassland and chaparral to desert scrub, juniper woodlands, and coniferous woodlands and forests. Rocky areas with some moisture, such as riparian zones, are favored.

**Reproduction:** This species breeds from April to June. A clutch of 2 to 6 eggs is laid in a burrow that the female excavates under a rock or log. She guards the eggs until they hatch, about 2 months later.

**Food Habits:** The western skink feeds on a wide variety of invertebrates, including beetles, grasshoppers, moths, flies, spiders, and earthworms.

**Ecology:** An active diurnal predator, skinks stalk prey in a cat-like fashion. They are inactive in cold weather.

**Comments:** Duration of winter inactivity varies with location. Western skinks are apparently absent from the Oregon coast north of Coos Bay.

*References:* Nussbaum et al. 1983, Tanner 1957, 1988.

43

# Western Whiptail
## (Cnemidophorus tigris)

Order: Squamata
Family: Teiidae
State Status: None
Federal Status: None
Global Rank: G5
State Rank: S4
Length: 12 in (30 cm)

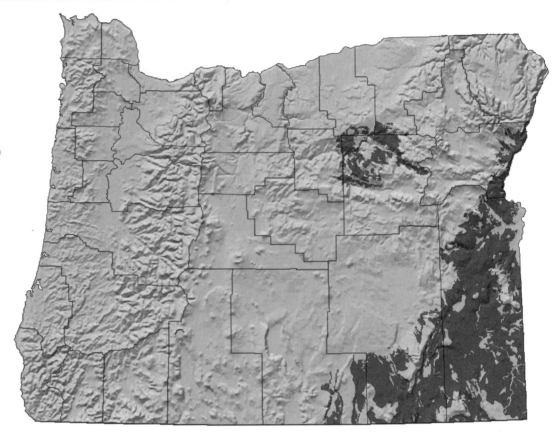

**Global Range:** Reaches the northern limit of its range in Oregon. It is found south to Baja California and northern Mexico and east to the Rocky Mountains.

**Habitat:** The western whiptail is found in eastern Oregon deserts and semiarid shrublands. It is most common in flat, sandy areas and along dry washes.

**Reproduction:** Breeding season begins in early June. Clutches of 1 to 4 eggs are laid in late June or early July and hatch in early to mid-August.

**Food Habits:** Western whiptails are primarily insectivorous. In a food habits study in southeastern Oregon, they ate caterpillars, crickets, grasshoppers, and beetles. They also eat spiders, scorpions, and other lizards.

**Ecology:** These lizards are inactive in winter and also during the hottest part of summer. They begin winter hibernation in early September. They are known to dig up prey with their front feet.

**Comments:** A population of the parthenogenetic plateau striped whiptail (*Cnemidophorus velox*) has been introduced to Cove Palisades State Park, Jefferson County, Oregon.

*References:* Hendricks and Dixon 1984, Maya and Malone 1989, Pianka 1970.

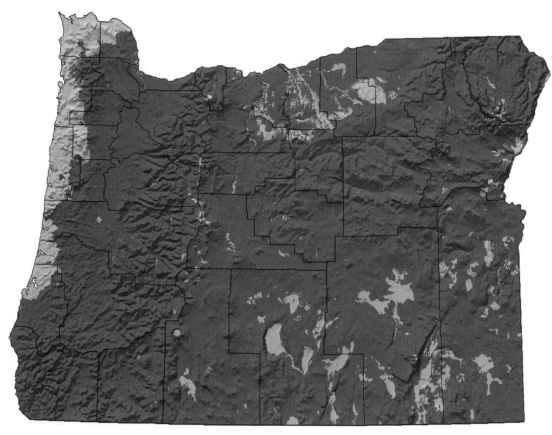

## Rubber Boa
### (Charina bottae)

*Order: Squamata*
*Family: Boidae*
*State Status: None*
*Federal Status: None*
*Global Rank: G5*
*State Rank: S4*
*Length: 24 in (61 cm)*

**Global Range:** Western North America from southern British Columbia south to southern California. Found from the Coast Ranges and Sierra Nevada east through the northern Great Basin to the Rocky Mountains in Montana, Colorado, and Utah.

**Habitat:** The rubber boa occurs in a variety of habitats, from desert scrub, foothill woodlands, and grasslands through deciduous and coniferous forests. In the Oregon coast ranges, it is found commonly in forest clearings that contain rotting stumps and logs. It is absent from the immediate vicinity of the coast north of Coos Bay.

**Reproduction:** This species bears live young. Mating occurs from April to late May or early June, with young usually born in late summer. Litter size averages 4 (range 2-8).

**Food Habits:** Rubber boas are constrictors and eat small mammals, especially young mice and shrews. There are two reports of them eating snakes.

**Ecology:** Rubber boas are usually found under logs or rocks. They are mostly active from dusk to dawn, but are also commonly active during the day in spring and early fall. Although seldom encountered, they can be common in appropriate habitat.

**Comments:** Forest clear-cuts provide habitat for this species. Most reptiles are inactive in the colder parts of the year, but rubber boas are active from February to early November, and are sometimes seen in January and, rarely, December.

*References:* Hoyer 1974, Nussbaum et al. 1983, Nussbaum and Hoyer 1974.

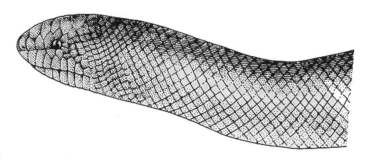

**45**

# Racer
## (Coluber constrictor)

*Order: Squamata*
*Family: Colubridae*
*State Status: None*
*Federal Status: None*
*Global Rank: G5*
*State Rank: S4?*
*Length: 36 in (91 cm)*

**Global Range:** Widespread throughout North and Central America, from southern British Columbia to Maine, and south to Florida and southern California. Central populations extend south from Texas to Guatemala.

**Habitat:** The racer is found in a variety of open habitats, including sagebrush flats, juniper woodlands, chaparral, and meadows. It avoids dense forests, high mountains, and very dry areas, and seeks cover under rocks, logs, or dense shrubs.

**Reproduction:** The racer breeds from April to June, laying a clutch of 3 to 7 eggs, which hatch 6 to 9 weeks later in late summer. Eggs are deposited in rotten logs, abandoned rodent burrows, under debris, or in stable talus. Racers are sexually mature in 2 to 3 years.

**Food Habits:** This species feeds on lizards, smaller snakes, frogs, toads, small mammals, birds and their eggs, and some insects. Young racers eat crickets, grasshoppers, and other insects.

**Ecology:** Racers are active diurnal predators. They may climb trees while hunting. They are inactive in winter, and often over-winter in communal dens, which can be some distance from their summer home range.

**Comments:** Some herpetologists have proposed that the western racer is a distinct species (*Coluber mormon*), but this has not been generally accepted.

*References:* Brown and Parker 1976, Corn and Bury 1986a, Fitch and Brown 1981, Greene 1984.

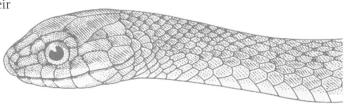

## Sharptail Snake
### *(Contia tenuis)*

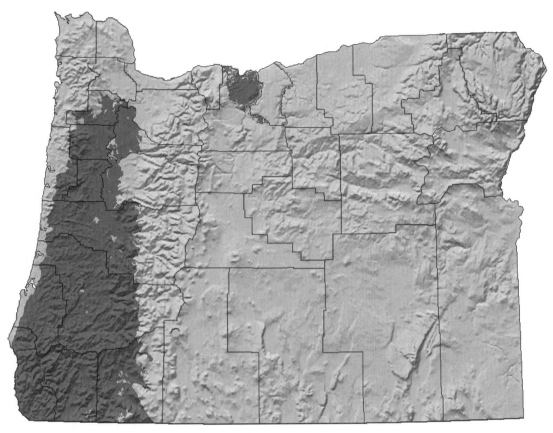

*Order: Squamata*
*Family: Colubridae*
*State Status: Sensitive*
*Federal Status: None*
*Global Rank: G5*
*State Rank: S3*
*Length: 12 in (30 cm)*

**Global Range:** Southern British Columbia south to central California in both coastal mountains and the Sierra Nevada. Isolated populations occur on Vancouver Island and in Washington.

**Habitat:** This snake is found in moist areas in coniferous forest, deciduous woodlands, chaparral, and grasslands. It frequents open grassy areas at forest edges and usually occurs under the cover of logs, rocks, fallen branches, or talus.

**Reproduction:** The female lays a clutch of 3 to 9 eggs in early summer. Incubation period is thought to be about 3 months.

**Food Habits:** The sharptail snake appears to specialize in feeding on slugs.

**Ecology:** This snake is active at lower temperatures (10°-17° C) than most reptiles and is active from late February to November in the Willamette Valley. In northern Oregon, these snakes are found at low elevations; however, they occur at higher elevations in the Siskiyou Mountains.

**Comments:** Although some herpetologists feel it is rare and declining in the Willamette Valley, the sharptail snake may be more common than thought.

*References:* Cook 1960, Marshall 1992.

# Ringneck Snake
## (Diadophis punctatus)

*Order: Squamata*
*Family: Colubridae*
*State Status: None*
*Federal Status: None*
*Global Rank: G5*
*State Rank: S4?*
*Length: 20 in (51 cm)*

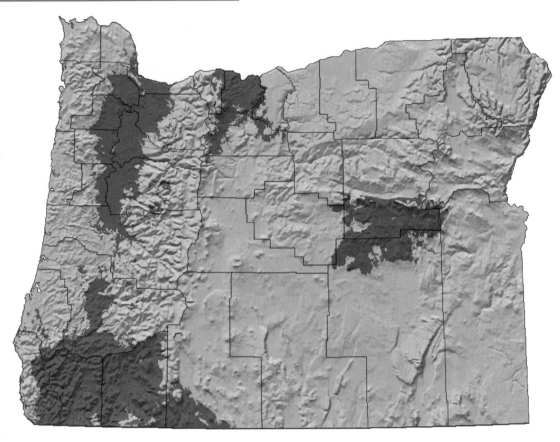

*References:* Blanchard 1942, Fitch 1975, Parker and Brown 1974.

**Global Range:** Widespread from the southwestern United States east to the Atlantic Coast and south into Mexico. It occurs along the Pacific Coast north to Vancouver, B.C., and farther north in the Canadian Rockies to central British Columbia, but is absent from most of the desert and interior West of the United States.

**Habitat:** The ringneck snake requires moist microhabitats such as downed logs, rocks, or stumps. It is found in a variety of vegetation types, but is most closely associated with pine-oak woodlands and moist canyon bottoms. It also can be abundant in Willamette Valley grasslands.

**Reproduction:** It breeds from March to July, laying a clutch of 3 eggs (range 1-10) in moist soil or under logs or rocks. The eggs hatch in late summer.

**Food Habits:** These snakes feed mainly on small lizards, snakes, and salamanders, and also slugs, earthworms, frogs, and insects.

**Ecology:** Ringneck snakes are active at night. In the Northwest, they may be seen from March to November, although the active season varies from year to year in response to the weather.

**Comments:** There are only a few records from east of the Cascade Mountains in Oregon, including a sight record near Troy on the lower Grande Ronde River by Alan D. St. John. Indirect evidence (tooth morphology and feeding behavior) suggests this snake is venomous, but it poses no known danger to humans.

*Order: Squamata*
*Family: Colubridae*
*State Status: None*
*Federal Status: None*
*Global Rank: G5*
*State Rank: S3*
*Length: 18 in (46 cm)*

**Global Range:** Arid western North America from central Washington south through Mexico. Generally west of the Rocky Mountains but east as far as central Texas. It occurs along the Pacific Coast from central California south through Baja California.

**Habitat:** In the Northwest, the night snake frequents arid desert scrub habitats near rocky outcrops or rimrock. It takes refuge in talus slopes or rocky crevices during the day.

**Reproduction:** A clutch of about 4 eggs (range 3-9) is deposited in late June or early July. Eggs hatch in about 2 months. Night snakes are sexually mature in about 1 year.

**Food Habits:** Night snakes tend to feed on cold-blooded prey, especially lizards and their eggs, frogs, toads, salamanders, large insects, and small snakes.

**Ecology:** These snakes are primarily nocturnal, and are more active on relatively cool nights of early summer.

**Comments:** The saliva of night snakes is toxic to small reptiles and amphibians, although these snakes pose no known threat to humans.

*References:* Diller and Wallace 1986, Nussbaum et al. 1983.

# Common Kingsnake
## *(Lampropeltis getula)*

*Order: Squamata*
*Family: Colubridae*
*State Status: Sensitive*
*Federal Status: None*
*Global Rank: G5*
*State Rank: S2*
*Length: 39 in (99 cm)*

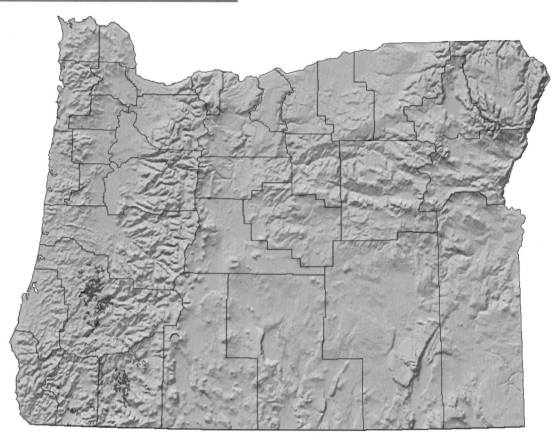

**Global Range:** From the Pacific to Atlantic coasts, reaching the northern limit of its range in Oregon and along the mid-Atlantic seaboard. It occurs south through Mexico.

**Habitat:** This snake is most common in thick vegetation along watercourses, but ranges into farmland, chaparral, and deciduous and mixed coniferous woodlands in the Rogue and Umpqua river valleys of southwestern Oregon.

**Reproduction:** Although the reproductive biology of Northwestern populations has not been studied, in other areas clutches of about 9 (range 2-12) eggs are laid in July. Eggs are usually laid in loose, well-aerated soil. The incubation period is just over 2 months.

**Food Habits:** Common kingsnakes usually feed on other snakes, but have been known to take small turtles, birds and their eggs, frogs, lizards, reptile eggs, and some small mammals.

**Ecology:** These snakes are active from April through October. They take cover under rocks, downed wood, or other debris. Newly hatched young are active and aggressive. Adults can kill other snakes as large as themselves.

**Comments:** This uncommon snake is known in Oregon from about 17 records. It is prized by collectors, and has a discontinuous distribution.

*References:* Blaney 1977, Marshall 1992, Nussbaum et al. 1983.

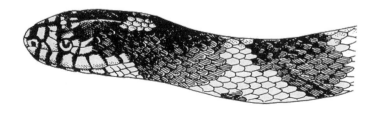

*Order: Squamata*
*Family: Colubridae*
*State Status: Sensitive*
*Federal Status: None*
*Global Rank: G4*
*State Rank: S3*
*Length: 30 in (76 cm)*

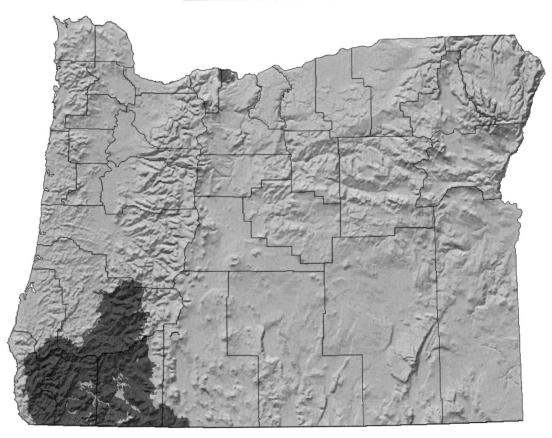

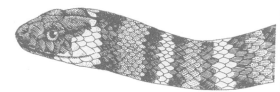

**Global Range:** Coastal and interior mountains from southwestern Oregon to Baja California. There are isolated populations along the Columbia River just east of the Cascade Mountains in Washington. Nussbaum et al. (1983) report an unverified population near the town of Maupin, Wasco County, Oregon.

**Habitat:** This species is found in pine forests, oak woodland, and chaparral of southwestern Oregon valleys. It is usually found in, under, or near rotting logs in open wooded areas near streams.

**Reproduction:** Breeding season is in April and May. Clutches of about 6 (range 3-8) eggs are laid in rotting logs or loose soil in June and July, and hatch in about 2 months.

**Food Habits:** The California mountain kingsnake preys upon snakes, lizards, birds and their eggs, and some small mammals.

**Ecology:** It is usually active in the daytime, but may become nocturnal during hot weather. Somewhat cold-tolerant, it can occur at higher elevations in mountains.

**Comments:** This rare snake is sought by collectors. There is questionable evidence for the presence of populations in Wasco County, Oregon. The taxonomic status of the Washington populations requires clarification.

*References:* Marshall 1992, Nussbaum et al. 1983, Zweiffel 1974.

# Striped Whipsnake
*(Masticophis taeniatus)*

*Order: Squamata*
*Family: Colubridae*
*State Status: None*
*Federal Status: None*
*Global Rank: G5*
*State Rank: S4*
*Length: 63 in (160 cm)*

**Global Range:** Central Washington south through the Great Basin and desert Southwest into central Mexico.

**Habitat:** In the Northwest, this snake is found in grasslands, sagebrush flats, rocky stream courses, and canyon bottoms. Elsewhere it also frequents juniper and pine-oak woodlands. In southwestern Oregon, it is found in dry, bushy areas close to rocks.

**Reproduction:** These snakes mate in late April and May, laying a clutch of 3 to 10 eggs in late June or early July. Eggs hatch about 2 months later, in late August to early September. Communal nest sites are often located in abandoned rodent burrows.

**Food Habits:** Young feed primarily on lizards and insects. Adults also take snakes, small mammals, young birds, and insects.

**Ecology:** This species is inactive during winter, hibernating in a den which may be as far as 1 kilometer from the summer home range. It is active during the daytime.

**Comments:** This is a highly active snake with keen eyesight. It is difficult to observe or capture, streaking away if a person approaches within 15-20 meters.

*References:* Bennion and Parker 1976, Nussbaum et al. 1983, Parker and Brown 1980.

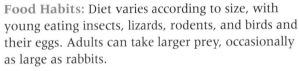

# Gopher Snake
## *(Pituophis melanoleucus)*

*Order: Squamata*
*Family: Colubridae*
*State Status: None*
*Federal Status: None*
*Global Rank: G5*
*State Rank: S5*
*Length: 40 in (102 cm)*

**Global Range:** This ubiquitous snake occurs from southern Canada south to Mexico. It is found throughout the United States, with the exception of some high mountains and humid coastal forests in Washington and Oregon.

**Habitat:** It occurs in a wide variety of habitats, from deserts and grasslands to woodlands and open forests. It frequents agricultural regions, especially where there is brushy cover such as fence rows.

**Reproduction:** This species mates from March to June. Clutches of 3 to 9 eggs are laid in early summer, often in a talus slope. The eggs hatch in 50 to 79 days. Males are sexually mature in 1 to 2 years, females in 3 to 5 years.

**Food Habits:** Diet varies according to size, with young eating insects, lizards, rodents, and birds and their eggs. Adults can take larger prey, occasionally as large as rabbits.

**Ecology:** This species often enters rodent burrows in search of prey. It can be locally common. In Utah, home range was estimated at 1-2 hectares with a population density of 0.3-1.3 per hectare. Generally, this snake is diurnal, but can be active at night in warm weather. It is inactive in cold weather, spending the winter in dens.

**Comments:** Young are aggressive when they hatch, but probably remain near their nest, and overwinter without feeding. This snake is adapted to brushy or weedy regrowth following disturbance, and is an important predator of rodents.

*References:* Diller and Johnson 1988, Parker and Brown 1980, Sweet and Parker 1990.

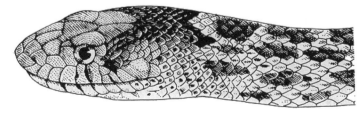

# Ground Snake
## (Sonora semiannulata)

*Order: Squamata*
*Family: Colubridae*
*State Status: Sensitive*
*Federal Status: None*
*Global Rank: G5*
*State Rank: S2*
*Length: 12 in (30 cm)*

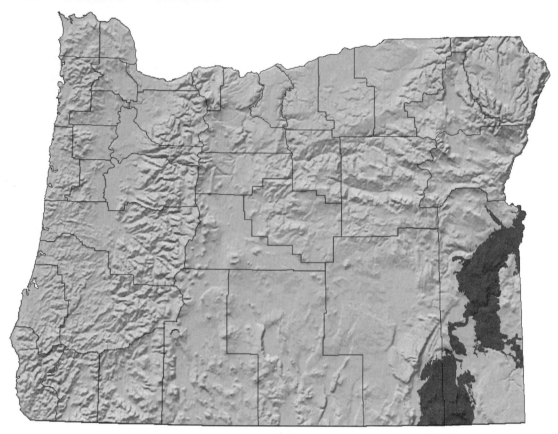

**Global Range:** A southwestern species, found from Oklahoma and Texas south into northern Mexico and west to the Mojave and western Great Basin deserts. Peripheral to Oregon, found only in the arid lowlands of Malheur County.

**Habitat:** The ground snake is found in arid desert scrub vegetation with sandy soil, usually under surface objects or in areas with some surface moisture, such as the edges of washes.

**Reproduction:** Little is known about the reproductive biology of Northwestern populations, but elsewhere this species lays a clutch of 4 to 6 eggs in early summer. It reaches sexual maturity in its second year.

**Food Habits:** This snake feeds on small arthropods such as spiders, scorpions, centipedes, crickets, and grasshoppers. It also takes insect larvae.

**Ecology:** The ground snake is most active on warm nights.

**Comments:** Grooves on the outer side of the teeth of ground snakes indicate that the species may be poisonous, but they pose no known danger to humans.

*References:* Frost 1983, Nussbaum et al. 1983.

# Western Aquatic Garter Snake
## (Thamnophis couchii)

*Order: Squamata*
*Family: Colubridae*
*State Status: None*
*Federal Status: None*
*Global Rank: G4*
*State Rank: S4?*
*Length: 30 in (76 cm)*

**Global Range:** The Oregon garter snake, subspecies *hydrophilus*, is found in the Siskiyou, Klamath, North Coast, and Cascade mountains of southwestern Oregon and northern California. The species, *T. couchii*, occurs south to Baja California.

**Habitat:** This highly aquatic snake is found in wet meadows, riparian areas, marshes, and moist forests near rivers, streams, lakes, and ponds. It requires streams with thick riparian vegetation (for escape) and exposed boulders for basking.

**Reproduction:** This live-bearing snake mates in the spring (March to June), and gives birth to litters of about 3-12 young from August to October.

**Food Habits.** It takes aquatic prey such as small fish and fish eggs, salamanders, tadpoles, frogs, toads, earthworms, and leeches.

**Ecology:** This snake is active primarily in the daytime. It is inactive during periods of extreme heat or cold. It hunts along water edges, and feeds in the water.

**Comments:** Some herpetologists consider the Oregon garter snake (*T. c. hydrophilus*) a separate species, *Thamnophis hydrophilus*. *Thamnophis couchii* is sometimes called the Sierra garter snake. Storm and Leonard (1995) assign this snake to *Thamnophis atratus* (Pacific Coast aquatic garter snake).

*References:* Fitch 1984, Nussbaum et al. 1983, Stebbins 1985.

# Western Terrestrial Garter Snake
## (Thamnophis elegans)

*Order: Squamata*
*Family: Colubridae*
*State Status: None*
*Federal Status: None*
*Global Rank: G5*
*State Rank: S5*
*Length: 24 in (61 cm)*

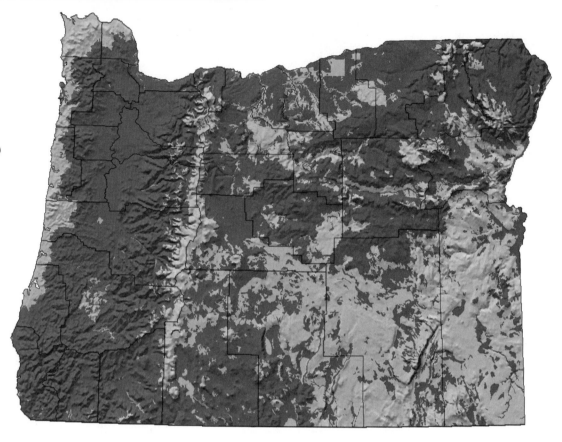

**Global Range:** Widespread in western North America from central British Columbia to southern California. Ranges east to the western Great Plains and northeastern New Mexico, with an isolated population in northern Baja California.

**Habitat:** This species is found in a variety of habitats. To make matters more confusing, four subspecies are found in Oregon (the Klamath, mountain, coast, and wandering garter snakes), each of which has somewhat different habitat preferences. All can be found in moist areas such as marshes and lake or stream margins, but two (mountain and coast) may occur some distance from water.

**Reproduction:** This snake mates in the spring (March-June) and bears litters of about 4-20 live young from late July to September.

**Food Habits:** The diet varies among subspecies: the more aquatic forms (Klamath, wandering) feed on fish, frogs, tadpoles, and leeches, which are eaten in the water. Terrestrial forms (mountain, coast) take frogs and toads, but also lizards, small mammals, salamanders, and slugs. Only the Klamath subspecies eats prey exclusively in the water.

**Ecology:** The aquatic subspecies escape into water.

**Comments:** The wandering garter snake is considered a separate species (*T. vagrans*) by some herpetologists. It occurs near marshes or water but feeds both in the water and on land. It is found in the mountains and water courses of eastern Oregon and in the vicinity of Columbia County. Suffice it to say, the taxonomy of this species needs investigation.

*References:* Nussbaum et al. 1983, Tanner and Lowe 1989.

# Northwestern Garter Snake
## (*Thamnophis ordinoides*)

*Order: Squamata*
*Family: Colubridae*
*State Status: None*
*Federal Status: None*
*Global Rank: G5*
*State Rank: S5*
*Length: 24 in (61 cm)*

**Global Range:** Southern British Columbia and Vancouver Island inland to the Cascade Mountains and south to northwestern California.

**Habitat:** This snake is found in meadows and at the edges of clearings in forests. It prefers areas with dense vegetation but, when basking, can be found in open areas or on talus slopes. It occurs in wooded areas on the floor of the Willamette Valley. This garter snake is commonly found in suburban areas and city parks.

**Reproduction:** This species mates in the spring and gives birth to a litter of about 10 (range 3-20) young from June to August.

**Food Habits:** The Northwestern garter snake feeds mainly on slugs and earthworms, but also takes insects, small salamanders, frogs, fish, small mammals, and possibly nestlings of ground-nesting birds.

**Ecology:** This common species is primarily terrestrial and is rarely found in or near water. It hunts on the ground surface, and is active during warm, sunny weather. Studies in the Willamette Valley indicate that only 3 out of 4 females breed in any given year.

**Comments:** The Northwestern garter snake is inactive during the winter, hibernating in talus slopes or deep in rock crevices, perhaps emerging on warm winter days in lowland areas.

*References:* Gregory 1978, Kirk 1979, Nussbaum et al. 1983.

# Common Garter Snake
## (Thamnophis sirtalis)

*Order: Squamata*
*Family: Colubridae*
*State Status: None*
*Federal Status: None*
*Global Rank: G5*
*State Rank: S5*
*Length: 48 in (122 cm)*

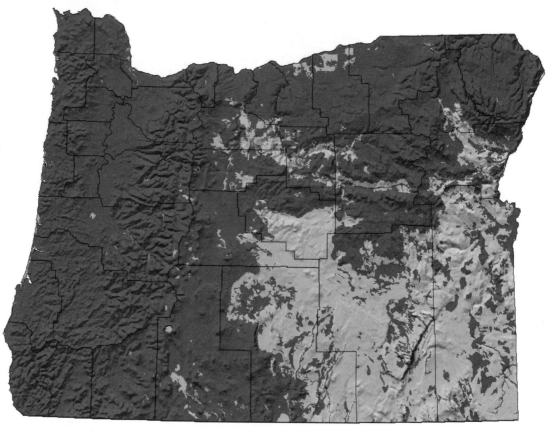

**Global Range:** Widespread throughout most of southern Canada and the United States, but absent from most of the arid region from west Texas to the Mojave and Great Basin deserts.

**Habitat:** While this snake frequents wet meadows and forest edges, it occurs in a variety of habitats far from water, including open valleys and moist coniferous forests.

**Reproduction:** Common garter snakes mate from March to June. Litters of about 16 (range 3-80) young are born from late July to September. Sexual maturity is reached at 2 to 3 years of age.

**Food Habits:** Smaller snakes eat earthworms, but adults feed on a variety of vertebrate prey, including frogs, toads, salamanders, birds, fish, reptiles, and small mammals. Invertebrates, including slugs and leeches, are also eaten. This snake can feed on the poisonous rough-skinned newt without ill effect.

**Ecology:** In most of Oregon, this snake usually is inactive in winter. In the low valleys of western Oregon it may emerge during warm periods. The timing of breeding and hibernation is variable depending on local climate.

**Comments:** This species can be extremely abundant, and, next to the Northwestern garter snake, is our most frequently encountered snake.

*References:* Fitch 1965, 1980, Lawson 1987.

*Order: Squamata*
*Family: Viperidae*
*State Status: None*
*Federal Status: None*
*Global Rank: G5*
*State Rank: S4*
*Length: 31 in (79 cm)*

**Global Range:** Generally, this aptly named rattlesnake is limited to the western half of the United States, from just north of the Canadian border south to Baja California del Sur and northern Mexico proper. It is found as far east as Kansas and Nebraska.

**Habitat:** Although they occur in a wide variety of habitat types, from deserts and chaparral to open forests, western rattlesnakes usually occur near rocks, cliffs, or downed logs. They overwinter in dens, which are usually located on south-facing rocky hillsides exposed to sunshine.

**Reproduction:** Rattlesnakes mate soon after emerging from winter hibernation, and are live-bearing. Females give birth to litters of 3-12 young in alternate years, in September and October. Larger females give birth to larger litters and litter size also varies among the two subspecies found in Oregon, with the Great Basin subspecies having larger litters.

**Food Habits:** Western rattlesnakes feed mainly on small mammals, including mice, gophers, squirrels, and rabbits, but will also take birds, lizards, and amphibians.

**Ecology:** They are most active at night in hot weather, but more diurnal in cool weather. Western rattlesnakes congregate in winter dens and may spend 7 or more months in hibernation.

**Comments:** Rattlesnakes are venomous, and their bites can be fatal to humans. The danger is probably exaggerated, since few people actually die of rattlesnake bites, but it is safest to leave them unmolested.

*References:* Diller and Johnson 1988, Diller and Wallace 1984, Klauber 1972, Macartney 1989, Macartney and Gregory 1988.

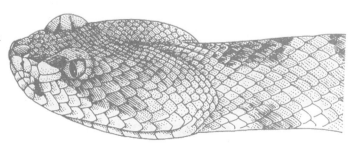

# Breeding
# Birds

# Pied-billed Grebe
## (Podilymbus podiceps)

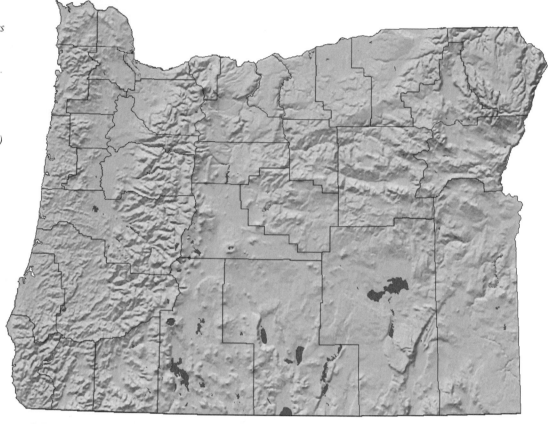

*Order: Podicipediformes*
*Family: Podicipedidae*
*State Status: None*
*Federal Status: None*
*Global Rank: G5*
*State Rank: S5*
*Length: 13.5 in (34 cm)*

**Global Range:** Breeds from central Canada south through Central America and into South America as far south as central Chile and southern Argentina.

**Habitat:** This is an aquatic species that breeds at the edge of open water in freshwater lakes, ponds, sluggish rivers, and marshes.

**Reproduction:** The nest is a floating or partly submerged mass of green and decayed aquatic vegetation. Most clutches of 4 to 7 (range 2-10) eggs are laid by April. The precocial young hatch after a 20-27-day incubation period.

**Food Habits:** This grebe mainly eats animals, including insects, crayfish, fish, leeches, frogs, molluscs, salamanders, snails, and shrimp. Some plant material is also taken. All of its food comes from aquatic habitats.

**Ecology:** During the breeding season, the Pied-billed Grebe defends a territory with a radius of about 50 meters around its nest. As a result, small ponds usually have but a single breeding pair.

**Comments:** This bird is widespread and common in Oregon. It winters locally in open water.

*References:* American Ornithologists' Union 1983, Gabrielson and Jewett 1940, Johnsgard 1987.

## Horned Grebe
*(Podiceps auritus)*

*Order: Podicipediformes*
*Family: Podicipedidae*
*State Status: Sensitive*
*Federal Status: None*
*Global Rank: G5*
*State Rank: S2*
*Length: 14 in (36 cm)*

**Global Range:** In North America, this species breeds throughout most of Alaska and Canada and, locally, just south of the Canadian border. It also breeds in northern Eurasia.

**Habitat:** This grebe favors areas with much open water surrounded with emergent vegetation.

**Reproduction:** A nest of aquatic plant material is anchored to emergent vegetation. Breeding commences in May and eggs are laid through July. Clutches of 4 to 6 (range 3-8) young hatch in 22-25 days. There is a single brood per year.

**Food Habits:** The Horned Grebe has a diet of animals. The primary foods are fish and tadpoles, but it will also take aquatic insects, crustaceans, amphibians, molluscs, and leeches.

**Ecology:** This grebe is territorial during nesting season. Horned Grebes winter on open water, mainly along the coast. Predation may result in high mortality at the nest.

**Comments:** This species only recently (1958) began breeding locally along the edges of marshes in eastern Oregon.

*References:* Littlefield 1990, Marshall 1992.

# Red-necked Grebe
## (Podiceps grisegena)

Order: Podicipediformes
Family: Podicipedidae
State Status: Sensitive
Federal Status: None
Global Rank: G5
State Rank: S1
Length: 18 in (46 cm)

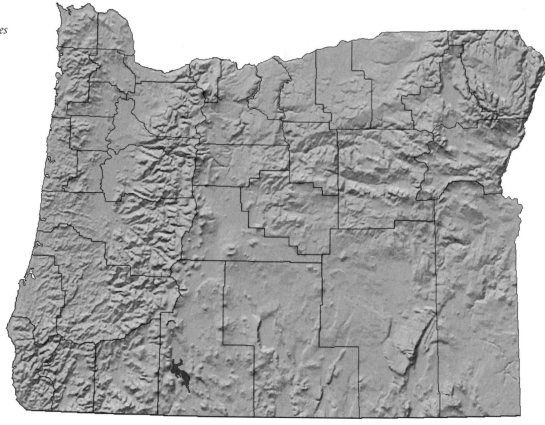

**Global Range:** This holarctic species breeds throughout most of Alaska and Canada from the Great Lakes westward. Some breed in the northern United States. It also breeds throughout northern Eurasia.

**Habitat:** This aquatic bird breeds in lakes and ponds, usually in forested areas. In Oregon, it is found in waters grown to hardstem bulrush intermixed with open water over 5 feet deep.

**Reproduction:** Breeding season begins in late April. There is a single clutch of about 5 eggs (range 2-8), which are incubated 22-27 days. The nest is built of rotting aquatic plants and is anchored to emergent vegetation.

**Food Habits:** Like other grebes, this bird is primarily a carnivore. It eats fish, aquatic insects, crustaceans, molluscs, amphibians, and annelids. Vegetable matter is sometimes consumed.

**Ecology:** Although the Red-necked Grebe is usually a solitary nester, it sometimes forms loose colonies. It winters along the coast. A minimum of 10 acres is needed for each nesting pair.

**Comments:** A single group of Red-necked Grebes has bred at Pelican Bay, Upper Klamath Lake, since at least 1945. In 1989, 28 adults and 10 juveniles were observed. There is no evidence the population is increasing or declining, though pollution in Upper Klamath Lake could be a threat. Recently (1993), it began breeding at Malheur National Wildlife Refuge, and it has bred occasionally at Howard Prairie Reservoir, Jackson County.

*References:* Gilligan et al. 1994, Kebbe 1958, Marshall 1992.

# Eared Grebe
*(Podiceps nigricollis)*

*Order: Podicipediformes*
*Family: Podicipedidae*
*State Status: None*
*Federal Status: None*
*Global Rank: G5*
*State Rank: S4*
*Length: 13 in (33 cm)*

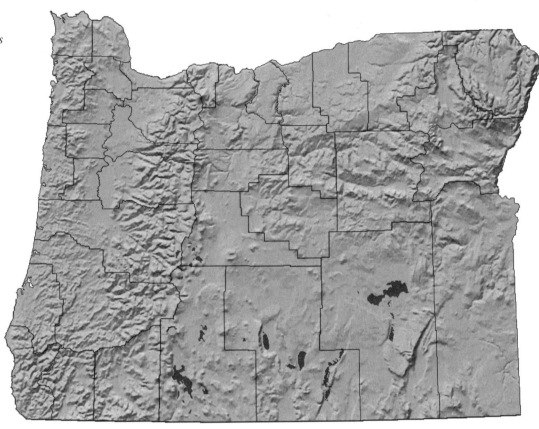

*References:* Gabrielson and
Jewett 1940, McAllister
1958.

**Global Range:** Breeds in western North America from southern Canada south to Arizona and New Mexico. It has a worldwide distribution, being found across Eurasia and in southern Africa.

**Habitat:** These grebes nest in the large, tule-fringed marshes of southeastern Oregon. They also use almost any open water for feeding after the breeding season.

**Reproduction:** Breeding season starts in late April and young are fledged by late September. One clutch (sometimes two) of 3 or 4 (range 1-8) eggs is incubated for 20-22 days in a floating nest made of a mat of reeds, usually hidden in vegetation. This grebe is a colonial nester.

**Food Habits:** The Eared Grebe feeds more heavily on aquatic insects than do other grebes. It will also take small fish, crustaceans, molluscs, spiders, and some terrestrial insects (grasshoppers, beetles, moths, flies, and insect larvae).

**Ecology:** On larger lakes, hundreds to thousands of these grebes will nest in dense colonies. They defend an area of about 2 feet around the nest. They will forage both on and beneath the water's surface. They winter both along the coast and inland, although most Oregon breeders winter out of state.

**Comments:** In winter, these grebes are more common in fresh water than salt water.

# Western Grebe
## *(Aechmophorus occidentalis)*

*Order: Podicipediformes*
*Family: Podicipedidae*
*State Status: None*
*Federal Status: None*
*Global Rank: G5*
*State Rank: S4?*
*Length: 27 in (61 cm)*

**Global Range:** Breeds in western North America from the prairie provinces of Canada south to California, Arizona, and New Mexico. There is also scattered breeding in Mexico.

**Habitat:** This water bird nests in the tule-fringed lakes and marshes of eastern Oregon. It winters along the coast and is common in bays and estuaries. It can also be found on major rivers during the winter.

**Reproduction:** This large grebe builds a floating nest of aquatic plant material, usually placed in the middle of thick, emergent vegetation. It is a colonial nester that begins breeding in early June. The clutch of 3 to 5 (range 1-10) young is fledged by October. Incubation lasts about 23 days.

**Food Habits:** The Western Grebe feeds primarily on fish, which it catches by diving. It also eats some insects, molluscs, crayfish, marine worms, and salamanders.

**Ecology:** The immediate vicinity of the nest is defended, but grebes nest in large colonies. Adults will carry the young around on their backs after hatching.

**Comments:** The Western Grebe was hunted for its plumes early in the century and nearly extirpated from Oregon. It has since recovered. Western Grebes have bred at Fern Ridge Reservoir, Lane County, since the early 1990s.

*References:* Gabrielson and Jewett 1940, Gilligan et al. 1994, Harrison 1978, Lawrence 1950.

*Order: Podicipediformes*
*Family: Podicipedidae*
*State Status: None*
*Federal Status: None*
*Global Rank: G5*
*State Rank: S4*
*Length: 27 in (61 cm)*

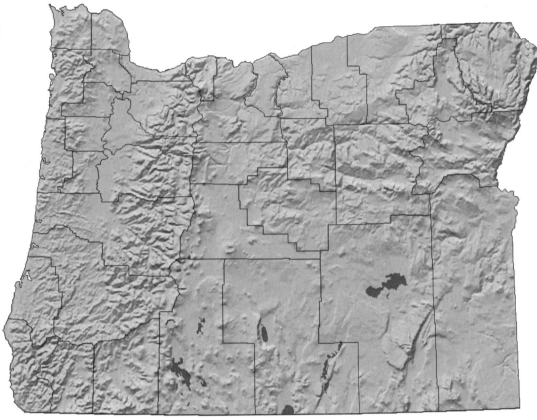

**Global Range:** This species has an identical range to the Western Grebe (western North America from Canada to New Mexico) but it is rare in the northern part of its range and as common as the Western Grebe in the southern portion of its range.

**Habitat:** Clark's Grebe breeds in inland lakes with emergent vegetation such as cattails and tules. It requires open water for foraging on the surface or diving below in pursuit of fish. It winters both along the seacoast and on major rivers.

**Reproduction:** This species builds a floating nest amongst reeds or tules. It breeds from July to October. A single clutch of 3 to 5 young hatches after being incubated by both sexes for 23 days.

**Food Habits:** Like the Western Grebe, Clark's Grebe is a specialist in fish, which make up over 80% of its diet. It will also eat other aquatic organisms like crayfish, amphibians, and aquatic insects. Where they nest together with Western Grebes, Clark's Grebes tend to feed in deeper water farther from shore.

**Ecology:** This species is often seen in association with the Western Grebe; these two grebes may even nest together in colonies and defend interspecific territories around the nest.

**Comments:** Until recently, Clark's Grebe was thought to be a lighter "color morph" of the Western Grebe, so information about the biology of the two species has long been combined. It was named after William Clark of the 1804-1806 Lewis and Clark Expedition. Clark's Grebe now breeds at Bolly Creek Reservoir, Malheur County.

*References:* American Ornithologists' Union 1983, 1985, Nuechterlein and Buitron 1989, Peterson 1990, Small 1994.

# Fork-tailed Storm-Petrel
## *(Oceanodroma furcata)*

*Order: Procellariiformes*
*Family: Hydrobatidae*
*State Status: Sensitive*
*Federal Status: None*
*Global Rank: G5*
*State Rank: S2*
*Length: 9 in (23 cm)*

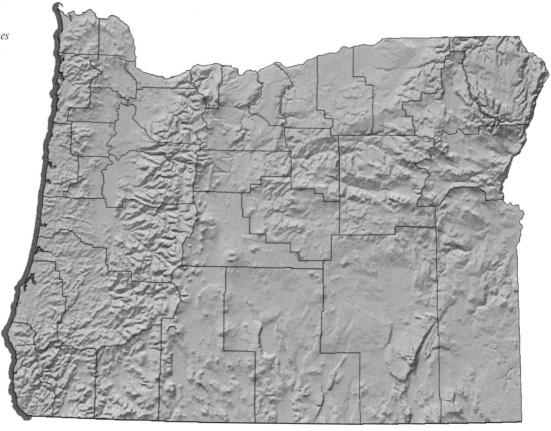

**Global Range:** Breeds around the rim of the northern Pacific Ocean, from the Kurile Islands in Asia, north and west to Alaska, and south as far as Baja California.

**Habitat:** When not flying over the open ocean in search of prey, it nests on offshore islands that have soil development. Its burrows are often located in grassy areas, but not where there are trees or shrubs.

**Reproduction.** This seabird digs its own burrow in which a single egg is laid in late June. Little is known about breeding specifics, save that the pair take turns incubating the egg. There is but one clutch per year.

**Food Habits:** The Fork-tailed Storm-Petrel feeds on marine plankton, small fish, and crustaceans.

**Ecology:** Storm-Petrels come ashore only to breed. Like most seabirds, they form colonies, usually on offshore islands that provide them some protection from terrestrial predators.

**Comments:** This is a rare species in Oregon, with estimates of the state's population varying between 100 and 1,000 birds. It nests on 6 or 7 offshore islands along the Oregon coast.

*References:* Browning and English 1972, Gilligan et al. 1994, Marshall 1992, Sanger 1972.

# Leach's Storm-Petrel
*(Oceanodroma leucorhoa)*

*Order: Procellariiformes*
*Family: Hydrobatidae*
*State Status: None*
*Federal Status: None*
*Global Rank: G5*
*State Rank: S5*
*Length: 8 in (20 cm)*

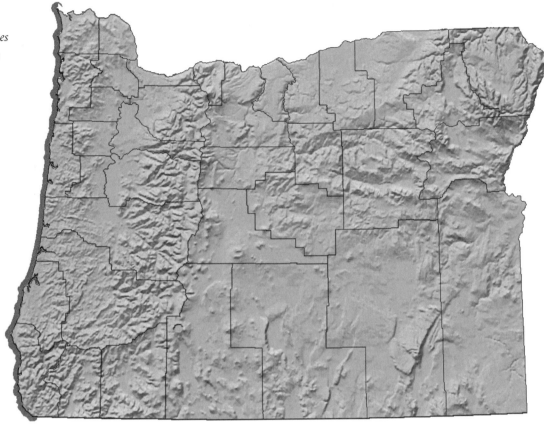

**Global Range:** Breeds along the shores of both the north Pacific and north Atlantic oceans. In nonbreeding season it wanders widely, as far south as northern South America.

**Habitat:** When not flying over the open ocean, Leach's Storm-Petrel comes ashore on grassy offshore islands with soil deep enough for burrow construction.

**Reproduction:** A single egg is laid in late May or June and is incubated by both parents for 41-42 days. Chicks are deserted about 40 days after hatching and take to sea when they are just over 2 months old. They do not breed until their fifth year.

**Food Habits:** Leach's Storm-Petrel obtains its food from the marine environment. Its diet consists of small fish, crustaceans, and squid.

**Ecology:** This petrel digs its own burrow and is colonial. At times, several pairs will have nest chambers off a single burrow. The same burrow may be used by a pair in successive years. Maximum life span is at least 24 years.

**Comments:** Predators, especially skunks, can do serious damage to colonies when they gain access to nesting sites. Several hundred thousand pairs nest on over a dozen rocks off the Oregon coast.

*References:* Gabrielson and Jewett 1940, Gilligan et al. 1994, Podolsky and Kress 1989, Warham 1991.

# American White Pelican
## (Pelecanus erythrorhynchos)

*Order: Pelecaniformes*
*Family: Pelicanidae*
*State Status: Sensitive*
*Federal Status: None*
*Global Rank: G3*
*State Rank: S1*
*Length: 60 in (152 cm)*
*Wingspread: 90 in (229 cm)*

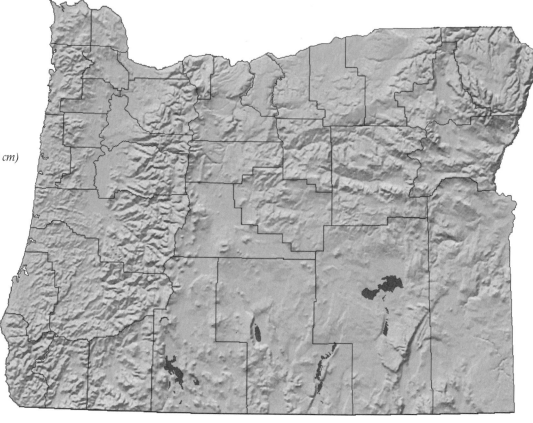

**Global Range:** Breeds in the interior of western and central Canada, south to northern California and Colorado. In winter, it wanders widely along coasts and over inland waters as far south as the Gulf of Mexico.

**Habitat:** American White Pelicans are found in inland lakes and marshes during breeding season. A predator-free island is required for nesting. During nonbreeding seasons, they may occur on almost any body of water, including ocean coasts.

**Reproduction:** The American White Pelican nests in colonies of up to several thousand birds. Breeding season begins in April or May and the usual clutch is 2 eggs. The incubation period is about 29 days.

**Food Habits:** Fish are the main item in the pelican's diet. The species of fish vary with location. At Malheur Lake they feed on carp. They will also take some crayfish and salamanders.

**Ecology:** The American White Pelican requires large quantities of fish during the breeding season. Individual birds may fly as far as 60 miles to feeding areas.

**Comments:** There are three breeding colonies in Oregon: Pelican Lake, Upper Klamath Lake, and Malheur Lake. In 1990, about 1,000 birds nested at Malheur Lake.

*References:* Marshall 1992, Paullin et al. 1988.

# Double-crested Cormorant
## *(Phalacrocorax auritus)*

*Order: Pelecaniformes*
*Family: Phalacrocoracidae*
*State Status: None*
*Federal Status: None*
*Global Rank: G5*
*State Rank: S5*
*Length: 30 in (76 cm)*

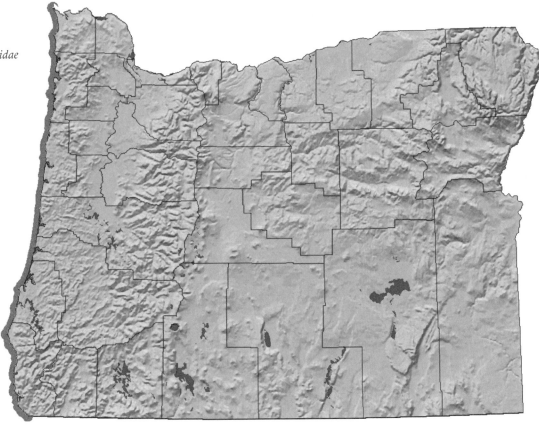

**Global Range:** Breeds along the coasts of North America, from Alaska and Newfoundland south to the Gulf Coast and northwestern Mexico. It is also found in colonies in the interior of the continent where the habitat is appropriate.

**Habitat:** The Double-crested Cormorant will nest on offshore rocks, on ledges of cliff faces overlooking water, or on isolated islands in interior lakes and marshes.

**Reproduction:** Breeding begins in April and young are fledged by late August. There is a single clutch of 3 or 4 (range 2-7) young each year.

**Food Habits:** Cormorants feed mostly on various types of fish, which they capture while diving as much as 15 meters deep. They also eat crustaceans, amphibians, marine annelids, and molluscs.

**Ecology:** The Double-crested Cormorant is a colonial breeder. It defends the immediate area around its nest. Nests may be as close as 1 meter apart. While foraging, these cormorants may fly as far as 10 miles from their nest site. The density of the population is correlated with the food supply.

**Comments:** This species is found in small colonies in the marshes of southeastern Oregon. About 2,000 birds nest in the Klamath Basin.

*References:* Gabrielson and Jewett 1940, Gilligan et al. 1994, Harrison 1978.

# Brandt's Cormorant
## *(Phalacrocorax penicillatus)*

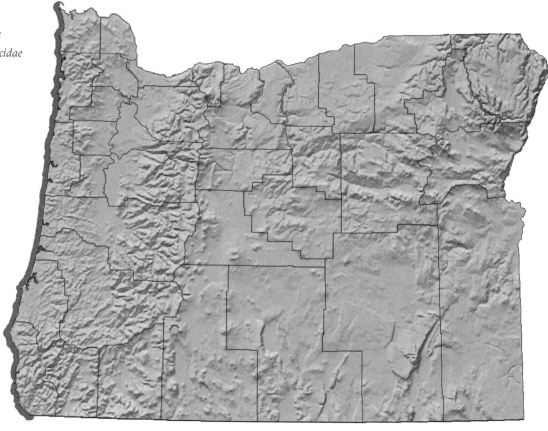

*Order: Pelecaniformes*
*Family: Phalacrocoracidae*
*State Status: None*
*Federal Status: None*
*Global Rank: G5*
*State Rank: S4*
*Length: 33 in (84 cm)*

**Global Range:** Brandt's Cormorant is a coastal bird of the eastern Pacific Ocean. There are some colonies in southeast Alaska, and it occurs regularly from Washington south to Baja California and the Gulf of California.

**Habitat:** Brandt's Cormorant forages in the ocean and brackish estuaries, seeking out rocky offshore islands or the cliffs of headlands for breeding. It often shares breeding islands with many other species.

**Reproduction:** The breeding season starts in June or July. The normal clutch is 4 eggs (range 3-6), and there is but one brood a year. The nest is made of seaweed and other plant material, and may be reused in subsequent years.

**Food Habits:** This cormorant feeds on small marine fish. It also takes some crabs and shrimp.

**Ecology:** This colonial cormorant is abundant throughout the year along the Oregon coast, although many birds that breed in Oregon winter out of state. It actively pursues fish during foraging dives.

**Comments:** There are no indications that cormorants prey on commercial or sport fish.

*References:* Gabrielson and Jewett 1940, Harrison 1978.

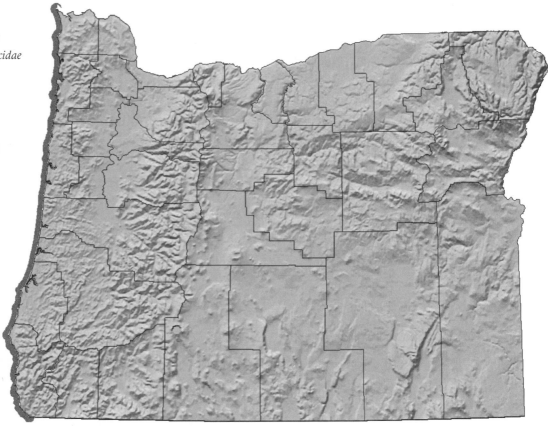

## Pelagic Cormorant
### *(Phalacrocorax pelagicus)*

*Order: Pelecaniformes*
*Family: Phalacrocoracidae*
*State Status: None*
*Federal Status: None*
*Global Rank: G5*
*State Rank: S5*
*Length: 26 in (66 cm)*

**Global Range:** Breeds along both the Asian and North American coasts of the northern Pacific Ocean. In North America, it is found as far south as Baja California.

**Habitat:** The Pelagic Cormorant forages over and in the ocean. It nests on island cliff ledges and in crevices in rocky areas along the coast, often in association with other cormorants and various seabirds.

**Reproduction:** The nesting season begins in June, and 3 to 5 eggs form the normal clutch. The nest is made of seaweed and other plant material. Both parents incubate the eggs for about 26 days.

**Food Habits:** Like other cormorants, this species eats small fish, crabs, shrimp, marine worms, and some amphipods.

**Ecology:** The Pelagic Cormorant is a colonial nester that rarely ventures far inland. It is present along the Oregon coast throughout the year.

**Comments:** This is the smallest of the three cormorant species found along the Oregon coast.

*References:* Gabrielson and Jewett 1940, Harrison 1978.

# American Bittern
## (Botaurus lentiginosus)

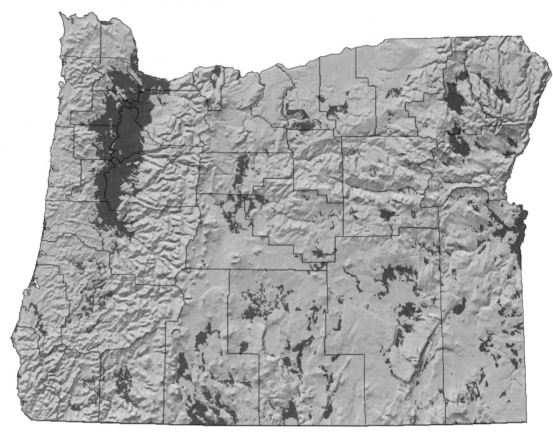

*Order: Ciconiiformes*
*Family: Ardeidae*
*State Status: None*
*Federal Status: None*
*Global Rank: G4*
*State Rank: S4*
*Length: 30 in (76 cm)*

**Global Range:** Breeds coast to coast from central Canada south to northern Mexico. It is not found in the arid Southwest.

**Habitat:** In Oregon, this is a bird of freshwater marshes with ample emergent vegetation. During the breeding season it is far more common in marshes on the east side of the Cascades.

**Reproduction:** The nest is usually a pile of vegetation at the edge of a marsh. Breeding season begins in late April. A clutch of 3 to 5 (range 2-7) eggs is incubated by the female for 24-28 days. The young are fledged by August.

**Food Habits:** The American Bittern feeds on a wide variety of animals that occur in its marshy environment, including frogs, small eels, fish, snakes, salamanders, crayfish, large insects and their larvae, and, occasionally, small mammals, birds, and bird eggs.

**Ecology:** Bitterns tend to be solitary nesters, although bittern nests may occur in a marsh from 50 to 150 feet apart. They are more common breeders in the marshes of eastern Oregon. In winter, a few remain on their breeding grounds, but they may also be found along the Oregon coast in marshy areas.

**Comments:** The American Bittern is cryptically colored and remains motionless in dense vegetation to avoid detection.

*References:* Gabrielson and Jewett 1940, Harrison 1978, Kushlan 1976, Root 1988.

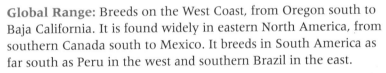

## Least Bittern
### *(Ixobrychus exilis)*

*Order: Ciconiiformes*
*Family: Ardeidae*
*State Status: Sensitive*
*Federal Status: SC*
*Global Rank: G5*
*State Rank: S1*
*Length: 13 in (33 cm)*

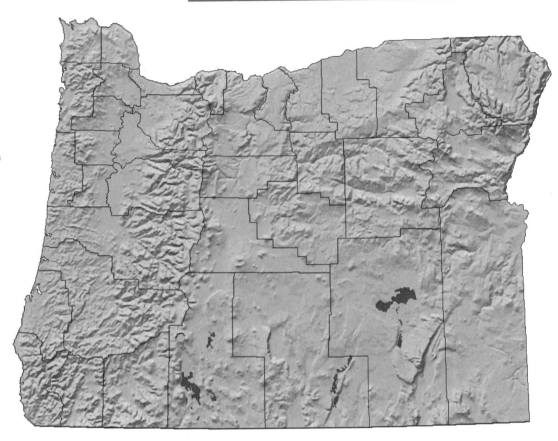

**Global Range:** Breeds on the West Coast, from Oregon south to Baja California. It is found widely in eastern North America, from southern Canada south to Mexico. It breeds in South America as far south as Peru in the west and southern Brazil in the east.

**Habitat:** In Oregon, this species breeds in freshwater cattail and bulrush marshes east of the Cascades.

**Reproduction:** The Least Bittern begins its breeding season around May. It builds a nest from aquatic vegetation and lays a clutch of 4 or 5 eggs. The female incubates the eggs for 17-20 days. There may be two clutches per year.

**Food Habits:** Like most herons, the Least Bittern eats animals, including small fishes, amphibians, crustaceans, insects, leeches, and slugs. It may occasionally take a small mouse.

**Ecology:** The Least Bittern is a solitary and secretive species that is rarely seen. It does not winter in Oregon.

**Comments:** In Oregon, the Least Bittern is at the northern limit of its range and is very rare. There are only six reports of this bird since 1981, although it is probably more common but has escaped detection.

*References:* Gabrielson and Jewett 1940, Harrison 1978, Marshall 1992.

# Great Blue Heron
## *(Ardea herodias)*

*Order: Ciconiiformes*
*Family: Ardeidae*
*State Status: None*
*Federal Status: None*
*Global Rank: G5*
*State Rank: S4*
*Length: 46 in (117 cm)*

**Global Range:** Breeds in coastal areas as far north as southeastern Alaska, but it occurs from coast to coast from southern Canada south through Mexico. It is also found on islands in the Caribbean Sea and on the northern coast of South America.

**Habitat:** This heron travels widely and uses many habitats in the course of a day. It does not use vast expanses of arid desert shrub or the interior of forests, but it can be found in nearly any meadow, grassland, marsh, riparian thicket, lake, river, or pond within every habitat type, including agriculture, pasture, and urban areas.

**Reproduction:** Typical nests are built of sticks with a depression lined with grasses and other soft material. They are commonly located in coniferous or deciduous trees, but also can be found on cliff ledges, or even on the ground in thick marsh vegetation. A clutch of 3 or 4 (range 1-8) eggs is laid in March or April, and the young fledge by September. There is one clutch per year.

**Food Habits:** This heron has a diet mostly consisting of fish and crustaceans. It will also eat amphibians, small reptiles, small birds and mammals, insects, and some aquatic plants and their seeds.

**Ecology:** Great Blue Herons are colonial birds that nest in groups of a few to several hundred pairs. They are found throughout the state all year, but migrate from areas where water is frozen in winter. They only defend territory immediately around the nest.

**Comments:** The Great Blue Heron is a common sight, even in the Portland metropolitan area. Individuals can sometimes become pests around fish hatcheries, but generally they eat nongame fish species.

*References:* Collazo 1981, Krebs 1974, Quinney and Smith 1980.

## Great Egret
*(Ardea alba)*

*Order: Ciconiiformes*
*Family: Ardeidae*
*State Status: None*
*Federal Status: None*
*Global Rank: G5*
*State Rank: S3*
*Length: 39 in (99 cm)*

**Global Range:** This species is truly cosmopolitan. It breeds on every continent except Antarctica, but it avoids high latitudes in North America and Europe, and is absent from the Sahara Desert in Africa.

**Habitat:** In Oregon, the Great Egret is a bird of the cattail and bulrush marshes east of the Cascades. It may forage in a number of habitat types around the nest site, including riparian and pasture.

**Reproduction:** Breeding season starts in early April, and young are fledged by late August. A single clutch of 3 to 5 (range 1-6) eggs is laid, and both sexes incubate the eggs for 23-24 days. Nests are built of sticks and are usually located in willows or other trees.

**Food Habits:** The Great Egret eats just about any animal it can catch in shallow water, marshes, or fields. This includes fish, frogs, salamanders, snakes, crayfish, aquatic insects, and small mammals.

**Ecology:** These egrets are colonial, and defend an area around the nest as far as they can reach. They do not winter on their breeding grounds in Oregon, but can be seen along the coast as nonbreeders. They can forage 5-10 miles from the nest. Great Egrets have nested within Great Blue Heron colonies at two locations near Coos Bay, Coos County.

**Comments:** Nesting colonies are easily disturbed. The largest colony in Oregon is at Malheur National Wildlife Refuge. Although colony size varies from year to year, the state breeding population is less than 1,000 birds scattered among 8 to 10 colonies. In mild years, a few have spent the winter in the Klamath Basin. The American Ornithologists' Union (1995) changed the Latin name of this species from *Casmerodius albus* to *Ardea alba*.

*References:* Gabrielson and Jewett 1940, Gilligan et al. 1994, Marshall et al. 1996, Wiese 1976.

# Snowy Egret
## (Egretta thula)

Order: Ciconiiformes
Family: Ardeidae
State Status: Sensitive
Federal Status: None
Global Rank: G5
State Rank: S2
Length: 24 in (61 cm)

**Global Range:** Breeds from Oregon south in the western United States, and from Maine south along the Atlantic Coast. It is found throughout South America as far south as central Argentina and southern Chile.

**Habitat:** Like most herons, this species is a bird of marshy areas. In breeding season, it is found in the cattail and bulrush marshes of southeastern Oregon. During winter, it can be found in marshy areas along the southern coast, mainly at Coos Bay.

**Reproduction:** Breeding season begins around late May. A clutch of 4 or 5 eggs is laid in a stick nest in a tree, or on the ground of an island inaccessible to predators. Both sexes incubate the eggs 18 days or longer.

**Food Habits:** Snowy Egrets feed on the typical heron menu of small fishes, frogs, lizards, snakes, crustaceans, worms, and insects. They forage in shallow water and in wet meadows.

**Ecology:** The Snowy Egret often nests in colonies with other water birds. It does not winter in its breeding range. Oregon is at the northern edge of the breeding range for this species, and its state population has fluctuated from zero to several hundred pairs.

**Comments:** Snowy Egrets have bred at a few sites in Oregon, including Malheur Lake, the Lower Silvies River Valley, Crump Lake, Pelican Lake, Summer Lake, and Upper Klamath Lake.

References: Littlefield 1990, Marshall 1992.

# Cattle Egret
## *(Bubulcus ibis)*

*Order: Ciconiiformes*
*Family: Ardeidae*
*State Status: None*
*Federal Status: None*
*Global Rank: G5*
*State Rank: SU*
*Length: 20 in (51 cm)*

*References:* Dinsmore 1973, Gilligan et al. 1994, Godfrey 1966, Janzen 1983, Jenni 1969, Zeiner et al. 1990a.

**Global Range:** Breeds from southern Canada south to northern Chile and Argentina and, in the Old World, from southern Europe east to Japan and south throughout Africa and Australia.

**Habitat:** Cattle Egrets are birds of open grasslands, pastures, irrigated croplands, and marshes. They will forage in flooded fields, wet meadows, and, elsewhere, coastal estuaries. They require trees or shrubs for nesting. They are often found in fields with grazing ungulates, hence their common name.

**Reproduction:** This species is a relatively recent addition to Oregon's avifauna, so knowledge of its breeding phenology is sketchy. Gilligan et al. (1994) report it has been present at Malheur National Wildlife Refuge from late April to September. A nest of twigs is built in a small tree or large shrub, usually near or over water. The clutch size is 4 or 5 (range 2-6) eggs, which are incubated by both parents for 3 to 4 weeks. The young fledge by 2 months of age. Cattle Egrets often nest in mixed colonies with other species of herons and egrets.

**Food Habits:** Insects form the largest portion of the Cattle Egret's diet, but earthworms, amphibians, reptiles, and small mammals are also eaten. Studies have shown that birds that catch insects disturbed by grazing cattle obtain more food per unit effort than individuals foraging in the absence of cattle.

**Ecology:** Male Cattle Egrets defend a small territory around the nest. In the absence of cattle, Cattle Egrets may forage in groups, some moving ahead to stir up insects for following birds. In winter, they may be observed along the Oregon coast and in western interior valleys.

**Comments:** Cattle Egrets are native to the Old World and were first reported in Surinam in 1877. By 1940, they had reached Florida and have been expanding northward and westward in North America ever since. Because they are presumed to have reached South America without human intervention, most authorities consider them a native New World species.

# Green Heron
## (Butorides virescens)

*Order: Ciconiiformes*
*Family: Ardeidae*
*State Status: None*
*Federal Status: None*
*Global Rank: G5*
*State Rank: S4*
*Length: 17 in (43 cm)*

**Global Range:** In the Western Hemisphere, this species breeds from southern Canada south to southern Peru and central Argentina. It is also found in sub-Saharan Africa, Australia, China, Japan, southeast Asia, and islands of the South Pacific.

**Habitat:** This species breeds in deciduous riparian woodlands or in the tops of coniferous trees growing near water. It will breed in trees growing along the banks of lakes and ponds, as well as those along stream banks, and seems to prefer to nest in willows.

**Reproduction:** Breeding season commences about mid-April, and young are fledged by September. The clutch of 4 or 5 (range 3-9) eggs is incubated by both parents for 19-21 days. It is likely to be single-brooded in Oregon. The nest is a platform of sticks and twigs, and is usually over water.

**Food Habits:** Most of the Green Heron's diet comes from the water. It eats small fish, frogs, crayfish, aquatic insects, and annelids. It occasionally eats some terrestrial species like mice, small snakes, and snails.

**Ecology:** The Green Heron usually nests alone or in small groups, but rarely in large colonies. The pair will defend only a small area around the nest. A few may winter in the Willamette Valley, southwestern Oregon, and on the south coast.

**Comments:** This heron was rare early in the century, but seems to have become more common in Oregon since 1940. Banks et al. (1987) include *B. virescens*, the Green Heron, in *Butorides striatus*, the Green-backed Heron, however the American Ornithologists' Union (1993) again recognizes the Green Heron as a separate species.

*References:* Gladson 1981, Larrison 1940.

## Black-crowned Night-Heron
*(Nycticorax nycticorax)*

*Order: Ciconiiformes*
*Family: Ardeidae*
*State Status: None*
*Federal Status: None*
*Global Rank: G5*
*State Rank: S4*
*Length: 25 in (64 cm)*

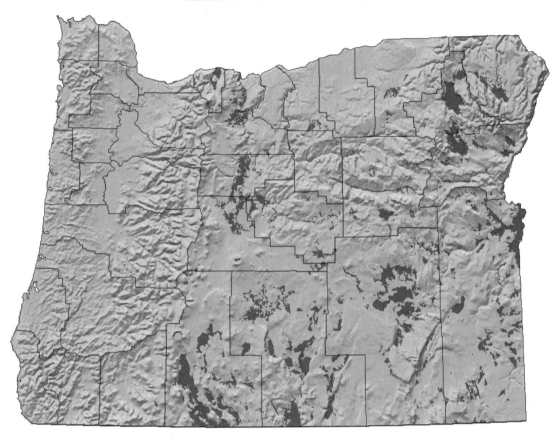

**Global Range:** Except for Antarctica and the arctic portions of the Northern Hemisphere, this species breeds, at least locally, throughout the world. It has even colonized remote oceanic islands in the South Pacific, and breeds on the Falkland Islands in the South Atlantic.

**Habitat:** The Black-crowned Night-Heron is an aquatic bird that nests in trees near water, or, rarely, in tules or cattails growing in marshes. It uses almost

any type of tree, including conifers, and may build its stick nest as high as 160 feet off the ground.

**Reproduction:** Breeding season for this heron begins in April and concludes in September. A single clutch of 3 to 5 (range 1-7) eggs is incubated 24-26 days by both parents. It nests in small to large colonies.

**Food Habits:** The Black-crowned Night-Heron finds most of its prey in shallow waters, but also takes small mammals and birds on land. It consumes small fish, molluscs, crustaceans, amphibians, snakes, and aquatic insects.

**Ecology:** This heron is present locally throughout the year in southwestern Oregon, but breeds both west of the Cascades and in the marshes of eastern Oregon. It defends a territory around the nest. Although it is sometimes active during the day, it is mainly nocturnal.

**Comments:** At night, Black-crowned Night-Herons may forage as far as 10 miles from their daytime roost.

*References:* Gabrielson and Jewett 1940, Harrison 1978.

# White-faced Ibis
### (Plegadis chihi)

*Order: Ciconiiformes*
*Family: Threskiornithidae*
*State Status: None*
*Federal Status: SC*
*Global Rank: G5*
*State Rank: S3*
*Length: 24 in (61 cm)*

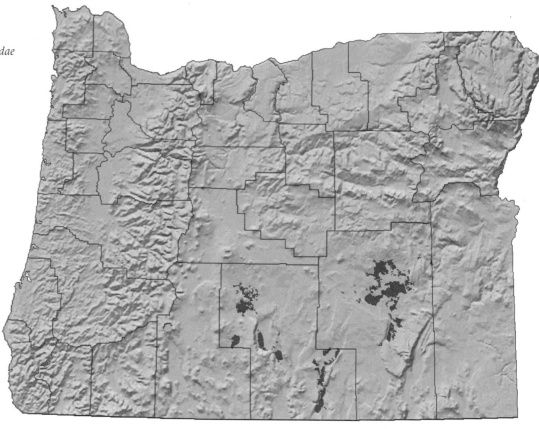

**Global Range:** The White-faced Ibis breeds in western North America from Oregon and North Dakota south to Mexico. It also breeds in South America south to central Chile and Argentina, but is absent from the Amazon Basin.

**Habitat:** Oregon's only ibis breeds in interior freshwater marshes. It usually nests among emergent hardstem bulrush, but it will feed in marshes, meadows, the edges of ponds, pastures, and irrigated alfalfa fields.

**Reproduction:** The nest is built from twigs, and usually is lined with green leaves. Egg-laying starts about the first of June, and a clutch of 3 or 4 eggs is incubated 21-22 days.

**Food Habits:** The White-faced Ibis forages in marshes and wet meadows for a variety of food items, but insects make up the bulk of its diet. It will also eat crayfish, leeches, molluscs, and worms. Rarely, small snakes, fishes, and amphibians are eaten.

**Ecology:** White-faced Ibises do not winter in Oregon, but after breeding they disperse widely to feed in agricultural areas and wetlands east of the Cascades. They often nest in association with Black-crowned Night-Herons. They usually forage in the company of other ibises.

**Comments:** The location of ibis colonies in eastern Oregon is related to runoff from winter precipitation. The number of breeding pairs has increased in recent decades and, as of 1990, there were about 4,000 pairs breeding in Oregon. Low water years of the early 1990s may have reversed this trend. Recent wet winters may result in increased numbers of breeding pairs in Oregon.

*References:* Ivey et al. 1988, Marshall 1992.

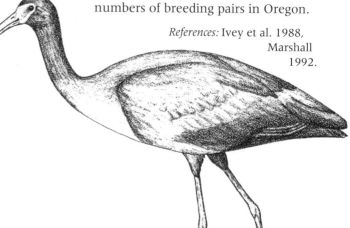

*Order: Anseriformes*
*Family: Anatidae*
*State Status: None*
*Federal Status: SC*
*Global Rank: G4*
*State Rank: S2*
*Length: 65 in (165 cm)*

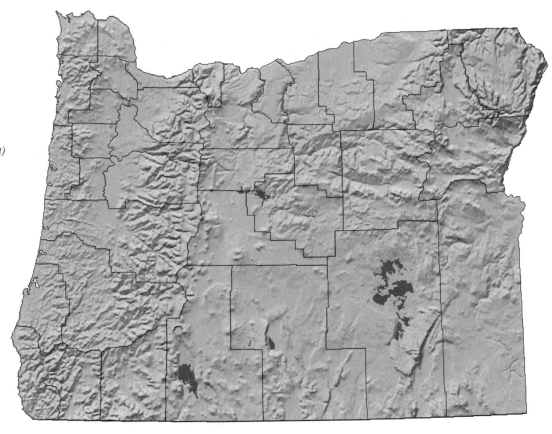

**Global Range:** Breeds primarily in southern Alaska and the Yukon Territory, but there are a few breeding populations as far south as Idaho. It probably is not a native breeding species in Oregon, although it does winter along the Pacific Coast.

**Habitat:** In Oregon, the Trumpeter Swan occurs in freshwater cattail and bulrush marshes around Malheur Lake. It nests on the shores of large inland lakes and marshes.

**Reproduction:** This swan builds a large nest (5 feet across) of aquatic vegetation (reeds, sedges, bulrush, etc.). Breeding commences in late April to early May, but young are not independent until the next spring. Clutches of 5 (range 2-9) eggs are incubated 33 days, mostly by the female.

**Food Habits:** Adults have a diet of various parts of aquatic plants, including roots, tubers, rhizomes, stems, and leaves. They will take some fish, molluscs, and insects. Hatchlings eat aquatic beetles and crustaceans for their first month, after which their diet shifts to vegetable matter.

**Ecology:** Trumpeter Swans often build their nests on muskrat houses. They rarely feed on land, and will dig into the substrate of shallow waters in search of roots and bulbs. In winter, birds from northern populations can occasionally be seen on the north coast and in the Willamette Valley.

**Comments:** The introduced population at Malheur National Wildlife Refuge consists of less than 20 pairs of swans. In 1987, 9 pairs nested. There is a new introduction program at Summer Lake. The Trumpeter Swan is listed as a game animal in Oregon.

*References:* Banko 1960, Gilligan et al. 1994, Harrison 1978.

# Canada Goose
## *(Branta canadensis)*

*Order: Anseriformes*
*Family: Anatidae*
*State Status: None*
*Federal Status: None*
*Global Rank: G5*
*State Rank: S5*
*Length: 40 in (102 cm)*

**Global Range:** Breeds throughout Alaska, Canada, and the northern tier of the lower 48 states. It winters from Oregon and the mid-Atlantic states south to southern Mexico.

**Habitat:** Canada Geese breed in a variety of habitats near water, including the shores of rivers, lakes, and reservoirs. They are most common in the cattail and bulrush marshes of eastern Oregon. They feed both in marsh and upland habitats, including meadow, pasture, and agricultural land.

**Reproduction:** The onset of breeding season depends on the climate, but eggs are present as early as March. A single clutch of 5 or 6 (range 2-11) young is produced each year. The female incubates the eggs about 25-30 days, and the young remain with the adults until the following spring. This species mates for life.

**Food Habits:** The Canada Goose is omnivorous. It will eat all parts of aquatic plants, grain in agricultural fields, grasses, and other crops. It also eats some insects, crustaceans, molluscs, snails, minnows, and tadpoles.

**Ecology:** Nests are depressions lined with grass, leaves, or other plant material and tend to occur in proximity to one another in suitable nesting habitat. In California, 31 nests were found in half an acre. Birds can forage up to 5 miles from the nest. After breeding season, they wander widely and can be seen almost anywhere there is open water, including urban areas.

**Comments:** In some regions, Canada Geese are considered pests in both agricultural fields and urban areas. They are a game species in Oregon. Only about 4,000 pairs breed in Oregon.

*References:* Bellrose 1980, Gabrielson and Jewett 1940, Gilligan et al. 1994, Harrison 1978.

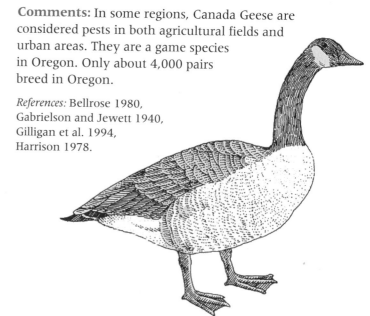

## Wood Duck
### (Aix sponsa)

*Order: Anseriformes*
*Family: Anatidae*
*State Status: None*
*Federal Status: None*
*Global Rank: G5*
*State Rank: S4*
*Length: 19 in (48 cm)*

**Global Range:** Breeds from southern Canada south to central California and the Gulf of Mexico; however, there is a gap between the West Coast populations and those found from the Great Plains eastward. It winters in the southern part of its breeding range, in Mexico, and on islands in the Caribbean.

**Habitat:** Wood Ducks nest in trees and are found anywhere suitable woodlands occur near water; they nest in riparian forests, in trees near lakes, and in marshes. Although usually associated with deciduous trees, they have been known to use cavities in conifers.

**Reproduction:** Breeding begins in April, and young are fledged by early August. This species uses natural cavities in trees, up to 65 feet above the ground, as nest sites. A clutch of 8 to 12 (range 3-15) eggs is incubated 27-37 days by the female.

**Food Habits:** About 90% of the diet is vegetable matter of all sorts, including seeds, green parts of aquatic plants, acorns, and other nuts. The animal component of the diet includes aquatic insects, dragonflies, beetles, crickets, and grasshoppers. Young are fed insects during their first weeks, then switch to vegetable matter.

**Ecology:** About 2 acres is needed for each pair of Wood Ducks. They are inconspicuous during breeding season. Nest trees must be within 1,200 feet of water, but are often directly over small streams or ponds. During winter, small groups can be found throughout the state.

**Comments:** The Wood Duck is a game species in Oregon. Wood Ducks will nest in artificial nest boxes in suitable locations.

*References:* Bellrose 1980, Bellrose et al. 1964, Gabrielson and Jewett 1940.

# Green-winged Teal
## *(Anas crecca)*

*Order: Anseriformes*
*Family: Anatidae*
*State Status: None*
*Federal Status: None*
*Global Rank: G5*
*State Rank: S5*
*Length: 14 in (36 cm)*

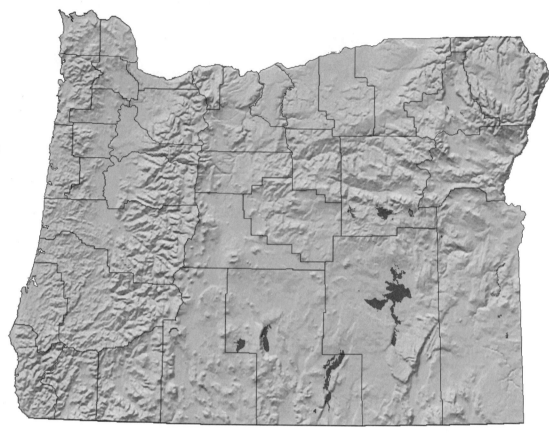

**Global Range:** The Green-winged Teal is a holarctic species that breeds throughout most of Eurasia, and coast to coast in North America from Alaska south to Oregon and, on the East Coast, to Maine. Occasional local breeding takes place as far south as southern California.

**Habitat:** These small ducks breed in freshwater marshes with ample emergent vegetation, within which they seek cover. They will occupy marshes within coniferous forests in the northern part of their range. In Oregon, they are best known from the inland marshes of the southeastern part of the state.

**Reproduction:** Green-winged Teal build nests of grasses lined with feathers, usually on an island in a lake, but sometimes on the shoreline, or as much as 200 feet from water, concealed by tall grass. The breeding season commences in April. A clutch of 8-12 (range 5-18) eggs is incubated for 21-23 days. Young are independent about 23 days after hatching.

**Food Habits:** Green-winged Teal feed on smaller seeds of aquatic plants, such as nutgrass, millet, water hemp, and sedges. They sometimes take larger seeds and green vegetation, especially tubers of sago pondweed, and they will also eat aquatic insects and molluscs. They forage on mudflats, shallow marshes, or flooded agricultural areas. Ducklings eat insects for their first 6 weeks.

**Ecology:** Green-winged and Blue-winged Teal have similar diets and often forage together. They are not overtly territorial. The home range in one study was 600 acres. In winter, Green-winged Teal can be seen along the coast, on the Columbia River and, rarely, on their breeding grounds in eastern Oregon.

**Comments:** The Green-winged Teal is a game bird in Oregon. It is noted for its ability to change direction rapidly and frequently in flight, making it a difficult target.

*References:* Bellrose 1980, Gabrielson and Jewett 1940, Harrison 1978.

# Mallard
## *(Anas platyrhynchos)*

*Order: Anseriformes*
*Family: Anatidae*
*State Status: None*
*Federal Status: None*
*Global Rank: G5*
*State Rank: S5*
*Length: 23 in (58 cm)*

**Global Range:** The most common and widely distributed duck in the Northern Hemisphere, breeding throughout most of North America, from Alaska to southern Mexico. It is also common in Eurasia.

**Habitat:** Mallards are "dabbling" ducks that do not dive when foraging. They frequent almost any type of open water with patches of emergent vegetation, including shallow lakes, ponds, slow-moving rivers, streams, and marshes.

**Reproduction:** Mallards build nests on the ground, concealed in vegetation close to water. In western Oregon, Mallards nest at the base of trees, or even in the notch of a tree limb, 10-12 feet above ground. A clutch of 8 to 10 (range 1-18) eggs is laid from early May to early June. The female incubates the eggs for 26-30 days. Young are flying by 2 months of age.

**Food Habits:** This duck has an almost exclusively vegetarian diet. Mallards feed on the roots, stems, and seeds of aquatic plants, such as bulrush, sedges, wild rice, and seeds of trees growing in wet areas. They will eat waste grain in agricultural fields after harvest. Animal matter is sometimes taken, and the young eat mainly aquatic insects.

**Ecology:** Mallards defend territories around the nest during breeding season. Density varies with the availability of breeding ponds, but is usually less than 10 pairs per square mile. In winter, Mallards can be seen nearly anywhere in Oregon, including urban areas.

**Comments:** The Mallard is a game species in Oregon. In 1958, almost 13,000,000 bred in North America, but trends have been downward since the early 1970s.

*References:* Bellrose 1980, Gabrielson and Jewett 1940.

# Northern Pintail
*(Anas acuta)*

*Order: Anseriformes*
*Family: Anatidae*
*State Status: None*
*Federal Status: None*
*Global Rank: G5*
*State Rank: S5*
*Length: 28 in (71 cm)*

**Global Range:** Breeds throughout North America, as far south as southern California in the West and the Great Lakes in the East. It also occurs widely in northern Eurasia.

**Habitat:** Northern Pintails nest in fairly open areas, usually within 40 yards of water (lakes, ponds, marshes). They may conceal the nest in low vegetation, but it is often merely a depression scraped in the ground, lined with dry vegetation and down. After hatching, the young are reared on ponds and lakes.

**Reproduction:** In Oregon, the breeding season begins with the arrival of pintails in March, and the young are fledged by September. Northern Pintails lay a clutch of 6 to 9 (range 3-14) eggs, which are incubated by the female for 23-25 days. Nestlings are tended by both parents, and fledge in about 7 weeks.

**Food Habits:** About 90% of the Northern Pintail's diet is vegetable material, but there is some seasonal variation, with females eating many aquatic invertebrates prior to egg-laying. Plants taken include seeds of bulrushes, sedges, pond weeds, and grasses.

**Ecology:** Nesting pairs are separated from one another, but territories are rarely aggressively defended. The home range during breeding season can be several thousand acres. Following breeding, Northern Pintails disperse throughout Oregon.

**Comments:** The Northern Pintail is a game species in Oregon. It is an abundant breeder in the alkaline marshes of the southeastern part of the state.

*References:* Bellrose 1980, Derrickson 1978, Gabrielson and Jewett 1940, Smith 1968.

## Blue-winged Teal
### (Anas discors)

*Order: Anseriformes*
*Family: Anatidae*
*State Status: None*
*Federal Status: None*
*Global Rank: G5*
*State Rank: S4*
*Length: 15 in (38 cm)*

**Global Range:** Breeds in North America from southern Alaska and most of Canada south to Texas, but not near the coast in most of the United States.

**Habitat:** In most of its range, the Blue-winged Teal breeds in marshes, lakes, ponds, and slow moving streams in prairies or deciduous woods. In Oregon, most breed in the alkaline marshes of the southeastern part of the state.

**Reproduction:** The nest is concealed in vegetation, usually within 40 yards of water. It is constructed of dead grasses and lined with down. The breeding season begins in May, and is completed by August. A clutch of about 10 (range 6-15) eggs is incubated by the female for 23-27 days. The young are flying 35-44 days after hatching.

**Food Habits:** Blue-winged Teal eat about 70% plant and 30% animal material. The plant portion of the diet includes seeds of grasses and sedges, the green parts of aquatic vegetation, and grain in agricultural fields. During breeding season, more animals are eaten, including snails, aquatic insects, and crustaceans.

**Ecology:** Males defend a small territory around the nest. Breeding home ranges are usually several hundred acres. Blue-winged Teal often feed in association with Green-winged Teal, but they prefer to feed in shallow water with a growth of pondweeds or other aquatic vegetation, while Green-winged Teal forage on mudflats.

**Comments:** The Blue-winged Teal is a game species in Oregon. This duck migrates south during the winter.

*References:* Bellrose 1980, Glover 1956, Wheeler 1965.

# Cinnamon Teal
## (Anas cyanoptera)

*Order: Anseriformes*
*Family: Anatidae*
*State Status: None*
*Federal Status: None*
*Global Rank: G5*
*State Rank: S5*
*Length: 16 in (41 cm)*

*References:* Bellrose 1980, Oring 1964.

**Global Range:** Breeds in western North America from southern Canada south through Mexico. It also is found in South America from Columbia, Peru, and southern Brazil south to Tierra del Fuego.

**Habitat:** This species nests in a cover of vegetation near water, sometimes in grasses or marsh vegetation at water's edge. Most nests are within 75 yards of a lake, marsh, pond, or sluggish creek. It prefers smaller, shallow bodies of water for feeding.

**Reproduction:** Breeding season begins in April, and young are usually fledged by August. The nest is built in a hollow in the ground, which is lined with grasses. A normal clutch of 8 to 12 (range 4-16) eggs is incubated for 21-25 days by the female. Young are fledged in 7 weeks.

**Food Habits:** Cinnamon Teal mostly consume the seeds and leaves of bulrush and pondweed, as well as the seeds of saltgrass, and a small amount of animal material (molluscs and insects).

**Ecology:** This duck is most abundant in the alkali marshes of southeastern Oregon, but it breeds throughout the state. It is territorial only when nesting, and defends an area within about 90 feet of the nest. Most winter to the south, but a few remain in western Oregon in some years.

**Comments:** The Cinnamon Teal is a game animal in Oregon. It is the only duck that breeds in both North and South America.

# Northern Shoveler
## *(Anas clypeata)*

*Order: Anseriformes*
*Family: Anatidae*
*State Status: None*
*Federal Status: None*
*Global Rank: G5*
*State Rank: S5*
*Length: 19 in (48 cm)*

**Global Range:** Breeds in North America from Alaska and Canada south to New Mexico. It also breeds on the East Coast south to Delaware, but its distribution is spotty in the East. It is found widely throughout Eurasia.

**Habitat:** Shovelers nest in grassy areas, 75 to 200 feet away from water. They frequent shallow, sometimes muddy, freshwater marshes, lakes, and ponds, that offer some cover of emergent vegetation.

**Reproduction:** The nest is placed in a depression in the ground and lined with grasses and down. A clutch of 8 to 12 eggs (range 5-15) is laid in April. Young are fledged by August. Females incubate the eggs for 23-25 days.

**Food Habits:** Shovelers forage by straining water or mud through comb-like filters in their bills. In deep water, they feed on plankton at the water's surface. In shallow water, they sweep the bottom for both plant seeds and small animals. About 25% of the diet is animal, consisting of clams, aquatic insects, and insect larvae.

**Ecology:** The Northern Shoveler defends a territory around the pond used by the pair during breeding season, but not around the nest. Densities depend on the distribution of breeding ponds, but range from 2-10 pairs per square mile. The Northern Shoveler can be found west of the Cascades in winter.

**Comments:** The Northern Shoveler is a game animal in Oregon. The shoveler's unique bill is adapted for straining zooplankton from the surface of deep waters, which they do in large numbers on Upper Klamath Lake each fall.

*References:* Bellrose 1980, Gabrielson and Jewett 1940.

# Gadwall
## (Anas strepera)

*Order: Anseriformes*
*Family: Anatidae*
*State Status: None*
*Federal Status: None*
*Global Rank: G5*
*State Rank: S5*
*Length: 21 in (53 cm)*

**Global Range:** The Gadwall is holarctic and breeds across Eurasia, from the United Kingdom to eastern Siberia, and south to the Mediterranean Sea and northern China. In North America, the Gadwall is found breeding from the Alaskan coast and central Canada, south to California in the West and the Great Lakes in the East.

**Habitat:** Gadwall nests are concealed by clumps of grasses or other vegetation in meadows or tall grasslands. Islands in freshwater marshes, lakes, and ponds are preferred sites, when available. The nest is usually within 100 yards of water and frequently just a few feet from open water.

**Reproduction:** Breeding begins in May, and young are fledged by August. A clutch of about 10 (range 5-20) eggs is incubated about 28 days by the female. The nest is a scooped-out depression in the ground, lined with grasses and down.

**Food Habits:** The main food items of Gadwalls are green parts (stems and leaves) of submerged aquatic plants, such as sago pondweed, parrot-feathers, sedges, widgeon grass, and salt grass. They rarely feed on grasses, but they do eat some seeds of aquatic plants and, occasionally, grain from agricultural fields. Aquatic invertebrates are added to the diet during breeding season.

**Ecology:** The Gadwall defends a radius of a few hundred feet around the nest, but home ranges (averaging 67 acres) of many pairs overlap. Some Gadwalls remain along the coast and on open water in eastern Oregon over the winter.

**Comments:** The Gadwall is a game species in Oregon. It is most abundant in the alkaline marshes of southeastern Oregon.

*References:* Bellrose 1980, Gates 1962, Serie and Swanson 1976.

# American Wigeon
## (Anas americana)

*Order: Anseriformes*
*Family: Anatidae*
*State Status: None*
*Federal Status: None*
*Global Rank: G5*
*State Rank: S5*
*Length: 20 in (51 cm)*

**Global Range:** The American Wigeon is a bird of higher latitudes. It breeds from Alaska and Canada south to northern New Mexico in the West and northern New York in the East. It occurs from coast to coast in the north, but only breeds inland in the western United States.

**Habitat:** This duck nests in clumps of grass or other vegetation 50-400 yards away from water. It prefers large marshes and lakes that have some exposed shoreline.

**Reproduction:** American Wigeons use a slight depression, lined with grass and down, for a nest. They begin breeding in May, and young are fledged by August. The normal clutch is about 10 (range 3-13) eggs, which are incubated 22-24 days by the female. Young develop quickly, and fly at about 43 days of age.

**Food Habits:** American Wigeons feed on the leaves and stems of aquatic plants. They will turn to seeds of aquatic plants in the absence of green vegetation. Sometimes they graze grass in pastures. During the breeding season, both adults and ducklings eat some aquatic invertebrates.

**Ecology:** Each pair of American Wigeons needs about a half acre of habitat. These ducks will steal aquatic vegetation brought to the surface by coots, Redheads, and Canvasbacks. In winter, this species can be found along the Oregon coast and in the Willamette Valley and locally elsewhere.

**Comments:** The American Wigeon is a game species in Oregon.

*References:* Bellrose 1980, Gabrielson and Jewett 1940.

93

# Canvasback
## (Aythya valisineria)

*Order: Anseriformes*
*Family: Anatidae*
*State Status: None*
*Federal Status: None*
*Global Rank: G5*
*State Rank: S4*
*Length: 22 in (56 cm)*

**Global Range:** Breeds in northwestern North America, from Alaska and western Canada south to northern California and northern New Mexico. It occurs as far east as the Great Lakes.

**Habitat:** Canvasbacks nest in the cover of emergent vegetation such as tules, cattails, or bulrushes. The nest is sometimes built on floating vegetation. Usually, the nest will be directly over the water in a small to medium pond or lake (less than 1 acre), but it may be located on dry ground.

**Reproduction:** Breeding begins in May, and young are fledged by August. There is only one brood per year. Females incubate a clutch of about 10 (range 7-12) eggs for 23-29 days. Hatchlings require 10-12 weeks to fly. The nest is a bulky cup made of reeds, grasses, and other available vegetation.

**Food Habits:** The Canvasback eats both plant and animal material. Plants eaten include pondweed, sedges, and tubers, stems, and roots of aquatic vegetation. The seeds of grasses and water lilies are also eaten. The animal portion of the menu includes aquatic insects, small fish, and crustaceans.

**Ecology:** Canvasback males sometimes defend a territory around their mates, but frequently the home ranges (as large as 1,300 acres) of many pairs overlap. Canvasbacks and related diving ducks feed under the surface of the water, where they may dive up to 30 feet deep. Usual dives are only 3 to 12 feet deep, and submergent vegetation or animal matter is taken from the bottom.

**Comments:** The Canvasback is a game animal in Oregon. It winters along the Oregon coast, although a few remain in the Klamath Basin all year.

*References:* Bellrose 1980, Gabrielson and Jewett 1940.

# Redhead
### *(Aythya americana)*

*Order: Anseriformes*
*Family: Anatidae*
*State Status: None*
*Federal Status: None*
*Global Rank: G5*
*State Rank: S4*
*Length: 20 in (51 cm)*

**Global Range:** Redheads have a similar distribution to Canvasbacks, but do not breed quite as far north. While a few breed in eastern Alaska, the breeding range is largely south of the Northwest Territories in Canada and in the western and central United States, as far south as southern California in the West and northern Pennsylvania in the East.

**Habitat:** Redheads breed around large freshwater marshes and lakes. Like those of the Canvasback, most nests are over water and concealed in dense stands of emergent vegetation, but sometimes they nest on islands, or even on dry land.

**Reproduction:** Breeding season begins in May, and young are fledged by mid-August. The nest is a mass of reeds, grasses, and other marsh vegetation, with a central depression lined with down and feathers. A clutch of 9 to 12 (range 4-16) eggs is incubated 24-28 days by the female.

**Food Habits:** About 90% of the Redhead's diet is plant material, including leaves, tubers, and seeds of pondweed, algae, sedges, grass, bulrushes, and other aquatic plants. Animal material is eaten in the breeding season, and includes small fish, tadpoles, snails, crustaceans, insects, and insect larvae.

**Ecology:** Redheads do not obviously defend a territory. They forage by diving, but often will feed in water so shallow they do not need to dive. Dives up to 10 feet deep are common. Female Redheads will lay their eggs in nests of other ducks, including other Redheads. They may be seen along the coast during winter.

**Comments:** The Redhead is a game animal in Oregon. It was more common in the early part of the century.

*References:* Alexander 1983, Bellrose 1980, Gabrielson and Jewett 1940.

# Ring-necked Duck
## (Aythya collaris)

*Order: Anseriformes*
*Family: Anatidae*
*State Status: None*
*Federal Status: None*
*Global Rank: G5*
*State Rank: S3*
*Length: 17 in (43 cm)*

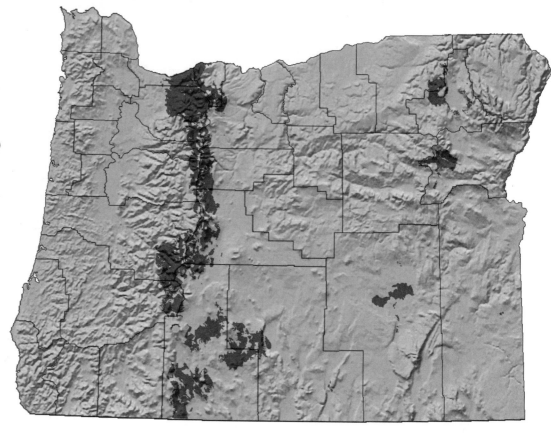

**Global Range:** Breeds from eastern Alaska across central Canada. It ranges as far south as northern California and western Pennsylvania. There are scattered breeding localities south of the main range in the United States.

**Habitat:** The Ring-necked Duck nests in thick emergent vegetation bordering lakes, ponds, marshes, and swamps, preferring to nest in marshes in wooded settings. It rarely nests on dry land, and often the nest is on floating mats of marsh vegetation.

**Reproduction:** The clutch size is usually 8 to 10 eggs (range 5-14), which are incubated by the female for 25-29 days. The young fly about 7 weeks after hatching. The breeding season lasts from May to August.

**Food Habits:** About 80% of the Ring-necked Duck's diet is plant material, chiefly seeds and tubers of aquatic plants like pondweeds, sedges, bulrush, and duckweed. They also consume plant leaves and stalks. Aquatic animals like snails, insects, insect larvae, and fresh water mussels are also eaten. Ducklings eat animal material for the first 2 weeks of life, then begin switching to plant seeds.

**Ecology:** Ring-Necked Ducks feed in shallower waters than other diving ducks, rarely diving more than 6 feet deep. During breeding season, the male will defend a small area around the female. Nests are spaced 500 to 1,000 yards apart. In winter, these ducks can be seen on major rivers, along the coast, and, locally, on open inland lakes, sometimes in large numbers.

**Comments:** The Ring-necked Duck is a game species in Oregon.

*References:* Bellrose 1980, Gabrielson and Jewett 1940, Marshall and Deubbert 1965.

*Order: Anseriformes*

*Family: Anatidae*

*State Status: None*

*Federal Status: None*

*Global Rank: G5*

*State Rank: S3*

*Length: 16 in (41 cm)*

**Global Range:** Breeds from Alaska through Canada, west of the Great Lakes, and south to northeastern California, Colorado, and northern Ohio.

**Habitat:** In its habitat preference, the Lesser Scaup more resembles a dabbling than a diving duck. It selects nest sites on dry land, which can be 150 to nearly 2,000 feet from water. It prefers to nest near lakes at least 10 feet deep and several acres large, especially lakes set in grasslands with some trees along the edge.

**Reproduction:** The nest is a depression, hidden in vegetation, and lined with meager amounts of plant material and, usually, some down. The clutch of 8 to 12 (range 6-15) eggs is incubated 22-27 days by the female. The Lesser Scaup begins breeding in May, and young are fledged by August.

**Food Habits:** Lesser Scaup dive deeper than other freshwater diving ducks, to depths of up to 40 feet (more frequently 10-25 feet deep). Up to 40% of their diet is animal food, including snails, clams, crustaceans, aquatic insects, and some fish. Plants eaten include seeds, leaves, and stalks of pondweed, wild celery, sedges, and bulrush. The diet shifts more to animal food in the summer.

**Ecology:** The male defends a small area around the female, and breeding pairs are spaced along the shoreline of marshes and lakes. Densities can reach over 20 pairs per acre on islands. In winter, Lesser Scaup can be found on almost any body of open water throughout the state.

**Comments:** The Lesser Scaup is a game animal in Oregon, but they eat enough animal material, especially along the coast, to give them an unpalatable flavor.

*References:* Bellrose 1980, Gabrielson and Jewett 1940, Hines 1977.

# *Harlequin Duck*
## *(Histrionicus histrionicus)*

*Order: Anseriformes*
*Family: Anatidae*
*State Status: Sensitive*
*Federal Status: SC*
*Global Rank: G4*
*State Rank: S2*
*Length: 16 in (41 cm)*

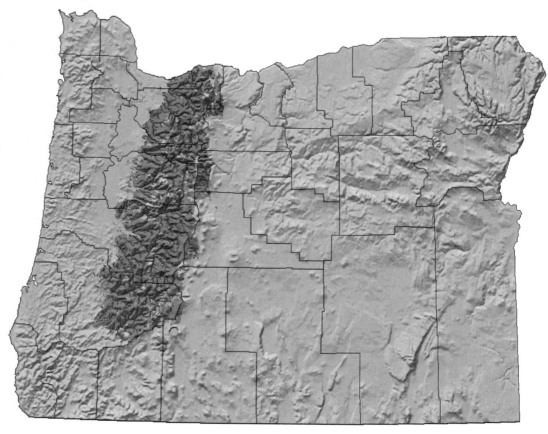

**Global Range:** The Harlequin Duck has an unusual distribution. It occurs from Iceland and Greenland west to eastern Canada. It is absent from the central part of North America, and the "western" population ranges from eastern Siberia east through Alaska and south to the Sierra Nevada of California and the mountains of southwestern Colorado.

**Habitat:** In the northwestern United States, the Harlequin Duck breeds along relatively low-gradient, slower-flowing reaches of mountain streams in forested areas. It is easily disturbed, and seeks out the most remote streams for breeding. It uses swift waters and rapids during other seasons.

**Reproduction:** A nest is built of leaves, lichen, and feathers, lined with down, and located under bushes or other cover alongside streams. A clutch of 6 to 8 (range 5-10) eggs is incubated 27-32 days by the female. Breeding begins in late April, and young are fledged by August.

**Food Habits:** On its breeding range, the Harlequin Duck feeds primarily on aquatic insects and their larvae, which are found on stream bottoms. A few fish are also eaten. Overall, about 98% of its diet is animal material. In winter, it is found off rocky ocean shorelines, where it eats crustaceans, molluscs, and some insects.

**Ecology:** The Harlequin Duck defends only a small territory around the female. Densities can be as high as one pair every 2 to 4 linear stream miles. It feeds by day, diving in swift flowing mountain streams and searching for food among rocks on the bottom.

**Comments:** While a game animal in Oregon, the total breeding population in the state is estimated to be about 100 pairs. A 1993 survey found Harlequin Ducks present along 31 streams on the west slope of the Cascade Mountains.

*References:* Anderson 1988, Bellrose 1980, Cassirer et al. 1991, Marshall 1992, Marshall et al. 1996.

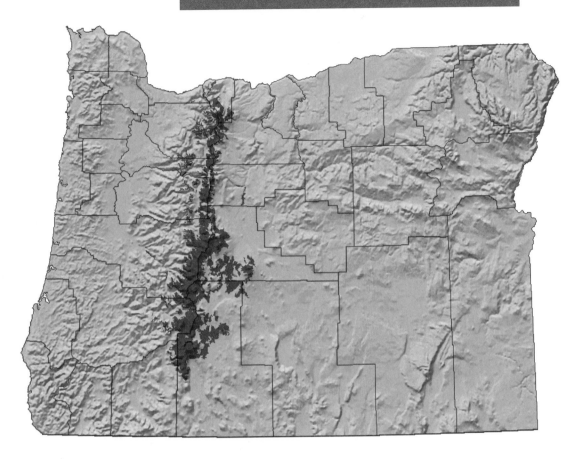

## Barrow's Goldeneye
### *(Bucephala islandica)*

*Order: Anseriformes*
*Family: Anatidae*
*State Status: Sensitive*
*Federal Status: None*
*Global Rank: G5*
*State Rank: S3*
*Length: 22 in (56 cm)*

**Global Range:** A small segment of the population breeds in Iceland, southern Greenland, and northern Labrador. Most Barrow's Goldeneyes breed in western North America, from Alaska and British Columbia south to the Sierra Nevada of California and the Rocky Mountains of Montana and Wyoming.

**Habitat:** Barrow's Goldeneyes nest around the edges of high mountain lakes, which are usually ringed with a thick growth of emergent vegetation. They are usually cavity nesters, and use abandoned woodpecker nests or natural holes in such trees as Douglas-fir, ponderosa pine, quaking aspen, and willow.

**Reproduction:** Breeding begins in May, and young are fledged by early to mid-August. A clutch of 9 or 10 (range 4-15) eggs is incubated about 1 month by the female. After hatching, the precocial young are led to water, where they commence feeding.

**Food Habits:** Animals make up 78% of the Barrow's Goldeneye's diet. They take many types of aquatic insects, including damsel- and dragonfly larvae, crustaceans, and molluscs. They also eat both seeds and vegetative parts of submerged vegetation, such as pondweed.

**Ecology:** A pair of Barrow's Goldeneyes needs a pond of at least 2 acres for breeding, and defends a territory around the nest. A few winter along the coast and, locally, east of the Cascade Mountains, mainly in the Klamath Basin and along the Columbia and Snake rivers.

**Comments:** The Barrow's Goldeneye is a game animal in Oregon. Probably fewer than 500 pairs breed in the state.

*References:* Bellrose 1980, Marshall 1992.

# Bufflehead
## (Bucephala albeola)

*Order: Anseriformes*
*Family: Anatidae*
*State Status: Sensitive*
*Federal Status: None*
*Global Rank: G5*
*State Rank: S2*
*Length: 14 in (36 cm)*

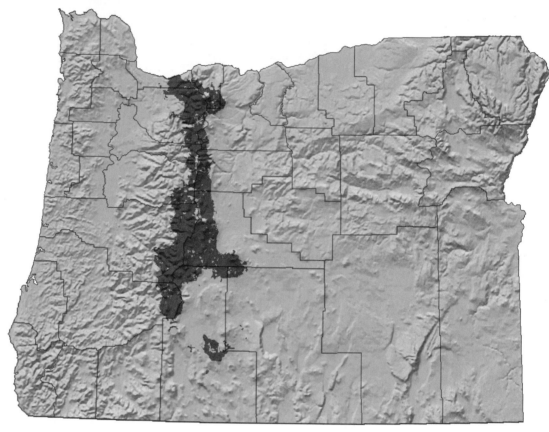

**Global Range:** This is a northern species that breeds from Alaska across Canada, and south to Oregon, northern California, and Wisconsin.

**Habitat:** Like Barrow's Goldeneye, the Bufflehead nests near mountain lakes surrounded by open woodlands containing snags. In many areas, the preferred nest trees are aspen, but it will also nest in ponderosa pine or Douglas-fir.

**Reproduction:** In Oregon, most Buffleheads nest in artificial nest boxes. Nesting begins in late April and early May, and young are fledged by early August. The female incubates a clutch of about 9 (range 5-16) eggs for 28-33 days. The young can fly about 55 days after hatching.

**Food Habits:** Buffleheads eat both animal and plant material, but primarily the former. During the breeding season, aquatic insects and their larvae are the most important item in the diet. They also eat the seeds of pondweeds and bulrushes. In winter, they feed on crustaceans and molluscs along the coast and snails in inland waters.

**Ecology:** Buffleheads defend a territory around the brood, which results in the spacing of family groups around the lake shore. After the breeding season, Buffleheads can be found on open waters throughout the state, along major rivers, and along the coast.

**Comments:** The Bufflehead is a game species in Oregon. Only several hundred pair are thought to breed in the state.

*References:* Bellrose 1980, Marshall 1992, Stern et al. 1987.

# Hooded Merganser
*(Lophodytes cucullatus)*

*Order: Anseriformes*
*Family: Anatidae*
*State Status: None*
*Federal Status: None*
*Global Rank: G5*
*State Rank: S4*
*Length: 18 in (46 cm)*

**Global Range:** In eastern North America, breeds from Hudson Bay to nearly to the Gulf of Mexico. In the West, breeds from southern Alaska through southern British Columbia, and in the mountains of Idaho, Montana, and Oregon.

**Habitat:** Hooded Mergansers are cavity nesters and eat aquatic animals: this limits their nesting habitat to wooded areas along the banks of streams or lakes. They require relatively clear water for foraging.

**Reproduction:** Nests are built in abandoned woodpecker holes or natural cavities in trees; they will also use artificial nest boxes. Breeding begins in late February, and young are fledged by July. A clutch of 10 or 12 (range 4-21) eggs is incubated 29-37 days by the female. Young fly at about 10 weeks of age.

**Food Habits:** The majority of the diet is composed of crustaceans and aquatic insects, such as caddisfly larvae and dragonfly nymphs. Hooded Mergansers also catch small fish and frogs and, rarely, eat seeds and roots of aquatic plants.

**Ecology:** The Hooded Merganser requires daylight and clear water for hunting. It competes with Wood Ducks for nest sites, and the two species will frequently lay eggs in the same nest cavity. It winters primarily along the coast and on open water west of the Cascades.

**Comments:** The Hooded Merganser is a game species in Oregon. This is the only merganser restricted to North America.

*References:* Bellrose 1980, Gabrielson and Jewett 1940, Morse et al. 1969.

# Common Merganser
## (Mergus merganser)

*Order: Anseriformes*
*Family: Anatidae*
*State Status: None*
*Federal Status: None*
*Global Rank: G5*
*State Rank: S4*
*Length: 25 in (64 cm)*

**Global Range:** This is a holarctic species that occurs widely in northern Eurasia, where it is called the Goosander. In North America, it breeds coast to coast from southern Alaska through Canada, and south into the northern United States. In the East, it does not regularly breed much south of the Great Lakes, but in the West it is found as far south as the mountains of New Mexico and Arizona.

**Habitat:** The Common Merganser usually nests in tree cavities along the forested edges of lakes, streams, and marshes. Like the Hooded Merganser, it will use a variety of tree species for nesting. Even hollows in cliff faces are used when tree cavities are absent. Ground nesting is uncommon.

**Reproduction:** Breeding season begins in April, and young are fledged by September. A clutch of 9 to 12 (range 6-17) eggs is incubated by the female for 28-32 days. Young are capable of flight at about 2 months of age.

**Food Habits:** Common Mergansers are far more piscivorous than Hooded Mergansers. They eat a wide variety of fish, depending on their availability. Near salmon or trout hatcheries, they can become pests. Other items in the diet include fish eggs, aquatic invertebrates, amphibians, crustaceans, molluscs, and, rarely, the roots and stems of aquatic plants. Hatchlings are fed aquatic invertebrates.

**Ecology:** Common Merganser pairs are generally spaced several miles apart along rivers. They require relatively clear waters, since they forage by pursuing fish in waters which are usually less than 6 feet deep. The species is widely distributed, but occurs at low density throughout Oregon. It can be found wintering on any body of open fresh water.

**Comments:** The Common Merganser is a game species in Oregon. In most areas, it primarily consumes nongame fish.

*References:* Bellrose 1980, Gabrielson and Jewett 1940.

## Ruddy Duck
*(Oxyura jamaicensis)*

*Order: Anseriformes*
*Family: Anatidae*
*State Status: None*
*Federal Status: None*
*Global Rank: G5*
*State Rank: S4*
*Length: 14 in (36 cm)*

**Global Range:** This duck has a wide distribution in the Western Hemisphere. It breeds from central and eastern Canada, south throughout most of the interior West and Great Plains into Mexico, the Caribbean islands, and through the Andes, south to southern Chile and western Argentina. It has been introduced to the United Kingdom.

**Habitat:** The Ruddy Duck breeds on freshwater marshes, lakes, and ponds where there is a dense growth of emergent vegetation to conceal the nest.

**Reproduction:** Nests are built of aquatic vegetation, and can be either up to 100 yards from water or built over the water in bulrushes and cattails. Nests in lakes are partly floating, being built up from the

bottom, and anchored to surrounding plants. Clutches usually contain 6 to 10 eggs (range 4-17). The incubation period is 24 days.

**Food Habits:** About three-quarters of the diet is plant material, but this varies with the season. Ruddy Ducks eat seeds, tubers, and foliage of submergent plants like pondweed, sedges, muskgrass, wild celery, and bulrushes. They also eat some insects, insect larvae, molluscs, and crustaceans.

**Ecology:** Male Ruddy Ducks defend a small territory around the female. Females with broods usually restrict their activities to an area a few hundred meters across. The male remains with the female and brood after the eggs hatch.

**Comments:** The Ruddy Duck is a game species in Oregon, but, along the coast, its diet of mussels and other animals makes its flesh unpalatable. The Ruddy Duck breeds locally in the Willamette Valley and the lower valleys of eastern Oregon. It winters on the coast and on other open waters both east and west of the Cascade Mountains.

*References:* Bellrose 1980, Gabrielson and Jewett 1940, McKnight 1974, Siegfried 1976.

**103**

# Turkey Vulture
## *(Cathartes aura)*

*Order: Falconiformes*
*Family: Cathartidae*
*State Status: None*
*Federal Status: None*
*Global Rank: G5*
*State Rank: S5*
*Length: 28 in (71 cm)*
*Wingspread: 60 in (152 cm)*

**Global Range:** The Turkey Vulture is widely distributed in the Western Hemisphere. It breeds from southern Canada, south through the United States and Central America, to the southern tip of South America.

**Habitat:** Although the Turkey Vulture is most typical of open rangeland, including meadows, pastures, and desert shrub communities, it frequents nearly every habitat in Oregon from sea level to alpine regions, travelling widely in search of carrion. It does not descend into dense forest or breed at very high elevations.

**Reproduction:** Breeding season begins in April, and eggs are laid by early May. The clutch of 2 (range 1-3) eggs is incubated by both parents for 38-41 days. Turkey Vultures nest in caves, cliff ledges, or hollow logs.

**Food Habits:** Although it will sometimes take ripe or rotting fruit, its main menu is vertebrate carrion, including road kills, dead livestock, amphibians, reptiles, and fish. It also eats bird eggs and nestlings.

**Ecology:** Nests are usually isolated from one another, but there is no known territorial defense. This species often roosts and feeds in large groups. The home range is very large. An average distance between roosting and feeding is at least 5 miles.

**Comments:** This species is resistant to botulism. It winters south of Oregon, but an occasional individual may wander up from coastal California during warm spells.

*References:* Coleman and Fraser 1989, Gabrielson and Jewett 1940, Wilbur and Jackson 1983.

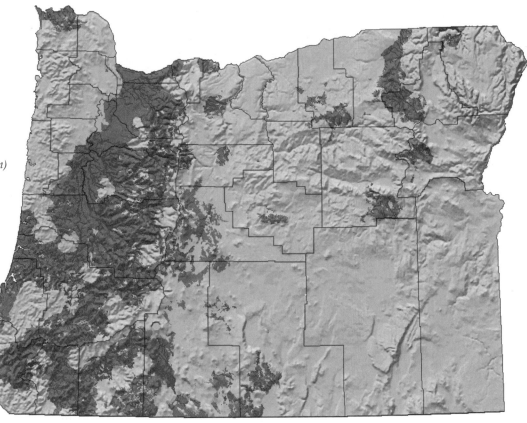

# Osprey
## *(Pandion haliaetus)*

*Order: Falconiformes*
*Family: Accipitridae*
*State Status: None*
*Federal Status: None*
*Global Rank: G5*
*State Rank: S4*
*Length: 24 in (61 cm)*
*Wingspread: 60 in (152 cm)*

**Global Range:** The distribution of the Osprey is essentially worldwide, except that it winters, but does not breed, in South America. In North America, it breeds from Alaska and Canada south to Baja California, Texas, and Florida. It is absent from the Great Plains, the Midwest, and much of the Southeast.

**Habitat:** The Osprey (also called the fish hawk) is specialized for catching fish. As a result, it nests in areas within easy reach of lakes and rivers. It also requires suitable nest sites such as large, dead trees or artificial nesting platforms.

**Reproduction:** Breeding season begins in April, shortly after Ospreys arrive in Oregon. Young are fledged by August. There is a single clutch of 3 (range 2-4) chicks per year. The female incubates the eggs for about 38 days, while the male provides food. Pairs often return to the same large stick-nest in successive years.

**Food Habits:** Most of the Osprey's diet consists of fish, which it captures by plunging feet-first into the water. Its claws are round in cross section, and taper to a sharp point, which assists holding slippery prey. It selects slow-moving fish species that swim near the surface. It sometimes eats other types of vertebrate prey (birds, reptiles, small mammals) and, rarely, feeds on invertebrates.

**Ecology:** The Osprey is usually solitary in Oregon, but sometimes forms colonies where food is concentrated. Nests usually overlook water, but are sometimes 5 or 6 miles away from feeding areas. It is mainly a diurnal predator. Bald Eagles may steal fish from Ospreys.

**Comments:** Osprey numbers fell in mid-century due to DDT contamination, but numbers in Oregon are now recovering. It can be seen at many lakes and reservoirs in the high Cascades, along the coast, and along major rivers.

*References:* DeGraaf et al. 1991, Gilligan et al. 1994, Harrison 1978.

# White-tailed Kite
## (Elanus leucurus)

*Order: Falconiformes*
*Family: Accipitridae*
*State Status: None*
*Federal Status: None*
*Global Rank: G5*
*State Rank: S1*
*Length: 16 in (41 cm)*
*Wingspread: 42 in (107 cm)*

*References:* DeGraaf
et al. 1991,
Eisenmann 1971,
Gilligan et al. 1994,
Hawbecker 1942.

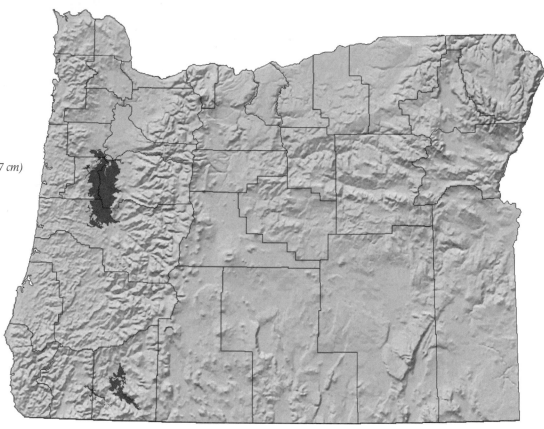

**Global Range:** If kites of the genus *Elanus* are considered a single species, *E. caeruleus* (American Ornithologists' Union 1983, Banks et al. 1987), their distribution includes Australia, Africa, southern Eurasia, South America, and regions of North America near the coast from Oregon and Florida, south through Central America. As recognized here, the New World form is a separate species, *E. leucurus* (American Ornithologists' Union 1993).

**Habitat:** The White-tailed Kite is a rare breeder in Oregon, but is likely expanding its range north from California, where it can be quite common in lower-elevation grasslands, agricultural areas, meadows, oak and riparian woodlands, marshes, and wetlands. It requires trees or tall shrubs for nesting.

**Reproduction:** White-tailed Kites build a stick-nest about 2 feet across and line it with plant material. There is usually a single clutch of 4 or 5 (range 3-6) eggs that the female incubates for 30-32 days, while the male brings food. Breeding season begins early, about mid-February, with young fledged by July.

**Food Habits:** In California, and likely in Oregon, this kite specializes in feeding on field voles (genus *Microtus*) where they are abundant. It will eat a wide variety of other vertebrates (small mammals, snakes, lizards, frogs) and invertebrates (large insects such as grasshoppers, beetles, and crickets).

**Ecology:** Where prey is abundant, a pair will forage over about 20 acres. Nesting pairs often occur in proximity to one another where nest trees are uncommon. They hunt by hovering above open areas and descending on prey with outstretched legs. Like most raptors, they are diurnal.

**Comments:** This kite has nested near Medford and Corvallis in the Willamette Valley. A pair raised 4 young in 1992 from a nest around Fern Ridge Reservoir. In winter, it is uncommon along the southern coast and western interior valleys.

## Bald Eagle
### *(Haliaeetus leucocephalus)*

*Order: Falconiformes*
*Family: Accipitridae*
*State Status: Threatened*
*Federal Status: Threatened*
*Global Rank: G4*
*State Rank: S3*
*Length: 36 in (91 cm)*
*Wingspread: 90 in (229 cm)*

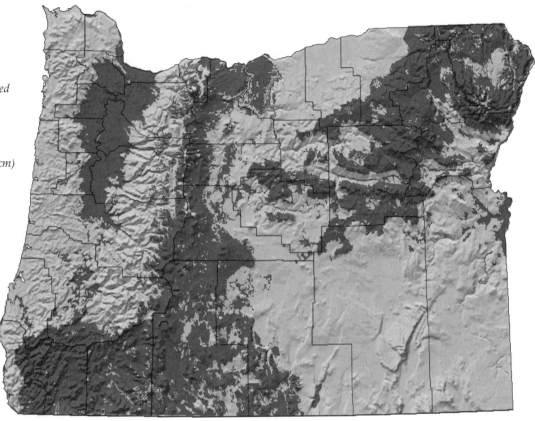

**Global Range:** Breeds from Alaska and Canada south to California and Florida. It is rare in the interior United States.

**Habitat:** Bald Eagles are associated with coasts, rivers, lakes, and marshes. They also require nearby tall trees or cliffs for nesting. In Oregon, they are most common around large inland lakes (such as Upper Klamath Lake) and marshes, along the coast, and along the Columbia River.

**Reproduction:** A Bald Eagle nest is usually a large platform, constructed with sticks in a tall tree, and within half a mile of water. The usual clutch size is 2. Eggs are incubated by both parents for 35-46 days. Breeding season begins early (January), and young are flying at about 3 months of age.

**Food Habits:** Bald Eagles feed mainly on fish, whether dead or alive. They also eat carrion, various water birds, and small mammals. They steal fish from Ospreys when they can. At inland wintering grounds they feed on dead waterfowl.

**Ecology:** The Bald Eagle defends a territory for a few hundred yards around the nest. Nests are usually spaced about a mile apart. In winter, they may travel up to 10 miles between roost trees and feeding areas.

**Comments:** About 250 pairs of Bald Eagles nested in Oregon in 1995, but during the winter their number is supplemented by migrants from farther north. Some 1,000 winter in the Klamath area alone.

*References:* DeGraaf et al. 1991, Gabrielson and Jewett 1940, Gilligan et al. 1994, Marshall et al. 1996.

**107**

# Northern Harrier
## (Circus cyaneus)

*Order: Falconiformes*
*Family: Accipitridae*
*State Status: None*
*Federal Status: None*
*Global Rank: G5*
*State Rank: S5*
*Length: 21 in (53 cm)*
*Wingspread: 44 in (112 cm)*

**Global Range:** This species, formerly called the Marsh Hawk, breeds throughout the Northern Hemisphere, from northern Eurasia south to the Mediterranean Sea in the Old World, and in North America from Alaska and Canada south to southern California in the West, and Virginia in the East.

**Habitat:** Northern Harriers frequent open country such as grassland, marshes, agricultural areas, and wet meadows. They occur in desert areas if there is some water nearby. They also frequent coastal headlands.

**Reproduction:** Breeding begins in late March, and young are fledged in September. A normal clutch of 5 (range 3-12) eggs is incubated 30-32 days by the female, who is fed by the male during this time. There is a single brood per year. The nest is usually built on dry ground, often near shrubs, or in good cover of grasses. Sometimes they will nest in reeds or bushes over water, in which case the nest is more substantial.

**Food Habits:** The Northern Harrier glides low over open ground when searching for food. Most of what it finds are small mammals, but it also eats lizards, small snakes, birds, crayfish, fish, a few insects, and carrion.

**Ecology:** It defends a territory around the nest, and may range up to 5 miles or more from the nest or roost while hunting. Breeding home ranges can be over 2,000 acres. Northern Harriers may hunt over the same area several days in a row. Most activity is in the morning and late afternoon.

**Comments:** The Northern Harrier is a permanent resident in Oregon, but wanders widely from breeding grounds during the fall and winter, even above timberline. It does not use forested areas.

*References:* DeGraaf et al. 1991, Gilligan et al. 1994.

*Order: Falconiformes*
*Family: Accipitridae*
*State Status: None*
*Federal Status: None*
*Global Rank: G5*
*State Rank: S4*
*Length: 11 in (28 cm)*
*Wingspread: 13 in (33 cm)*

**Global Range:** Widely distributed in the Western Hemisphere. It breeds from Alaska and Canada south through the highlands of Mexico and Central America, on some islands of the Caribbean Sea, and in the mountains of South America as far south as northern Argentina and Uruguay.

**Habitat:** Sharp-shinned Hawks are birds of forests and woodlands. They are associated with coniferous forests, but also use mixed and deciduous forests, riparian woodlands, and even juniper or oak woodlands.

**Reproduction:** The nest is built of twigs placed on a branch near a tree trunk or in the fork of a limb. The female incubates the clutch of 4 to 5 (range 3-8) eggs for 30-32 days while she is fed by the male. The male continues to bring food to the female and nestlings after the eggs hatch. Breeding season starts in April and young are fledged by late August.

**Food Habits:** The Sharp-shinned Hawk and other members of the genus *Accipiter* tend to specialize in eating other birds. The "Sharpie" will even take birds larger than itself, but mostly preys on small songbirds it takes in flight while cruising through foliage. It also eats a few small mammals, reptiles, and insects.

**Ecology:** This is an uncommon hawk, and pairs usually nest several miles apart, or one nest every 7,000 acres. A territory of a few hundred acres is defended around the nest. This species has reverse sexual dimorphism, meaning the females are larger than the males. Several explanations have been advanced, including the female's need to defend the young from the male, and the smaller male's tendency to bring smaller prey items to feed chicks.

**Comments:** The short, broad wings of the Sharp-shinned Hawk and its relatives allow rapid maneuvering through thick foliage.

*References:* DeGraaf et al. 1991, Harrison 1978.

# Cooper's Hawk
## *(Accipiter cooperii)*

*Order: Falconiformes*
*Family: Accipitridae*
*State Status: None*
*Federal Status: None*
*Global Rank: G4*
*State Rank: S4*
*Length: 16 in (41 cm)*
*Wingspread: 33 in (84 cm)*

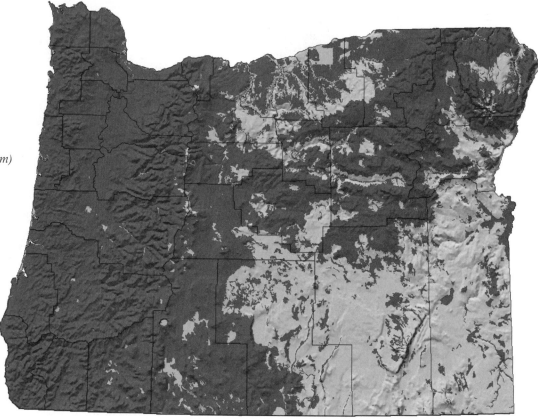

**Global Range:** Breeds coast to coast from southern Canada through almost all of the United States, and as far south as northern Baja California and northern Mexico.

**Habitat:** This species prefers coniferous forests and woodlands, but also can be found in mixed and deciduous forests, as well as riparian, juniper, or oak woodlands.

**Reproduction:** The nest is made of sticks and twigs, lined with bark, and located, usually, in the branch of a limb. The usual clutch of 4 or 5 eggs (range 2-6) is incubated and brooded by the female, while the male brings food. Breeding season stretches from late March to August, when the young are fledged.

**Food Habits:** About 80% of the diet consists of small to medium-size birds, caught on the wing or pounced upon from a hidden perch. It takes flickers and other woodpeckers, Mourning Doves, meadowlarks, robins, jays, quail, and other birds. The remainder of the diet consists of both small mammals and insects. Although the old nickname of "chicken hawk" doesn't really fit, it has been known to raid poultry farms.

**Ecology:** Nests are usually not closer than 2 miles to one another, and the home range is several hundred acres. Like the Sharp-shinned Hawk, the female is larger than the male. It tends to exclude the Sharp-shinned Hawk from its home range.

**Comments:** There is a size overlap between male Cooper Hawks and female Sharp-shinned Hawks. Because of their similar coloration, they are extremely difficult to tell apart.

*References:* DeGraaf et al. 1991, Gabrielson and Jewett 1940.

# Northern Goshawk
### (Accipiter gentilis)

*Order: Falconiformes*
*Family: Accipitridae*
*State Status: Sensitive*
*Federal Status: SC*
*Global Rank: G4*
*State Rank: S3*
*Length: 23 in (58 cm)*
*Wingspread: 33 in (84 cm)*

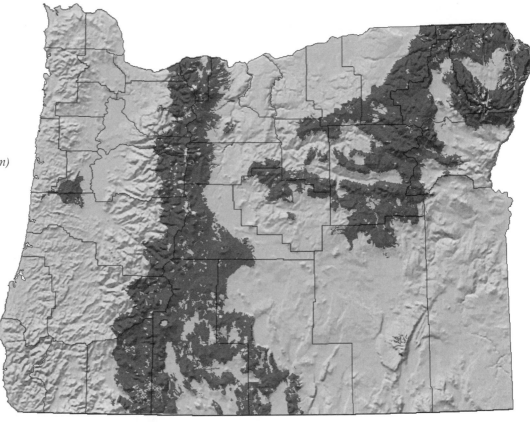

**Global Range:** The Goshawk is a holarctic species that breeds across Eurasia, as far south as the Mediterranean Sea and China. In North America, it breeds from Alaska across Canada, and south in the West to the Sierra Nevada of California and the mountains of Arizona and New Mexico. In the eastern United States, it extends south in the Appalachian Mountains to North Carolina, but it is absent from the plains and low valleys of the middle of the continent.

**Habitat:** Goshawks are forest birds. In Oregon, this usually means coniferous forests, but they will use quaking aspen groves on desert mountain ranges. These are large accipiters and prefer large patches of late-successional forests with large trees and considerable canopy closure. In ponderosa pine woodlands of the Blue Mountains, they are found in more open situations.

**Reproduction:** Goshawk nests are built of sticks, and are usually located in the fork of a limb, or on a large limb near the trunk of a tree, from 20 to 80 feet high. The usual clutch of 3 or 4 (range 1-5) eggs is incubated 32-34 days by the female. During incubation and brooding, the male provides food.

**Food Habits:** Goshawks include both birds and mammals in their diet. They take prey up to the size of grouse and rabbits, but some insects are also eaten. Among their prey are quail, owls, small hawks, ducks, thrushes, squirrels, and shrews.

**Ecology:** Goshawks defend territories of several hundred acres. A study in Oregon found nests separated by about 3.4 miles. They prefer to nest on north-facing slopes near some water. They hunt below the forest canopy. In winter, they can be seen in the coast ranges.

**Comments:** In Oregon, the loss of older forests is thought to have caused a decline of this species.

*References:* Marshall 1992, Reynolds et al. 1982, Reynolds and Wight 1978.

111

# Swainson's Hawk
## (Buteo swainsoni)

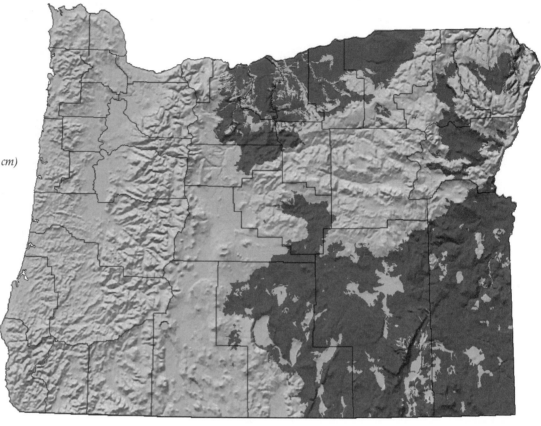

*Order: Falconiformes*
*Family: Accipitridae*
*State Status: Sensitive*
*Federal Status: None*
*Global Rank: G4*
*State Rank: S3*
*Length: 21 in (53 cm)*
*Wingspread: 51 in (130 cm)*

**Global Range:** Breeds in western North America, from Alaska south to northern Mexico and, at least formerly, as far east as Illinois and Missouri. It winters primarily on the pampas of South America.

**Habitat:** This is a hawk of open country. In eastern Oregon it uses grasslands, sagebrush flats, juniper woodlands, larger meadows, and grasslands within forested mountains. It requires some type of tree, often a willow or juniper, for nesting.

**Reproduction:** The nest is fairly typical for raptors, built of twigs and lined with leaves or other plant material. Breeding begins in mid-April and is completed by early August. A clutch of 2 (range 1-4) eggs is incubated about a month by both sexes.

**Food Habits:** At least during the breeding season, this hawk preys primarily on small mammals, especially ground squirrels, pocket gophers, and deer mice. It will take some birds living in open country (e.g., Horned Larks, meadowlarks). At other times of the year, its diet includes more insects and an occasional snake or lizard. It can even capture bats on the wing.

**Ecology:** A pair has a territory of about 1 to 3 square miles, which is defended not only against

other Swainson's Hawks but also against its main competitor, the Red-tailed Hawk. It will hunt in hay and alfalfa fields. Fire suppression, conversion of native grasslands, and encroachment of western juniper all modify habitat in favor of Red-tailed Hawks.

**Comments:** Swainson's Hawk was once common in eastern Oregon, but has declined since about 1950. Fewer than 800 pairs are estimated to still nest in Oregon.

*References:* Bechard 1982, Gilligan et al. 1994, Janes 1987, Marshall 1992.

# Red-tailed Hawk

## *(Buteo jamaicensis)*

*Order: Falconiformes*
*Family: Accipitridae*
*State Status: None*
*Federal Status: None*
*Global Rank: G5*
*State Rank: S5*
*Length: 22 in (56 cm)*
*Wingspread: 52 in (132 cm)*

**Global Range:** Breeds from southern Alaska, south and east through Canada and the United States to the highlands of Central America and Panama. It is found from coast to coast and is our most common *Buteo.*

**Habitat:** Red-tailed Hawks use nearly every open habitat in Oregon, from agriculture and grasslands to the woodlands and meadows of eastern mountains. They do not hunt over or beneath the closed canopy of dense forests.

**Reproduction:** Breeding begins in March and young are fledged by early September. The nest is a platform of sticks and twigs lined with green vegetation, often high above ground (over 100 feet) in the tallest tree bordering an open area. The normal clutch has 2 or 3 eggs (range 1-5), which are incubated 28-35 days, mostly by the female, who is fed by the male.

**Food Habits:** Red-tailed Hawks feed primarily on small mammals, up to the size of rabbits, which they detect while soaring at considerable heights. Ground squirrels, deer mice, and voles are staples, but they also eat small birds, reptiles, some insects and, rarely, carrion.

**Ecology:** This common hawk competes with other buteos in eastern Oregon, where nesting territories range from one half to 3 square miles. During the winter they center their activity in lower valleys.

**Comments:** Because there is much variation in coloration among adults and between adults and juveniles, it is difficult to distinguish among *Buteo* species.

*References:* DeGraaf et al. 1991, Gilligan et al. 1994, Rothfels and Lein 1983.

# Ferruginous Hawk
### (Buteo regalis)

*Order: Falconiformes*
*Family: Accipitridae*
*State Status: Sensitive*
*Federal Status: SC*
*Global Rank: G4*
*State Rank: S3*
*Length: 23 in (58 cm)*
*Wingspread: 54 in (137 cm)*

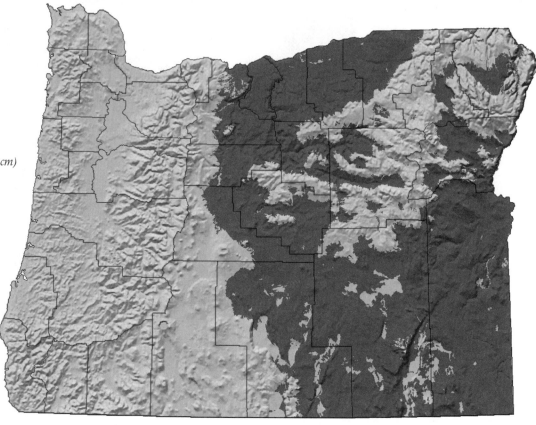

**Global Range:** This hawk has a fairly small range in western North America. It breeds from southern Canada, south to northern California, and south and east to west Texas and the Dakotas.

**Habitat:** Like most *Buteos*, the Ferruginous Hawk soars over open country (grassland, desert steppe, juniper woodland). It requires either ledges on cliffs, isolated trees, or riparian woodland for nesting, although there are reports of ground nesting from Malheur County. It uses agricultural land and pasture less than other species in the genus.

**Reproduction:** The nest, which is reused in subsequent years, is a large structure of sticks and other debris, lined with leaves or other fine plant material. Breeding season begins in mid-April. The usual clutch of 3 to 4 eggs is incubated about 28 days, mostly by the female, while the male brings her food. Young fledge about 2 months after hatching.

**Food Habits:** This hawk primarily preys on mammals found in its arid environment, including jackrabbits, ground squirrels, pocket gophers, and kangaroo rats. It will also eat birds and reptiles.

**Ecology:** Ferruginous Hawks hunt while soaring, or from a perch. There can be as many as 8 to 10 nests per 40 square miles. The home range of males is about 3 square miles.

**Comments:** This hawk was once fairly common in Oregon, but it has declined with the conversion of grasslands to agriculture over much of its range. It is also sensitive to disturbance while nesting. During winter, a few birds are occasionally seen in Lake County and surrounding parts of south central Oregon.

*References:* Marshall 1992, Schmutz 1989, Wakeley 1978, White and Thurow 1985.

# Golden Eagle
### (Aquila chrysaetos)

*Order: Falconiformes*
*Family: Accipitridae*
*State Status: None*
*Federal Status: None*
*Global Rank: G5*
*State Rank: S4*
*Length: 36 in (91 cm)*
*Wingspread: 84 in (213 cm)*

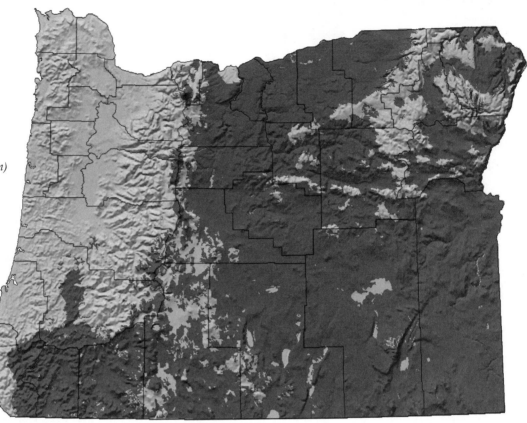

**Global Range:** The Golden Eagle is a holarctic species whose breeding range in Eurasia extends south to northern Africa. It is found from the United Kingdom to Japan in the Old World. In North America, it breeds from Alaska across Canada. In the West, it is found in open country south into Mexico, while it occurs only as far south as New York in the East.

**Habitat:** Golden Eagles are found in both open country (grassland, pasture, desert scrub) and in open coniferous forests and woodlands. They range over open areas above timberline, and usually require ledges on cliffs for nesting.

**Reproduction:** Nests are built of sticks and lined with leaves, mosses, twigs, and fur, usually on a cliff, but sometimes in a large tree. Breeding season begins in late March. There are usually 2 eggs (range 1-3), which are incubated mainly by the female for 43-45 days. Young are independent about 4 months after hatching.

**Food Habits:** Rabbits and hares are the mainstay of the Golden Eagle's diet, although other mammals such as ground squirrels, woodrats, and mice are also taken. Birds (grouse, owls, hawks, magpies, and others), snakes, and some carrion are occasionally eaten.

**Ecology:** The Golden Eagle hunts for prey by soaring several hundred feet above the ground. In eastern Oregon there are an estimated 4 to 5 pairs per 40 square miles. Golden Eagles have overlapping home ranges. Although they may breed quite high in the mountains, they seek lower valleys in winter.

**Comments:** They occasionally are seen feeding on young deer, sheep, or pronghorn. Some of these instances may reflect feeding on carrion, while others are kills.

*References:* Collopy 1984, DeGraaf et al. 1991, Olendorff 1976.

**115**

# American Kestrel
## *(Falco sparverius)*

*Order: Falconiformes*
*Family: Falconidae*
*State Status: None*
*Federal Status: None*
*Global Rank: G5*
*State Rank: S5*
*Length: 11.5 in (29 cm)*
*Wingspread: 22 in (56 cm)*

**Global Range:** The American Kestrel is widely distributed in the New World, except in dense tropical moist forests. It breeds from Alaska south through Canada, the United States, and Mexico, into Central America, and farther south in South America to Tierra del Fuego. It is not found in the Amazon Basin.

**Habitat:** American Kestrels use a wide variety of open to semi-open habitats, including grasslands, deserts, juniper woodlands, meadows and clear-cuts in forests, marshes, agricultural fields, and even urban areas. They perch on utility poles and wires in otherwise treeless areas.

**Reproduction:** The preferred nest is a woodpecker hole or natural cavity in a tree, but an American Kestrel will make do with covered rock ledges or nest boxes. The breeding season begins in April, and young are fledged by August. A clutch of 4 or 5 (range 3-7) eggs is incubated 29-30 days by the female. Young are tended by both parents and are independent in 4 or 5 weeks.

**Food Habits:** Despite its earlier common name of Sparrow Hawk, the American Kestrel feeds mainly on insects (grasshoppers, beetles, crickets). When insects are seasonally low, it feeds on small mammals and some birds. Rarely, it will eat lizards, snakes, frogs, spiders, and earthworms.

**Ecology:** American Kestrel density varies with the food supply. Two studies found territory sizes of 109 and 130 hectares. It is sometimes attacked by larger raptors. When foraging, it will either dart out from a perch or hover above an open area, hoping to spot an exposed mouse. It sometimes caches its food, especially in winter.

**Comments:** Away from the coast, the American Kestrel is our most common raptor, and has probably benefited from human activities.

*References:* Collopy 1977, Cruz 1976, Gilligan et al. 1994, Raphael 1985.

*Order: Falconiformes*
*Family: Falconidae*
*State Status: Endangered*
*Federal Status: Endangered*
*Global Rank: G4*
*State Rank: S1*
*Length: 18 in (46 cm)*
*Wingspread: 45 in (114 cm)*

**Global Range:** On every continent but Antarctica. In the New World, breeds locally in Alaska and the western United States, but is absent from much of middle and eastern North America, reappearing in the temperate portions of South America, mostly in central and southern Chile and Argentina.

**Habitat:** The most critical habitat components for Peregrine Falcons are suitable nest sites, usually cliffs, overlooking fairly open areas with an ample food supply. They nest along seacoasts, near marshes, and even in cities, but are not well suited to life in forest interiors. They were once called "duck hawks," and they usually nest or roost near a marsh, lake, or coast where waterbirds are plentiful.

**Reproduction:** Peregrines prefer to nest on inaccessible ledges high on rocks or cliffs. They begin breeding in March, and young are fledged by late August. The usual clutch is 3 to 4 eggs (range 2-6), which are incubated about 29-33 days, during which time the male brings food to the female.

**Food Habits:** This falcon attacks birds up to the size of herons, on the wing. It is well known for its high-speed dives, which end by striking the prey with closed talons. Rarely, small mammals, insects, or fish are eaten.

**Ecology:** Territory size depends on orientation of the nest site. Nests may be quite close if not visible to one another. Areas from 100 yards to a mile from the nest may be defended. The population density is probably limited by the availability of nest sites. The home range varies from 25 to 100 square miles. Great Horned Owls are a serious predator.

**Comments:** This species was much reduced due to pesticide accumulation that caused eggshell thinning. It is now recovering, and about 40 pairs are known to nest in Oregon. The U.S. Fish and Wildlife Service has proposed changing its status from endangered to threatened. Illegal collecting of Peregrine Falcon nestlings for use in falconry is a major threat to the recovering population. During migration, Peregrine Falcons can be seen throughout Oregon.

*References:* Gilligan et al. 1994, Marshall et al. 1996, Peakall 1990, Steidl et al. 1991.

**117**

# *Prairie Falcon*
## *(Falco mexicanus)*

*Order: Falconiformes*
*Family: Falconidae*
*State Status: None*
*Federal Status: None*
*Global Rank: G4G5*
*State Rank: S4*
*Length: 19 in (48 cm)*
*Wingspread: 40 in (102 cm)*

**Global Range:** The Prairie Falcon is a bird of the arid western United States. Its breeding range just enters Canada to the north and Mexico to the south. Eastern Colorado and western North Dakota mark its eastern limit.

**Habitat:** Prairie Falcons are mostly birds of arid deserts and open grasslands. They will use plateau grasslands and alpine meadows high on mountain ranges, provided their cliff-face nesting sites are available.

**Reproduction:** This falcon prefers to nest on a covered ledge on rock outcrops or cliff faces as high as 400 feet above surrounding open country. The normal clutch is 4 or 5 eggs (range 3-6), which the female incubates for 29-33 days while being fed by the male. Eggs are laid in April and May, and young leave the nest about 6 weeks after hatching.

**Food Habits:** The Prairie Falcon is an opportunistic feeder that takes small mammals up to the size of jackrabbits (deer mice, ground squirrels), birds (ducks, quail, doves, sparrows and other small birds), lizards, and grasshoppers.

**Ecology:** Prairie Falcons can be quite dense where prey is common. One study found 101 pairs along 72 kilometers of cliffs overlooking the Snake River. They hunt from perches or make swift dives on prey from cruising altitudes of 50 to 300 feet. They retreat from higher elevations in winter.

**Comments:** Although most Prairie Falcons nest east of the Cascades, a few breed in the Rogue River Valley of southwestern Oregon. A few birds winter in the Willamette Valley.

*References:* DeGraaf et al. 1991, Gilligan et al. 1994.

# Gray Partridge
*(Perdix perdix)*

*Order: Galliformes*
*Family: Phasianidae*
*State Status: None*
*Federal Status: None*
*Global Rank: G5*
*State Rank: SE*
*Length: 13 in (33 cm)*

*References:*
Gabrielson and
Jewett 1940,
Gilligan et al. 1994,
Hupp et al. 1988,
Weigand 1980.

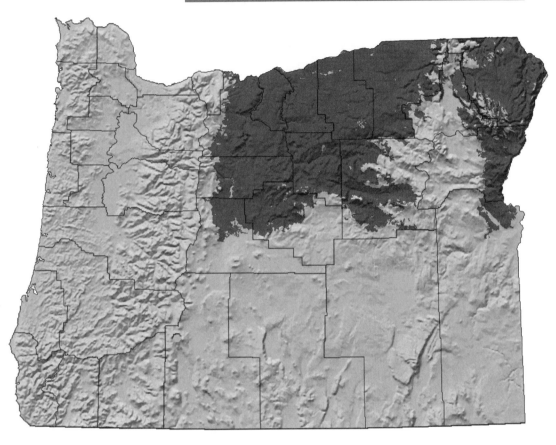

**Global Range:** The Gray Partridge is native to Europe and Asia, from the United Kingdom east through northern Russia and south to southern Europe, Turkey, and Mongolia. It has been introduced in North America from the Atlantic provinces of Canada west to southern British Columbia and south to northern Nevada, Indiana, and New York.

**Habitat:** The Gray Partridge is a bird of cultivated fields, grasslands, meadows, and pasture. At least some sparse brushy cover is required, such as is often found along the margins of agricultural fields. In Oregon, it is often found in wheat fields, but sometimes occurs in sagebrush or grassland a considerable distance from agriculture.

**Reproduction:** The breeding season starts in late May. The nest is a shallow depression scraped in the ground, usually in thick grass or under a bush. The clutch usually contains about 10 eggs (range 6-20), which the female incubates for 3 to 4 weeks. Like most gallinaceous birds, the young are precocial.

**Food Habits:** Young Gray Partridges feed on insects for their first few weeks, but then turn to an adult diet of grain seeds (wheat, corn, barley, oats), grasses, and forbs. Leaves of alfalfa, clover, dandelion, and a variety of forbs supplement the diet. In the absence of cultivation, the natural diet is likely to have consisted of grass seeds and leaves of native grasses and forbs.

**Ecology:** The Gray Partridge is relatively sedentary; coveys travel no more than about a quarter of a mile. In late summer, two or more family groups coalesce into coveys of 10-30 birds. The Gray Partridge is resident throughout the year, surviving in areas where there may be considerable snowfall.

**Comments:** This species was introduced to western Oregon in 1900, and later to eastern Oregon. Today it survives only east of the Cascades, where it is a game species. This species is also called the Hungarian Partridge, shortened by hunters to "Hun."

**119**

# Chukar
## (Alectoris chukar)

Order: Galliformes
Family: Phasianidae
State Status: None
Federal Status: None
Global Rank: G5
State Rank: SE
Length: 14 in (36 cm)

**Global Range:** Native to Eurasia from southeastern Europe east to northern China. It has been introduced into North America from southern British Columbia east to Montana and south to northern Baja California and northern New Mexico. It is also introduced and established in Hawaii.

**Habitat:** The Chukar prefers sparsely vegetated rocky canyons, slopes, and hillsides and can be found in sagebrush flats, grasslands, and open juniper canyonlands. It is not found in dense forests nor in areas subject to heavy accumulations of snowfall.

**Reproduction:** Chukars are ground nesters that line a slight depression under a rock or bush with grass or twigs. The breeding season varies with climate, but by June a clutch of 8 to 20 eggs has been laid. The female incubates the eggs, which hatch in just over 3 weeks. The male may incubate the first clutch if the female renests. Young are precocial and remain in a family group until coveys of up to 40 birds are formed in the fall.

**Food Habits:** The Chukar is primarily a granivore, eating the seeds of grasses and forbs. About 5% of the adult diet consists of insects, and the young eat a higher proportion of insects. Some fruits and the leaves of plants are also consumed.

**Ecology:** Chukars are not thought to be territorial, but males may defend an area around the female. Densities of one bird per 10 acres have been reported. They require daily access to water in hot, dry weather, which limits their distribution in arid regions. They may migrate downslope as far as 10 miles in winter.

**Comments:** Chukars were introduced to Oregon in 1951 and have become successfully established, aided by the spread of introduced Eurasian cheatgrass, a favored food. They are game birds in Oregon.

*References:* Christensen 1970, Godfrey 1966, Ryser 1985, Zeiner et al. 1990a.

## Ring-necked Pheasant
### *(Phasianus colchicus)*

*Order: Galliformes*
*Family: Phasianidae*
*State Status: None*
*Federal Status: None*
*Global Rank: G5*
*State Rank: SE*
*Length: 33 in (84 cm)*

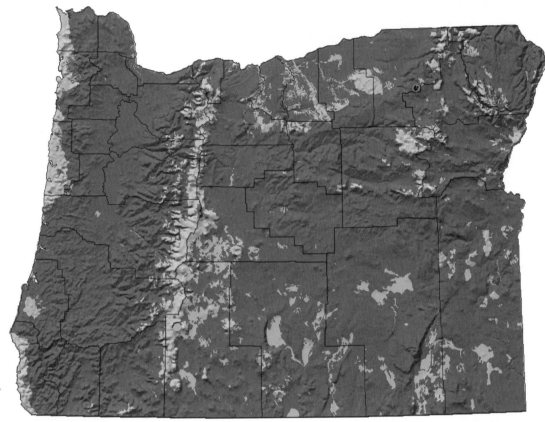

*References:* Eklund 1942, Gabrielson and Jewett 1940, Gilligan et al. 1994, Leopold 1959, Zeiner et al. 1990a.

**Global Range:** Native to Eurasia, from central Russia east to China and south to Iran. It is also found in Japan, although some authors think the Japanese Pheasant is a separate species. It is widely introduced in North America, from southern Canada to northern Mexico. It is also introduced in Europe, New Zealand, and Hawaii.

**Habitat:** Ring-necked Pheasants are associated with agricultural fields that provide some cover in the form of tall grasses or shrubby margins. They are most abundant in the wheat-growing areas of the Columbia Plateau, but also can be found in the Willamette Valley and irrigated croplands in Malheur County. They are not found in deserts, high mountains, or dense forests and are nearly always associated with human-modified landscapes.

**Reproduction:** A clutch of a dozen eggs (range 7-16) is laid in a shallow depression lined with grass or leaves. The female incubates for just over 3 weeks. There is a single brood per year, but females will renest if the first clutch is lost. Young are precocial and fly at 2 weeks of age. Leopold (1959) speculated that ground moisture is required to keep the eggs moist during incubation, excluding pheasants from arid regions in the absence of water.

**Food Habits:** Ring-necked Pheasants feed on grain (wheat, corn, rice) or weed seeds. They also eat green vegetation, such as clover, alfalfa, and native forbs, as well as a few insects. The young eat insects after hatching, then switch to the adult diet.

**Ecology:** Males are territorial during the breeding season, defending a territory several acres in size. Each male mates with several females (which are smaller) in his territory. In autumn, loose flocks of up to 40 birds are formed, especially around areas with abundant food. Ring-necked Pheasants roost in trees at night. They are nonmigratory and usually confine their activities to a few square miles.

**Comments:** Ring-necked Pheasants were introduced to the Willamette Valley in 1881. In agricultural areas, they can cause crop damage. They are a game species in Oregon.

**121**

# Spruce Grouse
## (Dendragapus canadensis)

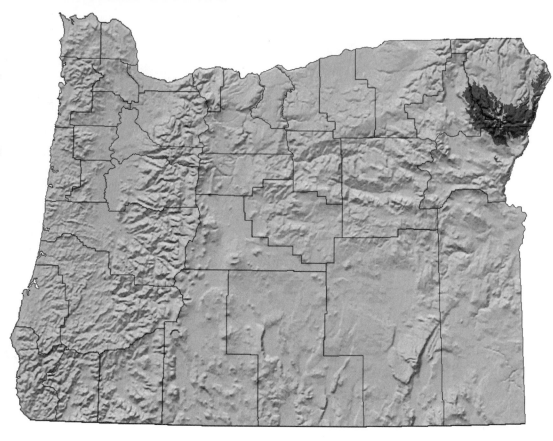

*Order: Galliformes*
*Family: Phasianidae*
*State Status: Sensitive*
*Federal Status: None*
*Global Rank: G5*
*State Rank: S3*
*Length: 16 in (41 cm)*

**Global Range:** The Spruce Grouse is a resident of boreal coniferous forests, from Alaska across Canada to the Atlantic Ocean, barely entering the northernmost tier of the conterminous United States.

**Habitat:** In the Wallowa Mountains of northeastern Oregon, the Spruce Grouse is found in mixed coniferous forests dominated by lodgepole pine, subalpine fir, and Engelmann spruce. It typically seeks out sites in the forest with a dense shrub understory, often of huckleberry. It prefers mature forests with large trees.

**Reproduction:** The breeding season begins around May. The nest is merely a hidden depression, scraped in the ground, and lined with fine vegetable material, leaves, and some feathers. The clutch of 6 or 7 (range 5-10) eggs is incubated by the female for 23-24 days.

**Food Habits:** Foods tend to change with seasonal availability. In winter, the Spruce Grouse survives on the needles of lodgepole pine, western larch, and Engelmann spruce. Insects, such as grasshoppers, become important in the spring, and fruits and berries are eaten later in the season. Some seeds are also eaten.

**Ecology:** Males defend a territory of 10-15 acres, and have a home range of 250-750 acres. Following breeding, Spruce Grouse migrate to higher elevations.

**Comments:** The Spruce Grouse has always been rare in Oregon. Wildfires and logging both destroy habitat and increase its exposure to hunting. Although listed as a game species, the hunting season is closed. Many are mistaken for Blue Grouse and shot anyway.

*References:* Gilligan 1994, Marshall 1992, Zwickel et al. 1974.

*Order: Galliformes*
*Family: Phasianidae*
*State Status: None*
*Federal Status: None*
*Global Rank: G5*
*State Rank: S4*
*Length: 17 in (43 cm)*

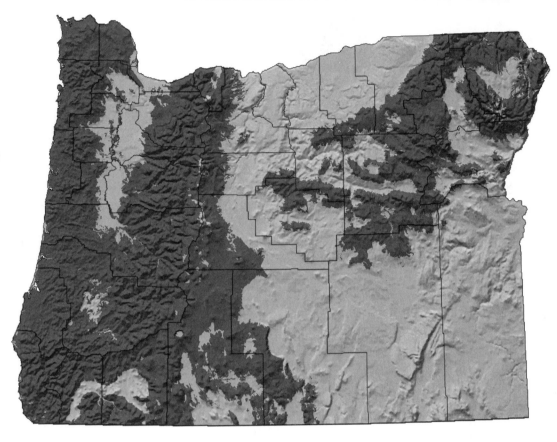

**Global Range:** Permanent residents from the Yukon Territory south through British Columbia, and into the mountainous western United States, occurring as far south as southern California and the mountains of eastern Arizona, and as far east as southeastern Colorado.

**Habitat:** Blue Grouse are birds of coniferous forests, usually dominated by Douglas-fir or true firs (genus *Abies*). Within those forests, they seek out areas with thickets of deciduous trees or shrubs. In winter, they move upslope to more open coniferous forests, and in spring, they move to the lower edge of the forest, where there is cover of deciduous trees and shrubs.

**Reproduction:** The nest is a scrape filled with grass and leaves, built in cover at the forest edge, and usually near water. A clutch of 6 to 8 (range 3-16) eggs is incubated by the female for about 26 days. Blue Grouse begin breeding in April, and young are fledged by September.

**Food Habits:** During winter, Blue Grouse depend on the needles and buds of conifers, including Douglas-fir, lodgepole pine, and hemlock. The young are fed insects and adults feed largely on berries

(bearberries, blueberries, serviceberries) when they are in season. Some insects are also eaten.

**Ecology:** The Blue Grouse is a solitary species that has a relatively small territory (about 2 to 10 acres) and home range (around 50-200 acres). It requires access to free water for drinking. Females move upslope with young shortly after hatching, and the flock remains together until late summer.

**Comments:** The Blue Grouse is a game species in Oregon. It occurs above timberline in late summer and fall.

*References:* Bendell and Elliott 1966, DeGraaf et al. 1991, Pekins et al. 1991.

# Ruffed Grouse
## (Bonasa umbellus)

*Order: Galliformes*
*Family: Phasianidae*
*State Status: None*
*Federal Status: None*
*Global Rank: G5*
*State Rank: S4?*
*Length: 17 in (43 cm)*

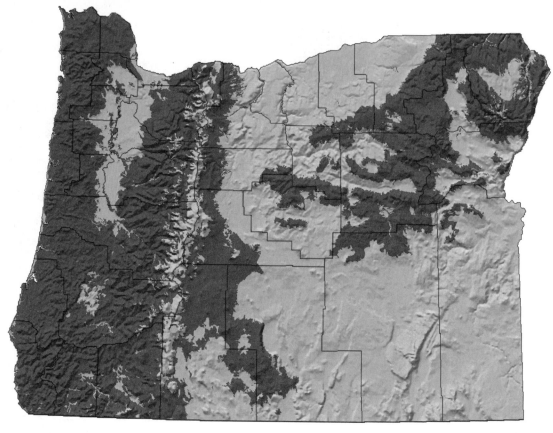

**Global Range:** Breeds from Alaska, across Canada, to the Atlantic Coast. In the West, it occurs south to northern California and Utah, and in the East, it follows the Appalachian Mountains south to northern Georgia.

**Habitat:** This is a species of forested areas. It seems to require the presence of some deciduous component to its habitat, such as willow and alder bottoms or young aspen stands; it is not found in pure stands of coniferous trees. It makes use of the brushy regrowth in cut-over areas.

**Reproduction:** In Oregon, breeding at lower elevations can begin in April, and young are fledged by late August. The female incubates a clutch of 9 to 12 (range 6-15) eggs for about 23-24 days. There is one clutch a year. The nest is a scrape in the ground, usually under cover and near a stream, lined with leaves or conifer needles.

**Food Habits:** The diet of the Ruffed Grouse varies from season to season. In spring, newly hatched young are fed insects and spiders, and adults eat about 30% insects along with various wild berries, seeds of trees, shrubs, and forbs, and blossoms, leaves and buds. In winter, however, buds and twigs of deciduous trees make up the main part of the diet, but tips of conifer twigs are also eaten.

**Ecology:** Male Ruffed Grouse defend a territory of 10-30 acres in breeding season. In habitats where food is abundant, this grouse can reach densities of about 30 birds per square mile. The subspecies in the Blue Mountains begins breeding later in the year (July) than coastal populations.

**Comments:** The Ruffed Grouse is a game species in Oregon.

*References:* Servello and Kirkpatrick 1987, Small and Rusch 1989, Thompson and Fritzell 1989.

124

# Sage Grouse
## (Centrocercus urophasianus)

*Order: Galliformes*
*Family: Phasianidae*
*State Status: Sensitive*
*Federal Status: SC*
*Global Rank: G5*
*State Rank: S3*
*Length: 28 in (71 cm)*

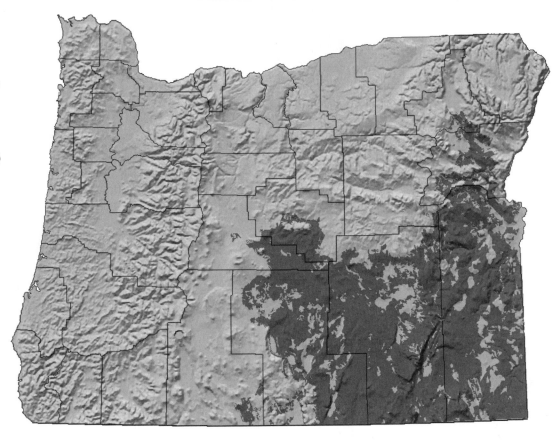

**Global Range:** Now has a local and reduced distribution in the central part of western North America. It is found sparingly from eastern Washington to North Dakota. It barely enters southern Canada above Montana, and occurs as far south as southern Utah.

**Habitat:** The Sage Grouse is well named, for it is found only in areas dominated by big sagebrush. It prefers areas where big sagebrush covers 15% to 50% of the ground. It also uses more open areas (known as leks), where many males congregate, for courtship display.

**Reproduction:** The breeding season begins in late March. The female incubates a clutch of 7 or 8 (sometimes up to 13) eggs for 25-27 days. The nest, usually found under the cover of sagebrush, is a depression lined with fine plant material.

**Food Habits:** The Sage Grouse feeds exclusively on the leaves of big sagebrush (preferring Wyoming big sagebrush to mountain big sagebrush) during winter. While continuing to eat sagebrush leaves in other seasons, it also eats leaves, blossoms, and buds of other plants, as well as a few insects, like ants and grasshoppers.

**Ecology:** Large numbers of males (up to 400) gather in open areas and display to females. This species always had fluctuating population densities, but it has been scarce since mid-century, likely because of loss or damage to its sagebrush habitat from grazing, wild fires, and conversion to grassland or agriculture.

**Comments:** Although the Sage Grouse is a game species in Oregon, the season is closed in much of the state.

*References:* Connelly et al. 1988, 1991; Marshall 1992.

# Wild Turkey
## (Meleagris gallopavo)

Order: Galliformes
Family: Phasianidae
State Status: None
Federal Status: None
Global Rank: G5
State Rank: SE
Length: 48 in (122 cm)

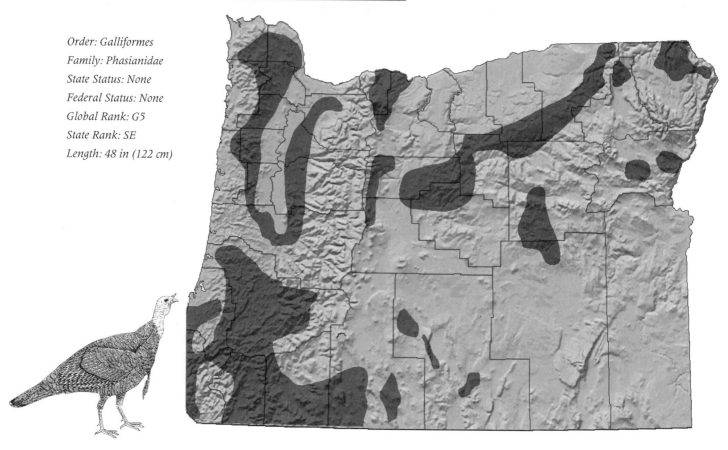

**Global Range:** Native from Arizona north and east across the Midwest to southeastern Canada and New England. Original range extended south through the southern United States to southern Mexico, at least as far south as northern Veracruz. Today, it has disappeared from much of its former range, although locally common. It has been widely introduced in the western United States, including local areas of California, Nevada, Utah, Washington, Wyoming, and Oregon, and into southern Canada, Hawaii, and New Zealand.

**Habitat:** Wild Turkeys inhabit open woodlands and riparian areas. They are most often found in oak or mixed oak-conifer woodlands. They prefer open forests with frequent grassy openings and hilly terrain and are usually found near water.

**Reproduction:** Wild Turkeys breed from March through August. The nest is a shallow depression lined with leaves, under cover of a bush or near a tree or fallen log. The clutch of 10 to 12 (range 5-20) eggs is incubated by the female. Males accumulate a harem of 5 or more females. The eggs hatch in about 4 weeks, and young remain with the female over the winter. Young can fly by 2 weeks of age.

**Food Habits:** Wild Turkeys eat a wide variety of seeds, fruit, grain, nuts, and arthropods. Where available, acorns are a favorite winter food. They consume buds, green leaves, and, occasionally, some small vertebrates (frogs, salamanders, lizards, snakes). Their diet varies with food availability, and may include juniper, madrone, manzanita, and blackberries.

**Ecology:** The Wild Turkey lives in flocks of the same sex except during the breeding season. Female flocks tend to be larger (over 20) than male flocks (usually under 10). During breeding season, males defend a harem, which may spread out over many acres. The home range is variously estimated to be from one to several square miles. Flocks may move over 20 miles to winter ranges. It roosts in trees at night.

**Comments:** Wild Turkeys were domesticated by the Aztecs before the arrival of Columbus. They were introduced to the West Coast as long ago as 1877, though not all introductions were successful. They are a game species in Oregon.

*References:* Dickson 1992, Gilligan et al. 1994, Leopold 1959, Rumble and Anderson 1992, Zeiner et al. 1990a.

# California Quail
## (Callipepla californica)

*Order: Galliformes*
*Family: Phasianidae*
*State Status: None*
*Federal Status: None*
*Global Rank: G5*
*State Rank: S4*
*Length: 10 in (25 cm)*

**Global Range:** Found along the west coast of North America, from southern British Columbia south to the tip of Baja California. Occurs from the Pacific Coast east to Utah, but it is believed that all populations north of southern Oregon and east of California are introduced.

**Habitat:** The original habitat of native California Quail in southern Oregon was probably lower valleys, oak woodland, chaparral, and grassland with scattered brushy areas. The species now occurs widely in the sagebrush country of eastern Oregon, and around fields and agriculture throughout the rest of the state. It is not found in dense coniferous forests or at high elevations.

**Reproduction:** Breeding season begins in April, and hatchlings stay in family groups until fall. The nest is a scrape in the ground concealed in cover, and lined with grass or leaves. The clutch of 12-17 (range 6-28) eggs is incubated by the female for 21-23 days. In more moderate climates, there may be a second brood.

**Food Habits:** The diet changes seasonally. During summer, the California Quail eats mostly seeds and a few insects. In winter, seeds of forbs and grasses become the primary food items. Green vegetation and a small amount of insects are eaten in the spring, and wild berries are added to the diet when available.

**Ecology:** Males defend a territory around the female during breeding season. Females with their broods may range over 10 to 20 acres. Larger coveys form in the winter, and birds may disperse several miles when the coveys break up in the spring. Coveys roost in dense thickets at night, 15-25 feet off the ground.

**Comments:** The California Quail is a game bird in Oregon. It is rarely farther than 1,200 feet from water.

*References:* Gabrielson and Jewett 1940, Gilligan et al. 1994, Leopold 1977.

## Mountain Quail
### (Oreortyx pictus)

*Order: Galliformes*
*Family: Phasianidae*
*State Status: None*
*Federal Status: 3C*
*Global Rank: G5*
*State Rank: S4?*
*Length: 11 in (28 cm)*

**Global Range:** Resident from the Puget Sound area of Washington south to northern Baja California. It occurs no farther east than central Idaho (where it is now rare) and the mountains of northern Nevada.

**Habitat:** While the Mountain Quail can be found high in the mountains, it is not a bird of dense coniferous forests, preferring open forests and woodlands with an ample undergrowth of brushy vegetation. It also inhabits thickets of chaparral and riparian woodland, meadow edges in forests, and brushy regrowth following timber harvest. It winters at the lower edges of forests, sometimes traveling on foot 20-40 miles from its breeding habitat.

**Reproduction:** The nest is a shallow depression on the ground, concealed in thick cover, and lined with grass and leaves. Nests are usually within a half mile of water. The female incubates the clutch of 7 to 10 (range 5-15) eggs for 24-25 days. The male is nearby and may share incubation of the eggs and brooding of the hatchlings.

**Food Habits:** During spring and summer, leaves, buds, flowers, and bulbs make up most of the diet. Some berries and insects such as grasshoppers, beetles, and ants are also eaten. In winter, seeds of a variety of plants, including acorns, make up most of the diet.

**Ecology:** Breeding territories range from 5 to 50 acres. The female and young do not travel far following hatching, often staying within 1 square mile. Coveys of 3 to 20 birds form in the fall and break up in late winter, prior to the breeding season.

**Comments:** The Mountain Quail is a game animal in Oregon. This species is reported to have declined recently in the mountains of eastern Oregon.

*References:* Brennan et al. 1987, DeGraaf et al. 1991, Gilligan et al. 1994.

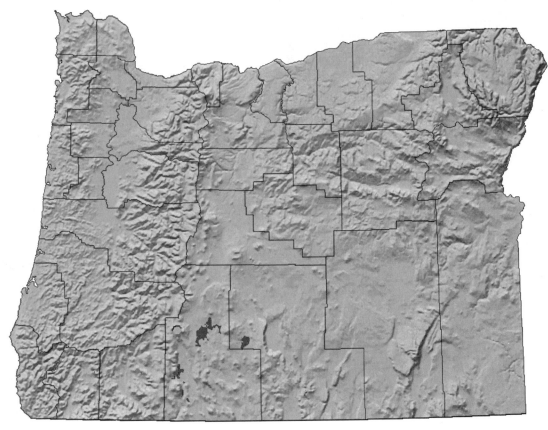

# *Yellow Rail*
## *(Coturnicops noveboracensis)*

*Order: Gruiformes*
*Family: Rallidae*
*State Status: Sensitive*
*Federal Status: None*
*Global Rank: G4*
*State Rank: S1*
*Length: 7 in (18 cm)*

**Global Range:** Breeds from central and eastern Canada south to New England and the Great Lakes region. An isolated subspecies is a local resident in the Toluca Valley, Mexico. The Oregon populations are extralimital and were thought to have disappeared early this century.

**Habitat:** The Yellow Rail inhabits freshwater marshes and wet meadows with a growth of sedges, usually surrounded by willows, and often with standing water up to a foot deep during the breeding season.

**Reproduction:** The Yellow Rail begins nesting in Oregon at least by May. The nest is a cup, built of marsh vegetation, and attached to emergent plants above water level. A clutch of 7-12 eggs is incubated by both sexes for 16-20 days.

**Food Habits:** This rail eats small invertebrates (insects and molluscs), seeds, and some leafy green vegetation. Little is known about the specifics of the diet.

**Ecology:** The Yellow Rail is very secretive, and little is known about its habits in Oregon. It is mainly detected through its vocalizations during breeding

season. Most Yellow Rails winter in the southeastern United States, but the winter residence of the Oregon populations is unknown. Wintering Yellow Rails have recently turned up at Humboldt Bay, California, and these may represent birds breeding in south central Oregon.

**Comments:** The Yellow Rail is listed as a game species in Oregon, although it is not present during the fall. The total state breeding population is probably less than 100 pairs. Its presence in Oregon was "rediscovered" in 1982. Some taxonomists consider the Yellow Rail to be conspecific with *C. exquisita* of Asia.

*References:* Gilligan et al. 1994, Marshall 1992, Stern et al. 1993.

# Virginia Rail
## *(Rallus limicola)*

*Order: Gruiformes*
*Family: Rallidae*
*State Status: None*
*Federal Status: None*
*Global Rank: G5*
*State Rank: S4*
*Length: 10 in (25 cm)*

*References:* DeGraaf et al. 1991, Gilligan et al. 1994, Ripley 1977.

**Global Range:** Discontinuous distribution throughout the Western Hemisphere. Breeds from central Canada south to California and Virginia, reappears as a breeder in central Mexico, also breeds in Colombia, Peru, and southern Chile and Argentina.

**Habitat:** Like most rails, this one is a bird of marshes. It is most common in freshwater marshes, but also inhabits marshes of coastal estuaries. It requires areas of shallow water with good cover of emergent vegetation, like cattails, bulrushes, and sedges.

**Reproduction:** Breeding begins in early April, and young are fledged by mid-August. The nest is a large cup woven of marsh plants, built above the water, and fastened to emergent vegetation. Both sexes incubate the single clutch of 7 to 12 (range 4-13) eggs for about 16-20 days. Young quickly wander from the nest, but return at night for brooding.

**Food Habits:** About 60% of the diet is larval and adult insects, but it also takes earthworms, small fish, frogs, crayfish, some seeds of grasses and aquatic plants, and occasional berries.

**Ecology:** The Virginia Rail defends an area around the nest ranging from a tenth to a half acre. It is a secretive bird, and many authors suggest it may be more common than usually thought. In winter, it can be found along the coast, in the marshes of western Oregon, and, locally, east of the Cascade Mountains.

**Comments:** The Virginia Rail is a game bird in Oregon.

## Sora
### *(Porzana carolina)*

*Order: Gruiformes*
*Family: Rallidae*
*State Status: None*
*Federal Status: None*
*Global Rank: G5*
*State Rank: S4*
*Length: 9 in (21 cm)*

**Global Range:** The Sora's breeding range in North America is much like that of the Virginia Rail, but the Sora does not occur south of the Mexican border. It breeds from the Yukon Territory through central Canada to the Atlantic Ocean, and south to California in the West and Maryland in the East.

**Habitat:** The Sora uses marshes, emergent vegetation along the edges of lakes and streams, and wet meadows with clumps of sedges, cattails, or other aquatic vegetation.

**Reproduction:** This rail begins breeding in late April or early May in Oregon. The nest is a small basket constructed from marsh vegetation, concealed in emergent vegetation, and usually built over water. The clutch of 8 to 12 (range 5-18) eggs is incubated by both sexes for about 16-20 days.

**Food Habits:** The Sora's diet varies by season. In spring and summer, its main food items are molluscs and other aquatic invertebrates, but it eats many seeds and leaves of aquatic plants in fall and winter.

**Ecology:** Soras defend a territory of one-half to 1 acre. When not breeding, they will sometimes forage in the open on beaches or in flooded grain fields. Most migrate south for the winter, but some Soras winter in Oregon, mainly on the coast and in southwestern valleys. The species is regularly seen wintering in the Rogue River Valley.

**Comments:** The Sora is the state's most common rail. It is a game species in Oregon.

*References:* Contreras 1992, DeGraaf et al. 1991, Gilligan et al. 1994, Harrison 1978.

# American Coot
## *(Fulica americana)*

*Order: Gruiformes*
*Family: Rallidae*
*State Status: None*
*Federal Status: None*
*Global Rank: G5*
*State Rank: S5*
*Length: 14 in (36 cm)*

*References:* DeGraaf et al. 1991, Gilligan et al. 1994, Ripley 1977.

**Global Range:** Breeds from central Canada south to Costa Rica from coast to coast. A few breed farther north in Alaska.

**Habitat:** The coot breeds in freshwater marshes and on the marshy edges of lakes or slow moving streams.

**Reproduction:** American Coots lay their eggs in May and early June. The coot builds several nests hidden in emergent vegetation (cattails, bulrushes) within a few feet of water or over water. They are constructed from aquatic vegetation, and serve as sites for courtship, egg laying, and brooding young. Both sexes incubate the clutch of 6 to 10 (range 4-17) eggs for 21-24 days.

**Food Habits:** About nine-tenths of the coot's diet consists of green parts of aquatic vegetation. The young are fed aquatic insects, insect larvae, and small crustaceans. Other animal food includes snails, worms, tadpoles, small fish, and bird eggs.

**Ecology:** Coots defend territories of about an acre during breeding season, but at other times are usually found in flocks. They are permanent residents and, during winter, concentrations in the hundreds can be found along the Columbia River and on larger lakes, marine bays, and estuaries.

**Comments:** American Coots are the most common nesting water bird in Oregon. They are a game species in Oregon, but few hunters deliberately take them because their meat is strongly flavored.

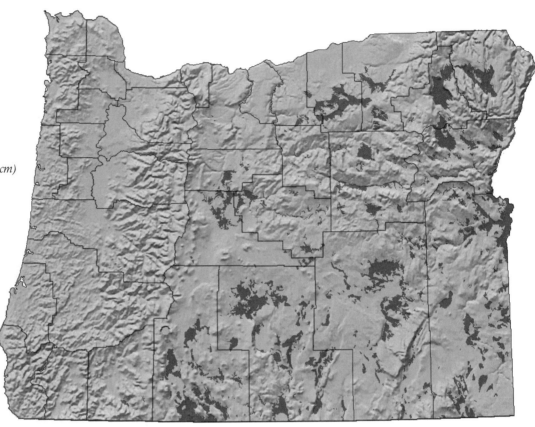

# Sandhill Crane
## (Grus canadensis)

*Order: Gruiformes*
*Family: Gruidae*
*State Status: Sensitive*
*Federal Status: None*
*Global Rank: G5*
*State Rank: S3*
*Length: 44 in (112 cm)*
*Wingspread: 90 in (229 cm)*

*References: Gilligan et al. 1994, Littlefield and Paullin 1990, Marshall 1992, Mullins and Bizeau 1978.*

**Global Range:** Some Sandhill Cranes breed in eastern Siberia, but most are found in North America, from Alaska and northern Canada south to Oregon and the Great Lakes region.

**Habitat:** Sandhill Cranes will nest in marshes and wet meadows or in drier grasslands and pastures, including irrigated hay meadows.

**Reproduction:** This crane arrives on its breeding grounds about March, and young fledge by August. The nest is a depression in the soil, lined with grass and feathers, or merely a mass of vegetation in shallow water. Both sexes incubate the clutch of 2 (range 1-3) eggs for 28-30 days. The young remain with their parents over the winter.

**Food Habits:** Sandhill Cranes eat a fairly broad range of food items, but more of the diet consists of plants than animals. They eat waste grain in agricultural areas in the fall, as well as seeds, berries, tubers, roots, green leaves, and shoots. They eat animals caught in marshes or grasslands, including large insects, reptiles, frogs, small mammals, birds, and bird eggs.

**Ecology:** During breeding season, Sandhill Cranes are territorial and defend an area around the nest that varied from 3 to 168 acres in an Oregon study. In 1984, the breeding density at Sycan Marsh, Lake County, was one pair per 180 acres. Almost all of Oregon's Sandhill Cranes migrate to the Central Valley of California during winter, but a small number may overwinter on Sauvie Island in some years.

**Comments:** About 1,000 pairs of Sandhill Cranes breed in Oregon, accompanied by another 500 nonbreeding individuals. In recent years the population has been stable in most areas, but eggs and young are often killed when hay fields are mowed early in the season. The Greater Sandhill Crane (*G. c. tabida*) is the subspecies nesting in Oregon.

**133**

# Snowy Plover
*(Charadrius alexandrinus)*

*Order: Charadriiformes*
*Family: Charadriidae*
*State Status: Threatened*
*Federal Status: Threatened (coastal populations)*
*   3C (interior populations)*
*Global Rank: G4*
*State Rank: S2*
*Length: 7 in (18 cm)*

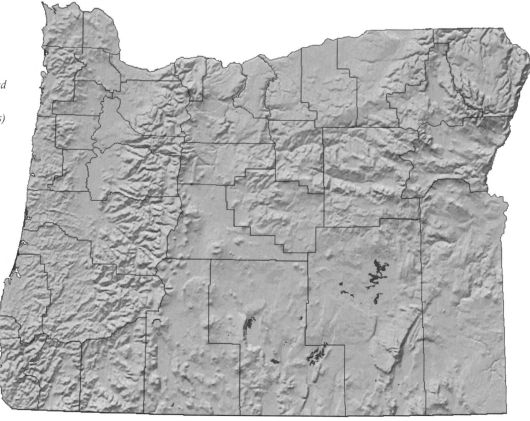

**Global Range:** Current taxonomy (American Ornithologists' Union 1983) considers Snowy Plovers from the coasts of Eurasia, Africa, South America, and Australia as one species. In North America, the Snowy Plover breeds along the Pacific Coast from Washington to Baja California, along the Gulf Coast from Florida to Mexico, and from interior Oregon and California east, locally, to Texas.

**Habitat:** Along the coast, the Snowy Plover nests on sand spits near river outlets and on level sandy beaches. In the Great Basin, this plover nests around the bare margins of alkaline lakes.

**Reproduction:** The breeding season is generally from mid-May to July, with inland populations nesting later. The nest is a shallow depression that may be unlined or lined with some plant material. Both sexes incubate the clutch of 3 (range 2-4) eggs about 24-25 days.

**Food Habits:** Like many shorebirds, the Snowy Plover's diet consists of small animals it finds along ocean beaches or on the margins of alkaline lakes. These include small crustaceans, beetles, shoreflies, flies, marine worms, and, rarely, small fish. Inland, brine flies are an important food.

**Ecology:** Snowy Plovers that breed in eastern Oregon winter along the California and Baja California coast, while some coastal birds remain on their breeding grounds throughout the year. Along the Oregon coast, less than one bird nests for every hectare of suitable habitat.

**Comments:** Less than 200 Snowy Plovers still nest along the southern Oregon coast, and the inland population is thought to number less than 1,000.

*References:* Gilligan et al. 1994, Herman et al. 1988, Marshall et al. 1996, Page et al. 1985.

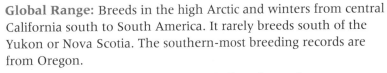

# Semipalmated Plover
## *(Charadrius semipalmatus)*

*Order: Charadriiformes*
*Family: Charadriidae*
*State Status: None*
*Federal Status: None*
*Global Rank: G5*
*State Rank: S1B?*
*Length: 7.5 in (19 cm)*

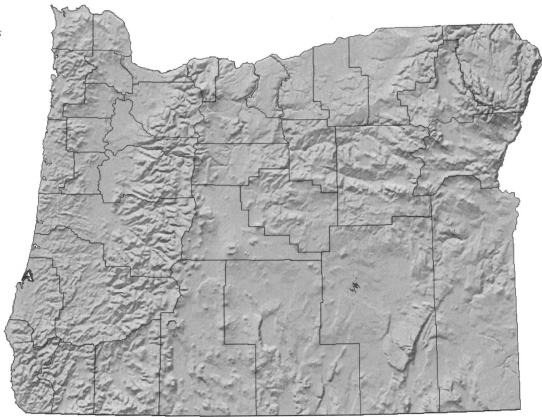

**Global Range:** Breeds in the high Arctic and winters from central California south to South America. It rarely breeds south of the Yukon or Nova Scotia. The southern-most breeding records are from Oregon.

**Habitat:** In the Arctic, it nests in a shallow depression on open, lichen-covered tundra. Specifics on the Oregon nests are not available, but they may be similar to those for the Snowy Plover.

**Reproduction:** The usual clutch of 4 (range 2-4) eggs is incubated by both sexes for 23-25 days. There is one clutch a year.

**Food Habits:** The Semipalmated Plover has an animal diet of insect larvae, beetles, grasshoppers, ants, spiders, marine worms, molluscs, and small crustaceans.

**Ecology:** Little is known about this very rare breeder in its Oregon range. It is a common bird along the coast during spring and fall migrations.

**Comments:** A single pair produced young at Malheur National Wildlife Refuge over 3 years in the late 1980s and, more recently, one or two pairs bred near Coos Bay, along the Oregon Coast.

*References:* DeGraaf et al. 1991, Gilligan et al. 1994.

# *Killdeer*
## (*Charadrius vociferus*)

*Order: Charadriiformes*
*Family: Charadriidae*
*State Status: None*
*Federal Status: None*
*Global Rank: G5*
*State Rank: S5*
*Length: 10 in (25 cm)*

**Global Range:** Breeds from Alaska south, coast to coast, through Canada and the United States into central Mexico. It also breeds on several Caribbean Sea islands and on the west coast of South America (Peru and northwestern Chile).

**Habitat:** The Killdeer is fairly unselective about its breeding habitat: almost any open area near water will do. This includes meadows, pastures, rangeland where grass is short, the shores of lakes or rivers, and agricultural areas.

**Reproduction:** Killdeer begin breeding as early as March, and young are fledged by mid-August. The nest, such as it is, is a depression in the ground that is either unlined or has some nearby plant material placed in it. The clutch of 4 (range 3-5) eggs is incubated about 24-30 days by both parents. Young can fly at 25 days of age. There may be two broods per year.

**Food Habits:** Most of the diet consists of insects, such as beetles, grasshoppers, ants, caterpillars, and dragonflies. Some spiders, worms, crayfish, snails and an occasional seed are also eaten.

**Ecology:** The male of the pair carries out territorial displays around the nest. Nesting pairs are generally several hundred feet apart. Killdeer are sometimes active on nights when the moon is bright.

**Comments:** The development of agricultural lands in Oregon has probably created more habitat for Killdeer. The southern Willamette Valley is one of the major wintering areas for the species in North America.

*References:* Gabrielson and Jewett 1940, Gilligan et al. 1994, Lenington and Mace 1975.

## Black Oystercatcher
### (Haematopus bachmani)

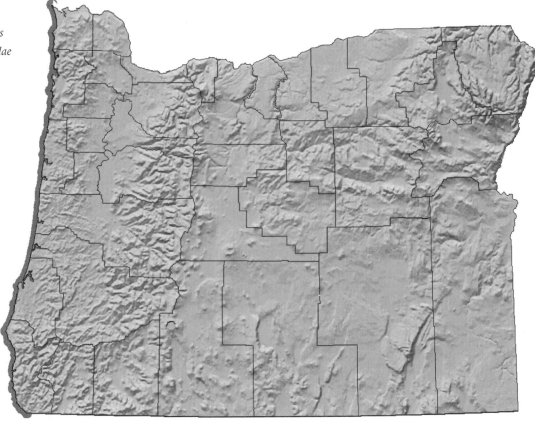

*Order: Charadriiformes*
*Family: Haematopodidae*
*State Status: None*
*Federal Status: None*
*Global Rank: G5*
*State Rank: S4*
*Length: 17 in (43 cm)*

*References:* Gilligan
et al. 1994,
Harrison 1978,
Jewett et al. 1953.

**Global Range:** This is a coastal bird found along the rim of the eastern Pacific Ocean, from the Aleutian Islands south to central Baja California. Another common name is the American Black Oystercatcher. Some taxonomists consider the oystercatchers of eastern North America and Eurasia to be one species, the Pied Oystercatcher.

**Habitat:** The Black Oystercatcher is a rather unusual bird, in that it is associated only with the intertidal environment. It nests either on offshore islands or rocky shorelines and cliffs, and finds its food in the rocky intertidal area or, rarely, on sandy beaches alongside streams entering the ocean.

**Reproduction:** The nest is a shallow depression that, usually, is lined with little more than a few pebbles or other debris. It may be near the high-water level or high on a rock. The breeding season begins in late May or June. Both sexes incubate the clutch of 2 or 3 (range 1-4) eggs for 26-27 days. There is one brood per year, but it will replace lost clutches.

**Food Habits:** Despite its name, oysters are not its main food item. It does eat oysters, but more often feeds on mussels, limpets, chitons, abalone, and other molluscs. In sandy areas, it will probe for marine worms. Young are fed insects before turning to shellfish for food.

**Ecology:** Black Oystercatchers form small flocks in winter, but rarely travel more than 30 miles from their breeding areas. They reuse the same nests in successive years.

**Comments:** A 1988 study concluded that only about 350 Black Oystercatchers live along the Oregon Coast.

**137**

# Black-necked Stilt
## *(Himantopus mexicanus)*

*Order: Charadriiformes*
*Family: Recurvirostridae*
*State Status: None*
*Federal Status: None*
*Global Rank: G5*
*State Rank: S4*
*Length: 15 in (38 cm)*

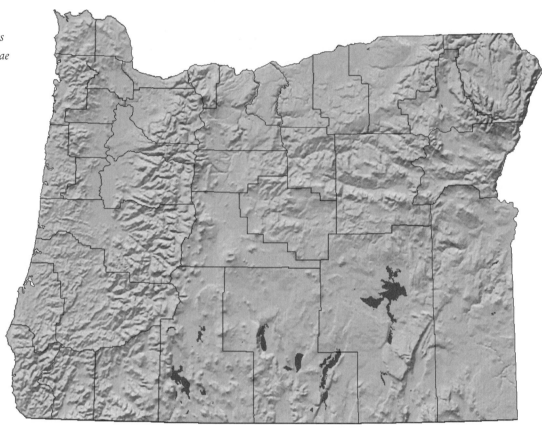

**Global Range:** Breeds in parts of North America and throughout most of South America. It is found on the Atlantic Coast from Delaware to Florida, and on islands of the Caribbean. In western and central North America, it breeds from eastern Washington to Baja California, and east to Kansas and Louisiana on the Gulf Coast.

**Habitat:** This shorebird nests along the edges of shallow, freshwater, saltwater, or alkaline marshes, lakes, and sloughs. It will use wet meadows and irrigated or flooded fields for feeding.

**Reproduction:** In Oregon, the Black-necked Stilt usually lays its clutch of 2 or 3 eggs in the last part of May. Both sexes incubate the clutch for 25-26 days. The hatchlings are independent by the time they are a month old. This species usually nests in small colonies. The nest is a small scrape, which may be built up with plant material.

**Food Habits:** Black-necked Stilts forage in shallow, often muddy, water for a variety of animal prey, including insect larvae, caddisflies, snails, crayfish, small fish, and seeds of aquatic plants. In wet meadows, they take some terrestrial insects, such as grasshoppers and beetles.

**Ecology:** This is a social species that nests in groups of up to 50 individuals, often seen foraging in close proximity to one another and with other shorebirds.

**Comments:** The Black-necked Stilt is very abundant in the large marshes of southeastern Oregon during years with adequate rainfall.

*References:* DeGraaf et al. 1991, Gilligan et al. 1994, Hamilton 1975.

# American Avocet
## (Recurvirostra americana)

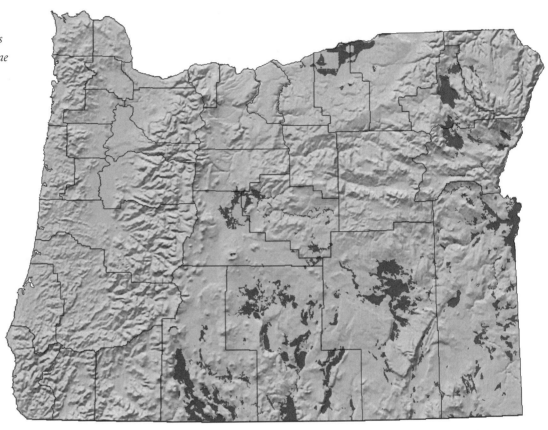

*Order: Charadriiformes*
*Family: Recurvirostridae*
*State Status: None*
*Federal Status: None*
*Global Rank: G5*
*State Rank: S4*
*Length: 18 in (46 cm)*

**Global Range:** Breeds widely in western North America, from south-central Canada south to Mexico. It is found from the San Francisco Bay area east to coastal Texas and Minnesota.

**Habitat:** American Avocets nest in open areas alongside freshwater or alkaline lakes and marshes. Their breeding habitat usually has scattered clumps of grass or other low vegetation. They require some shallow water for feeding, which they find in marshes, mudflats, ponds, and coastal estuaries.

**Reproduction:** Avocets arrive in Oregon in April, and lay eggs in late May or June. They usually nest in small colonies. Their nest is a simple scrape lined with some vegetation or pebbles. The clutch of 4 (range 3-5) eggs is incubated 22-24 days by both parents. Young become independent about 6 weeks after hatching.

**Food Habits:** The American Avocet has a thin, upturned bill that is swept from side to side over the bottom of shallow water. With it, the avocet catches insects (including dragonfly nymphs, beetles, flies, and water boatmen), shrimp and other small crustaceans, seeds of aquatic plants, snails, and small fish.

**Ecology:** Avocets nest and feed in small groups, but defend a territory around both the nest and a feeding area. In one instance, nests were anywhere from 7 to 138 feet apart along a dike. They commonly feed while walking alongside other avocets and in the company of other shorebirds, like Black-necked Stilts or Willets.

**Comments:** American Avocets do not winter in Oregon, but may occur in large numbers on summer breeding grounds. Over 50,000 were estimated during an August census at Summer Lake, Lake County.

*References:* Gibson 1971, Gilligan et al. 1994, Hamilton 1975.

**139**

# Greater Yellowlegs
### (Tringa melanoleuca)

*Order: Charadriiformes*
*Family: Scolopacidae*
*State Status: None*
*Federal Status: None*
*Global Rank: G5*
*State Rank: S1*
*Length: 14 in (36 cm)*

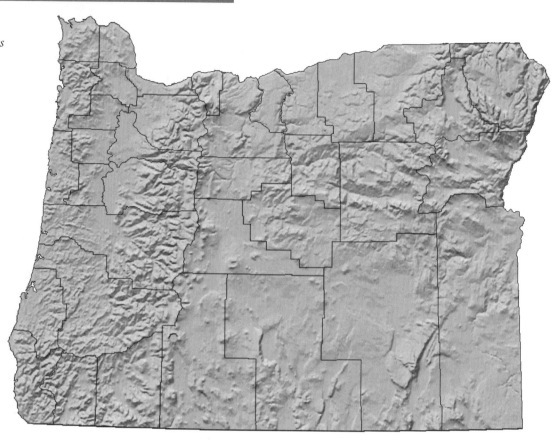

**Global Range:** Breeds in Alaska and most of Canada, except along its southern border. A few wayward pairs have nested in Oregon. It winters along the U.S. coasts and south to the southern tip of South America.

**Habitat:** Like the Solitary Sandpiper, the Greater Yellowlegs prefers boggy areas where there are some coniferous trees and open water in the neighborhood.

**Reproduction:** The Greater Yellowlegs nests in a slight depression on the ground that may or may not have a meager lining of vegetation. The usual clutch of 4 eggs is incubated for 23 days. Young fledge in about 18-20 days. In the short summer of the north country, it begins breeding in late May and young are fledged by mid-July.

**Food Habits:** This shorebird actively hunts animal food in the water, even chasing small fish. Its usual diet is aquatic invertebrates, such as insects and their larvae, molluscs, crustaceans, worms, tadpoles, and an occasional berry.

**Ecology:** During migration, this species can be seen nearly anywhere in Oregon where there is shallow water for feeding.

**Comments:** The only Oregon breeding locality is Downy Lake, in the Wallowa Mountains. Breeding occurred at least 4 times in recent years. In winter, it is common along the coast and, locally, inland.

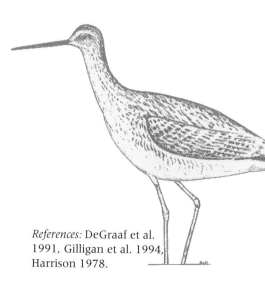

*References:* DeGraaf et al. 1991, Gilligan et al. 1994, Harrison 1978.

# Solitary Sandpiper
*Tringa solitaria*

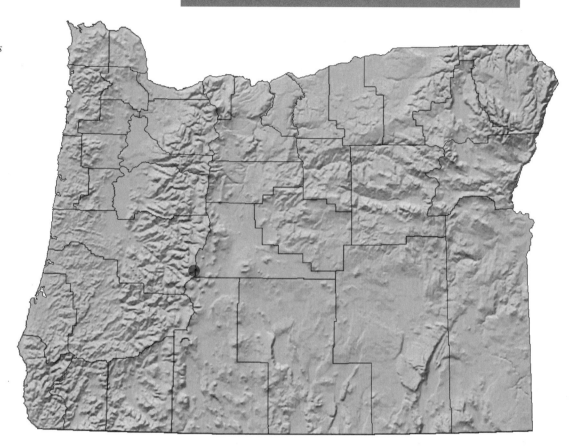

*Order: Charadriiformes*
*Family: Scolopacidae*
*State Status: None*
*Federal Status: None*
*Global Rank: G5*
*State Rank: S1*
*Length: 9 in (23 cm)*

**Global Range:** Breeds mainly in Alaska and Canada, just entering the United States to breed in northern Minnesota. Its breeding in Oregon represents an extralimital occurrence.

**Habitat:** In the Arctic, this sandpiper inhabits boggy areas, selecting clumps of coniferous trees near lakes or ponds for nesting. Appropriately, the records of Oregon breeding have been at Gold Bog Lake, in the high Cascades of Lane County.

**Reproduction:** Details of Oregon breeding are lacking, but seasonal phenology at higher elevations may be similar to that in its normal Arctic range. Breeding begins in late May or June, and it lays eggs in an abandoned passerine nest. A normal clutch has 4 (range 3-5) eggs. Little else is known about its reproductive biology.

**Food Habits:** The Solitary Sandpiper eats just about any small animal it encounters in shallow water or wet meadows. Food items include insects, small molluscs, worms, crustaceans, spiders, and small frogs.

**Ecology:** Not much information is available on this sandpiper. It winters south of the United States in Central and South America, as far south as Peru, Uruguay, and Argentina.

**Comments:** Gilligan et al. (1994) report that pairs with young juveniles were seen in the 1980s.

*References:* DeGraaf et al. 1991, Gilligan et al. 1994, Godfrey 1966.

141

# Willet
## *(Catoptrophorus semipalmatus)*

*Order: Charadriiformes*
*Family: Scolopacidae*
*State Status: None*
*Federal Status: None*
*Global Rank: G5*
*State Rank: S4*
*Length: 16 in (41 cm)*

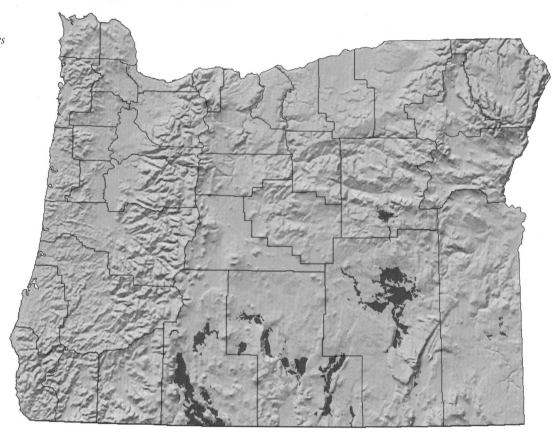

**Global Range:** Breeds in western North America from southern Canada to California and east to Iowa. Also breeds along the Atlantic and Gulf coasts from New Brunswick to Texas, as well as on many Caribbean Sea islands.

**Habitat:** In Oregon, the Willet breeds along the margins of inland alkaline marshes, lakes, and wet meadows. During the breeding season, it is usually found within several hundred yards of water.

**Reproduction:** Willet nests consist of a slight hollow in the ground, sometimes lined with grass or other vegetation. It begins breeding in late April or May. A clutch of 4 (range 3-5) eggs is incubated by the female for about 22 days. Some birds leave the breeding grounds as early as late June.

**Food Habits:** The Willet has a fairly mixed diet, which it obtains by gleaning the surface of the ground or probing its bill into mud. It eats insects (both terrestrial and aquatic), crustaceans, molluscs, and worms. Vegetable matter eaten includes seeds, shoots, leaves of grasses, and waste grain from agricultural fields.

**Ecology:** Willets are semi-colonial nesters, with pairs nesting within a few hundred feet of one another.

Their foraging areas are distinct from their nest sites. Several thousand breed at Malheur National Wildlife Refuge.

**Comments:** Willets are commonly seen in association with large plovers or other shorebirds. They tend to perch in trees or other elevated sites more than their relatives do.

*References:* Gilligan et al. 1994, Howe 1974, Stenzel et al. 1976.

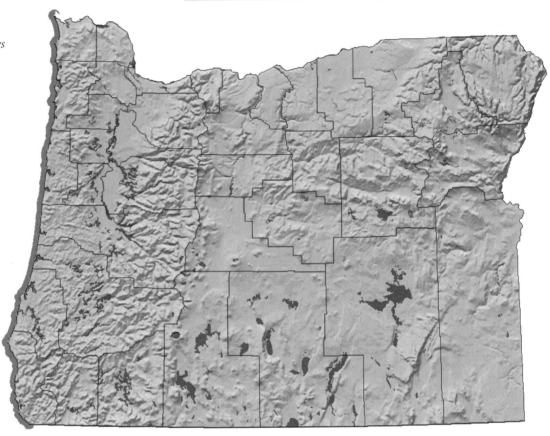

# Spotted Sandpiper
### (Actitis macularia)

*Order: Charadriiformes*
*Family: Scolopacidae*
*State Status: None*
*Federal Status: None*
*Global Rank: G5*
*State Rank: S4*
*Length: 7 in (18 cm)*

**Global Range:** Breeds throughout most of North America north of Mexico, except for the southeastern United States.

**Habitat:** This sandpiper breeds in a variety of habitats, but it must be near freshwater. It can be found from sea level to above timberline near lakes, streams, ponds, or even temporary pools. It uses both open and wooded habitats.

**Reproduction:** Eggs are laid from May to July in slight depressions lined with dead leaves, grass, moss, or other vegetation. The normal clutch is usually 4 (range 3-5) eggs and is incubated by the male (later clutches assisted by the female) for 20-24 days. Young can fly when they are 16-18 days of age.

**Food Habits:** The Spotted Sandpiper hunts for its food at water's edge or in shallow water. Its diet consists of insects (flies, grasshoppers, crickets, insect larvae, beetles), small crustaceans, molluscs, spiders, and even small fish.

**Ecology:** This species is polyandrous, and females may lay multiple clutches, which are tended by different males. Spotted Sandpipers can be quite common in good habitat. One report found 43 pairs in an 18-acre study site.

**Comments:** Spotted Sandpipers display a characteristic bobbing action as they walk.

*References:* DeGraaf et al. 1991, Gilligan et al. 1994, Oring et al. 1983.

# Upland Sandpiper
## *(Bartramia longicauda)*

*Order: Charadriiformes*
*Family: Scolopacidae*
*State Status: Sensitive*
*Federal Status: None*
*Global Rank: G5*
*State Rank: S1*
*Length: 11 in (28 cm)*

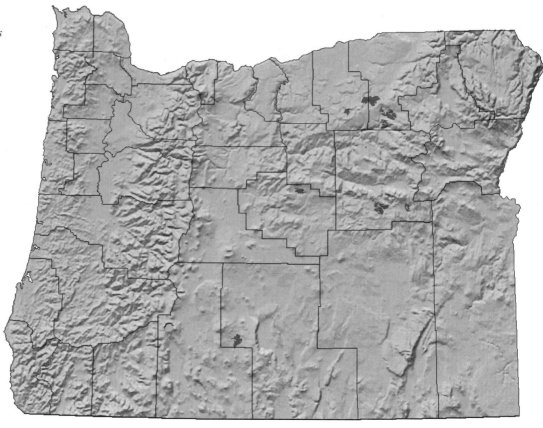

**Global Range:** This relatively large sandpiper breeds from Alaska south and east through Canada to the eastern United States, as far south as Texas and Maryland. West of the Rockies, very small numbers breed at scattered locations in the Northwest. It winters in South America.

**Habitat:** In Oregon, the Spotted Sandpiper nests in partly flooded meadows and grasslands, usually with a fringe of trees, and often in the middle of higher-elevation sagebrush communities. Meadows favored by this sandpiper are little grazed and have some growth of forbs.

**Reproduction:** Spotted Sandpipers have a short breeding season in Oregon, arriving in early May and leaving in late July. The usual clutch is 4 eggs, laid in a slight depression in the ground and concealed by a cover of grass. Both sexes incubate the eggs about 21 days. Young reach adult size in about a month.

**Food Habits:** This sandpiper forages in open meadows for its favorite foods, grasshoppers and crickets. It also eats ants, berries, and seeds of grasses and forbs.

**Ecology:** This shorebird may perch in coniferous trees or snags surrounding the nesting site. Little is known of its social interactions in Oregon, but several dozen pairs are attracted to the same local patches of habitat.

**Comments:** Upland Sandpipers (once called Upland Plovers) breed at about 8 locations in Oregon. The estimated state breeding population is less than 100 birds, most of which breed in the Bear and Logan valleys of Grant County.

*References:* Gilligan et al. 1994, Marshall 1992, Stern and Rosenberg 1985.

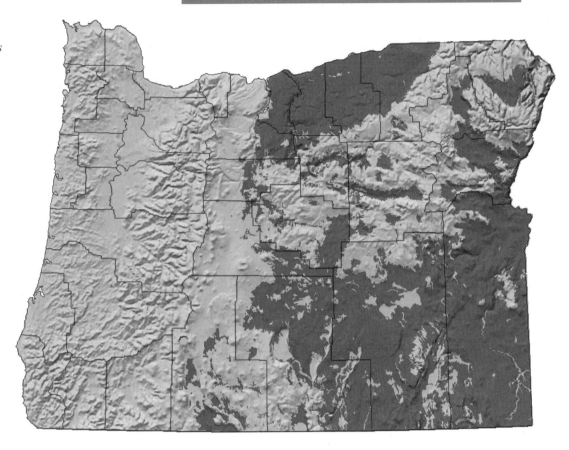

## Long-billed Curlew
*(Numenius americanus)*

*Order: Charadriiformes*
*Family: Scolopacidae*
*State Status: None*
*Federal Status: None*
*Global Rank: G5*
*State Rank: S3S4*
*Length: 23 in (58 cm)*

**Global Range:** Breeds in open valleys and flatlands of western North America, from southern Canada south to Nevada and Texas, and east to about the North Dakota-Minnesota border.

**Habitat:** Long-billed Curlews nest in open grasslands, prairies, and meadows, often near scattered shrubs and usually near water or wet meadows, but also in dry situations.

**Reproduction:** This curlew arrives in Oregon near the end of March. The nest is a depression in open grassland and is lined with grass. A clutch of 4 eggs (range 3-5) is incubated by both parents for 28-30 days.

**Food Habits:** Long-billed Curlews find a variety of food items in grasslands and nearby agricultural areas. They eat insects (beetles, insect larvae, grasshoppers), worms, crustaceans, nestling birds, bird eggs, small fishes, seeds, and berries.

**Ecology:** The home range can be up to 1,000 hectares. They will fly as far as 6 miles from their nest site to forage. Their nests can be damaged by grazing livestock.

**Comments:** This species was formerly more abundant in Oregon, but much habitat has been converted to agriculture.

*References:* Cochran and Anderson 1987, Pampush and Anthony 1993, Stenzel et al. 1976.

**145**

# Common Snipe
## (Gallinago gallinago)

*Order: Charadriiformes*
*Family: Scolopacidae*
*State Status: None*
*Federal Status: None*
*Global Rank: G5*
*State Rank: S4*
*Length: 11 in (28 cm)*

**Global Range:** Breeds throughout Europe, northern Asia, and South America. In North America, it breeds from Alaska and Canada south to northern California, Colorado, and Virginia on the East Coast.

**Habitat:** This shorebird breeds in wet meadows or shallow marshes. Because it feeds by probing saturated soil with its bill, it is normally restricted to areas with very damp or slightly submerged soils. Breeding sites have a cover of low grasses or sedges. Occasionally it will use open ground in cattail marshes.

**Reproduction:** Common Snipe arrive on their Oregon breeding grounds in March. The nest is a shallow cup of plant material, concealed in vegetation. The clutch of 4 (range 3-5) eggs is incubated by the female for 18-20 days. The young can fly 19-20 days after hatching.

**Food Habits:** Burrowing insect larvae and earthworms form the major part of the Common Snipe's diet. It also takes some food from the surface of the mud or marsh bottom. Other food items include small crustaceans, spiders, molluscs, grass seeds, and some small vertebrates.

**Ecology:** The Common Snipe defends a territory that shrinks from about 50 acres to just a few acres as the breeding season progresses. This species will make use of freshly plowed agricultural fields and pastures when feeding.

**Comments:** This species is the only shorebird that is a game species in Oregon. It was formerly called Wilson's Snipe.

*References:* Gilligan et al. 1994, Mason and MacDonald 1976, Tuck 1972.

# Wilson's Phalarope
*(Phalaropus tricolor)*

*Order: Charadriiformes*
*Family: Scolopacidae*
*State Status: None*
*Federal Status: None*
*Global Rank: G5*
*State Rank: S4*
*Length: 9 in (23 cm)*

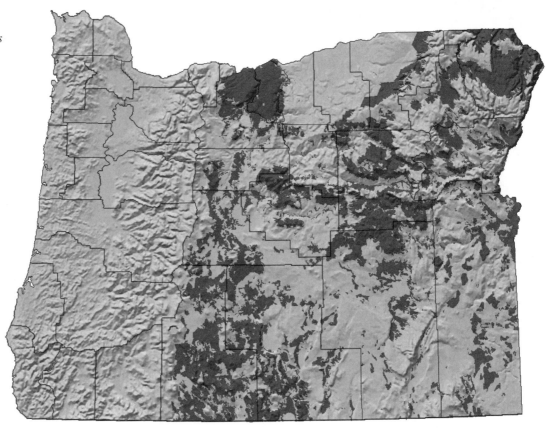

**Global Range:** Breeds in the western half of North America, from the Yukon south to California and Colorado, and east to Wisconsin. It also breeds in southeastern Canada and, locally, in New England.

**Habitat:** In Oregon, this species is found east of the Cascade Mountains, where it nests near freshwater or alkaline marshes, small ponds, lakes, and wet meadows.

**Reproduction:** Breeding season for Wilson's Phalarope begins in late April or early May. The nest is a depression in the ground in an area offering good cover of grass or other vegetation. The normal clutch of 4 (range 3-4) eggs is incubated by the male for 19 days. This species displays sequential polyandry: females may lay a second clutch with a new mate.

**Food Habits:** Wilson's Phalarope hunts on muddy shores and shallow waters, where it finds insect larvae, brine flies, seeds of aquatic plants, brine shrimp, other small crustaceans, and diving beetles.

**Ecology:** This shorebird nests in loose colonies. The female will defend the male—rather than a specific area—from other females. This species nests in marshes along lake margins quite high in the Cascade Mountains (to 5,000 feet). It spends its winters in South America, often departing Oregon in mid-summer.

*References:* Colwell 1986, Gilligan et al. 1994, Howe 1975, Jehl 1988.

# Franklin's Gull
## (Larus pipixcan)

*Order: Charadriiformes*
*Family: Laridae*
*State Status: Sensitive*
*Federal Status: None*
*Global Rank: G5*
*State Rank: S1*
*Length: 15 in (38 cm)*

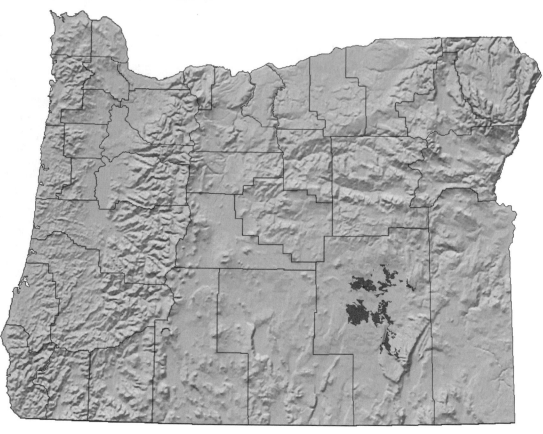

**Global Range:** Breeds in north-central North America, from the Canadian prairie provinces south into the Great Plains, Montana, Idaho, Utah, and northern California. It usually winters along the coasts of Central America and western South America.

**Habitat:** In Oregon, this species nests in or near alkaline hardstem bulrush marshes. It flies to meadows, irrigated pasture, and agricultural areas to feed.

**Reproduction:** This colonial nester arrives on its breeding ground in late April or May. The nest consists of plant material, either floating on the water or anchored to emergent vegetation just above the water. A clutch of 3 (range 2-4) eggs is incubated by both parents for 18-20 days. The young fly at about 1 month of age.

**Food Habits:** Franklin's Gull feeds mostly on insects, but also eats small fish, crustaceans, worms or grubs unearthed by plowing, and some grain.

**Ecology:** The size and location of Oregon's Franklin's Gull colony changes depending on water conditions. After the breeding season, these gulls spread out and can be seen over a wide area of eastern Oregon. They leave the state by the 1st of October.

**Comments:** This species expanded its range into Oregon in the 1940s. It only breeds in a single colony at Malheur National Wildlife Refuge. About 360 pairs nested in 1990, but the colony has had more than 1,300 pairs in the past.

*References:* DeGraaf et al. 1991, Gilligan et al. 1994, Littlefield 1990, Littlefield and Thompson 1981, Marshall 1992.

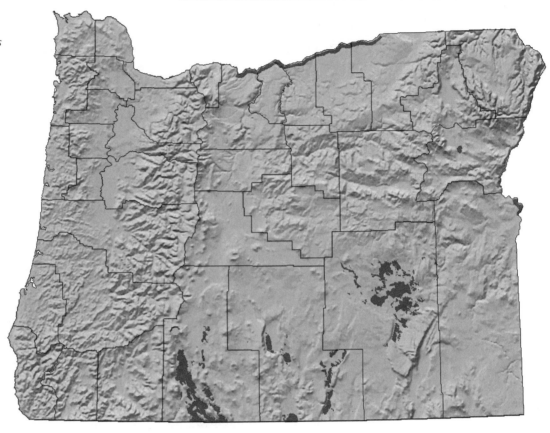

# Ring-billed Gull
## *(Larus delawarensis)*

*Order: Charadriiformes*
*Family: Laridae*
*State Status: None*
*Federal Status: None*
*Global Rank: G5*
*State Rank: S5*
*Length: 19 in (48 cm)*

*References:* Broadbrooks 1961, DeGraaf et al. 1991, Gilligan et al. 1994, Vermeer 1970.

**Global Range:** Breeds in the prairie provinces of Canada and in the northern United States.

**Habitat:** This is a colonial species that uses rocky islands or spits in large freshwater marshes, lakes, and rivers for nesting. It may fly at least 5 miles from the nest to forage in marshes, rivers, pastures, or other open habitats.

**Reproduction:** The breeding season lasts from March or April to mid-August, when young are fledged. The nest is a hollow on the ground, with a thin lining of whatever material is handy. Rarely, in other parts of the range, Ring-billed Gulls nest low in trees. A clutch of 3 (range 2-4) eggs is incubated 21-23 days by both sexes.

**Food Habits:** Like most gulls, the Ring-billed Gull has a varied diet. During the breeding season it eats insects, small fish, and grain. It will also eat small birds, bird eggs, carrion, and earthworms. It may also scavenge from garbage dumps.

**Ecology:** Ring-billed Gulls sometimes nest in mixed colonies with other birds, especially California Gulls. They defend a small area around the nest, usually only a few square meters in size.

**Comments:** There are colonies at marshes of southeastern Oregon and on islands in the Columbia and Snake rivers. Gilligan et al. (1994) report there has been a colony near Baker City for over 20 years.

**149**

# California Gull
## (*Larus californicus*)

*Order: Charadriiformes*
*Family: Laridae*
*State Status: None*
*Federal Status: None*
*Global Rank: G5*
*State Rank: S5*
*Length: 22 in (56 cm)*

*References:*
Gabrielson and
Jewett 1940,
Gilligan et al. 1994,
Vermeer 1970.

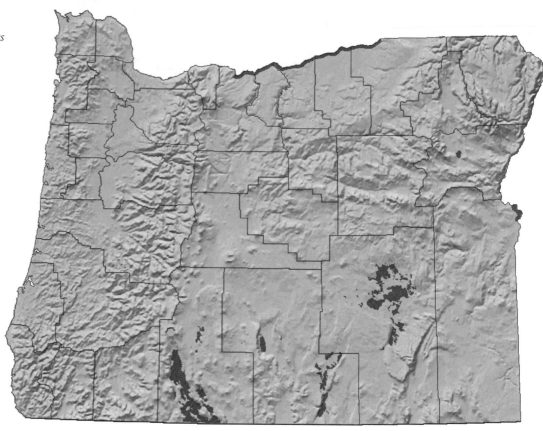

**Global Range:** Breeds from the prairie provinces of Canada south to northern California and Utah. It does not range as far east as the Ring-billed Gull, breeding only as far east as North Dakota.

**Habitat:** California Gull colonies are usually located on unvegetated islands in lakes or rivers. Sometimes colonies are on the shore, but these are subject to increased predation from mammalian carnivores.

**Reproduction:** California Gulls nest in colonies, often in the company of Ring-billed Gulls. They begin breeding from late April to early May, and leave colonies as soon as young are fledged, in late July or August. The nest is a scrape in the ground, lined with plant material and feathers. A clutch of 3 (range 2-5) eggs is incubated by both parents for 23-27 days.

**Food Habits:** Like most gulls, the California Gull includes many types of food in its diet. Insects and their larvae, young birds, and earthworms are important on the breeding grounds. It also eats small fish, bird eggs, small mammals, amphibians, berries, seeds, carrion, and scavenges for garbage.

**Ecology:** This gull nests in fairly dense colonies and only defends an area around the nest about as far as it can peck. Young wandering from the nest may be killed by other adults. It may forage for food a considerable distance (up to 20 miles) from the nest.

**Comments:** Although the exact number of California Gulls nesting in Oregon is not known, there are several large colonies in the southeastern marshes, and at least 5,000 gulls nest on islands in the Columbia River. Smaller numbers nest on islands in the Snake River near Ontario, Malheur County.

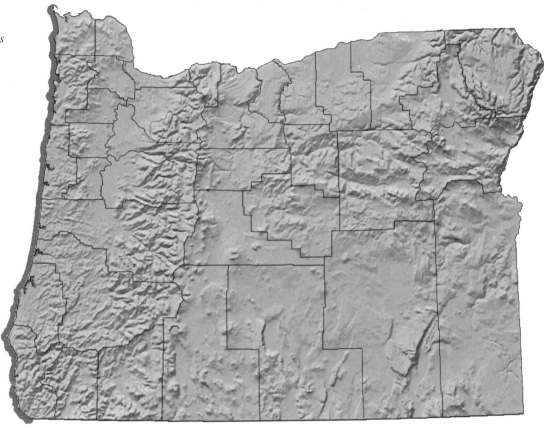

# Western Gull
## (Larus occidentalis)

*Order: Charadriiformes*
*Family: Laridae*
*State Status: None*
*Federal Status: None*
*Global Rank: G5*
*State Rank: S4*
*Length: 26 in (66 cm)*

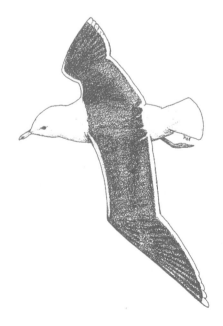

**Global Range:** Only breeds along the coast of the eastern Pacific Ocean, from southern British Columbia south to central Baja California.

**Habitat:** This is a marine gull that breeds on both offshore islands and rocky cliffs along the Oregon Coast. It also uses structures for nesting and, occasionally, will nest on grass-covered headlands.

**Reproduction:** The breeding season commences in late April, and young fly around mid-August. The nest is a cup of plant material, and may be used for several nesting seasons. The usual clutch is 3 eggs (range 1-4), which are incubated 24-29 days by both parents.

**Food Habits:** The Western Gull's food comes from the marine environment, estuaries, and the immediate shoreline. It eats small fish, clams, mussels, bird eggs, the young of other birds nesting nearby, sea urchins, starfish, squid, crustaceans, marine worms, and carrion. It will scavenge garbage or waste from fishing boats.

**Ecology:** While it lives in colonies, nests are usually at least 20 feet apart. It may fly at least 5 miles away from the nest while foraging.

**Comments:** In 1979, the state population was estimated to be about 5,000 pairs.

*References:* Coulter 1975, Gilligan et al. 1994, Scott 1971, Spear et al. 1986.

# Glaucous-winged Gull
### (Larus glaucescens)

*Order: Charadriiformes*
*Family: Laridae*
*State Status: None*
*Federal Status: None*
*Global Rank: G5*
*State Rank: S5*
*Length: 26 in (66 cm)*

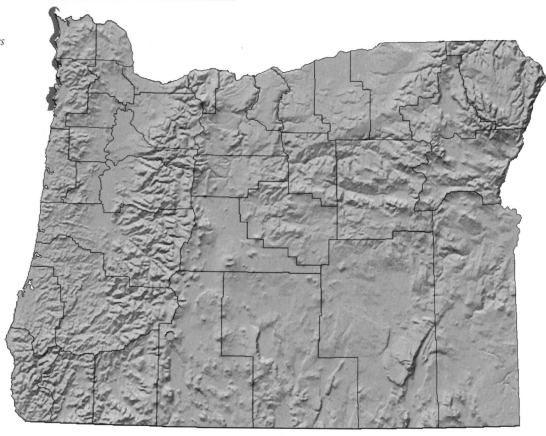

**Global Range:** Breeds along the coast of the Bering Sea and the North Pacific Ocean, from the Commander Islands east and south to at least Yaquina Bay, Oregon. A few pairs have nested on islands in the Columbia and Willamette rivers.

**Habitat:** This gull nests on offshore islands, rocky coastal cliffs, and, sometimes, in grass on level parts of headlands. It forages in the marine and intertidal environments.

**Reproduction:** Breeding begins in late May or June. The nest is a large cup of grass, seaweed, and feathers. The typical clutch of 3 (range 2-4) eggs is incubated about 26 days. It often nests in association with other seabirds, and frequently hybridizes with the Western Gull.

**Food Habits:** Glaucous-winged Gulls eat just about any animal material they find. They take small fish, barnacles, molluscs, sea urchins, bird eggs, carrion, and animal waste discarded from fishing boats.

**Ecology:** This gull may nest colonially or singly. A small territory (about 2 feet in diameter) is defended around the nest. In the morning, it feeds up to several miles out to sea, and roosts in the afternoon.

Flocks sometimes follow fish runs (smelt, salmon) up rivers far inland.

**Comments:** Less than 400 pairs of Glaucous-winged Gulls breed in Oregon, most on an island in the mouth of the Columbia River.

*References:* Gilligan et al. 1994, Hoffman et al. 1978, Reid 1988.

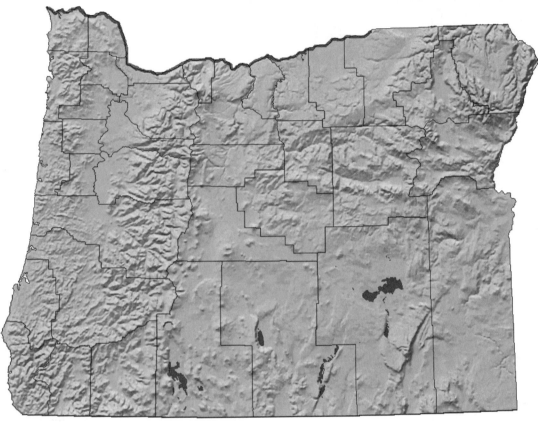

## Caspian Tern
### *(Sterna caspia)*

*Order: Charadriiformes*
*Family: Laridae*
*State Status: None*
*Federal Status: None*
*Global Rank: G5*
*State Rank: S4?*
*Length: 21 in (53 cm)*

**Global Range:** Breeds in scattered areas throughout the world. It is found along coasts from Africa to Australia, and throughout the temperate parts of Eurasia. In North America, it breeds in local colonies along both coasts, and from northern Alberta south to Baja California and the Gulf Coast.

**Habitat:** The Caspian Tern nests on flat sandy or gravelly areas on islands, or on the margins of lakes, rivers, and marshes. It is always near water, and forages both in nearby water bodies and on prey exposed in nearby open areas.

**Reproduction:** Breeding begins in May, and the last young are fledged by September. The nest is a simple depression in the sand, and may be lined with any convenient plant material. The clutch of 2 or 3 eggs (rarely only one ) is incubated 20-22 days by both parents. Caspian Terns usually nest in association with California and Ring-billed Gulls.

**Food Habits:** This is a large tern that primarily feeds on fish 3 to 10 inches long. It dives into the water and catches fish just below the surface. It also eats crustaceans, molluscs, insects, bird eggs, and the nestlings of other birds.

**Ecology:** This tern usually nests in tightly-packed colonies. Nests can be only a few feet apart, but a territory is maintained around the nest. Caspian Terns can fly great distances (in one case 37 miles) from the nest to feed. If no suitable nesting sites are available, due to water conditions at inland marshes, they will defer breeding that year.

**Comments:** Oregon's largest Caspian Tern colony (over 200 birds) is found on Three Mile Island, in the Columbia River, Morrow County.

*References:* Cuthbert 1985, 1988; Gill 1976, Gill and Mewaldt 1983, Gilligan et al. 1994.

**153**

# Forster's Tern
## *(Sterna forsteri)*

*Order: Charadriiformes*
*Family: Laridae*
*State Status: None*
*Federal Status: None*
*Global Rank: G5*
*State Rank: S4*
*Length: 14 in (36 cm)*

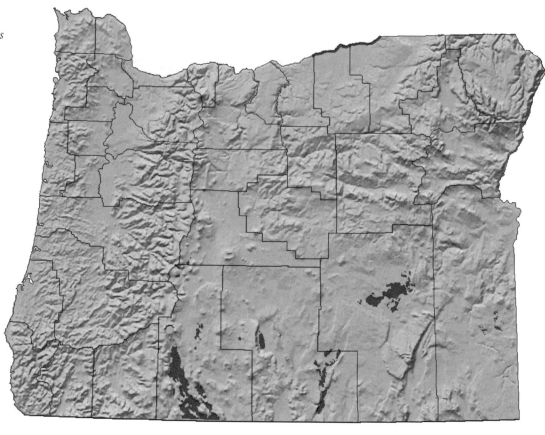

**Global Range:** Breeds in widely separated colonies in the interior and along the coasts of North America. It can be found from coastal British Columbia to southern California, from central Canada locally to the Gulf Coast (Tamaulipas to Louisiana), and along the Atlantic Coast.

**Habitat:** This tern breeds on lakes and marshes (in floating nests) and, sometimes, on mud or sand flats near water.

**Reproduction:** Breeding season begins in May, and these terns depart Oregon in October. The nest depends partly on the colony site. It can be a cup of aquatic vegetation built on a floating mat of marsh plants, or it may sit on a floating log or muskrat house. The clutch of 2 or 3 (range 1-4) eggs is incubated for 23-25 days by both parents.

**Food Habits:** Forster's Tern catches its food over or in the water. It will catch insects in flight, or take them from the surface of the water. It also dives for small fish, and will eat dead fish or frogs found in the water. Rarely, it eats eggs of other birds nesting nearby.

**Ecology:** Forster's Terns nest in small, loose colonies. Nests are often within a few feet of one another.

They actively defend their nests against intruders and, generally, are intolerant of other birds. American Coots, however, have been known to parasitize their nests.

**Comments:** Most Forster's Tern colonies in Oregon are in the alkaline marshes of the southeastern part of the state, but there is a colony on Three Mile Island in the Columbia River.

*References:* Bergman et al. 1970, Gilligan et al. 1994, Salt and Willard 1971.

**154**

*Order: Charadriiformes*

*Family: Laridae*

*State Status: None*

*Federal Status: SC*

*Global Rank: G4*

*State Rank: S3*

*Length: 10 in (25 cm)*

*References:* Bergman et al. 1970, DeGraaf et al. 1991, Gilligan et al. 1994.

**Global Range:** The breeding distribution of the Black Tern is holarctic. It breeds from Scandinavia to Siberia, south to the Mediterranean and Aral seas. In North America, it breeds from north-central Canada south to central California. Its southern limits move north across the country and, on the East Coast, it does not breed south of New England. It winters in South America and Africa.

**Habitat:** The Black Tern usually nests in or on emergent vegetation in alkaline lakes and freshwater marshes, or in marshy areas along rivers or ponds. It forages within a few hundred meters of its nest on both open ground (wet meadows, pastures, agricultural fields) and in the water.

**Reproduction:** Breeding begins in April or May, and some adults leave the breeding marshes by late July. The nest is a platform of vegetation that may be floating and attached to reeds, or placed on marshy ground. The Black Tern will use abandoned nests of other water birds. The clutch of 3 (range 2-4) eggs is incubated about 20-22 days by both sexes.

**Food Habits:** The Black Tern is more insectivorous than other Oregon terns. It catches insects resting on marsh vegetation or on the ground. It eats dragonflies, grasshoppers, moths, flies, and crickets, as well as aquatic invertebrates like crayfish and small molluscs. It eats a few small fish and tadpoles.

**Ecology:** Black Terns are only loosely colonial, with several pairs occupying the same marsh, nesting less than 100 feet apart. They defend their nests against intruders and, when young are fledged, at about 4 weeks of age, they help their parents defend a feeding territory.

**Comments:** These terns breed at Malheur National Wildlife Refuge and other desert marshes. They also breed at marshes bordering mountain lakes as high as 4,000 feet on the east slope of the southern Cascade Mountains. A few pairs have nested at Fern Ridge Reservoir, Lane County. The Black Tern population of North America declined from 1966 to 1985.

155

# Common Murre
## *(Uria aalge)*

*Order: Charadriiformes*
*Family: Alcidae*
*State Status: None*
*Federal Status: None*
*Global Rank: G5*
*State Rank: S5*
*Length: 17 in (43 cm)*

**Global Range:** The Common Murre is strictly a seabird that breeds along the arctic, subarctic, and temperate coasts of the Northern Hemisphere. In North America, it breeds along the Pacific Coast, from Alaska to central California, and along the Atlantic Coast, from Labrador to Newfoundland.

**Habitat:** Common Murres nest on ledges of coastal cliffs and on sides and tops of rock stacks and other offshore islands. They forage in the marine and estuarine environments.

**Reproduction:** Eggs are usually laid in early June. Common Murres do not construct a nest, but lay a single egg on rocky substrate. Both sexes incubate the egg 28-35 days. The egg is held between the adult's feet during incubation. The young leave the nest for the sea about 3 weeks after hatching, but cannot fly for another 3 weeks.

**Food Habits:** Common Murres find all of their food at sea. They eat fish up to 7 inches long (sand lance, herring, anchovies, rockfish), squid, shrimp, and mysid crustaceans. They may forage 10 miles from the nest, and are able to dive to depths of over 200 feet in pursuit of prey.

**Ecology:** Common Murres nest in very close proximity to one another, often less than one body length away, and only become aggressive if another bird comes too close. They forage during the day and are sensitive to disturbance at the colony.

**Comments:** There are 60 known Common Murre colonies along the Oregon Coast. In 1988, the state population was estimated to be nearly three-quarters of a million birds.

*References:* Bayer et al. 1991, Cairns et al. 1987, Gilligan et al. 1994, Zeiner et al. 1990a.

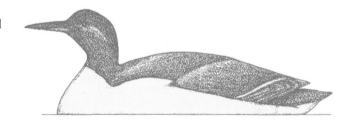

# Pigeon Guillemot
*(Cepphus columba)*

*Order: Charadriiformes*
*Family: Alcidae*
*State Status: None*
*Federal Status: None*
*Global Rank: G5*
*State Rank: S5*
*Length: 14 in (36 cm)*

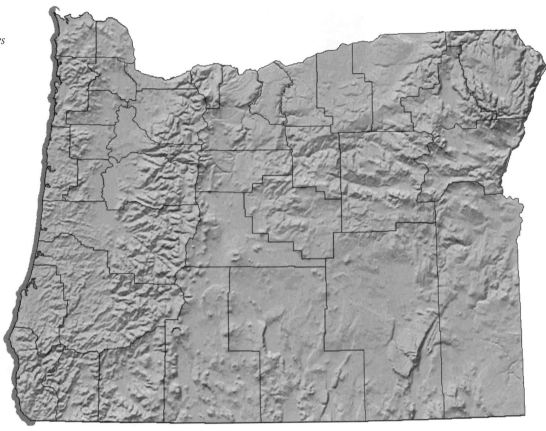

**Global Range:** Breeds along the coasts of the Bering Sea and the northern Pacific Ocean, from the Kurile Islands in Asia, north and east along the coast of Alaska, and south as far as central California and Santa Barbara Island.

**Habitat:** Like many marine birds, the Pigeon Guillemot nests on the ledges of coastal cliffs, on offshore islands, and rock stacks. It also will nest on artificial structures (docks, bridges).

**Reproduction:** Although this species is not colonial, many nest in close proximity because of limited nesting sites. The nest is usually located in a covered depression. Sometimes Pigeon Guillemots excavate their own burrow in soil. A pile of small stones may keep the egg in place. The usual clutch is 2 eggs (one egg 9% of the time), which are laid from late April to June and incubated about 28 days by both male and female. The young fledge when they are 34-40 days old.

**Food Habits:** Small fish (sand lance, smelt, sculpins, blennies) form the bulk of their diet, especially while breeding. They also eat some crustaceans (small crabs, shrimp), molluscs, and marine annelids.

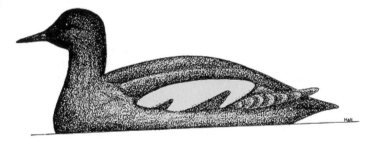

**Ecology:** The Pigeon Guillemot is a late breeder, and often is still feeding young on the nest in August, long after other seabirds disperse to the pelagic environment. Most breeding sites support fewer than 20 birds. They carry out most of their foraging within 3 miles of shore.

**Comments:** One estimate placed the number of nesting Pigeon Guillemots in Oregon at just under 5,000 birds.

*References:* Gilligan et al. 1994, Sealy 1990, Zeiner et al. 1990a.

# Marbled Murrelet
### (*Brachyramphus marmoratus*)

*Order: Charadriiformes*
*Family: Alcidae*
*State Status: Threatened*
*Federal Status: Threatened*
*Global Rank: G4*
*State Rank: S2*
*Length: 10 in (25 cm)*

**Global Range:** Breeds near the Pacific Coast, from the Kurile Islands (possibly Japan and Korea) eastward along the Siberian coast and offshore islands, to Alaska and then south to central California. Recent taxonomic work suggests that the Asian and North American races are different species (Friesen et al. 1996).

**Habitat:** In the Arctic parts of its range, the Marbled Murrelet nests on the ground, but Oregon populations nest in large trees in older forests (Douglas-fir, western red cedar, western hemlock, and Sitka spruce) or forests with old-growth characteristics, usually within 50 miles of the coast. It forages in the marine environment, usually within 2 kilometers of the shoreline and is known to visit its nesting habitat during the nonbreeding season.

**Reproduction:** Nests are depressions on large limbs near the trunk of a large-diameter tree, and are usually lined with moss or conifer needles. Nests usually have overhead cover, sometimes a growth of mistletoe, and are an average of 200 feet above the ground. One egg is laid, between April and mid-June. The incubation period is estimated to be 28 days. Young are fledged by early September.

**Food Habits:** The Marbled Murrelet feeds on small fish, especially during the breeding season, and marine invertebrates. Fish eaten include sand lance, Pacific herring, anchovies, smelt, capelin, seaperch, sardines, and rockfish. They also eat small shrimp-like crustaceans (euphausids, mysids, gammarids) and squid.

**Ecology:** Several pairs of Marbled Murrelets will nest in the same stand of trees, perhaps more due to limited habitat availability than any colonial tendencies. Of nests with known outcomes (n=32) 72% failed, about half due to predation (mostly by Common Ravens and Steller's Jays).

**Comments:** Loss and fragmentation of nesting habitat, leading to nest failure, is thought to be responsible for a yearly 4-7% decline in populations in the western United States. There may be over 8,000 pairs of Marbled Murrelets off the Oregon coast, including nonbreeding adults. If current rates of decline continue, the species may disappear from Oregon over the next several decades.

*References:* Marshall 1988, 1992; Marshall et al. 1996, Ralph et al. 1995, Singer et al. 1991, U.S. Fish and Wildlife Service 1995.

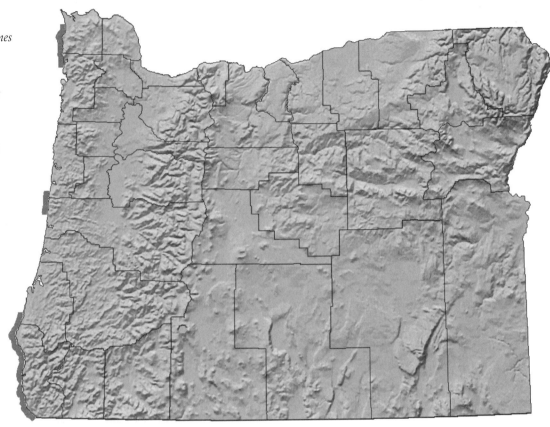

## Cassin's Auklet
### *(Ptychoramphus aleuticus)*

*Order: Charadriiformes*
*Family: Alcidae*
*State Status: None*
*Federal Status: None*
*Global Rank: G4*
*State Rank: S4*
*Length: 9 in (23 cm)*

**Global Range:** Breeds along the West Coast of North America from the Aleutian Islands south to central Baja California.

**Habitat:** Cassin's Auklet uses a few offshore islands for breeding in Oregon. It forages in the marine environment.

**Reproduction.** Cassin's Auklet nests in a chamber under rocks or digs its own burrow, 2-6 feet long, in the soil. The nest chamber is bare or nearly so. Egg laying may last from March to July. Both parents incubate the single egg about 37 days, and hatchlings leave the burrow by 50 days of age.

**Food Habits:** The main item in the diet seems to be euphausid crustaceans. They also eat amphipods, copepods, larval squid, crab larvae, and some small fish.

**Ecology:** This species spends its days at sea and only comes to the burrow at night, perhaps to reduce discovery by predaceous Western Gulls; as a result, little is known about its biology. It will defend its burrow from conspecifics. This species is found alone or in small groups on the open ocean by day.

**Comments:** Only 240 Cassin's Auklets are estimated to breed on four offshore islands in Oregon.

*References:* Gilligan et al. 1994, Harrison 1978, Zeiner et al. 1990a.

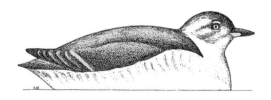

# Rhinoceros Auklet
## (Cerorhinca monocerata)

*Order: Charadriiformes*
*Family: Alcidae*
*State Status: None*
*Federal Status: None*
*Global Rank: G5*
*State Rank: S4?*
*Length: 15 in (38 cm)*

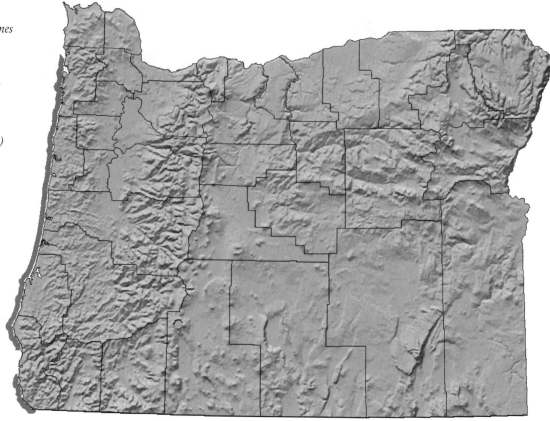

**Global Range:** Breeds along the coasts of the northern Pacific Ocean. In Asia, it occurs from Sakhalin Island south to Japan and Korea. In North America, it breeds from southern Alaska to northern California.

**Habitat:** The Rhinoceros Auklet makes use of the tops and slopes of offshore islands and coastal headlands. These areas can be bare or forested, but must have some soil development. It forages in the ocean.

**Reproduction:** This species digs its own burrow, usually 5-15 feet in length. Both sexes take part in the excavation, which can take 2 weeks. It begins breeding in May, and both parents incubate the single egg for 31-33 days.

**Food Habits:** Unlike Cassin's Auklet, the Rhinoceros Auklet feeds mostly on fish, but will take crustaceans and squid when fish are scarce. Sand lance, caught by underwater pursuit, are the most important item in the diet.

**Ecology:** Like Cassin's Auklet, this species mostly enters and leaves its burrow at night. The location of breeding colonies is probably determined by the

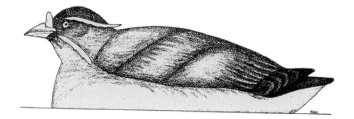

presence of undisturbed sites with soft, easily dug soil that are near productive ocean fisheries.

**Comments:** There are 7 known colonies of Rhinoceros Auklets in Oregon, of which 5 are on coastal headlands. The estimated state breeding population is about 1,000 birds.

*References:* Gilligan et al. 1994, Wilson and Manuwal 1986, Zeiner et al. 1990a.

## Tufted Puffin
*(Fratercula cirrhata)*

*Order: Charadriiformes*
*Family: Alcidae*
*State Status: None*
*Federal Status: None*
*Global Rank: G5*
*State Rank: S3*
*Length: 15 in (38 cm)*

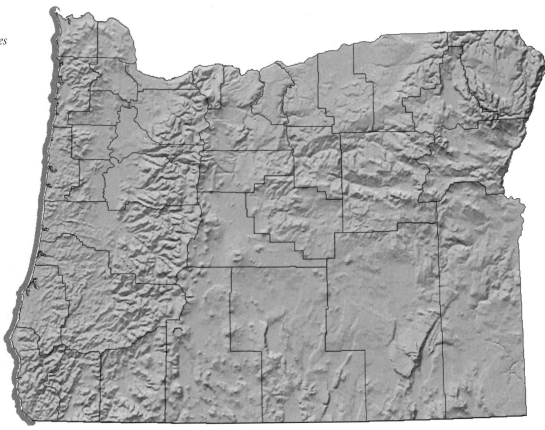

**Global Range:** Breeds along the coasts of the Bering Sea and northern Pacific Ocean, south to northern Japan in Asia and central California in North America.

**Habitat:** Tufted Puffins dig burrows on slopes or turf-covered headlands of offshore islands and coastal bluffs. Soil in which they can dig is a factor determining which sites they use, although sometimes they nest in rock crevices. They forage in the marine environment.

**Reproduction:** These puffins dig relatively shallow burrows, and line the nest chamber with dry grass and feathers. They arrive at breeding colonies around April and return to sea in August or September. Both parents incubate the single egg for 30 days. Young travel to the sea about a month after hatching. It is unclear whether or not they remain in the company of their parents during the transition to marine life.

**Food Habits:** The Tufted Puffin eats medium-size fish like smelt, sea perch, and herring. It also feeds on some squid, crustaceans, and marine annelids.

**Ecology:** Tufted Puffins are colonial and often nest in the same areas as the Rhinoceros Auklet. They are diurnal and visit their burrows in the daytime. They may fly 5 miles or more while foraging for fish to feed the nestling.

**Comments:** There are an estimated 3,300 pairs of Tufted Puffins breeding in Oregon, two-thirds of which are on Three Arch Rocks National Wildlife Refuge. Four other large colonies and 26 smaller colonies are known.

*References:* Baird 1991, Gilligan et al. 1994, Price and Simons 1986, Zeiner et al. 1990a.

# Rock Dove
## *(Columba livia)*

*Order: Columbiformes*
*Family: Columbidae*
*State Status: None*
*Federal Status: None*
*Global Rank: G5*
*State Rank: SE*
*Length: 13 in (33 cm)*

**Global Range:** Native to Eurasia, from Norway east to northern China and south to the Sahara Desert, Iran, India, and Burma. Widely introduced to major cities throughout the world, including all of North and South America and Hawaii.

**Habitat:** Natural populations nest on cliffs and ledges, along canyon walls, in caves, and at desert oases. More commonly, they nest on ledges of buildings, bridges, statues, and other urban structures. In agricultural areas, they frequently take up residence in barns. Although most common in the cities and towns of western Oregon, some populations nest on cliffs in arid country east of the Cascade Mountains.

**Reproduction:** Rock Doves begin nesting in March and may rear two or three broods by September. The nest is a thin layer of plant material deposited on a ledge. Both parents incubate the clutch of 2 (range 1-3) eggs for about 18 days. Young fledge in about a month.

**Food Habits:** Natural foods include seeds, grain, forbs, grasses, and berries. In urban or agricultural settings they consume nearly any dry edible material (e.g., popcorn, bread scraps, grain spilled at shipping facilities).

**Ecology:** Rock Doves congregate into breeding or foraging colonies. The pair will defend a small area around the nest. They rarely perch in trees, and forage on the ground. Rock Doves are not migratory.

**Comments:** Due to their litter and droppings and to the fact they may harbor diseases, Rock Doves are considered a pest species. They are easy prey for falcons, including the Peregrine Falcon, which sometimes shares their urban environment.

*References:* Gabrielson and Jewett 1940, Gilligan et al. 1994, Godfrey 1966, Kindschy 1964, Zeiner et al. 1990a.

# Band-tailed Pigeon
## (Columba fasciata)

Order: Columbiformes
Family: Columbidae
State Status: None
Federal Status: None
Global Rank: G5
State Rank: S4
Length: 15 in (38 cm)

References:
Gabrielson and
Jewett 1940,
Gilligan et al. 1994,
Wight et al. 1967.

**Global Range:** Breeds from southwestern British Columbia south to Baja California and east to Colorado, New Mexico, and again south to Central America. Also resident from Costa Rica south through the temperate parts of South America to Peru and northwestern Argentina, but some taxonomists consider these populations a different species.

**Habitat:** In Oregon, Band-tailed Pigeons are birds of coniferous or mixed coniferous-deciduous forests. They are often associated with forests and woodlands containing oaks, but use dense coniferous forests in western Oregon. They do not breed in higher-elevation forests (above 4,000 feet) of the western Cascade Mountains.

**Reproduction:** This pigeon nests at the tops of trees in a shallow cup of twigs. Most eggs are laid from early May to early June. The clutch of one egg (rarely 2) is incubated by both parents for 18-20 days. The nest is usually located near water.

**Food Habits:** The Band-tailed Pigeon's diet consists mainly of nuts and berries, including acorns, hazelnuts, and berries of madrone, salal, elderberry, salmonberry, manzanita, chokecherry, and sumac. It also eats some insects, waste grain, oak flowers, and madrone, lupine, and pine seeds.

**Ecology:** A breeding territory of a tenth to a half mile is defended around the nest. They may travel several miles for food or water. After the breeding season, they form flocks and may migrate upslope in search of food or mineral deposits. They are diurnal, and most travel south for the winter, but a few remain in western Oregon.

**Comments:** The Band-tailed Pigeon is a game species in Oregon. Populations are thought to be declining. Naturalized colonies of Band-tailed Pigeons are established on rocky cliffs in eastern Oregon (e.g., along the Columbia and Owyhee rivers, at Smith Rocks State Park, and along the Catlow Rim).

# Mourning Dove
## (Zenaida macroura)

*Order: Columbiformes*
*Family: Columbidae*
*State Status: None*
*Federal Status: None*
*Global Rank: G5*
*State Rank: S5*
*Length: 12 in (30 cm)*

**Global Range:** Breeds coast to coast from southern Canada to northern Mexico and also, locally, in Central America south to Panama. It has been introduced to the island of Hawaii.

**Habitat:** The Mourning Dove uses a wide range of habitats in Oregon. It is common around agricultural lands, pastures, woodlands, and suburbs in the valleys of western Oregon. It also makes use of chaparral, oak, juniper, ponderosa pine woodlands, and arid desert communities of eastern Oregon, provided there is some access to water. It does not breed in high elevation or dense forests.

**Reproduction:** Mourning Doves typically lay two or more clutches during their breeding season, which lasts from April to September in Oregon. The usual clutch is 2 eggs and is incubated 14-16 days by both sexes. The male is on the nest, usually built of twigs 10-20 feet high in a tree, during the day. The young grow rapidly and are flying in 2 weeks.

**Food Habits:** The diet is almost exclusively (98%) made up of seeds and grain. They consume seeds from many weedy herbs and shrubs, as well as some fruits and nuts and, rarely, small snails or insects.

**Ecology:** Mourning Doves defend a small territory around their nest, but the home ranges of several pairs often overlap. They feed at distances of up to 3 miles from the nest. After the breeding season, these doves form small flocks.

**Comments:** The Mourning Dove is a game animal in Oregon. Pairs of this species mate for life.

*References:* Baskett et al. 1993, Gilligan et al. 1994, Zeiner et al. 1990a.

# Yellow-billed Cuckoo
### *(Coccyzus americanus)*

*Order: Cuculiformes*
*Family: Cuculidae*
*State Status: Sensitive*
*Federal Status: None*
*Global Rank: G5*
*State Rank: S1*
*Length: 12 in (30 cm)*

**Global Range:** Breeds coast to coast from southern Canada south to Central America and the Caribbean. Now rare or absent from much of the western U.S. Winters in South America. In Oregon, formerly bred in riparian forests along the Columbia River at least as far west as Sauvie Island, near Portland.

**Habitat:** Yellow-billed Cuckoos are birds of thick, closed-canopy riparian forests with an understory of dense brush. These riparian forests are usually composed of various species of willows and cottonwoods, especially black cottonwoods along the rivers of eastern Oregon. Studies in California have suggested that patches of suitable habitat must be at least 37 acres in size and include over 7.5 acres of closed canopy riparian forest (Marshall 1992).

**Reproduction:** Their nest of twigs, vines, grass and other plant material is usually placed 4 to 10 feet above the ground in thick brush (often willow). The clutch of 3 to 4 (range 1-5) eggs is incubated by both parents for about 2 weeks. Young leave the nest after about a week, while they are still flightless.

**Food Habits:** Although their nest is usually placed in willows, Yellow-billed Cuckoos seem to feed among cottonwoods. They eat mainly caterpillars, including tent caterpillars, although they eat other insects (including grasshoppers and cicadas), fruit, and, occasionally, small lizards and frogs.

**Ecology:** Reproductive success is thought to increase in years with outbreaks of tent caterpillars. Records indicate that cuckoos arrive in Oregon in mid-May and depart for their wintering grounds by the end of September. Elsewhere they nest in orchards, and in August 1980 one was seen feeding young at a residence in La Grande, Union County.

**Comments:** Although many cuckoos are brood parasites, this species builds its own nest, though it has been known to lay eggs in nests of other species, including American Robins, Red-winged Blackbirds, and Mourning Doves. While a few pairs of this species probably breed in eastern Oregon during most years, no areas have been documented to be regularly occupied; therefore we have not mapped breeding locations for this species. The decline of Oregon's Yellow-billed Cuckoo population is probably related to the loss of large stands of riparian forests along our major rivers. Marshall (1992) suggests that the replacement of willow by introduced Himalayan blackberry in the riparian forests of western Oregon may create unsuitable habitat for this species. Pesticide contamination on the wintering grounds may also cause eggshell thinning.

*References:* Banks 1988, Gaines and Laymon 1984, Gilligan et al. 1994, Laymon and Halterman 1987, Marshall 1992, Zeiner et al. 1990a.

# Barn Owl
## (Tyto alba)

*Order: Strigiformes*
*Family: Tytonidae*
*State Status: None*
*Federal Status: None*
*Global Rank: G5*
*State Rank: G4?*
*Length: 17 in (43 cm)*
*Wingspread: 39 in (99 cm)*

**Global Range:** The Barn Owl is absent from the high latitudes of the Northern Hemisphere, but otherwise breeds throughout the world (southern Eurasia, Africa, South America, Australia). It does not breed much farther north than southern British Columbia and New England in North America. Some believe the Barn Owl of Australia is a separate species.

**Habitat:** Barn Owls are most suited to open country with an abundance of rodents. As a result, they do well in the agricultural areas of western Oregon valleys and the Columbia Basin. They can, however, also be found in oak, juniper, or ponderosa pine woodlands and in the arid, open deserts of southeastern Oregon, where canyons and caves offer nest sites.

**Reproduction:** This owl usually nests in hollow trees, ledges on cliffs, or caves, but will use structures as well. Nests are often reused in successive years, and the accumulated debris is the only nesting material. A clutch of 4 to 7 (range 3-11) eggs is incubated by the female for 32-34 days, during which time she is fed by the male.

**Food Habits:** The Barn Owl is a carnivore that tends to specialize in small nocturnal mammals, including various mice, gophers, moles, shrews, kangaroo rats, and rabbits. It sometimes takes unusual prey items such as bats, insects, frogs, lizards, crayfish, and small birds.

**Ecology:** Only a short distance (less than 50 feet) around the nest is defended, but the home range can be a square mile or more. After the breeding season, Barn Owls may roost together. Oregon is near the northern limit of its range, and many migrate south in the winter. Like most owls, the Barn Owl is a nocturnal hunter.

**Comments:** Barn Owls have been known to starve over the winter in eastern Oregon, due to low temperatures or heavy snow cover.

*References:* Gilligan et al. 1994, Marti 1974, Marti and Wagner 1985.

## Flammulated Owl
### (Otus flammeolus)

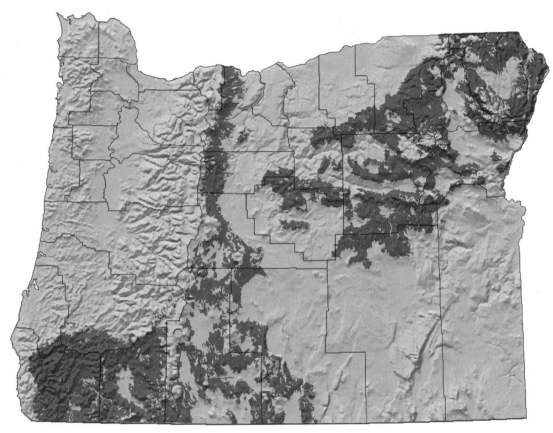

*Order: Strigiformes*
*Family: Strigidae*
*State Status: Sensitive*
*Federal Status: None*
*Global Rank: G4*
*State Rank: S4*
*Length: 7 in (18 cm)*

**Global Range:** This is an owl of western North America that reaches its northern limit in southern British Columbia. It occurs east to the Rocky Mountains and south to central Mexico. It winters in Mexico and Central America.

**Habitat:** In Oregon, the Flammulated Owl is closely associated with open forests that have a ponderosa pine component. It uses other forest types elsewhere (Douglas-fir in British Columbia; Jeffrey Pine in California). It requires fairly large trees for roosting that are adjacent to patches of grassland or meadow, where it forages.

**Reproduction:** Flammulated Owls nest in abandoned woodpecker holes or natural tree cavities. Their breeding season begins in May, and most disappear to the south by October. There are usually 3 (range 2-4) eggs in the clutch, which are incubated 21-22 days by the female, who is fed by the male. The young fledge about 3 weeks after hatching.

**Food Habits:** This small owl is mainly an insectivore, preferring grasshoppers and moths, but also eating beetles, crickets, spiders, scorpions, and, rarely, a small mammal or bird.

**Ecology:** The home range of male Flammulated Owls in one Oregon study was 25 acres. They forage along ecotones between forests and grasslands. They tend to nest in ponderosa pine trees even in mixed forests.

**Comments:** If no abandoned woodpecker holes are found, this owl will displace a flicker from an occupied nest. Loss of mature ponderosa pine woodland and individual large ponderosa pines has resulted in loss of habitat for this species in Oregon.

*References:* Bull and Anderson 1978, Bull et al. 1990, Franzreb and Ohmart 1978.

# Western Screech-Owl
### (Otus kennicottii)

*Order: Strigiformes*
*Family: Strigidae*
*State Status: None*
*Federal Status: None*
*Global Rank: G5*
*State Rank: S4?*
*Length: 9 in (23 cm)*

**Global Range:** Resident from the south Alaskan coast and coastal British Columbia south to the area of Mexico City and east to Montana and the Rio Grande Valley of Texas.

**Habitat:** Western Screech-Owls are birds of woodlands and forests. They occupy suburban areas; oak, juniper, and riparian woodlands; chaparral; and, less commonly, coniferous forests. They require the presence of trees at least 1 foot in diameter for nesting, so are confined to towns and riparian areas in the arid parts of eastern Oregon.

**Reproduction:** This owl is usually a "secondary cavity nester," meaning that it nests in holes in trees excavated by other birds, although it sometimes uses holes in cliffs. A typical clutch has 4 or 5 (range 2-8) eggs, and is incubated for 26 days by the female, while she is fed by the male.

**Food Habits:** Unlike its relative the Flammulated Owl, this owl feeds primarily on rodents (especially voles). It also eats quite a few small birds, some insects, and other invertebrates. Sometimes unusual food items such as fish, amphibians, reptiles, and crayfish turn up in the diet.

**Ecology:** Studies have found densities of about one bird per square mile for this owl. Males defend a territory around nesting and feeding areas. It does not breed in high elevation forests (above 4,000 feet) of the Cascade Mountains.

**Comments:** This is the most common small owl in western Oregon, and can often be found in city parks.

*References:* Gilligan et al. 1994, Hayward and Garton 1988, Murray 1976, Van Camp and Henny 1975.

*Order: Strigiformes*
*Family: Strigidae*
*State Status: None*
*Federal Status: None*
*Global Rank: G5*
*State Rank: S5*
*Length: 24 in (61 cm)*
*Wingspread: 54 in (137 cm)*

*References:* Knight and Erickson 1977, McGillivray 1989, Zeiner et al. 1990a.

**Global Range:** Found almost everywhere in the Western Hemisphere except the high Arctic. It breeds from central Alaska south to Tierra del Fuego. Strangely, it is absent from most islands in the Caribbean Sea.

**Habitat:** This versatile owl can be found in nearly every habitat type, although it needs large trees or ledges on cliffs for nesting. It may forage far from the nest and, as a result, uses habitats from deserts and grasslands to woodlands, forests, and suburbs. It is often found near water, so riparian forests make especially favorable habitat. It prefers open areas to dense forests, and humans have accommodated this preference through clear-cutting.

**Reproduction:** Great Horned Owls nest in tree cavities, caves, ledges, and abandoned stick nests of hawks or crows. The normal clutch has 2 or 3 (range 1-5) eggs, and is incubated for 26-35 days by the female, fed by the male. Although young fledge in about 10 weeks, they are fed by adults for an additional time.

**Food Habits:** The majority of the Great Horned Owl's diet consists of small to medium-size mammals (mice, voles, squirrels, rabbits, skunks, and porcupines), but it also takes a fair number of birds, including ducks, quail, and even other raptors and owls. Some insects, fish, reptiles, and amphibians are also eaten.

**Ecology:** Great Horned Owls frequently perch high (up to 70 feet) in trees, and can be heard vocalizing in territorial defense. Territories are usually several hundred acres, and home ranges are over a thousand acres. They often hunt at the edges between open areas and forests or woodlands. Birds breeding at higher elevations spend winter in valleys.

**Comments:** The Great Horned Owl is a major predator of the threatened Spotted Owl in the forests of western Oregon, aided by increased edge created by timber harvest.

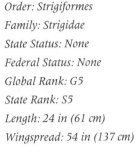

# Northern Pygmy-Owl
### *(Glaucidium gnoma)*

*Order: Strigiformes*
*Family: Strigidae*
*State Status: Sensitive*
*Federal Status: None*
*Global Rank: G5*
*State Rank: S4?*
*Length: 7 in (18 cm)*
*Wingspread: 15 in (38 cm)*

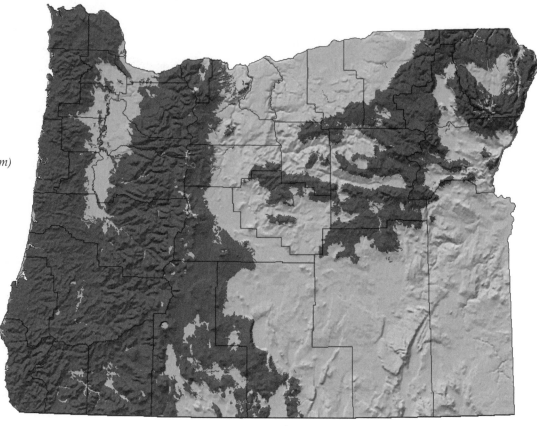

**Global Range:** The Northern Pygmy-Owl is a forest bird of western North America and Central America. It breeds from northern British Columbia south to the interior highlands of Honduras and is found as far east as New Mexico and west Texas.

**Habitat:** This owl is found in both coniferous and mixed coniferous-deciduous forests. It will occupy dense, moist forests (Douglas-fir, western hemlock, western red cedar), riparian woodlands, and also drier woodlands (ponderosa pine). It tends to hunt in open areas within the forest matrix, and has been seen using clear-cuts. It also occurs in chaparral in southwestern Oregon.

**Reproduction:** The Northern Pygmy-Owl uses abandoned woodpecker holes for its nest. Nest trees are usually near a meadow or clearing. The normal clutch is 3 to 5 (range 3-7) eggs, which are incubated by the female for about 29 days, while the male brings her food. The young are ready to leave the nest in another month.

**Food Habits:** This owl has a relatively broad diet for a small raptor. It feeds on small mammals (mice, chipmunks), large insects (grasshoppers, beetles, and crickets), other invertebrates, reptiles, amphibians,

and birds. During the breeding season it takes relatively more birds, some larger than itself.

**Ecology:** Unlike most owls, this one is active during daylight hours (most active around dawn and dusk), explaining its ability to take a variety of birds. It is often mobbed by passerines. During winter, it will move down from the mountains of eastern Oregon to juniper or aspen groves.

**Comments:** Forest practices that remove snags containing old woodpecker holes may be reducing available nest sites for this species in Oregon.

*References:* Bull et al. 1987, DeGraaf et al. 1991, Gilligan et al. 1994, Marshall 1992, Zeiner et al. 1990a.

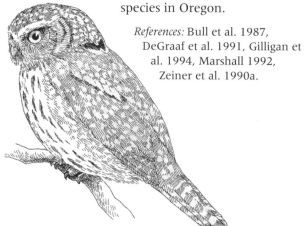

## Burrowing Owl
*(Athene cunicularia)*

*Order: Strigiformes*
*Family: Strigidae*
*State Status: Sensitive*
*Federal Status: SC*
*Global Rank: G4*
*State Rank: S3*
*Length: 10 in (25 cm)*
*Wingspread: 24 in (61 cm)*

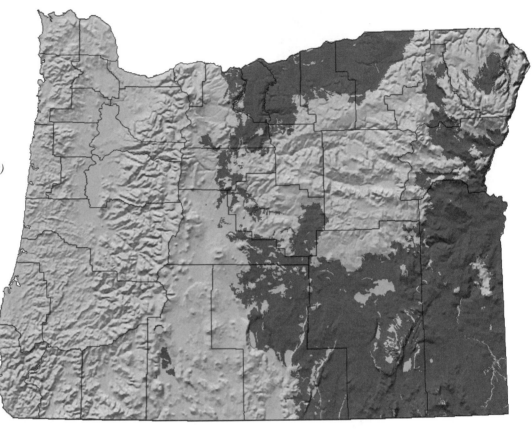

**Global Range:** Found throughout most of western North America, from southern Canada south to central Mexico. It is found as far east as eastern Texas. It reappears in South America in arid habitats from Colombia south to northern Tierra del Fuego.

**Habitat:** This unusual owl lives in open deserts, grasslands, fields, and pastures. It will use roadsides and airports. In Oregon, it is most common in the sagebrush steppe of the southeastern part of the state, but also occurs in arid parts of the Columbia Basin.

**Reproduction:** Burrowing Owls usually nest in modified burrows made by ground squirrels or badgers. The burrow can be 5 to 10 feet long, ending in a chamber lined with debris. A clutch of 5 or 6 (range 4-11) eggs is incubated by the female for 27-30 days, while the male brings food.

**Food Habits:** The main items on the Burrowing Owl's menu are rodents (pocket mice, sagebrush voles), followed by insects (including grasshoppers). It will also eat bats, shrews, small birds, crayfish, reptiles, and amphibians.

**Ecology:** Burrowing Owl nests tend to be spaced several hundred feet apart, but only a short distance (33 feet) around the entrance is defended. They are diurnal in the breeding season, becoming more nocturnal in winter. In the Columbia Basin, they tend to use old badger holes. The badger is also one of their chief predators.

**Comments:** Burrowing Owls were once considered common in the Rogue River Valley, but they have disappeared from the valleys of southwestern Oregon. A few spend the winter in western Oregon.

*References:* Gilligan et al. 1994, Green and Anthony 1989, Marshall 1992, Maser et al. 1971.

# Spotted Owl
## *(Strix occidentalis)*

*Order: Strigiformes*
*Family: Strigidae*
*State Status: Threatened*
*Federal Status: Threatened*
*Global Rank: G3*
*State Rank: S3*
*Length: 18 in (46 cm)*
*Wingspread: 45 in (114 cm)*

**Global Range:** From southwestern British Columbia south along the coast to southern California, and south in the Sierra Nevada and Rocky Mountains to the mountains of central Mexico.

**Habitat:** Southern subspecies occur in oak communities in shaded canyons, but in Oregon the Spotted Owl successfully breeds only in late-successional mixed coniferous forests, usually dominated by Douglas-fir. It prefers larger forest stands (more than 1,200 acres) with multiple layers and a closed canopy. Nests are usually within 400 meters of small streams. At higher elevations in the Cascade Mountains, forests of mountain hemlock, true fir, lodgepole pine, or ponderosa pine may be used.

**Reproduction:** The Spotted Owl breeding season begins in late March. Nests are in a tree hollow or on a large horizontal limb, as high as 200 feet above the ground. The clutch of 2 (range 2-4) eggs is incubated by the female, who is fed by the male. Young fly about 6 weeks after hatching, and breed at 2 or 3 years of age.

**Food Habits:** Major food items include northern flying squirrels, red tree voles, western red-backed voles, and dusky-footed woodrats. They also eat some small birds, bats, and insects.

**Ecology:** Spotted Owls have large home ranges (larger farther north in their range), and nests are about 1 to 2 miles apart in good habitat. They are nocturnal hunters. The current level of recruitment is insufficient to maintain the population. Predation by Great Horned Owls, especially on dispersing young, is a major mortality factor that has probably increased as forest fragmentation creates more edge.

**Comments:** The Spotted Owl has been eliminated from much of the lower-elevation forest of western Oregon through timber harvest. The long-term viability of the remaining populations is uncertain. Surveys through 1992 estimated that about 2,000 pairs nested in Oregon.

*References:* Carey et al. 1990, 1992, Doak 1989, Gilligan et al. 1994, Hunter et al. 1995, Lamberson et al. 1992, 1994, Marshall et al. 1996, Noon and Biles 1990.

## Barred Owl
### *(Strix varia)*

*Order: Strigiformes*
*Family: Strigidae*
*State Status: None*
*Federal Status: None*
*Global Rank: G5*
*State Rank: SU*
*Length: 22 in (56 cm)*

**Global Range:** Resident throughout most of the eastern United States, and across southern Canada to British Columbia. They range south in the mountains of the Pacific Northwest, possibly entering northern California. Their range in the Northwest is expanding rapidly. A disjunct population occurs in central Mexico.

**Habitat:** The Barred Owl is a forest species, but as one would expect for a species with such a wide distribution, it lives in different types of forest in different regions. In the Northwest it is found in coniferous and mixed forests dominated by Douglas-fir and ponderosa pine. Like the closely related Spotted Owl, it tends to occur in forests with old-growth characteristics.

**Reproduction:** Barred Owls nest in tree cavities, tops of snags, and abandoned crow or raptor nests. The breeding season begins in March, and 2 or 3 (range 2-4) eggs are incubated by the female for about 28 days. Young are tended by their parents for some time after fledging.

**Food Habits:** The Barred Owl is a generalist carnivore that eats small mammals (shrews, mice, chipmunks, foxes, rabbits, opossums), birds (other owls, grouse, quail, flickers, crows, jays), insects, crayfish, spiders, fish, amphibians, and reptiles. It forages over clearings in forests.

**Ecology:** Barred Owls are usually found near streams or lakes. In the Midwest, the home range is 1 or 2 square miles, and is used exclusively by members of a pair.

**Comments:** The Barred Owl was first sighted in eastern Oregon just over 20 years ago and has rapidly colonized the state. Its niche requirements are very similar to those of the Spotted Owl, except that it makes use of forest clearings. It may someday displace the Spotted Owl, with which it occasionally hybridizes, in Oregon.

*References:* Gilligan et al. 1994, Nicholls and Warner 1972, Taylor and Forsman 1976.

# Great Gray Owl
*(Strix nebulosa)*

*Order: Strigiformes*
*Family: Strigidae*
*State Status: Sensitive*
*Federal Status: None*
*Global Rank: G5*
*State Rank: S4*
*Length: 27 in (69 cm)*
*Wingspread: 57 in (145 cm)*

*References:* Bull and Henjum 1990, DeGraaf et al. 1991, Franklin 1988, Gilligan et al. 1994, Marshall 1992, Zeiner et al. 1990a.

**Global Range:** The Great Gray Owl is a holarctic species of high latitudes. It is found in the taiga of Eurasia, from Scandinavia to Siberia, and in North America from Alaska to Hudson Bay. It enters the northern tier of states in several places, but its southernmost outpost is in the central Sierra Nevada of California. It is expanding its range in Oregon.

**Habitat:** Great Gray Owls forage over open areas, such as meadows or clear-cuts, in coniferous forests. They are found in mixed coniferous, ponderosa pine, and lodgepole pine forests. They most frequently are found in old-growth forests on north-facing slopes.

**Reproduction:** The breeding biology of this owl has not been studied in Oregon, but elsewhere the breeding season begins in March. Great Gray Owls nest in tree cavities, on top of snags, or in abandoned raptor, corvid, or squirrel nests. The normal clutch of 3 to 5 (up to 9) eggs is incubated by the female for 28-29 days. Young are independent in about 4 or 5 months.

**Food Habits:** The Great Gray Owl's diet consists mostly of small mammals, especially voles and pocket gophers, but it will also eat birds as large as grouse.

**Ecology:** In an Idaho study, the home range of a pair of Great Gray Owls was 2.6 square kilometers. In Oregon, larger home ranges (30-60 square miles) have been recorded. A territory of over 100 acres is defended. An Oregon study found 6 pairs and 9 individual males in a 46-square-kilometer study area. Great Horned Owl predation can be a major mortality factor. Great Gray Owls move to lower elevations during the winter, even into agricultural areas.

**Comments:** Like the Barred Owl, this species is expanding its range in Oregon, taking advantage of openings in forests created by timber harvest practices. Young leave the nest before they can fly, and need leaning trees or a dense canopy to climb to the ground.

174

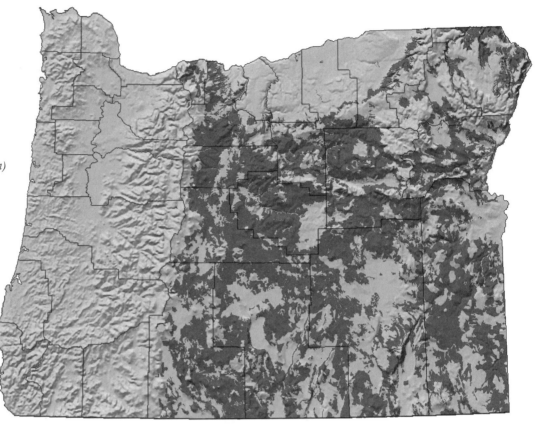

## Long-eared Owl
### (Asio otus)

*Order: Strigiformes*
*Family: Strigidae*
*State Status: None*
*Federal Status: None*
*Global Rank: G5*
*State Rank: S4?*
*Length: 15 in (38 cm)*
*Wingspread: 39 in (99 cm)*

**Global Range:** Breeds throughout the Northern Hemisphere. In the Old World, it breeds from the United Kingdom to Siberia, and south to North Africa and Korea. In North America, it breeds coast to coast, from central Canada south to California in the West and Virginia in the East.

**Habitat:** In Oregon, the Long-eared Owl is found in open coniferous and mixed coniferous-deciduous forests at lower elevations, as well as in juniper and riparian woodlands. It prefers areas of mixed forest and open country.

**Reproduction:** The breeding season begins in April, and most young are fledged by late July. The Long-eared Owl rarely builds its own nest, but uses the abandoned nest of a corvid or raptor, usually 15 to 30 feet high. Rarely, it may nest on the ground. The clutch of 4 or 5 (range 3-8) eggs is incubated for 25-30 days by the female. Young are independent about 2 months after hatching.

**Food Habits:** The Long-eared Owl's main prey items are small mammals (voles, deer mice, bats, shrews, rabbits), but it also eats small birds, amphibians, small snakes, and some insects.

**Ecology:** During breeding season, Short-eared Owls have a home range of about 100-300 acres. Their nests are subject to predation by raccoons and other arboreal mammals, and they compete for food with Northern Harriers. Although they are usually nocturnal, they may hunt at dawn and dusk. In winter, many Long-eared Owls may share a common roost.

**Comments:** While common in parts of eastern Oregon, there are fewer than 5 nesting records from the Willamette Valley (Gilligan et al. 1994:172).

*References:* Bull et al. 1989, Gilligan et al. 1994, Marks 1984.

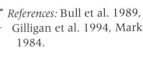

# Short-eared Owl
## (Asio flammeus)

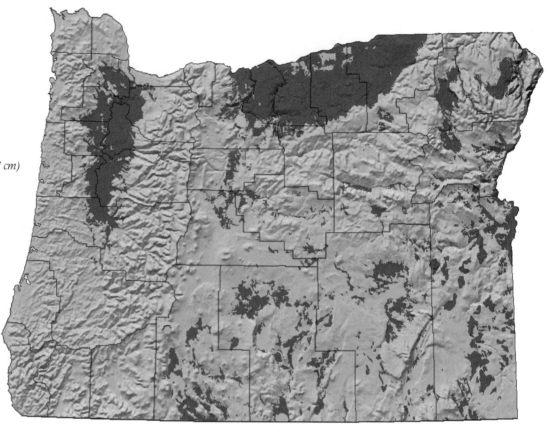

*Order: Strigiformes*
*Family: Strigidae*
*State Status: None*
*Federal Status: None*
*Global Rank: G5*
*State Rank: S4?*
*Length: 16 in (41 cm)*
*Wingspread: 42 in (107 cm)*

**Global Range:** The Short-eared Owl is a holarctic species that breeds in Eurasia, from Iceland and the United Kingdom east to Siberia and south to southern Europe and Manchuria. In North America, it breeds in the Alaskan and Canadian Arctic, south to California and New Jersey.

**Habitat:** This is an owl of open terrain that uses marshes (most commonly), grasslands, dunes, agricultural fields, pastures, and hay meadows. It closely resembles the Northern Harrier in its habitat preferences and was once colloquially called the "Marsh Owl."

**Reproduction:** The nesting season begins in April, and young are fledged by August. The nest is a depression in the ground with little or no lining. A clutch of 4 to 8 (range 3-14) eggs is incubated by the female for 24-28 days.

**Food Habits:** Short-eared Owls feed mainly on small terrestrial mammals (voles, deer mice), but also will catch bats, birds (shorebirds, sparrows), reptiles, and large insects.

**Ecology:** This owl is more diurnal than most, foraging both at night and in the daytime. Territory size varies with food supply. There may be

anywhere from 0.6-6.0 pairs per square kilometer. In winter, it may roost communally. It forages by flying low, usually into the wind, and dropping on prey.

**Comments:** This owl is most common around the marshes of eastern Oregon, but a few nest in the valleys of western Oregon.

*References:* Clark 1975, Gilligan et al. 1994, Marti 1974.

Order: *Strigiformes*
Family: *Strigidae*
State Status: *Sensitive*
Federal Status: *None*
Global Rank: *G5*
State Rank: *S4?*
Length: *10 in (25 cm)*

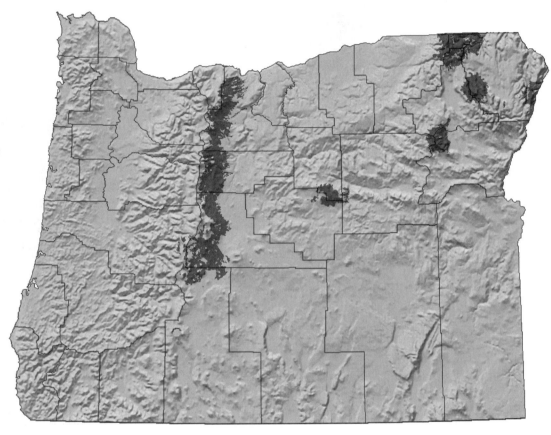

**Global Range:** A holarctic species, the Boreal Owl is found in Eurasia, from the northern tree line to the mountains of southern Europe and Japan. In North America, it breeds throughout most of Alaska and Canada. Oregon is the southern limit of its range on the West Coast. It is also known from the high Rocky Mountains of Colorado.

**Habitat:** In Oregon, this species frequents high-elevation (above 5,000 feet) forest communities, frequently dominated by Engelmann spruce and subalpine fir, but often mixed with lodgepole pine, ponderosa pine, Douglas-fir, or quaking aspen. Little is known of its Oregon habitat preferences, but elsewhere it nests in both coniferous and deciduous forests and in alder thickets and hunts over open meadows at the edges of forests.

**Reproduction:** Again, information comes from study of out-of-state populations, where the breeding season begins in mid-April to late May. The nest is in a natural hole in a tree or an old woodpecker hole. The clutch of 2 to 4 eggs is incubated by the female for 25-36 days. Young fledge a month after hatching, and are independent when 5 or 6 weeks old.

**Food Habits:** The Boreal Owl's main food items are small mammals (meadow voles, red-backed voles, shrews, deer mice, pocket gophers, flying squirrels), but they eat some birds, insects, amphibians, and small reptiles.

**Ecology:** In an Idaho study, the home range was 1,100 to 1,500 hectares, but only the immediate area of the nest site was defended. This owl may forage either by day or night, but usually roosts in dense cover in the daytime.

**Comments:** This owl was virtually unknown in Oregon prior to 1987, possibly because its habitat is snowed in during the early breeding season. It is possible that it is relatively common in its isolated, subalpine habitat. The Boreal Owl is associated with older forests, and its habitat in Oregon is likely to decline in the future.

*References:* DeGraaf et al. 1991, Gilligan et al. 1994, Marshall et al. 1996, Whelton 1989.

# Northern Saw-whet Owl
## *(Aegolius acadicus)*

*Order: Strigiformes*
*Family: Strigidae*
*State Status: None*
*Federal Status: None*
*Global Rank: G5*
*State Rank: S4?*
*Length: 8 in (20 cm)*
*Wingspread: 18 in (46 cm)*

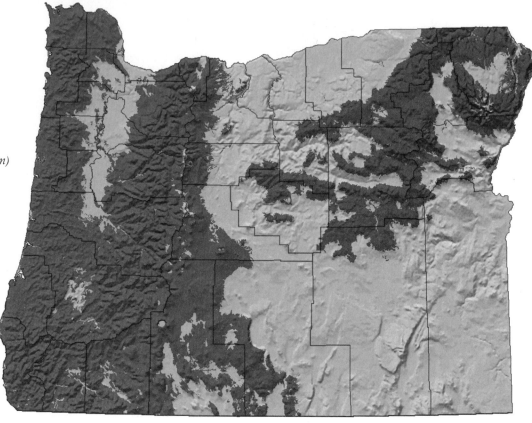

**Global Range:** Restricted to North America, where it breeds from coastal Alaska south to southern California, and east across southern Canada to the Atlantic Coast. It is found in the Appalachian Mountains as far south as North Carolina. Disjunct populations breed in the mountains of central Mexico.

**Habitat:** The Northern Saw-whet Owl lives in dense coniferous or broad-leaved forests. In Oregon, little use is made of oak or juniper woodlands. It hunts in open areas in or near forests and breeds in high-elevation forests.

**Reproduction:** The breeding season starts in April or May. The nest is usually a cavity in a tree, often an old woodpecker hole. The normal clutch of 5 to 7 eggs (range 4-7) is incubated for 26-28 days by the female.

**Food Habits:** This owl tends to eat mainly small terrestrial mammals (shrews, deer mice, chipmunks, squirrels, rabbits) but will also catch small passerine birds (juncos, sparrows, warblers), large insects (beetles, grasshoppers), bats, and frogs.

**Ecology:** The home range is several hundred acres, and a territory is defended within that area (no information on its size). The density of breeding pairs is no more than a few pairs per square kilometer.

**Comments:** During winter, these owls are thought to move down from higher elevations to valleys, where they favor riparian thickets.

*References:* Forsman and Maser 1970, Gilligan et al. 1994, Marks et al. 1989.

# Common Nighthawk
## (Chordeiles minor)

*Order: Caprimulgiformes*
*Family: Caprimulgidae*
*State Status: None*
*Federal Status: None*
*Global Rank: G5*
*State Rank: S5*
*Length: 9 in (23 cm)*

*References:* Armstrong 1965, Caccamise 1974, Gilligan et al. 1994, Zeiner et al. 1990a.

**Global Range:** Breeds in North America as far north as the Yukon and Nova Scotia, and south as far as southern California and eastern Mexico to Panama. It winters in South America, as far south as northern Argentina.

**Habitat:** Common Nighthawks forage over a wide variety of habitats. While they nest in open areas, including forest clearings, they forage over nearly every habitat in Oregon. They range from ocean dunes to high-elevation forests, including chaparral, grasslands, agricultural land, pastures, dense forest, juniper woodland, arid desert communities, and around cities and towns. In desert areas, they are usually found close to canyons or outcroppings.

**Reproduction:** The breeding season begins late, around early June. The Common Nighthawk nests on bare ground. A clutch of 2 eggs (sometimes only one) is incubated by the female for around 19 days. The male brings food to the nestlings. Young fledge at 23 days, and are independent in a month.

**Food Habits:** Common Nighthawks forage over many different habitats, catching flying insects on the wing. A variety of flying insects are taken, including ants, mosquitoes, grasshoppers, flies, moths, and caddisflies.

**Ecology:** Although called the nighthawk, this species also forages during the daytime, usually early in the morning or late in the afternoon. A territory around the nest and another one for feeding, ranging from 10 to 60 acres, is defended. The breeding home range is less than a square mile.

**Comments:** This species once nested on roofs in Portland, but is declining in the Willamette Valley.

# Common Poorwill
## *(Phalaenoptilus nuttallii)*

*Order: Caprimulgiformes*
*Family: Caprimulgidae*
*State Status: None*
*Federal Status: None*
*Global Rank: G5*
*State Rank: SU*
*Length: 7.5 in (19 cm)*

**Global Range:** The Common Poorwill is a species of western North America. It breeds from extreme southwestern Canada south to Baja California and central Mexico. It is found as far east as South Dakota and central Texas. Winters are spent from California south to Mexico.

**Habitat:** Common Poorwills use a variety of habitats, but are more particular than Common Nighthawks. They are found in sagebrush steppe and juniper woodlands in eastern Oregon, as well as higher up in ponderosa pine woodlands. They occur less commonly in chaparral and oak woodlands in southwest Oregon. They forage over open areas, and forest clear cutting may be providing them an opportunity to expand their range in Oregon.

**Reproduction:** Like their relative the Common Nighthawk, Common Poorwills lay their eggs on the ground, often beneath a bush. Their breeding season begins in May. A clutch of 2 eggs is incubated by both sexes.

**Food Habits:** The Common Poorwill catches most of its food on the wing. It eats flying insects such as moths, grasshoppers, locusts, beetles, and chinch bugs. It sometimes takes insects on the ground.

**Ecology:** Little is reported about their activities. They are thought to have a breeding home range of 100 acres in California. Hawks and owls sometimes catch Common Poorwills, which are more nocturnal in their activities than Common Nighthawks. Common Poorwills have been seen drinking water while skimming over the surface.

**Comments:** In warmer parts of its range, this bird sometimes overwinters in a state of hibernation, rather than migrating south.

*References:* Fears 1975, Gilligan et al. 1994, Horn and Marshall 1975.

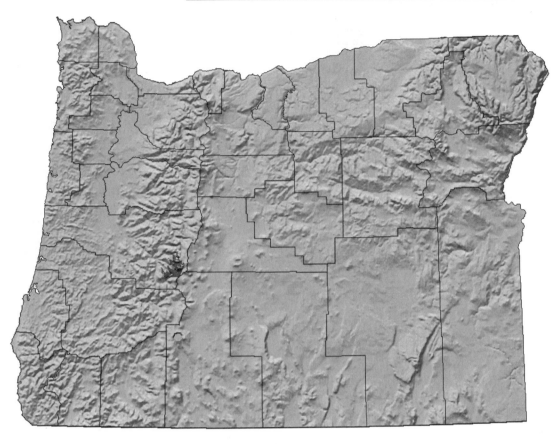

## Black Swift
*(Cypseloides niger)*

*Order: Apodiformes*
*Family: Apodidae*
*State Status: Sensitive*
*Federal Status: None*
*Global Rank: G4*
*State Rank: SU*
*Length: 7 in (18 cm)*

**Global Range:** The Black Swift has a scattered distribution in western North America and Central America. It breeds from southern Alaska south to California and east to Colorado and Utah. It also is found locally from Mexico south to Costa Rica, and on some Caribbean Sea islands.

**Habitat:** Black Swifts nest in cliff faces near or behind waterfalls. In western North America, these situations are usually in deep canyons in wooded areas.

**Reproduction:** Breeding season is likely to begin in June, and these swifts are on their way south by September. The species nests in small colonies. It builds a nest cup of mosses, ferns, and other handy plant matter. A single egg is laid. The difficulty in observing this species prevents a detailed knowledge of its breeding biology.

**Food Habits:** Like other swifts, the Black Swift catches insects in flight. It takes flying ants, termites, beetles, bees, flies, aphids, midges, and an occasional dispersing spider.

**Ecology:** Black Swift colonies generally consist of between 5 and 15 pairs. In other parts of the range, they use a variety of habitats, including sea cliffs and caves. They winter in Central America and on some Caribbean Sea islands.

**Comments:** Black Swifts were discovered during the breeding season in Oregon at Salt Creek Falls, Lane County. There are other sites in Oregon that qualify as breeding habitat but, even at the Salt Creek Falls site, it has not been possible to confirm breeding (eggs or nestlings seen). This is partly due to the inaccessibility of the nesting area.

*References:* Gilligan et al. 1994, Knorr 1961, Marshall 1992.

**181**

# Vaux's Swift
## (Chaetura vauxi)

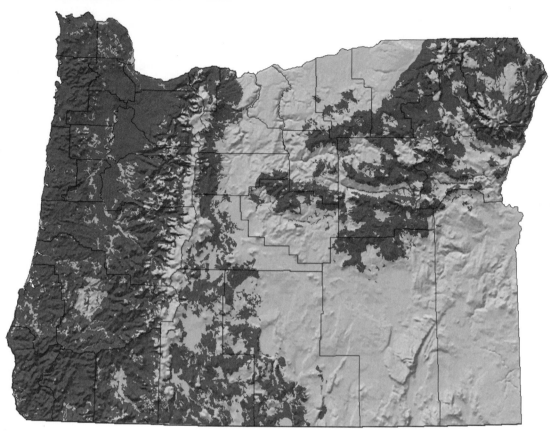

Order: Apodiformes
Family: Apodidae
State Status: None
Federal Status: None
Global Rank: G5
State Rank: S5
Length: 4.5 in (11 cm)

**Global Range:** Vaux's Swift has a disjunct distribution. Northern populations breed from southern Alaska south to central California, and east to the Rocky Mountains of Montana. The southern populations breed from southern Mexico south to northern South America.

**Habitat:** This swift is found in coniferous forests with large, usually hollow, trees that serve as nesting sites. It is known from Douglas-fir, ponderosa pine, Jeffrey pine, and mixed coniferous forests. Southern populations occupy broad-leaved forests.

**Reproduction:** Nests are often located inside a hollow tree, but are also found in abandoned woodpecker holes or snags. The nest is a mass of conifer needles and twigs. As with most other swifts, its nest is held together with hardened saliva. The clutch of 4 to 6 (range 3-7) eggs is incubated 18-20 days by both parents. Young fledge about a month after hatching. The breeding season begins in April.

**Food Habits:** This swift forages over the forest canopy, but more often over openings in the forest, or over water. It catches flying insects such as mosquitoes, gnats, small beetles, and flies. It can also skim aquatic insects from the surface of lakes or streams.

**Ecology:** Vaux's Swift nests alone or in a colony. Occasionally, it will nest in chimneys in urban areas. Most nesting occurs below 4,500 feet. During post-breeding dispersal it may occur in cities and suburbs.

**Comments:** This swift is more common in old-growth forests, presumably because of the availability of nest trees. Loss of older forests to timber harvest may reduce overall habitat for this bird in Oregon.

*References:* Baldwin and Zaczkowski 1963, Gabrielson and Jewett 1940, Gilligan et al. 1994, Lundquist and Mariani 1991.

*Order: Apodiformes*
*Family: Apodidae*
*State Status: None*
*Federal Status: None*
*Global Rank: G5*
*State Rank: S4?*
*Length: 6.5 in (17 cm)*

# White-throated Swift
*(Aeronautes saxatalis)*

**Global Range:** This is a bird of western North America and Central America. It breeds from southern British Columbia south through Mexico to Honduras. It occurs as far east as western Texas and Nebraska.

**Habitat:** The White-throated Swift specializes in nesting high on cliff faces. In forested mountains, it occupies cliffs above timberline. It also uses canyon walls in deserts. It forages widely over open country near the nest.

**Reproduction:** White-throated Swifts' breeding season begins about early May, and they migrate south as early as September. Their nest is a cup of feathers, plantdown, or other materials, gathered while flying. There are usually 4 or 5 eggs (range 3-6).

**Food Habits:** This swift has a diet of flying insects, including flies, beetles, flying ants, wasps, bees, true bugs, and leafhoppers.

**Ecology:** The White-throated Swift nests in small colonies, and returns to the same site in successive years. It forages considerable distances from the nest. Peregrine Falcons are a significant predator.

Occasionally, this swift will nest on the sides of tall buildings in urban areas.

**Comments:** The family name, "Apodidae," means "without feet." While swifts do have feet, they are not adapted to perching and the White-throated Swift carries out most of its activities, including mating, on the wing.

*References:* DeGraaf et al. 1991, Gilligan et al. 1994, Zeiner et al. 1990a.

# Black-chinned Hummingbird
### (Archilochus alexandri)

*Order: Apodiformes*
*Family: Trochilidae*
*State Status: None*
*Federal Status: None*
*Global Rank: G5*
*State Rank: S4*
*Length: 4 in (10 cm)*

**Global Range:** Breeds in western North America, from central Texas and northern Mexico north to southern British Columbia. In the northern part of their range, they are found as far east as northwestern Montana.

**Habitat:** In Oregon, the Black-chinned Hummingbird is found in riparian woodlands, wooded canyons, open ponderosa pine woodlands, and mountain chaparral of the eastern part of the state. They also can be found breeding in parks and small towns; they will frequently build a nest in a tree or shrub over a creek or dry creek bed.

**Reproduction:** This species arrives in Oregon by late May. The nest is a small woven cup, placed in a low tree or tall shrub. The clutch of 2 (range 1-3) eggs is incubated by the female for 13-16 days. The young are independent about a month after hatching. There may be more than one brood per year.

**Food Habits:** Like most hummingbirds, the Black-chinned Hummingbird is adapted to feeding on the nectar of flowers. It also eats some insects, found on foliage or caught on the wing. Although it nests in

fairly arid habitats, it must nest in restricted areas, like riparian woodland or canyons, where there is an adequate supply of nectar-producing plants.

**Ecology:** The male has little to do with raising the young. Females may nest in proximity to one another, due to limited habitat. This is described by some authors as "semicolonial" nesting.

**Comments:** This hummingbird is said to preferentially nest in oak trees.

*References:* DeGraaf et al. 1991, Gilligan et al. 1994, Zeiner et al. 1990a.

# Anna's Hummingbird
## *(Calypte anna)*

*Order: Apodiformes*
*Family: Trochilidae*
*State Status: None*
*Federal Status: None*
*Global Rank: G5*
*State Rank: S4?*
*Length: 4 in (10 cm)*

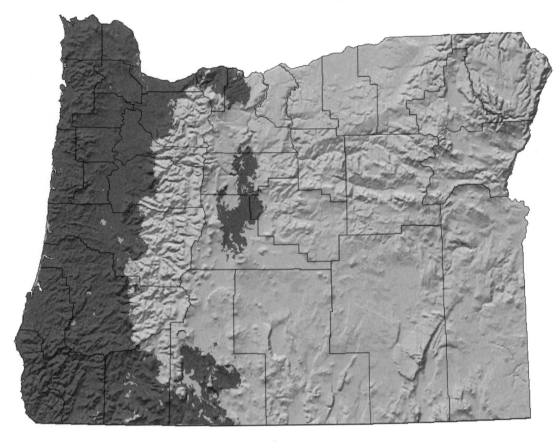

**Global Range:** Found on the west coast of North America, from southern British Columbia to northern Baja California. In the north, it breeds west of the Cascade-Sierra Nevada mountains, but in the south it breeds as far east as Arizona, and possibly west Texas.

**Habitat:** Anna's Hummingbird nests in chaparral-covered hillsides and canyons, sparse forests with open canopies, and residential and agricultural areas.

**Reproduction:** This hummingbird nests early in the year. Eggs were laid before February 1st in a 1986 nest observed in Medford, Jackson County. It builds a small cup nest of fine plant material and feathers. The female incubates the clutch of 2 (range 1-3) eggs for 16-17 days. There may be more than one brood per year.

**Food Habits:** Anna's Hummingbird feeds on nectar, especially from early-blooming currants and other berry-producing species. It also eats pollen, small insects, and sap.

**Ecology:** Male Anna's Hummingbirds defend a territory of up to a quarter of an acre, through aerial displays and dives. The territory usually includes rich food sources.

**Comments:** Most of our information on Anna's Hummingbird comes from studies in California, since the species was unknown to Oregon until 1958. It has rapidly extended its range north and is now well established. Growth of suburbs may have benefited this species.

*References:* Gilligan et al. 1994, Paczolt 1987, Powers 1987, Stiles 1973.

# Calliope Hummingbird
## (Stellula calliope)

Order: Apodiformes
Family: Trochilidae
State Status: None
Federal Status: None
Global Rank: G5
State Rank: S4?
Length: 3 in (8 cm)

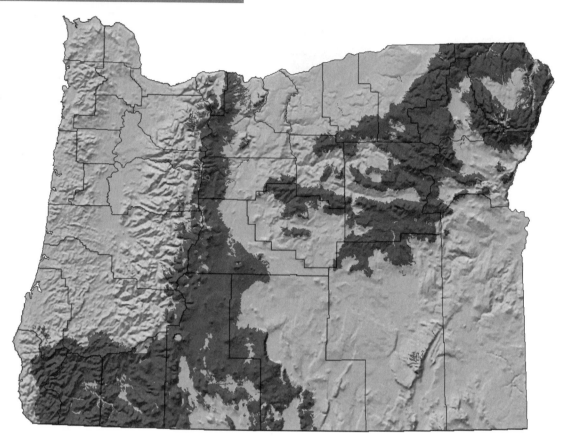

**Global Range:** This mountain hummingbird breeds from central British Columbia south to the Sierra San Pedro Mártir of northern Baja California. It extends its range as far east as Wyoming and Utah, but does not occupy the mountains of Arizona or New Mexico.

**Habitat:** The Calliope Hummingbird nests in riparian areas and open forests at the edges of meadows or alder, willow, or aspen thickets. It prefers forested areas with open canopies near water or wet meadows, where there is an abundance of flowering plants.

**Reproduction:** Calliope Hummingbirds return to Oregon in late April and early May. They usually nest in coniferous trees, selecting a branch 2 to 70 feet high. The single clutch of 2 eggs is incubated for about 15 days by the female. Young are fed until they leave the nest, at about 3 weeks of age.

**Food Habits:** This hummingbird has a diet of nectar, small insects, and spiders. It takes the nectar from manzanita, currant, gooseberry, paintbrush, columbine, penstemon, trumpet gilia, and elephant head flowers.

**Ecology:** One study found 4 territorial males in just over 6 acres of habitat. Males abandon the breeding areas while the females are still incubating eggs. Following the short breeding season, the species moves to higher elevations, possibly to seek late-flowering plants.

**Comments:** This species may breed in the open forests of the Siskiyou Mountains of southwestern Oregon.

*References:* Armstrong 1987, Calder 1971, Gilligan et al. 1994.

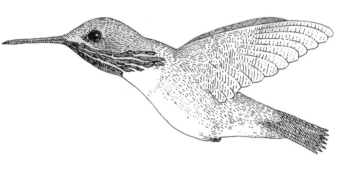

# Rufous Hummingbird
## (Selasphorus rufus)

*Order: Apodiformes*
*Family: Trochilidae*
*State Status: None*
*Federal Status: None*
*Global Rank: G5*
*State Rank: S4*
*Length: 3.5 in (9 cm)*

**Global Range:** Rufous Hummingbirds breed fairly far north for the family. They range from southern Alaska south to northern California and east to Yellowstone, on the Idaho-Wyoming border.

**Habitat:** In Oregon, this hummingbird occupies open forest edges near meadows and riparian thickets in mountainous areas, and scattered riparian woodlands in the lower, arid northeastern parts of the state. It will breed high in the mountains where meadows, thickets, and coniferous woodlands combine to provide good access to flowering plants.

**Reproduction:** The breeding season begins in April in warmer parts of western Oregon, but may be delayed until July at higher elevations. The nest is a cup made of plant down covered with moss or lichen, and often wrapped with spider webs. The female incubates the clutch of 2 (range 1-3) eggs. Nests may be close together, usually in branches of conifers, where there is a good food supply. Young fledge at about 3 weeks of age.

**Food Habits:** This hummingbird feeds on the nectar of flowers (various berries, penstemons, paintbrushes), and eats a few insects and spiders. Early in the year, it feeds on nectar from the early-appearing white flowers of madrone.

**Ecology:** Males defend small feeding territories (about 20 x 20 feet). Following the breeding season, but prior to migration, Rufous Hummingbirds move upslope. Females of this species tend to return to the same nest in subsequent years.

**Comments:** Rufous Hummingbirds have been observed drinking while hovering over water, or while sitting in water and dipping the bill.

*References:* Gass et al. 1976, Gilligan et al. 1994, Horvath 1964.

**187**

## Allen's Hummingbird
### (Selasphorus sasin)

*Order: Apodiformes*
*Family: Trochilidae*
*State Status: None*
*Federal Status: None*
*Global Rank: G5*
*State Rank: S3?*
*Length: 3.5 in (9 cm)*

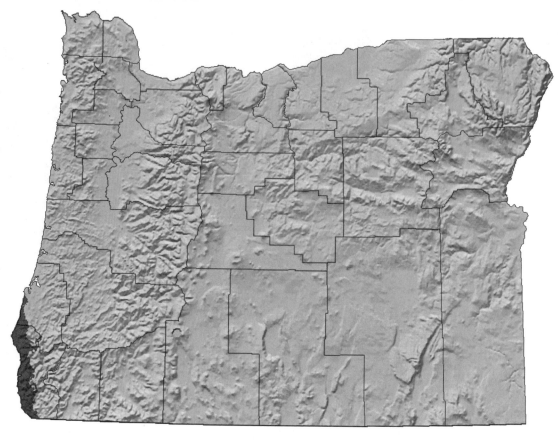

**Global Range:** Allen's Hummingbird has a relatively restricted range. It is found along the coast and on coastal slopes of mountains from Coos County, Oregon, south to Los Angeles County, California. It extends slightly farther inland up the valleys of the Klamath River system in northwestern California. It spends its winters in Mexico.

**Habitat:** This hummingbird breeds in coastal scrub, riparian thickets in moist canyon bottoms, and in the brushy edges, usually near meadows, of redwood and other coniferous forests. It also breeds in suburban gardens.

**Reproduction:** Breeding season begins in April, and most travel south by October. The nest is a cup made of moss and other plant material, lined with feathers, and placed in a branch of a tall shrub or on the limb of a tree. The clutch of 2 eggs is incubated by the

female for 16-22 days. In other areas it is double brooded, and may be in the temperate climate of southwestern Oregon.

**Food Habits:** Allen's Hummingbird takes nectar from the flowers of a number of plants, often with red flowers and trumpet-shaped corollas to accommodate its bill. It will also catch insects in flight, or glean insects and spiders from foliage.

**Ecology:** The nesting density is related to the quality of the food supply and can be quite high (several pairs per acre) in good habitat. Males defend a breeding territory early in the season, and females defend a feeding territory until eggs are laid.

**Comments:** This species is at the northern edge of its range in Oregon. Growth of suburbs around the towns along the southern Oregon coast may benefit Allen's Hummingbird. Most information on this species comes from California studies.

*References:* Gilligan et al. 1994, Legg and Pitelka 1956, Pitelka 1951a, Zeiner et al. 1990a.

## Belted Kingfisher
*(Ceryle alcyon)*

*Order: Coraciiformes*
*Family: Alcedinidae*
*State Status: None*
*Federal Status: None*
*Global Rank: G5*
*State Rank: S4*
*Length: 13 in (33 cm)*

**Global Range:** Breeds throughout most of the United States and Canada south of the high Arctic. It occurs from central Alaska and northern Canada south to southern California, northern Arizona and New Mexico, and central Florida. It winters from the southern United States south to northern South America.

**Habitat:** As its name implies, the Belted Kingfisher catches fish (among other things) in rivers and lakes and is therefore always near water, both inland and along the coast. Since it digs its own nest burrow, it is not associated with any particular terrestrial vegetation cover type.

**Reproduction:** This species digs a burrow several feet deep in a natural embankment or road cut. Its breeding season starts in late April or early May. A clutch of 6 to 8 (range 5-14) eggs is incubated by both parents for 23-24 days. Young leave the nest about 5 weeks after hatching.

**Food Habits:** Small fish (3-4 inches long) make up about 90% of the kingfisher's diet. Other adult foods include crayfish, amphibians, mice, molluscs, lizards, and fruit. The young are fed insects.

**Ecology:** Kingfishers often perch over water looking for prey. They defend a territory around the perch, and are usually spaced one-half to 2 miles apart along rivers or the shoreline. They usually feed within a mile of their nest, but have been known to fly as far as 5 miles for food.

**Comments:** Belted Kingfishers can become pests around fish hatcheries.

*References:* Brooks and Davis 1987, DeGraaf et al. 1991, Kilham 1974a.

**189**

# Lewis' Woodpecker
## (Melanerpes lewis)

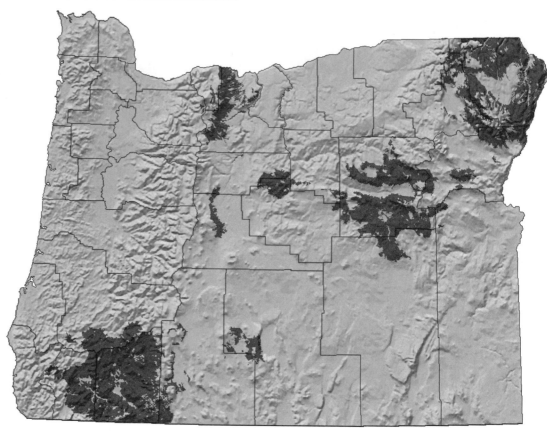

*Order: Piciformes*
*Family: Picidae*
*State Status: Sensitive*
*Federal Status: None*
*Global Rank: G5*
*State Rank: S4*
*Length: 11 in (28 cm)*

**Global Range:** The Lewis' Woodpecker is a bird of western North America. It breeds from central British Columbia south to central California and New Mexico. The eastern edge of its breeding range barely enters southern South Dakota. In winter, it occurs as far south as northern Mexico.

**Habitat:** This woodpecker is associated with open forests, often at lower elevations. It nests in Oregon white oak woodlands, ponderosa pine woodlands, mixed oak-pine woodlands, and cottonwood riparian woodlands of the river valleys of eastern Oregon.

**Reproduction:** The Lewis' Woodpecker can excavate its own nest chamber, usually in a dead or decayed tree, but prefers to use an existing abandoned woodpecker hole. It begins breeding in late April and May. Both male and female incubate the clutch of 6 to 7 (range 5-9) eggs for about 2 weeks. Young fledge about a month after hatching.

**Food Habits:** The diet varies by season. In spring and summer, it mostly eats insects (beetles and their larvae, ants, grasshoppers, flies, and tent caterpillars) and spiders. The diet turns to acorns and berries (currants, cherries, serviceberry, poison oak) in the fall. It stores acorns for winter consumption, usually under the bark or in crevices of trees, or in cracks of utility poles.

**Ecology:** Lewis' Woodpeckers defend a territory around the nest and feeding area. One study found the average territory was about 15 acres. They catch flying insects by hawking, and also glean insects from bark and foliage. Following breeding season, they move upslope, looking for berries and insects.

**Comments:** This species has declined in numbers in Oregon, the result of loss of nesting and food storage trees, and increased competition for nest cavities from introduced European Starlings. In Malheur County, Lewis' Woodpeckers breed at scattered locations where there are suitable riparian forests.

*References:* Bock 1970, Bock and Brennan 1987, Gilligan et al. 1994, Marshall 1992.

## Acorn Woodpecker
### (*Melanerpes formicivorus*)

*Order: Piciformes*
*Family: Picidae*
*State Status: None*
*Federal Status: None*
*Global Rank: G5*
*State Rank: S3?*
*Length: 9.5 in (24 cm)*

*References:* Gilligan
et al. 1994,
Gutierrez and
Koenig 1978,
Koenig and
Mumme 1987,
MacRoberts and
MacRoberts 1976,
Marshall 1992,
Roberts 1979.

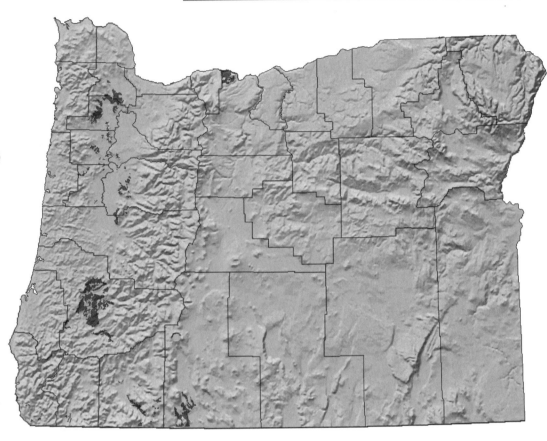

**Global Range:** Acorn Woodpeckers reach the northern limit of their distribution in Washington County, Oregon. They are found south through most of California, into northern Baja California, from Arizona and New Mexico south through Mexico and Central America, to the northern Andes of Colombia.

**Habitat:** Acorn Woodpeckers live where oaks grow, and occur both in oak savannas and in open oak-conifer woodlands. In Oregon, they inhabit Oregon white oak communities, with or without co-occurring ponderosa pine. Other coniferous and broad-leaved trees are often present.

**Reproduction:** This woodpecker is unusual because it lives in groups of 2-16 birds, out of which only a single female lays eggs. Group members are related, and cooperatively raise the offspring. They nest in cavities, which they dig themselves. The clutch of 4 to 6 (range 4-10) eggs is incubated by most members of the group for about 14 days. Young fledge in about a month, and most remain with the group.

**Food Habits:** The Acorn Woodpecker's main food item is acorns, which are eaten green in summer and stored for winter food. Storage trees may hold thousands of acorns. It will also hawk insects in the spring, and feed on sap and bast (the cambium layer under the bark of trees) during winter. Fruits and other seeds are also eaten.

**Ecology:** Each cooperatively breeding group defends a territory around its acorn storage trees. Territories are usually around 10-20 acres in size. Acorn Woodpeckers are permanent residents whose territory size may change from year to year, depending on the abundance of acorns.

**Comments:** Although still common in parts of its Oregon range, this woodpecker is dependent on large oak trees, which are disappearing because of clearing, firewood harvest, and invasion by Douglas-fir in response to fire suppression. There is a small, isolated population near The Dalles, Wasco County.

191

# Red-naped Sapsucker
### (Sphyrapicus nuchalis)

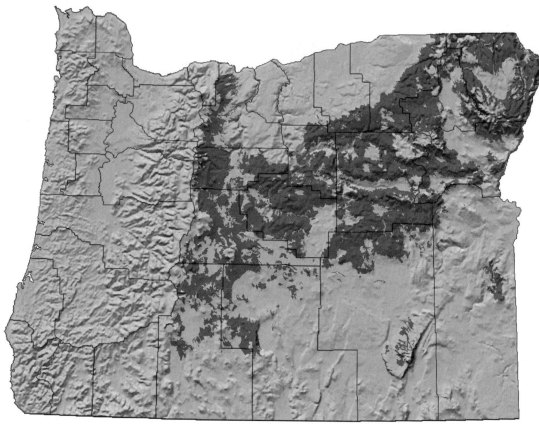

*Order: Piciformes*
*Family: Picidae*
*State Status: None*
*Federal Status: None*
*Global Rank: G5*
*State Rank: S4*
*Length: 8.5 in (22 cm)*

**Global Range:** Breeds from southern British Columbia and Alberta south to Nevada, northern Arizona, and southern New Mexico.

**Habitat:** This sapsucker inhabits a variety of coniferous forest communities within which there are stands of quaking aspen. In mountains, it also uses riparian woodlands of willow and other deciduous trees.

**Reproduction:** The breeding season for this woodpecker begins in late April or May. The nest is a cavity excavated in a tree; aspens are the preferred nest tree, but ponderosa pines are also used. The clutch of 4 or 5 (range 4-7) eggs is incubated for about 2 weeks by both parents. The young fledge in about a month but are fed by their parents for another 2 weeks.

**Food Habits:** The Red-naped Sapsucker drills a series of holes in compact rows, usually in deciduous trees, from which it eats cambium and sap that flows from the wound. During the breeding season, it also eats insects, especially ants, attracted to the sap. It flies out from perches to catch flying insects. It also gleans foliage for caterpillars, beetles, and spiders, and eats fruit when available.

**Ecology:** Nests are spaced from 300 to 1,500 feet apart in the forest, indicating a territory of from 2-15 acres. Its sapwells are defended not only from other Red-naped Sapsuckers, but also from hummingbirds, warblers, and even chipmunks. A few birds overwinter in Oregon, but most depart for southern California and Mexico by October.

**Comments:** This species was formerly considered a subspecies of the Yellow-bellied Sapsucker (*S. varius*) of central and eastern North America (American Ornithologists' Union 1985).

*References:* Crockett and Hadlow 1975, Dobkin et al. 1995, Gilligan et al. 1994, Howell 1952, Johnson and Johnson 1985, Zeiner et al. 1990a.

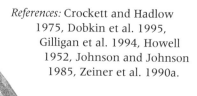

*Order: Piciformes*
*Family: Picidae*
*State Status: None*
*Federal Status: None*
*Global Rank: G5*
*State Rank: S4*
*Length: 9 in (23 cm)*

**Global Range:** Breeds in the Pacific Coast region of North America. It is found as far north as south-central Alaska, and is resident from western British Columbia to northern California. It also breeds farther south in the Sierra Nevada and southern California mountain ranges, and winters south to northern Baja California.

**Habitat:** The Red-breasted Sapsucker is western Oregon's equivalent to the Red-naped Sapsucker. It is found in coniferous forests or mixed coniferous-broad-leaved forests, within which grow stands of deciduous trees, especially along streams or next to meadows. It also visits fruit and nut orchards.

**Reproduction:** This sapsucker digs its own nest cavity in a variety of trees, preferring alder, birch, cottonwood, aspen, or willow, but also using coniferous species. The breeding season varies with elevation, but usually commences in April. The clutch of 4 or 5 (range 3-7) eggs is incubated by both sexes for 12-14 days.

**Food Habits:** Like other sapsuckers, the Red-breasted Sapsucker drills a grid pattern of holes in the trunks of deciduous trees, eating the cambium and sap, as well as insects that are caught in flight over meadows and lakes. It eats berries and other fruits when they are available.

**Ecology:** Red-breasted Sapsuckers defend a territory of up to 15 acres against both conspecifics and other species (warblers, hummingbirds) attracted to its network of sapwells. Its nest cavities are important to secondary cavity nesters, who do not excavate their own nest holes.

**Comments:** This species was once considered a subspecies of the Yellow-bellied Sapsucker. It is sometimes considered an orchard pest, but its holes cause little damage to fruit or nut trees.

*References:* Gilligan et al. 1994, Johnson and Zink 1983, Lawrence 1967.

# Williamson's Sapsucker
## (Sphyrapicus thyroideus)

*Order: Piciformes*
*Family: Picidae*
*State Status: Sensitive*
*Federal Status: None*
*Global Rank: G5*
*State Rank: S4*
*Length: 9.5 in (24 cm)*

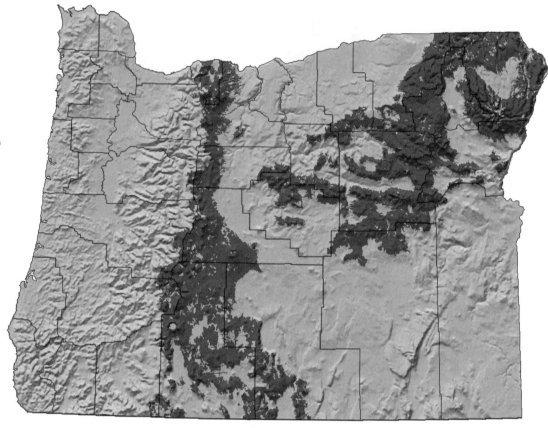

**Global Range:** This western sapsucker breeds from southern British Columbia south to the mountains of southern California and northern New Mexico. It is found as far east as the Rocky Mountains of Colorado.

**Habitat:** Williamson's Sapsucker distinguishes itself from our other sapsuckers by its use of mature, higher-elevation coniferous forests for nesting and feeding. In Oregon, it prefers open ponderosa pine forests, but may also use lodgepole pine, red fir, grand fir, subalpine spruce, Douglas-fir, and aspen forests. It may breed in riparian thickets within coniferous forest mosaics.

**Reproduction:** Williamson's Sapsucker begins breeding in May. It digs its own nest cavity, usually in the soft wood of a diseased or dead conifer. Pairs may return to the same tree in successive years, but excavate a new nest each year. The clutch of 5 or 6 (range 3-7) eggs is incubated 12-14 days by both parents. Young fledge in 4 or 5 weeks.

**Food Habits:** This sapsucker drills its holes in coniferous trees (ponderosa pine, lodgepole pine, true fir), as well as in aspen. It eats sap and cambium, as well as insects (like ants), which are the primary food for the nestlings. It also feeds on spruce budworm larvae and moths, and gleans some other insects. When they are available, it eats berries.

**Ecology:** An area of 10 to 20 acres is defended, most intensely near the nest. Nest holes are placed from 5 to 60 feet above the ground. Most Williamson's Sapsuckers leave the state in winter.

**Comments:** It sometimes interbreeds with the Red-naped Sapsucker, and may defend its territory against other sapsucker species.

*References: Crockett and Hadlow 1975, Gabrielson and Jewett 1940, Gilligan et al. 1994, Marshall 1992.*

194

# Downy Woodpecker
## *(Picoides pubescens)*

*Order: Piciformes*
*Family: Picidae*
*State Status: None*
*Federal Status: None*
*Global Rank: G5*
*State Rank: S4*
*Length: 6.5 in (17 cm)*

**Global Range:** Breeds coast to coast, from Alaska and central Canada south to southern California and Florida. Absent from the deserts of southeastern California east to western Texas.

**Habitat:** These small woodpeckers use a wide range of wooded habitats, tending to prefer broad-leaved trees and riparian areas, but also use mixed coniferous-deciduous forests and, sometimes, coniferous forests. They may be found in suburbs and city parks.

**Reproduction:** The Downy Woodpecker excavates its own nest cavity in a dead or diseased tree trunk, from 4 to 50 feet high, often in a riparian forest with ample cover. Nesting begins in late April and May. The usual clutch of 4 or 5 (range 3-7) eggs is incubated by both sexes for 11-12 days.

**Food Habits:** Downy Woodpeckers dig for insects and insect larvae in tree trunks or under flakes of bark. They also glean insects from leaves and twigs of shrubs. Occasionally, they fly out to catch insects on the wing. They also eat some fruits, seeds, and acorns.

**Ecology:** Like most woodpeckers, Downy Woodpeckers announce their territorial limits by drumming on resonant parts of trees in the breeding season. Territories are usually smaller than 10 acres. This species remains abundant in Oregon through the winter. It does not breed in higher-elevation forests, but may move upslope in the fall in search of food, only to seek shelter in valleys when winter arrives.

**Comments:** Although the Downy Woodpecker is found throughout the state, it is present in arid deserts and grasslands only where there are patches of trees, such as in riparian thickets or around towns.

*References:* DeGraaf et al. 1991, Jackson 1970, Zeiner et al. 1990a.

# Hairy Woodpecker
## (Picoides villosus)

*Order: Piciformes*
*Family: Picidae*
*State Status: None*
*Federal Status: None*
*Global Rank: G5*
*State Rank: S4*
*Length: 9.5 in (24 cm)*

**Global Range:** Breeds from southern Alaska through central Canada south, coast to coast, to southern California and Florida, continuing south, in the mountains and highlands of Mexico and Central America, to western Panama.

**Habitat:** This woodpecker inhabits coniferous forests, mixed coniferous-deciduous forests, and riparian woodlands at all elevations. Unlike its smaller relative, the Downy Woodpecker, it does not occur in isolated riparian habitats of arid southeastern Oregon. However, it too can be found in wooded suburbs and urban areas. It is more common in older forests, but readily uses burned areas and forest edges for foraging.

**Reproduction:** The Hairy Woodpecker digs its own nest in a dead or diseased tree. Its breeding season begins in mid-March to early April. The clutch of about 4 eggs (range 3-6) is incubated by both male and female (with the male taking the night shift) for 11-

12 days. The young become independent about 6 weeks after hatching.

**Food Habits:** Male Hairy Woodpeckers tend to dig larger insects from tree trunks away from the nest, while females glean smaller insects from the surface and under bark close to the nest. They forage on the ground. They also eat ants, caterpillars, spiders, millipedes, aphids, berries, and seeds, including acorns. Seeds can be an important winter food.

**Ecology:** In eastern Oregon, territories as large as 25 acres may be defended against many species of woodpeckers. The females may remain on the territory throughout the year, being joined by the male in the spring.

**Comments:** This woodpecker is common in burned-over areas or in concentrations of dead or insect-infested trees. It may be important in keeping bark beetle populations under control.

*References:* Gilligan et al. 1994, Kilham 1965, Lawrence 1967.

# White-headed Woodpecker
## (Picoides albolarvatus)

*Order: Piciformes*
*Family: Picidae*
*State Status: Sensitive*
*Federal Status: None*
*Global Rank: G5*
*State Rank: S3*
*Length: 9 in (23 cm)*

**Global Range:** The White-headed Woodpecker has a relatively small range in the Pacific states. It is resident from southern British Columbia south to northern California in the Coast Ranges and Cascade Mountains. It is found farther south in the Sierra Nevada and Transverse Ranges of California, nearly to the Mexican border. It occurs as far east as the mountains of north-central Idaho.

**Habitat:** In Oregon, this species is closely associated with ponderosa pine or ponderosa pine-mixed conifer forests. It requires large trees for foraging and snags for nesting, which are characteristic of older forests. In other parts of its range, it inhabits true fir, sugar pine, and Jeffrey pine forests.

**Reproduction:** Breeding season begins in April or May, and young are fledged by August. It digs its own nest cavity, usually in a dead tree. Both parents incubate the clutch of 4 or 5 (range 3-7) eggs for about 2 weeks.

**Food Habits:** This woodpecker has a diet that varies seasonally. In late spring and summer, it eats mostly insects, but in winter and early spring, it mainly feeds on seeds of ponderosa pine.

**Ecology:** An Oregon study found that White-headed Woodpeckers spent most of their time foraging in trees larger than 20 inches in diameter. The home range size was measured at 250-500 acres. Nest trees averaged 18 inches in diameter and were usually located at the edge of a clearing.

**Comments:** Loss of mature ponderosa pine forests has resulted in a severe decline of this species in the Blue Mountains of Oregon.

*References:* Gilligan et al. 1994, Ligon 1973, Marshall 1992, Milne and Hejl 1989.

197

# Three-toed Woodpecker
*(Picoides tridactylus)*

*Order: Piciformes*
*Family: Picidae*
*State Status: Sensitive*
*Federal Status: None*
*Global Rank: G5*
*State Rank: S3*
*Length: 9.5 in (24 cm)*

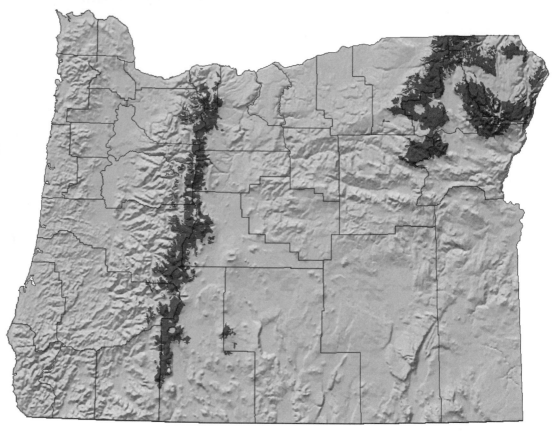

**Global Range:** Found throughout most of Alaska and Canada, but its range extends south only in the mountainous areas of the western conterminous United States. Its southern limit is the mountains of eastern Arizona and western New Mexico. It is also found in Eurasia, from northern Scandinavia to Siberia, and as far south as Mongolia and Japan.

**Habitat:** In Oregon, this species is closely associated with higher-elevation (above 4,500 feet) forests of grand fir-lodgepole pine, lodgepole pine, or lodgepole pine mixed with other conifers. Elsewhere in its wide distribution, it also lives in true fir and spruce forests. Even in mixed forests, it spends most time foraging on older lodgepole pine trees.

**Reproduction:** Breeding begins in mid- to late May. It excavates its own nest cavity in dead or diseased lodgepole pine trees which are at least 11 inches in diameter. The clutch of 4 (range 3-6) eggs is incubated for 2 weeks by both parents.

**Food Habits:** This woodpecker feeds mostly on wood-boring larvae of moths and beetles, which it captures by probing dead or decaying wood. It also gleans some insects (ants, caterpillars), and eats fruits, mast, and cambium.

**Ecology:** Pairs of this species may nest in close proximity to one another. The home range in Oregon ranged from 100 to over 700 acres. It often congregates in areas of insect outbreaks or in recently burned stands. In winter, it may be seen at lower elevations.

**Comments:** The loss of mature lodgepole pine forest habitat, essential to this naturally rare species, may lead to its decline in Oregon.

*References:* Bock and Bock 1974, Gilligan et al. 1994, Marshall 1992.

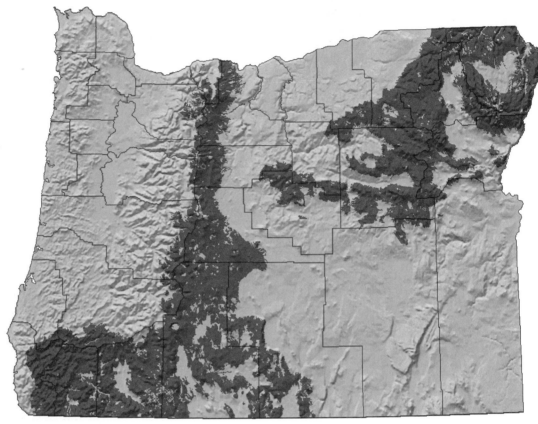

## Black-backed Woodpecker
*(Picoides arcticus)*

*Order: Piciformes*
*Family: Picidae*
*State Status: Sensitive*
*Federal Status: None*
*Global Rank: G5*
*State Rank: S3*
*Length: 9.5 in (24 cm)*

**Global Range:** Resident from southern Alaska across timbered Canada to Newfoundland. It ranges south to New York and the Upper Midwest, and, in the mountainous West, as far south as the central Sierra Nevada of California.

**Habitat:** This species inhabits a broader range of forest types than the Three-toed Woodpecker. It generally is found at lower elevations where the two species are sympatric, and lives in forests dominated by lodgepole pine or ponderosa pine, which may be mixed with other species (western larch, true firs, Engelmann spruce). It selects lodgepole pine, ponderosa pine, and western larch for its nest holes.

**Reproduction:** Breeding season begins in late April and early May. It excavates its own nest cavity in dead or diseased lodgepole pines, which are usually less than 20 inches in diameter. The clutch of 4 (range 2-6) eggs is incubated by both sexes for about 14 days. Nest trees are usually located at the edges of clearings, and near water.

**Food Habits:** About three-quarters of the Black-backed Woodpecker's diet consists of wood-boring beetles and their larvae. It also eats other beetles, and gleans ants and spiders from bark and foliage. It eats some fruit, acorns, and cambium, depending on the season.

**Ecology:** During breeding season, a pair requires about 100 acres around the nest; however this may vary with the food supply. In Oregon, one study found that the average home range was about 400 acres. This species spends most of its time foraging in mature lodgepole pine stands with many dead or dying trees.

**Comments:** Forestry practices that replace older forest stands (which provide food) with younger forests reduce habitat for this species.

*References:* Bock and Bock 1974, Gilligan et al. 1994, Marshall 1992, Short 1974.

**199**

# Northern Flicker
### (Colaptes auratus)

*Order: Piciformes*
*Family: Picidae*
*State Status: None*
*Federal Status: None*
*Global Rank: G5*
*State Rank: S5*
*Length: 13 in (33 cm)*

**Global Range:** While several species of flickers were once recognized, they now are considered varieties of the Northern Flicker, which breeds from Alaska and Canada south through North America to the highlands of Mexico and Central America.

**Habitat:** The Northern Flicker is an adaptable generalist, found in nearly every terrestrial habitat type in Oregon. It prefers open habitats and is rare in dense coniferous forests, although clear-cuts may be occupied. It is found in deserts, especially along riparian woodland, near agriculture, or in towns. It uses urban and suburban areas, but any open forest or woodland, especially with patches of broad-leaved trees, will be occupied.

**Reproduction:** Breeding season begins in April, and eggs are laid through late June. The Northern Flicker usually digs its own nest cavity in a tree, but sometimes, especially in eastern Oregon, it will dig a hole in a dirt embankment. The clutch of 6 to 8 (range 3-14) eggs is incubated by both parents for about 11-12 days. Young are independent about a month after hatching.

**Food Habits:** The Northern Flicker's diet is diverse and changes with the season. Just over half its diet consists of insects, and it especially favors ants. It searches the ground for other insects (beetles, grasshoppers, moths, insect larvae, caterpillars), and for seeds, berries, nuts, and acorns. In winter, it eats berries and fruits.

**Ecology:** Unlike most other woodpeckers, the Northern Flicker spends much of its time searching for food on the ground. It can dig insects from trees in traditional woodpecker fashion, but ants, its main prey item, are found on foot. It will also fly from a perch and catch insects on the wing. One study found it defended a breeding territory of 40 acres.

**Comments:** This is one of the most frequently seen and familiar birds in the state.

*References:* Anderson 1972, Gilligan et al. 1994, Kerpez and Smith 1990.

# Pileated Woodpecker
## *(Dryocopus pileatus)*

*Order: Piciformes*
*Family: Picidae*
*State Status: Sensitive*
*Federal Status: None*
*Global Rank: G5*
*State Rank: S4?*
*Length: 18 in (46 cm)*

*References:* Bull 1987, Bull and Jackson 1995, Gilligan et al. 1994, Marshall 1992, Mellen et al. 1992.

**Global Range:** The Pileated Woodpecker is a bird of the eastern United States, Canada, and the mountains and forests of the Northwest, extending south into the Coast Ranges and Sierra Nevada of northern and central California.

**Habitat:** In the western part of its range, the Pileated Woodpecker is associated with forest habitats that have large trees, especially snags, for nesting and foraging. It will use both coniferous and deciduous trees, but it tends to be most common in old-growth Douglas-fir forests in western Oregon and old-growth ponderosa pine-mixed conifer forests in eastern Oregon.

**Reproduction:** The Pileated Woodpecker is a large bird that excavates a large hole (8 inches around and 22 inches deep) in trees and snags that are at least 2 or 3 feet in diameter. It begins breeding around April, and the clutch of 4 or 5 (range 3-8) eggs is incubated by both parents for about 12 days. The family group stays together for some time after the young fledge.

**Food Habits:** Carpenter ants, which it finds in decaying wood, are the main food item of this woodpecker. It will also eat larvae of wood-boring beetles, termites, berries, and acorns. More vegetable matter is taken in winter.

**Ecology:** Pileated Woodpeckers have large home ranges, over 1,000 acres. The male and female of a pair have home ranges that do not completely overlap. This species may forage in open areas, but it requires forests over 70 years old for nesting. It disperses after breeding, and may be seen in a variety of wooded situations, including wooded suburbs.

**Comments:** Conversion of older forests, with snags and downed logs, to even-aged, short-rotation tree plantations, is reducing habitat for this species in Oregon.

# Olive-sided Flycatcher
## *(Contopus borealis)*

*Order: Passeriformes*
*Family: Tyrannidae*
*State Status: None*
*Federal Status: SC*
*Global Rank: G4*
*State Rank: S4*
*Length: 7.5 in (19 cm)*

*References:* Gabrielson and Jewett 1940, Gilligan et al. 1994, Hagan and Johnson 1992, Tvrdik 1971.

**Global Range:** Breeds from Alaska across Canada and south to North Carolina in the East, and the mountains of northern Baja California, Arizona, and New Mexico in the West. It is absent from the Great Plains and from most of the southeastern United States.

**Habitat:** In Oregon, the Olive-sided Flycatcher is a bird of coniferous forests. It perches on tall trees or snags that provide an overview of its territory. As a result, it prefers forests with an uneven canopy, especially those with tall snags left over from fires. While open forests are preferred, it occupies a variety of forest types, from sea level to subalpine environments.

**Reproduction:** Its breeding season commences in May, and by October it has begun its journey south. The nest, a cup of plant material, is typical for passerine birds. The clutch of 3 or 4 eggs is incubated by the female for 16-17 days. The young are independent in 2 to 3 weeks.

**Food Habits:** Like most flycatchers, this one flies out from a perch to catch insects in flight. It eats bees, flying ants, flies, small beetles, mosquitoes, and any other small flying insect that approaches.

**Ecology:** Despite its conspicuous territorial advertisement call, surprisingly little is known about the spatial requirements of this species.

**Comments:** U.S. Fish and Wildlife Service Breeding Bird Survey data indicate this species is declining throughout the West.

# Western Wood-Pewee
## *(Contopus sordidulus)*

*Order: Passeriformes*

*Family: Tyrannidae*

*State Status: None*

*Federal Status: None*

*Global Rank: G5*

*State Rank: S4*

*Length: 6 in (15 cm)*

**Global Range:** Breeds throughout most of western North America, from as far north as south-central Alaska, south to the highlands of Mexico and Central America. It occurs as far east as western South Dakota and west Texas.

**Habitat:** Western Wood-Pewees are tree-living birds that breed nearly anywhere in Oregon where a stand of trees is to be found. This includes open coniferous forests and forest edges, broad-leaved and mixed forests and woodlands, including ponderosa pine and oak woodlands, trees in towns and suburbs, and, in eastern Oregon, riparian woodlands and aspen groves. Rarely, they breed in juniper woodlands or thickets of mountain mahogany.

**Reproduction:** The Western Wood-Pewee begins breeding in May, and most have left Oregon by the end of September. The clutch of 3 eggs (range 2-4) is laid in a cup woven of plant material, and incubated by the female for about 12 days.

**Food Habits:** This flycatcher takes most, but not all, of its food on the wing. It eats flying insects (bees, wasps, ants, flies, craneflies, moths, mayflies, dragonflies, lacewings, beetles, mosquitoes, and termites), but also some spiders and caterpillars. A few seeds and berries are also included in its diet.

**Ecology:** Western Wood-Pewees defend a territory of a few acres in most habitats, but territories can be more compact along favorable riparian woodlands. Although it breeds in high-elevation forests, it is relatively rare above 5,500 feet in the Cascade Mountains.

**Comments:** This species is a neotropical migrant, spending its winters in South America.

*References:* DeGraaf et al. 1991, Gilligan et al. 1994, Verbeek 1975.

# Willow Flycatcher
## *(Empidonax traillii)*

*Order: Passeriformes*
*Family: Tyrannidae*
*State Status: None*
*Federal Status: SC*
*Global Rank: G5*
*State Rank: S4*
*Length: 6 in (15 cm)*

**Global Range:** Breeds coast to coast, from southernmost Canada south to southern California, New Mexico, and Georgia. It winters in Mexico and Central America.

**Habitat:** This flycatcher is found in willows at the edges of streams flowing through meadows and marshes, but also breeds in thickets along the edges of forest clearings and, generally, in tall, brushy vegetation in the vicinity of water. It should not be expected in grasslands and desert valleys of eastern Oregon in the absence of these microhabitat elements, although it does use vegetation around springs and seeps in desert mountain ranges.

**Reproduction:** This is a late breeder, not laying eggs until June, presumably waiting for the peak abundance of flying insects. A cup nest is built low in a shrub or small tree. A clutch of 3 to 4 (range 2-4) eggs is incubated by the female for 12-15 days. Young fledge at about 2 weeks of age.

**Food Habits:** Willow Flycatchers have a fairly routine diet for their family, eating mostly flying insects, especially wasps, but including flies, mosquitoes, ants, beetles, bees, grasshoppers, and dragonflies. They glean some spiders, seeds, and berries from foliage.

**Ecology:** Willow Flycatchers defend a breeding territory of about 1 to 3 acres. A study in the Oregon coast range found a density of one bird per 2 acres in 10-year-old regenerating clear-cuts.

**Comments:** Although declining and a species of concern in California, the Willow Flycatcher is reported to be fairly common in Scotch broom and Himalayan blackberry thickets growing around the edges of farms in the foothills of the Willamette Valley (D. B. Marshall, personal communication). It was formerly considered conspecific with the Alder Flycatcher (*Empidonax alnorum*), which breeds in Alaska and Canada, north of the range of the Willow Flycatcher.

*References:* Ashmole 1968, Ettinger and King 1980, Gilligan et al. 1994, Prescott and Middleton 1988.

*Order: Passeriformes*
*Family: Tyrannidae*
*State Status: None*
*Federal Status: None*
*Global Rank: G5*
*State Rank: SU*
*Length: 5 in (13 cm)*

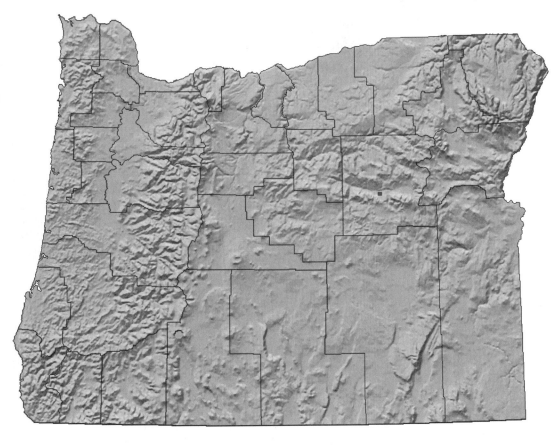

**Global Range:** Breeds from the Yukon east to the Atlantic Coast and south to Montana and Virginia. Its Oregon breeding locality is outside of the normal range of the species.

**Habitat:** The Least Flycatcher is resident along brushy edges of open forests, often near a lake or stream. It also occupies agricultural areas, parks, and suburban gardens. Given its limited occurrence in Oregon, it is impossible to generalize about its habitat use in our state.

**Reproduction:** The small cup-shaped nest is placed on a tree branch at varying heights. In other parts of the range, it begins breeding in May or June. The clutch of 4 (range 3-6) eggs is incubated by the female for 14-16 days. The young are independent about 2 weeks after hatching.

**Food Habits:** Like other *Empidonax* flycatchers, this species primarily eats insects caught by sallying forth from a perch. It will also glean some food (e.g., insects, spiders) from foliage. It eats some seeds and small berries.

**Ecology:** All sources indicate this flycatcher is partial to open deciduous or mixed deciduous-coniferous forests. This may explain its absence from coastal British Columbia and the coastal forests of the Pacific Northwest.

**Comments:** Gilligan et al. (1994) confirm one breeding locality for this flycatcher in Oregon, Clyde Holliday State Wayside in Grant County, where it nested in 1995 (A. Contreras, personal communication). The continued presence of the Least Flycatcher as a breeding member of Oregon's avifauna is uncertain.

*References:* DeGraaf et al. 1991, Gilligan et al. 1994, Godfrey 1966, Sherry and Holms 1988.

# Hammond's Flycatcher
## *(Empidonax hammondii)*

*Order: Passeriformes*
*Family: Tyrannidae*
*State Status: None*
*Federal Status: None*
*Global Rank: G5*
*State Rank: S4*
*Length: 5.5 in (14 cm)*

*References:* Beaver and Baldwin 1975, Gilligan et al. 1994, Mannan 1984, Manuwal 1970, Sakai and Noon 1991.

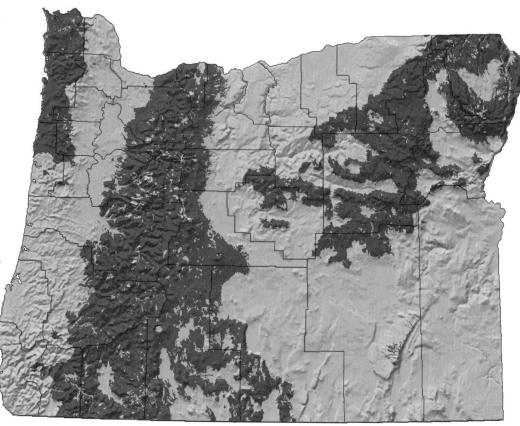

**Global Range:** Breeds from central Alaska south along the coast and interior mountains to the southern Sierra Nevada in California, the Rocky Mountains of northern New Mexico, and mountain ranges of the Intermountain West.

**Habitat:** This flycatcher occupies mid- to high-elevation coniferous forests. It seems to prefer dense, well-watered forests. It was found to be associated with older Douglas-fir/ponderosa pine forests in the interior West. It occupies all the forest types on the west slope of the Cascade Mountains, and may breed where dense forests occur in the valleys and slopes of the Coast Ranges. Small (1994) suggests this species is associated with red fir in California, and this may apply to populations in the red fir zone of the Siskiyou and southern Cascade mountains of Oregon.

**Reproduction:** Like most tyrannid flycatchers that depend on an abundance of insects, Hammond's Flycatcher begins its breeding season late (early June), and leaves for its wintering grounds in Mexico

and Central America by September. Its cup nest is placed on a fairly high (25-40 feet) tree limb. The clutch of 3 or 4 (range 2-5) eggs is incubated for about 15 days by the female. Young are not independent until about 3 weeks after hatching.

**Food Habits:** This flycatcher has a rather specialized foraging niche. It perches in the middle or upper parts of tall conifers or aspens, and flies out to catch insects in the openings between trees. Its food is almost entirely composed of flying insects (mosquitoes, moths, ants, beetles, bees, wasps, and others).

**Ecology:** Hammond's Flycatcher often selects areas near streams for nesting. It defends a territory of several acres around its nest. Depending on habitat, singing males of this species occur at densities of 5-20 per 100 acres.

**Comments:** In the northern part of its range, this species tends to nest in deciduous trees, but in the Northwest it is found in conifers, mainly at higher elevations.

## Dusky Flycatcher
*(Empidonax oberholseri)*

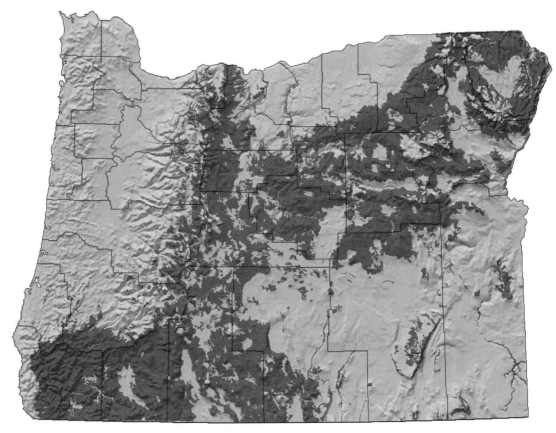

*Order: Passeriformes*
*Family: Tyrannidae*
*State Status: None*
*Federal Status: None*
*Global Rank: G5*
*State Rank: S4*
*Length: 6 in (15 cm)*

**Global Range:** The Dusky Flycatcher's range is similar to that of Hammond's Flycatcher. It breeds from the southern Yukon south in the western United States to southern California and northern New Mexico. It is found east as far as western South Dakota. It winters in Mexico and Central America.

**Habitat:** In contrast to Hammond's Flycatcher, the Dusky Flycatcher prefers drier, open forests and woodlands with a shrub understory, or simply patches of mountain chaparral. It also can be found in juniper woodland, and in isolated quaking aspen groves in sagebrush country. The brushy regrowth in clear-cuts is used by this species.

**Reproduction:** This species places its cup nest relatively low to the ground (4-7 feet) in a tree or in a shrub. It begins breeding in late May or early June, and young are fledged by late July. The clutch of 3 or 4 (range 2-6) eggs is incubated by both parents for 12-15 days. The young fledge in about 18 days.

**Food Habits:** This flycatcher flies low over thickets of shrubs in search of flying insects. It also gleans insects from leaves. Among others, it eats flies, mosquitoes, small moths, flying ants, and beetles.

**Ecology:** Two studies found territories of the Dusky Flycatcher to be between 3 and 6 acres. In the mountains of eastern Oregon, this flycatcher uses stands of mountain mahogany.

**Comments:** Aside from its preference for a different habitat, it is difficult to distinguish the Dusky Flycatcher from Hammond's Flycatcher in the field.

*References:* DeGraaf et al. 1991, Eckhardt 1977, Gilligan et al. 1994.

# Gray Flycatcher
### *(Empidonax wrightii)*

*Order: Passeriformes*
*Family: Tyrannidae*
*State Status: None*
*Federal Status: None*
*Global Rank: G5*
*State Rank: S4*
*Length: 6 in (15 cm)*

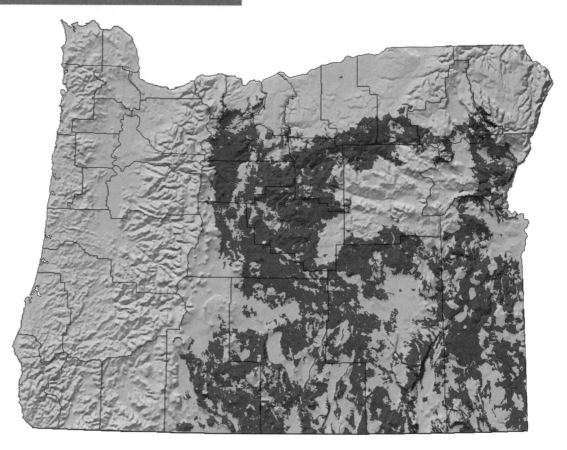

**Global Range:** The Gray Flycatcher is a bird of the arid interior West. It breeds from eastern Washington south to southern California and east to Colorado and New Mexico.

**Habitat:** The habitat preferences of the Gray Flycatcher resemble those of the Dusky Flycatcher, except that it prefers relatively treeless areas with tall sagebrush, bitterbrush, or mountain mahogany communities, but will also occupy these communities within open forests of ponderosa or lodgepole pine. It also lives in juniper woodland with a sagebrush understory. It is usually found below 6,000 feet.

**Reproduction:** The Gray Flycatcher begins breeding around early June and leaves for its wintering grounds in Arizona and Mexico by the end of September. Its nest cup is placed near the trunk of a good-sized bush. The clutch of 3 or 4 eggs is incubated by the female for 2 weeks. Young are independent of their parents at 1 month of age.

**Food Habits:** Appropriate to its preference for brushy vegetation, it both catches insects on the wing between bushes and catches them on the ground. Included in its diet are a variety of species ranging from small beetles to butterflies.

**Ecology:** This species is most abundant in extensive tracts of big sagebrush, often selecting areas along washes where the sagebrush is especially large. The territory has been reported to vary from 3 to 9 acres, and the home range seems to be about 10 acres.

**Comments:** About 20-30% of Gray Flycatcher nests in one Oregon study were parasitized by the Brown-headed Cowbird.

*References:* Friedmann et al. 1977, Gilligan et al. 1994, Johnson 1963, 1966; Zeiner et al. 1990a.

## Pacific-slope Flycatcher
*(Empidonax difficilis)*
## Cordilleran Flycatcher
*(Empidonax occidentalis)*

*Order: Passeriformes*
*Family: Tyrannidae*
*State Status: None*
*Federal Status: None*
*Global Rank: G5*
*State Rank: S4(S?)*
*Length: 6 in (15 cm)*

*Distribution map for Pacific-slope Flycatcher*

**Global Range:** The Pacific-slope Flycatcher breeds from coastal southeastern Alaska and British Columbia south to Baja California del Norte, usually west of the crest of the Cascade and Sierra Nevada mountains. The Cordilleran Flycatcher breeds from eastern Washington south to the highlands of Mexico, and east to the eastern foothills of the Rocky Mountains and west Texas (American Ornithologists' Union 1989). In Oregon, the Cordilleran Flycatcher breeds east of the crest of the Cascade Mountains (although a few have been recorded on the west slope of Mt. Hood). These species were formerly considered subspecies of the Western Flycatcher (*E. difficilis*) (American Ornithologists' Union 1989).

**Habitat:** In western Oregon, the Pacific-slope Flycatcher is found in every forest type from the coast to mid-elevations (about 4,000 feet) of the Cascade Mountains. On the east slope of the Cascade Mountains and in the Blue Mountains, the Cordilleran Flycatcher seeks out moister, denser parts of the forests in shaded canyon bottoms, often near streams.

**Reproduction:** These flycatchers have a short breeding season in Oregon, generally lasting from late May to late August. Nest sites are varied, but are usually shaded, and can be in a tree, woodpecker hole, under bark, or in a rock crevice. The clutch of 4

**209**

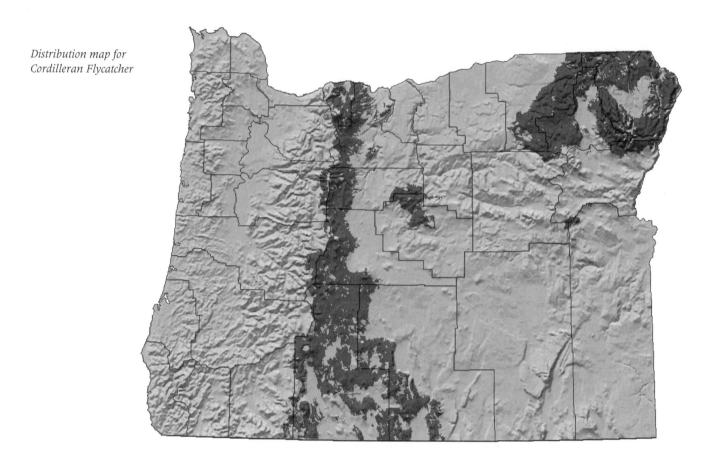

*Distribution map for Cordilleran Flycatcher*

(range 3-5) eggs is incubated by the female for 14-15 days. Young are independent in about a month.

**Food Habits:** The diet includes a mix of flying insects (e.g., ants, wasps, bees, flies, beetles, moths), as well as some invertebrates gleaned from foliage (caterpillars, budworms, spiders). A limited number of seeds and berries are also eaten.

**Ecology:** Densities of these flycatchers vary with habitat, but can reach one per 3 acres. Foraging is usually done in the lower to middle canopies of trees. Along the Oregon coast, they are most common in willow and alder thickets along stream courses.

**Comments:** There is little information about the ecological differences between these two species formerly known as the Western Flycatcher. The State Rank of the Cordilleran Flycatcher has not been determined.

*References:* Beaver and Baldwin 1975, Davis et al. 1963, Gilligan et al. 1994, Johnson 1980, Johnson and Marten 1988, Sakai and Noon 1991, Verbeek 1975.

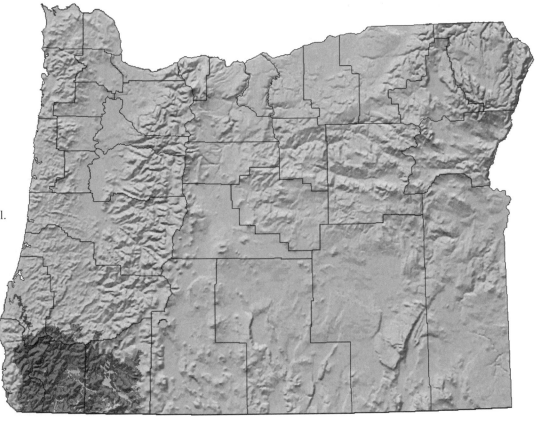

*Order: Passeriformes*

*Family: Tyrannidae*

*State Status: None*

*Federal Status: None*

*Global Rank: G5*

*State Rank: S3*

*Length: 6.5 in (17 cm)*

*References:* Gilligan et al. 1994, Verbeek 1975, Zeiner et al. 1990a.

**Global Range:** The Black Phoebe reaches the northern limit of its rather extensive range in southwestern Oregon. It is found south through most of cismontane California, and east to west Texas. Farther south, it breeds in the highlands of Mexico and Central America, and into South America, where it occurs in temperate areas as far south as northern Argentina.

**Habitat:** Black Phoebes are unusual flycatchers that build a nest of mud and plant material attached to a protected cliff face or rock outcrop. As a consequence, they are almost always found near water. They prefer riparian areas with thick vegetation and some vertical surface nearby. Bridges are now often used for nest sites, as well as thickets near ponds. They need open areas for foraging, so are absent from dense forests.

**Reproduction:** Information from other regions indicates the Black Phoebe's breeding season starts in March. A clutch of 4 or 5 (range 3-6) eggs is incubated by the female for 15-18 days. The young leave the nest in 3 weeks, and may continue to be fed by the male while the female incubates a second clutch.

**Food Habits:** Black Phoebes perch on low branches at the edge of open areas, often over water, and sally forth to catch flying insects. They eat beetles, bees, wasps, flying ants, flies, and moths. They also glean insects (e.g., caterpillars) from vegetation, and eat some fruit when it is available. They have been reported to catch small fish by diving into the water.

**Ecology:** Black Phoebes sometimes winter on their breeding range in Oregon. Males and females are thought to maintain separate feeding territories in winter. The same nest may be used year after year.

**Comments:** This species is apparently expanding its range northward along the southern Oregon coast. A substantial number of Black Phoebes winter in the lower Coquille River Valley, Coos County.

# Say's Phoebe
## *(Sayornis saya)*

*Order: Passeriformes*
*Family: Tyrannidae*
*State Status: None*
*Federal Status: None*
*Global Rank: G5*
*State Rank: S4?*
*Length: 7 in (18 cm)*

**Global Range:** Breeds in the western half of North America, from Alaska to central Mexico. It is not found farther east than the Dakotas.

**Habitat:** This phoebe is a bird of open, often arid, country. In Oregon, it is found in desert and grassland communities, breeding near a canyon, cliff face, or (more recently) farm buildings and towns. It nests in protected crevices in vertical surfaces. Nonbreeders may frequent broad, flat valleys covered with juniper, sagebrush, or salt desert shrub communities.

**Reproduction:** The breeding season for Say's Phoebes begins in late March, when they return to Oregon. The nest cup is made of plant material and other fibers, but not with mud. The clutch of 4 or 5 (range 3-7) eggs is incubated by the female for about 2 weeks. The young leave the nest in 2 weeks, and are fed by the male while the female re-nests.

**Food Habits:** Say's Phoebes feed on flying insects, especially grasshoppers, but including bees, wasps, flying ants, and flies. They also glean some spiders and caterpillars from vegetation, and eat berries when they are available.

**Ecology:** The territory size varies with habitat, but can be as large as 100 acres. This species winters in the desert southwest and in Mexico.

**Comments:** The species is probably nest-site limited in the arid deserts and grasslands of eastern Oregon. A few Say's Phoebes winter in Oregon, mainly in the Rogue River Valley and the Klamath Basin.

*References:* DeGraaf et al. 1991, Gilligan et al. 1994, Zeiner et al. 1990a.

# Ash-throated Flycatcher
## (Myiarchus cinerascens)

*Order: Passeriformes*
*Family: Tyrannidae*
*State Status: None*
*Federal Status: None*
*Global Rank: G5*
*State Rank: S4?*
*Length: 8 in (20 cm)*

**Global Range:** Breeds in the western United States and Mexico, from southeastern Washington south to Baja California, and in the central Mexican highlands. It occurs as far east as Texas, but does not cross the Continental Divide north of southern Wyoming.

**Habitat:** This is a bird of oak, juniper, and riparian woodlands. Near the northern edge of its distribution, it is found in Oregon white oak-ponderosa pine woodlands in Wasco County. In eastern Oregon, it occurs in wooded canyons and valleys.

**Reproduction:** Ash-throated Flycatchers have a short breeding season in Oregon, lasting from May to early September. Atypical for a flycatcher, this species is usually a cavity nester, placing its nest cup in a natural opening in a tree or in a woodpecker hole, sometimes evicting its original occupant. The normal clutch of 4 or 5 (range 3-7) eggs is incubated by the female for 15 days. Young leave the nest in about 16-17 days.

**Food Habits:** Like many flycatchers, this one flies out from a perch to catch flying insects (bees, wasps, beetles, ants, moths, grasshoppers) on the wing. It also gleans some spiders and caterpillars from leaves. Insects make up 92% of the diet, but it also eats some berries.

**Ecology:** The breeding territory varied between about 5 and 20 acres in several studies. In linear riparian habitat, the density can be quite high (50-150 per square mile), but this is misleading since the territorial males are spaced along the river. This species winters in Mexico and Central America.

**Comments:** In Oregon, the Ash-throated Flycatcher is most common in oak woodlands of the Rogue River Valley.

*References:* Gilligan et al. 1994, Harrison 1978, Zeiner et al. 1990a.

213

# Western Kingbird
## (Tyrannus verticalis)

*Order: Passeriformes*
*Family: Tyrannidae*
*State Status: None*
*Federal Status: None*
*Global Rank: G5*
*State Rank: S5*
*Length: 9 in (23 cm)*

**Global Range:** This is basically a bird of the western part of the conterminous United States. It breeds from extreme southern Canada to northern Mexico, and ranges from the Pacific states to east Texas and Wisconsin. It is absent from the immediate area of the Pacific Coast.

**Habitat:** The Western Kingbird is an inhabitant of open country with scattered trees. It will nest in isolated cottonwoods, trees around ranches in arid sagebrush, or alkali desert communities, as well as in oak woodland and grassland.

**Reproduction:** This flycatcher usually builds its large, cup-shaped nest 15-40 feet up a fairly large tree, but sometimes nests are placed in a bush. It begins breeding around May, and departs to the south in September. The clutch of 4 (range 3-7) eggs is incubated for 12-14 days.

**Food Habits:** Western Kingbirds sit on low perches and fly out to catch insects. They can often be seen foraging from the top strand of barbed wire fences. The diet is 90% insect (bees, wasps,

beetles, flies, grasshoppers), but they also eat spiders and some fruit.

**Ecology:** Although kingbird nests are often spaced about one-quarter of a mile apart, it is not unusual for 2 or more pairs to nest in the same tree. They actively defend their nests from crows, jays, and raptors. Their reproductive success is tied to insect abundance.

**Comments:** Western Kingbirds occasionally breed in deciduous woodland along the Columbia River, as far west as Columbia County.

*References:* Dick and Rising 1965, Gilligan et al. 1994, Hespenheide 1964.

# Eastern Kingbird
## (Tyrannus tyrannus)

*Order: Passeriformes*
*Family: Tyrannidae*
*State Status: None*
*Federal Status: None*
*Global Rank: G5*
*State Rank: S4*
*Length: 8 in (20 cm)*

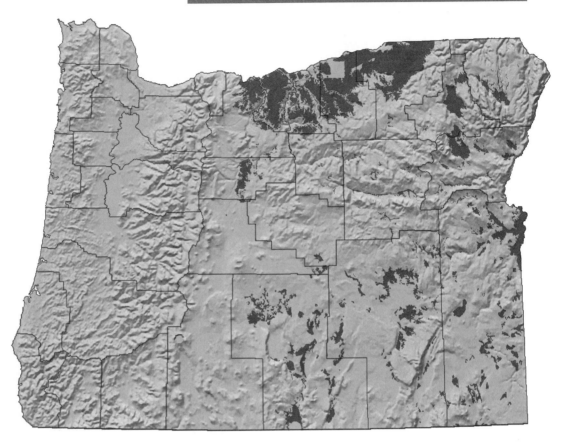

**Global Range:** Despite its name, the Eastern Kingbird breeds throughout most of Canada south of the Yukon, and across most of the United States, except for the Pacific Coast and the arid Southwest. It does not extend its breeding range south into Mexico, but it winters from southern Mexico to Central America.

**Habitat:** In Oregon, the Eastern Kingbird breeds in trees that border open fields and agricultural land. It is not a bird of coniferous forests or treeless desert valleys, but uses riparian woodlands, scattered cottonwoods, orchards, and trees along lake shores or roads for nesting.

**Reproduction:** Like most flycatchers, Eastern Kingbirds arrive in Oregon fairly late, in the last part of May, perhaps to insure a good supply of insects. The cup nest is usually in a tree or shrub, from 2 to 60 feet high. A clutch of 3 or 4 (range 3-5) eggs is incubated for 12-13 days by the female. Young are fed by both parents for over a month after hatching.

**Food Habits:** The Eastern Kingbird catches most of the insects it eats on the wing, but it will also alight on the ground to take insects exposed by plowing. It takes some insects from the surface of water. It eats a wide variety of berries, which it picks from bushes while hovering.

**Ecology:** Like the Western Kingbird, this species chases corvids and raptors from its nest. It also defends its nest from Western Kingbirds. It frequently nests in riparian vegetation alongside or over water. It does not nest in the same tree as other pairs of Eastern Kingbirds.

**Comments:** Although this is primarily a bird of eastern Oregon, Gilligan et al. (1994) cite a 1993 breeding record for the Eastern Kingbird along the Sandy River, Multnomah County.

*References:* Davis 1955, DeGraaf et al. 1991, Gilligan et al. 1994, Zeiner et al. 1990a.

# Horned Lark
## (Eremophila alpestris)

*Order: Passeriformes*
*Family: Alaudidae*
*State Status: Sensitive*
*Federal Status: None*
*Global Rank: G5*
*State Rank: S5*
*Length: 7 in (18 cm)*

**Global Range:** The Horned Lark has a cosmopolitan distribution, breeding throughout Eurasia, in North Africa, and in northern South America. In North America, it breeds from the arctic coasts of Alaska and Canada south to southern Mexico. It is absent from the extreme southeastern United States.

**Habitat:** Horned Larks are rather unusual birds because of their preference for nesting where there is little or no vegetation. In Oregon, this includes agricultural areas, pastures, grasslands (both native and introduced), sparsely vegetated desert shrublands, the margins of playas, and alpine areas.

**Reproduction:** The Horned Lark breeds from March through June, raising 2 or more broods. Its nest is a depression in the ground, with the top of the eggs coming to ground level. Pebbles may be placed around the nest rim. The clutch of 4 (range 2-7) eggs is incubated by the female for 10-14 days. Young fly within 3 weeks of hatching.

**Food Habits:** The Horned Lark's diet varies with the season, including insects (beetles and their larvae, caterpillars, grasshoppers, ants, wasps, leafhoppers), molluscs, and spiders in spring and summer, and seeds of grasses and forbs and waste grain in winter.

**Ecology:** While breeding, the Horned Lark's territory has been reported as 1-12 acres. They form flocks after the breeding season. In some areas they are considered agricultural pests.

**Comments:** Conversion of sagebrush to crested wheatgrass in eastern Oregon has probably increased habitat for this species. On the other hand, the bird has declined in the Willamette Valley in the last several decades. Only the Willamette Valley population is state listed as sensitive.

*References:* Beason and Franks 1974, Gilligan et al. 1994, Littlefield 1990, Marshall 1992, Medin 1990, Trost 1972, Verbeek 1967.

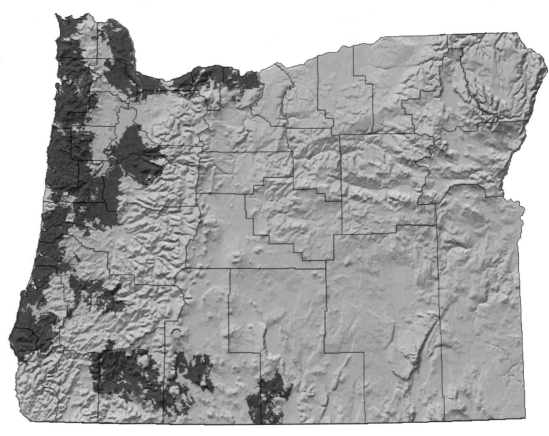

## *Purple Martin*
### *(Progne subis)*

*Order: Passeriformes*
*Family: Hirundinidae*
*State Status: Sensitive*
*Federal Status: None*
*Global Rank: G5*
*State Rank: S3*
*Length: 8 in (20 cm)*

**Global Range:** Breeds from central Canada south to the highlands of central Mexico. Its distribution is fairly continuous in the eastern United States, but it is absent from much of the arid West.

**Habitat:** Purple Martins require the juxtaposition of appropriate nesting habitat (holes in trees, nest boxes) with open areas for foraging. Such situations are often found near the shores of lakes or rivers, in unsalvaged tracts of forest killed by fire, near large meadows, and in cities and towns. They will use open forests or woodlands, but lack of nest trees keeps them from arid grasslands and deserts.

**Reproduction:** This swallow may nest singly or in colonies. The nest cavity is often an abandoned Northern Flicker or Pileated Woodpecker hole or an artificial nest box; those birds using a nest box tend to be most often observed. It begins breeding in late May or early June. The clutch of 4 or 5 (range 3-8) eggs is incubated by the female for 15-18 days. Young leave in about 1 month, but return to the natal nest hole to roost, and return to the same colony the following year to breed.

**Food Habits:** Purple Martins, like all swallows, catch most of their food on the wing. A few insects and spiders may be picked up from the ground, but they mainly eat flying ants, wasps, bees, mosquitoes, beetles, dragonflies, moths, butterflies, and grasshoppers.

**Ecology:** The Purple Martin defends only the entrance to its nest hole. The species is thought to be declining in Oregon because of diminishing nest sites, as snags in burned areas are cleared. It also faces competition for nest holes from introduced species (European Starlings, House Sparrows).

**Comments:** A 1977 survey found only 168 pairs breeding in Oregon.

*References:* Gilligan et al. 1994, Marshall 1992, Morton and Derrickson 1990, Richmond 1953, Zeiner et al. 1990a.

**217**

# Tree Swallow
## *(Tachycineta bicolor)*

*Order: Passeriformes*
*Family: Hirundinidae*
*State Status: None*
*Federal Status: None*
*Global Rank: G5*
*State Rank: S5*
*Length: 5.5 in (14 cm)*

**Global Range:** Breeds from northern Alaska and Canada south through most of the United States. It is absent from southern Arizona east through most of the Gulf Coast states and Georgia.

**Habitat:** Tree Swallows are closely associated with water, including marshes, farm ponds, lakes, and rivers. They also require trees with nesting holes near, over, or in the water. They forage over water or other open areas, and do not use dense forests, although meadows and ponds within a forest matrix are acceptable.

**Reproduction:** This species begins its breeding season in Oregon around early April, and the young are fledged by July. Tree Swallows nest in an abandoned woodpecker hole or a natural tree cavity. The usual clutch is 4 to 6 (range 3-7) eggs and is incubated for 13-16 days. Young leave the nest by 4 weeks of age.

**Food Habits:** Flying insects—including ants, beetles, flies, bees, wasps, moths, dragonflies, and grasshoppers—make up about 80% of the diet. The Tree Swallow will feed on seeds (bulrush, sedge, smartweed) or berries when insects are scarce.

**Ecology:** Tree Swallows nest in close proximity to one another, and defend only the nest hole. They will use nest boxes or holes in buildings if suitable tree holes are unavailable. This species winters from the southern edge of the United States south along coastal areas of the Caribbean Sea and the Pacific Coast of Mexico.

**Comments:** This swallow is common around the lakes of the high Cascade Mountains.

*References:* DeGraaf et al. 1991, Gabrielson and Jewett 1940, Gilligan et al. 1994.

# Violet-green Swallow
## (Tachycineta thalassina)

*Order: Passeriformes*
*Family: Hirundinidae*
*State Status: None*
*Federal Status: None*
*Global Rank: G5*
*State Rank: S5*
*Length: 5.5 in (14 cm)*

**Global Range:** Breeds only in western North America, from central Alaska south through the United States to the highlands of Mexico. It is found as far east as western South Dakota and western Texas.

**Habitat:** Although this species is similar in appearance to the Tree Swallow, it is much more versatile in use of habitats. It nests in abandoned woodpecker holes, but also in crevices of cliffs in desert areas and in similar niches in buildings. It is common in urban areas, agricultural regions, and in open forests and woodlands, such as the ponderosa pine forests of eastern Oregon. It forages over clearings, like meadows in forests or over open water, but is not strongly associated with aquatic habitats.

**Reproduction:** Breeding season for this species begins in late April or early May. The normal clutch of 4 to 5 (range 4-7) eggs is incubated for about 2 weeks. Young remain in the nest for 23-25 days, and may continue to be fed on the wing by either parent for some time after fledging.

**Food Habits:** The Violet-green Swallow is apparently a strict insectivore, feeding on flies, ants, leafhoppers, wasps, bees, moths, beetles, midges, mayflies, and dragonflies, all caught in flight. It has been reported to sometimes feed on accumulations of insects on the ground.

**Ecology:** Where nest sites are abundant, Violet-green Swallows may nest in loose colonies. Only the nest itself is defended as a territory. There are reports of densities ranging from 3 to 63 pairs per 100 acres.

**Comments:** This swallow nests from sea level, where it uses cliffs overlooking the ocean, to timberline.

*References:* Brawn and Balda 1988, DeGraaf et al. 1991, Gilligan et al. 1994.

# Northern Rough-winged Swallow
### (Stelgidopteryx serripennis)

Order: Passeriformes
Family: Hirundinidae
State Status: None
Federal Status: None
Global Rank: G5
State Rank: S4
Length: 5.5 in (14 cm)

**Global Range:** Breeds from central British Columbia east across the northern United States to New England and south throughout the United States to the highlands of Mexico and Central America.

**Habitat:** This swallow digs its own burrow in dirt banks, and uses all open habitat types, including open woodlands and deserts, where an appropriate bank is present. Many occur along stream banks, but it will nest up to one-half mile from water. It is found at the mouths of streams along the Oregon Coast, and as high as 5,000 feet in mountains.

**Reproduction:** It begins breeding in Oregon in April, and most leave by October. The clutch of 6 or 7 (range 4-8) eggs is laid in a chamber, lined with plant material, at the end of a burrow that can be up to 6 feet long. Only the female incubates the eggs for 15-16 days. Young are independent about 3 weeks after hatching.

**Food Habits:** The Northern Rough-winged Swallow eats mostly flying insects, taken on the wing. These include flies, moths, beetles, ants, bees, wasps, and dragonflies. It may pick insects from the surface of the water while in flight.

**Ecology:** This is not a colonial swallow. It may nest alone or, because of limited nesting habitat, several pairs may nest in the same bank, but usually no closer than about 5 meters apart. This species winters from the southern United States south to Panama.

**Comments:** This swallow will dig holes in road cuts, and it may have increased in numbers with the coming of the automobile.

*References:* DeGraaf et al. 1991, Gilligan et al. 1994, Lunk 1962, Zeiner et al. 1990a.

*Order: Passeriformes*
*Family: Hirundinidae*
*State Status: Sensitive*
*Federal Status: None*
*Global Rank: G5*
*State Rank: S4*
*Length: 5 in (13 cm)*

**Global Range:** This is a holarctic species that breeds in Eurasia, from the British Isles to Siberia, and south to the Mediterranean and Japan. In North America, it breeds from Alaska across Canada, and south to California, southern New Mexico, and South Carolina.

**Habitat:** The Bank Swallow, like the Northern Rough-winged Swallow, digs its nest tunnels in dirt embankments, and it has similar habitat requirements. It can be found in open habitat types (grasslands, desert scrub, agricultural areas, pastures) near a nesting site. Areas in proximity to water are preferred, which means it is locally distributed in arid eastern Oregon.

**Reproduction:** The breeding season begins in April, and young are fledged by July. The nest is dug by both adults in friable soil, ending in a chamber lined with plant material. The clutch of 4 or 5 (range 3-7) eggs is incubated for 12-16 days by both sexes. Young fledge in about 3 weeks.

**Food Habits:** Bank Swallows catch their food while flying over water,

pastures, or other open areas. Flies are an important item in their diet, but they also eat moths, wasps, bees, beetles, ants, termites, leafhoppers, aphids, mosquitoes, craneflies, and dragonflies.

**Ecology:** This species is a colonial nester, and hundreds of pairs nest in a suitable embankment. The immediate area of the nest entrance is defended, and tunnel entrances are as close as 6 to 8 inches from one another. By mid-September, Bank Swallows have left for their winter quarters in Panama and northern South America.

**Comments:** Gilligan et al. (1994) report a single colony west of the Cascade Mountains, on the Chetco River, Curry County. Marshall (1992) summarizes the information on 12 known active colonies in Oregon; however the species may be more common in eastern Oregon than these records indicate.

*References:* Beecher and Beecher 1979, DeGraaf et al. 1991, Gilligan et al. 1994, Marshall 1992, Peterson 1955, Zeiner et al. 1990a.

# Cliff Swallow
### (Hirundo pyrrhonota)

*Order: Passeriformes*
*Family: Hirundinidae*
*State Status: None*
*Federal Status: None*
*Global Rank: G5*
*State Rank: S5*
*Length: 5.5 in (14 cm)*

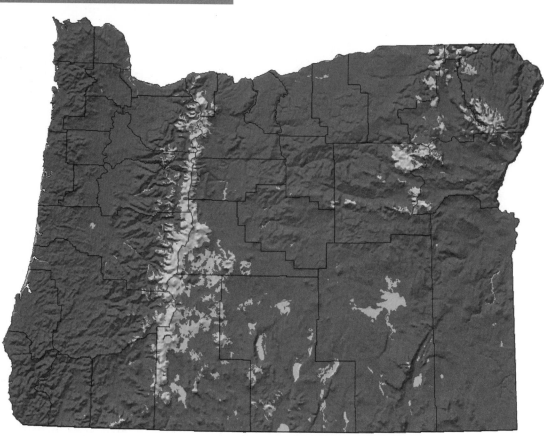

**Global Range:** Breeds throughout most of North America, from Alaska east across Canada and south to central Mexico and the Gulf Coast of Mexico. It is absent from the southeastern United States, and is declining in the Northeast.

**Habitat:** This is another bird of open habitats, but these include mountain canyons and open forests. Its primary requirements are a protected, nearly vertical, cliff on which to build its nest, and the proximity of a mud source. Water must be available nearby through the nesting season. It will use pastures, agricultural fields, freeway overpasses, and urban areas, in addition to natural communities.

**Reproduction:** This swallow is strongly colonial (colonies may contain thousands of birds), and builds conical mud nests. The breeding season begins in April in Oregon. A clutch of 4 or 5 (range 3-6) eggs is incubated, mostly by the female, for about 16 days. Young are independent in about 4 weeks.

**Food Habits:** The Cliff Swallow catches insects in flight, usually at a considerable height (100 or more feet above the surface). It requires a source of clean water for drinking. Insects caught include beetles, ants, wasps, bees, flies, moths, mosquitoes, and grasshoppers. Most foraging takes place a few hundred yards from the colony, but it may forage 2 to 4 miles away.

**Ecology:** Nest entrances are spaced about 4 inches apart. Colonies decline if unseasonably cold weather reduces the insect supply in spring or early summer. This species leaves Oregon for South America in July and August.

**Comments:** The Cliff Swallow has no doubt expanded its range in Oregon through the use of human structures as nest sites.

*References:* Behle 1976, Brown and Brown 1988, DeGraaf et al. 1991, Gilligan et al. 1994, Mayhew 1958.

# Barn Swallow
*(Hirundo rustica)*

*Order: Passeriformes*
*Family: Hirundinidae*
*State Status: None*
*Federal Status: None*
*Global Rank: G5*
*State Rank: S5*
*Length: 7 in (18 cm)*

*References:* DeGraaf et al. 1991, Gilligan et al. 1994, Snapp 1976.

**Global Range:** This is a cosmopolitan species. It breeds across Eurasia, south to the Mediterranean Sea and Japan. These populations winter in Africa, Australia, and the Pacific islands. In North America, it breeds from coastal Alaska across Canada and south to central Mexico. It is absent only from parts of Florida, Georgia, and South Carolina. Most North American breeders winter in South America.

**Habitat:** Barn Swallows use open forests and woodlands, as well as open habitats like grassland, pasture, agricultural areas, marshes, lakes, rivers, and suburbs. They require a protected surface for nest building, and a source of mud for nests, and are therefore restricted to places near water.

**Reproduction:** Barn Swallows have a longer breeding season than most of Oregon's other swallows, and may raise a second brood. They nest alone or in small groups, beginning breeding in April or May. The clutch of 4 or 5 (range 3-8) eggs is incubated, mostly by the female, for 14-16 days. Young fly in 3 to 4 weeks, but often remain with the parents to help raise the second brood.

**Food Habits:** Most food consists of insects caught on the wing, including flies, beetles, moths, ants, wasps, bees, crickets, dragonflies, leafhoppers, and grasshoppers. It may drink while skimming the surface of water. Rarely, it alights to eat some seeds or berries.

**Ecology:** Barn Swallows defend the nest and an area a few feet away. During breeding season, their foraging usually takes place within a few hundred yards of the nest. The pair bond may last for several breeding seasons, and pairs often use the same nest in successive years.

**Comments:** This is another swallow whose numbers have increased as human structures are used for nest sites.

# Gray Jay
## (Perisoreus canadensis)

*Order:* Passeriformes
*Family:* Corvidae
*State Status:* None
*Federal Status:* None
*Global Rank:* G5
*State Rank:* S4
*Length:* 12 in (30 cm)

*References:* Gilligan et al. 1994, Ha and Lehner 1990, Ouellet 1970, Rutter 1969.

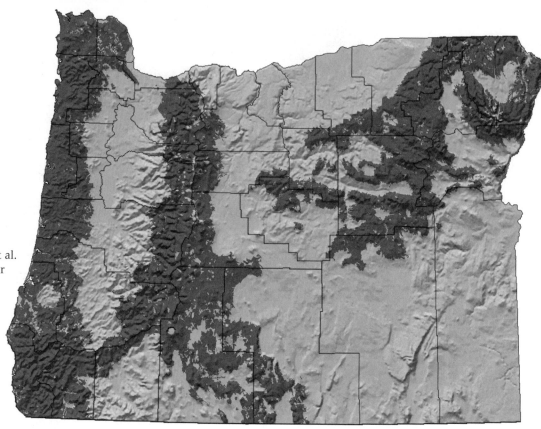

**Global Range:** Breeds throughout most of Alaska and Canada, and occurs as far south as northern New York in the East. In the West, it is resident in forested areas south to northern California and to the Rocky Mountains of northern New Mexico.

**Habitat:** Gray Jays are residents of coniferous forests, especially those forests above 2,000 feet. They are found in most types of forest communities up to timberline, including ponderosa pine, lodgepole pine, true fir, spruce, mixed conifer, and Douglas-fir forests. They prefer more open forests, and use patches of deciduous woods mixed in with coniferous forests.

**Reproduction:** The breeding season begins sometime in March, but varies with locality and climate. The nest cup of plant material is usually 5-30 feet high in a conifer. The clutch of 3 or 4 (range 2-5) eggs is incubated for 16-18 days by the female. The young leave the nest when just over 2 weeks old.

**Food Habits:** Gray Jays, like most corvids, have a broad diet that includes both plant and animal material. In the warmer parts of the year, they eat insects (grasshoppers, wasps, bees, beetles, and caterpillar eggs), small mammals, eggs and young of birds, and, when available, fruits and berries. They also eat lichens, fungi, carrion, and conifer seeds. They store small caches of food in the foliage of coniferous trees, attached by their sticky saliva.

**Ecology:** One study found the breeding season territory was 160-320 acres. This species tends to move down slope, probably in search of better foraging conditions, during winter. They adapt readily to human activities and seek food scraps at campgrounds. They probably benefit from forest edges created by timber harvest practices.

**Comments:** There are suggestions that this species has declined in Oregon's coastal forests in recent decades.

# Steller's Jay
*(Cyanocitta stelleri)*

*Order: Passeriformes*
*Family: Corvidae*
*State Status: None*
*Federal Status: None*
*Global Rank: G5*
*State Rank: S5*
*Length: 13 in (33 cm)*

*References:* Brown 1964, DeGraaf et al. 1991, Gilligan et al. 1994, U.S. Fish and Wildlife Service 1995.

**Global Range:** Resident from southeastern Alaska south through western North America to the highlands of Mexico and Central America, as far south as Nicaragua and as far east as Colorado and west Texas.

**Habitat:** This jay is most closely associated with various types of coniferous forests, from sea level to timberline. It also uses mixed coniferous-deciduous forests and can be found in small wooded patches in suburbs and cities. Rarely, it occupies dense juniper woodland, and lives on some isolated mountain ranges of eastern Oregon.

**Reproduction:** Steller's Jays begin nesting in late March in warmer parts of their Oregon range. Their nest cup of plant material is usually built in a conifer, and can be up to 100 feet off the ground. The usual clutch is 4 (range 2-6) eggs, which the female incubates for about 16 days. The family group stays together for over a month after the young fledge.

**Food Habits:** Like most jays, Steller's Jay is omnivorous, but about three-quarters of its diet is vegetable matter. It eats the fruit and seeds of many berries, pine seeds, acorns, and nuts. For the animal quarter of its diet, it eats insects (beetles, wasps, bees, grasshoppers, caterpillars, moths, sow bugs), spiders, eggs and young of small birds, frogs, snakes, and carrion. In suburbs, it may damage garden fruits and vegetables.

**Ecology:** The territory is limited to the area around the nest. The home ranges of many pairs of Steller's Jays will overlap where food is abundant, and they maintain order through a dominance hierarchy. They forage both on the ground and in trees, even probing the soil for insects. After breeding, this species tends to move upslope in search of berries and other food, but they move to the lowlands with the arrival of winter.

**Comments:** This jay favors forest edges, and forest fragmentation may increase its ability to prey on birds adapted to forest interior habitats, such as the Marbled Murrelet.

# Western Scrub-Jay

## (*Aphelocoma californica*)

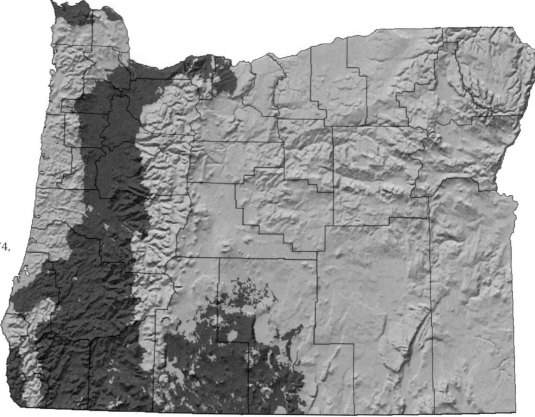

Order: *Passeriformes*
Family: *Corvidae*
State Status: *None*
Federal Status: *None*
Global Rank: *G5*
State Rank: *S5*
Length: *12 in (30 cm)*

References: Brown 1974,
Gilligan et al. 1994,
Pitelka 1951b.

**Global Range:** Reaches the northern limit of its distribution in southwestern Washington. From there, it is resident south through Oregon and California to the Mexican highlands, at least as far south as Veracruz. It is found as far east as central Texas. It was formerly considered conspecific with the Florida Scrub-Jay (*Aphelocoma coerulescens*) (American Ornithologists' Union 1995). The scrub-jay living on Santa Cruz Island, off the southern California coast, is now recognized as a distinct species, the Island Scrub-Jay (*Aphelocoma insularis*).

**Habitat:** The Western Scrub-Jay, as its name implies, favors drier habitat than Steller's Jay. In Oregon, it is found at lower elevations in chaparral, oak woodland, juniper woodland, orchards, agricultural, urban, and suburban areas, and at brushy edges of riparian woodland or thickets in pastureland. It occupies the edges of ponderosa pine-oak woodlands.

**Reproduction:** Breeding probably commences in Oregon in late March, with a cup nest built in a large bush or tree. The normal clutch is 4 or 5 (range 2-6) eggs, which are incubated by the female for 15-17 days, during which time she is fed by the male. Mates stay together for successive breeding seasons.

**Food Habits:** Western Scrub-Jays eat a variety of foods, but eat more plants than animals. They feed on acorns, juniper seeds, and an assortment of wild berries and cultivated fruits. They also eat insects (wasps, bees, butterflies, moths, caterpillars, cutworms, grasshoppers, crickets), spiders, scorpions, ticks, mites, snails, eggs and young of birds, mice, shrews, and lizards.

**Ecology:** Western Scrub-Jays are permanent residents that maintain their territory of 5 to 10 acres throughout the year; however, they tolerate the passage of vagrant scrub jays through their territory after the breeding season.

**Comments:** This species has been expanding its range in Oregon in recent decades, especially east and west along the Columbia River from Portland and in central Oregon.

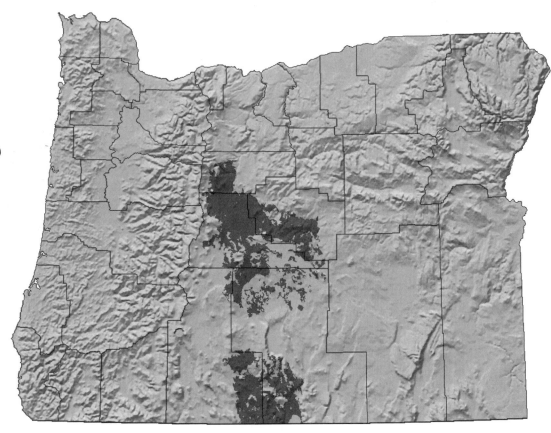

## Pinyon Jay
### (Gymnorhinus cyanocephalus)

*Order: Passeriformes*
*Family: Corvidae*
*State Status: None*
*Federal Status: None*
*Global Rank: G5*
*State Rank: S3S4?*
*Length: 11 in (28 cm)*

**Global Range:** The Pinyon Jay is a bird of the interior western United States. It breeds from central Oregon east to western South Dakota and south to southern New Mexico and the mountains of northern Baja California.

**Habitat:** Even though there are no pinyon trees in Oregon, the Pinyon Jay makes use of juniper and ponderosa pine woodlands for its breeding habitat. It is only found in south-central Oregon, even though seemingly suitable habitat occurs elsewhere east of the Cascade Mountains.

**Reproduction:** This species breeds in colonies, but usually only one bird will nest per tree. The nest cup made of plant material is placed on the limb of a tree 6-20 feet off the ground. In Oregon, breeding probably begins in mid-March. The clutch of 3 or 4 eggs is incubated by the female, while the male brings her food. Young usually remain with the flock over winter. Flocks wander widely during the post-breeding season in search of food.

**Food Habits:** Pinyon Jays eat the seeds of ponderosa pine and other conifers. They may communally cache conifer seeds for use in following years. They also eat other small seeds, fruit, berries, insects, and insect larvae. They eat bird eggs and nestlings, and glean grasshoppers, ants, and caterpillars from foliage.

**Ecology:** Flocks of Pinyon Jays have an established home range during the breeding season. The adults are apparently monogamous and remain paired through the nonbreeding season. Females do not breed until their second spring.

**Comments:** The nesting success of this species is dependent on a good supply of conifer seeds.

*References:* DeGraaf et al. 1991, Gilligan et al. 1994, Marzluff and Balda 1988, 1992.

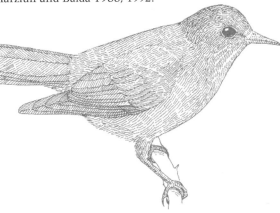

# Clark's Nutcracker
## (Nucifraga columbiana)

*Order: Passeriformes*
*Family: Corvidae*
*State Status: None*
*Federal Status: None*
*Global Rank: G5*
*State Rank: S4*
*Length: 12.5 in (32 cm)*

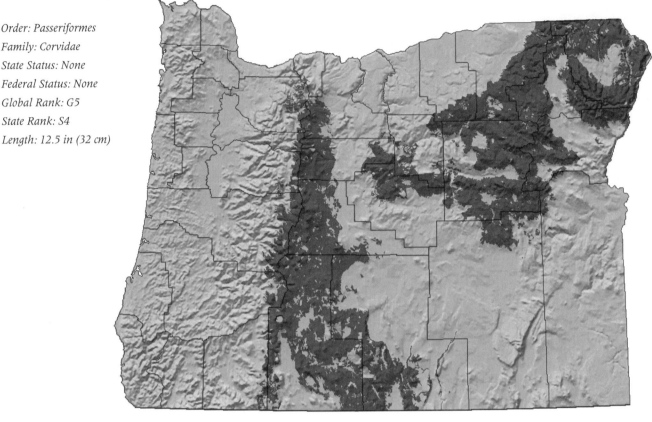

**Global Range:** Found in mountainous western North America, from southern British Columbia to the Mexican border; resident as far east as the Rocky Mountains of Colorado and New Mexico. Isolated populations occur in the mountains of Baja California and northern Mexico.

**Habitat:** This jay breeds in open coniferous forests, usually at higher elevations (above 4,000 feet). In Oregon, it is found in spruce and pine forests near timberline. Less often, it breeds in juniper and ponderosa pine woodlands east of the Cascade Mountains.

**Reproduction:** The nest is a cup of twigs on the limb of a conifer, sometimes as high as 150 feet off the ground. The clutch size is 2 or 3 (range 2-6) eggs. Both sexes incubate the eggs for about 17-18 days. The young leave the nest at about 24-28 days, but are fed by adults for a while after fledging.

**Food Habits:** Both adults and nestlings feed primarily on pine seeds, although they also eat insects, acorns, juniper berries, snails, some small mammals, carrion, and bird eggs and nestlings.

**Ecology:** Breeding begins early in the year (perhaps in March), even at high elevations. This is possible because they feed on pine seeds stored the previous fall. Following breeding, Clark's Nutcrackers wander widely in flocks at lower elevations.

**Comments:** Clark's Nutcracker is considered one of the oldest representatives of the New World subfamily of jays. Some have speculated that it formerly nested at lower elevations, and has been displaced from those habitats by more recently evolved and successful jays, like Steller's Jay and the Gray Jay.

*References:* Gilligan et al. 1994, Goodwin 1986, Mewaldt 1956, Tomback 1977, 1980; Van der Wall and Balda 1977.

*Order: Passeriformes*
*Family: Corvidae*
*State Status: None*
*Federal Status: None*
*Global Rank: G5*
*State Rank: S5*
*Length: 20 in (51 cm)*

**Global Range:** The Black-billed Magpie is a holarctic species that is resident from the British Isles to central Siberia and Japan, south to the Mediterranean and eastern China. In North America, it is found from southern Alaska south to central California and east to Quebec and western Kansas.

**Habitat:** This is a bird of open habitats that contain scattered trees. It is found in sagebrush and salt desert scrub, juniper woodlands, riparian woodlands, marshes, agricultural areas, pastures, and even rural towns. It may use the edges of forests and ascend high in unforested mountains, but it is absent from dense coniferous forests. It is usually found near water.

**Reproduction:** The breeding season begins in March, peaks in April, and eggs may still be found in July. A large stick nest is lined with plant material, and usually has a covering of twigs. The normal clutch of 7 to 9 (range 5-12) eggs is incubated for 17-18 days by the female. Young leave the nest in about a month.

**Food Habits:** Insects, grasshoppers in particular, are the main food item, but like most corvids, a wide variety of foods is eaten, including small mammals, bird eggs and nestlings, molluscs, spiders, reptiles, amphibians, fruit, and carrion.

**Ecology:** Black-billed Magpies are gregarious birds that may sometimes hunt grasshoppers cooperatively. There is no information to suggest a territory, although their close relative, the Yellow-billed Magpie (*Pica nuttalli*), defends a territory of about 1-5 acres throughout the year.

**Comments:** The Black-billed Magpie destroys the nests of some game birds and may peck the hides of livestock. It is therefore sometimes considered a pest by ranching and hunting interests.

*References:* Erpino 1968, Gabrielson and Jewett 1940, Gilligan et al. 1994, Reese and Kadlec 1985, Verbeek 1973, Zeiner et al. 1990a.

# American Crow
## (Corvus brachyrhynchos)

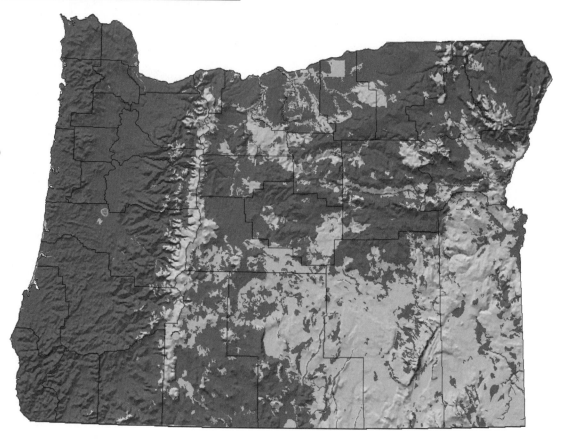

Order: *Passeriformes*
Family: *Corvidae*
State Status: *None*
Federal Status: *None*
Global Rank: *G5*
State Rank: *S5*
Length: *18 in (46 cm)*

**Global Range:** Breeds essentially coast to coast, from central Canada south to the Mexican border. It is absent from a few areas of the desert Southwest, and from the immediate area of the Pacific Coast south to northern Washington state.

**Habitat:** The American Crow roosts and nests in trees, but spends much time foraging in open, treeless areas. It is absent, except as a visitor, from large expanses of treeless deserts and grasslands, but is found in abundance in woodlands, agricultural areas, orchards, cities and towns, pastures, riparian areas, and the edges and open areas of forests. It generally does not use dense coniferous forests or high elevation habitats.

**Reproduction:** In Oregon, the breeding season begins in late March. Large stick nests are built in a tree, 10-75 feet high. The usual clutch of 4 or 5 (range 3-6) eggs is incubated for about 18 days. Young leave the nest at about 7 weeks of age.

**Ecology:** Nests of the American Crow are often found evenly spaced in a habitat patch, which tends to suggest territoriality in a loose colony. One study found 60 nests in just over 100 acres. Groups of crows will often mob large raptors (e.g., Red-tailed

Hawks); crows, in turn, are mobbed by smaller passerines.

**Comments:** Although resident throughout the year west of the Cascades, the American Crow does not typically winter east of the Cascades, except along the Columbia and Snake river systems, where winter roosts can be found.

*References:* Chamberlain-Auger et al. 1990, Gilligan et al. 1994, Goodwin 1986, Zeiner et al. 1990a.

230

# Common Raven
## (Corvus corax)

*Order: Passeriformes*
*Family: Corvidae*
*State Status: None*
*Federal Status: None*
*Global Rank: G5*
*State Rank: S4*
*Length: 24 in (61 cm)*
*Wingspread: 50 in (127 cm)*

**Global Range:** The Common Raven has a complicated distribution in North America, being resident throughout Alaska, most of Canada, the western part of the continent south to Nicaragua, and from New England south in mountains to Georgia. It is absent from the Great Plains, much of the Midwest, and the Southeast. In Eurasia, it is resident from Iceland to Siberia, south to the Mediterranean and Japan.

**Habitat:** Common Ravens are mobile birds that cover much area in their daily wanderings, and are capable of using almost every habitat type in Oregon, from the seacoast to high mountains. They nest in trees, cliffs, or structures, but can forage over desert valleys and grasslands. They are most common in open woodlands, but use forests of nearly every type, less frequently dense forests.

**Reproduction:** The breeding season begins in March, and young are fledged by July. The nest is a large mass of sticks and is lined with plant material. The usual clutch of 4 to 6 eggs is incubated for 20-21 days by the female,

while the male brings food. Young are independent in about a month and a half.

**Food Habits:** The Common Raven eats a varied diet, but is especially prone to scavenge carrion of various sorts. It also eats small vertebrates (mammals, reptiles, amphibians, bird nestlings and eggs), molluscs, fruit, mast, grain, and crustaceans.

**Ecology:** Common Ravens are more solitary than crows, and are territorial in the breeding season. Their breeding home ranges may be as large as several thousand acres.

**Comments:** Common Ravens are most abundant in southeastern Oregon, where marshes and cattle ranches provide a source of carrion.

*References:* Gilligan et al. 1994, Goodwin 1986, White and Tanner-White 1988.

# Black-capped Chickadee
## *(Parus atricapillus)*

*Order: Passeriformes*
*Family: Paridae*
*State Status: None*
*Federal Status: None*
*Global Rank: G5*
*State Rank: S5*
*Length: 5.5 in (14 cm)*

**Global Range:** Found in the northern part of North America, from southern Alaska and central Canada east to the Atlantic Coast and south to northern California in the West and the Appalachian highlands of North Carolina in the East.

**Habitat:** The Black-capped Chickadee is most common in deciduous woodlands and forests, but uses mixed coniferous-deciduous forests, and occasionally occurs in pure coniferous forests. It makes use of ornamental plants in cities and towns and, in eastern Oregon, is present in many riparian woodlands, but absent from vast expanses of desert.

**Reproduction:** Members of the titmouse family are cavity nesters, and the Black-capped Chickadee is no exception. They excavate their own cavities a few feet high in decaying trees. The normal clutch of 6 to 8 (range 5-13) eggs is incubated by the female for 12-14 days. The family group remains together for some time after the young fledge.

**Food Habits:** Chickadees glean insects from foliage, including flies, wasps, true bugs, plant lice, scale insects, leafhoppers, ants, and aphids. They also eat snails and small amphibians. During times when insects are scarce, they eat conifer seeds and berries.

**Ecology:** Black-capped Chickadees defend a territory of 5-20 acres, depending on locality. They form small foraging flocks during the winter. They often nest at the edges of forests.

**Comments:** This chickadee usually stays near its breeding site throughout the year. It is most abundant in valleys west of the Cascades.

*References:* Brewer 1963, Desrochers et al. 1988, Gilligan et al. 1994, Hill and Lein 1989, Smith 1991.

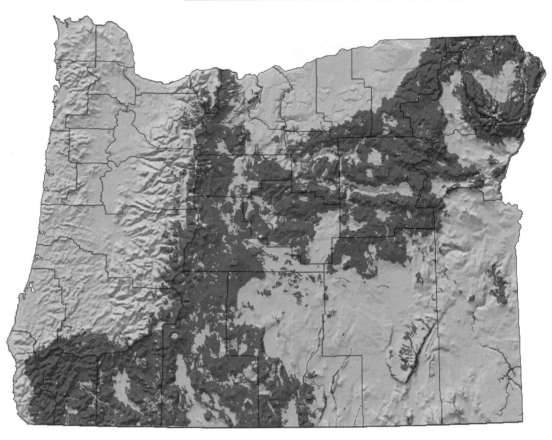

## Mountain Chickadee
### (Parus gambeli)

*Order: Passeriformes*
*Family: Paridae*
*State Status: None*
*Federal Status: None*
*Global Rank: G5*
*State Rank: S4*
*Length: 5.5 in (14 cm)*

**Global Range:** This is a chickadee of western North America. It is resident from northern British Columbia south to northern Baja California and east to Montana, New Mexico, and west Texas. It is absent from lower-elevation forests near the coast.

**Habitat:** Mountain Chickadees make their homes in mid- to high-elevation forests, mainly ponderosa pine, lodgepole pine, mountain hemlock, true fir (genus *Abies*), Engelmann spruce, and mixed conifer forests. They also use juniper woodland or quaking aspen stands within coniferous forests. They are usually found above the Douglas-fir zone.

**Reproduction:** The breeding season begins in May. This chickadee prefers to nest in abandoned woodpecker holes or natural tree cavities, but may dig its own nest hole. The clutch of 6 to 9 (range 5-12) eggs is incubated for 2 weeks. Young fledge in about 3 weeks. This species may produce 2 broods a year.

**Food Habits:** Mountain Chickadees glean most of their food from foliage. They eat insect eggs, larvae, and adults (moths, plant lice, beetles, flies, scale insects), and spiders. They also feed on some conifer seeds and buds, and eat some berries.

**Ecology:** At lower elevations, Mountain Chickadee territories are maintained year round, and vary from a few to 20 acres or more. These chickadees can be fairly dense in favorable habitat. In winter, they leave higher elevations for more moderate climate in lower woodlands.

**Comments:** This species is absent as a breeder from the desert valleys of eastern Oregon, being found only in mountain ranges (e.g., Hart, Steens, and other mountains).

*References:* Franzreb 1976, Gilligan et al. 1994, Hill and Lein 1989, Kluyver 1961, Laudenslayer and Balda 1976, Manolis 1977.

# Chestnut-backed Chickadee
## (Parus rufescens)

*Order: Passeriformes*
*Family: Paridae*
*State Status: None*
*Federal Status: None*
*Global Rank: G5*
*State Rank: S5*
*Length: 5 in (13 cm)*

*References:* Brennan and Morrison 1991, DeGraaf et al. 1991, Gabrielson and Jewett 1940, Gilligan et al. 1994, Stiles 1980.

**Global Range:** The Chestnut-backed Chickadee has an unusual distribution, being found mainly along the coastal forests of western North America, from southern Alaska to central California. For some reason, it is found inland from northern Washington east to the Rocky Mountains of northern Idaho and southern Alberta, and locally in the Blue Mountains of eastern Oregon.

**Habitat:** In Oregon, this species is found in moist, dense coniferous forests. It will use stands of deciduous trees in coniferous forests. In western Oregon, it is found in mid- to low-elevation Douglas-fir, lodgepole pine, western hemlock, western red cedar, and Sitka spruce forests. In the Blue Mountains, it is found in moist Douglas-fir, hemlock, and grand fir forests, usually at middle elevations.

**Reproduction:** This species begins its breeding season about April in western Oregon. It nests in an existing cavity in a tree, or will dig its own nest hole, lined with moss, fur, feathers, and plant fibers. The usual clutch is 6 or 7 (range 5-9) eggs. Both sexes feed the nestlings.

**Food Habits:** About two-thirds of the diet consists of insects and spiders, gleaned from foliage or extracted from decaying, downed wood. It eats beetles, hemipterans, caterpillars, wasps, moths, scale insects, ants, plant lice, treehoppers, spiders, and other invertebrates. The other third of the diet is made up of conifer seeds and berries.

**Ecology:** The Chestnut-backed Chickadee has territories of at least 3 acres. In winter, populations in eastern Oregon tend to move downslope to warmer forests. After the breeding season, it often forms multi-species flocks with other small passerines (juncos, kinglets, nuthatches, etc.).

**Comments:** This species is western Oregon's most common chickadee, often seen in urban and suburban parks and woodlots.

234

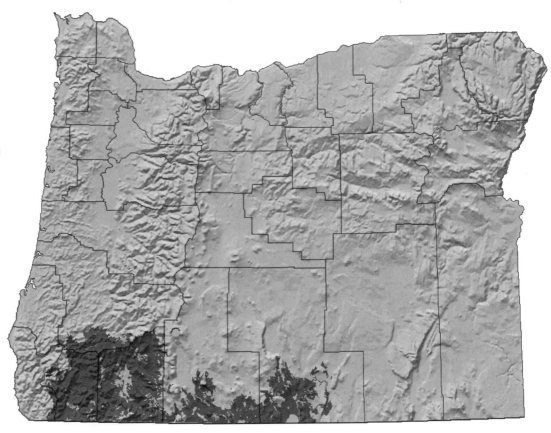

## Plain Titmouse
### *(Parus inornatus)*

*Order: Passeriformes*
*Family: Paridae*
*State Status: None*
*Federal Status: None*
*Global Rank: G5*
*State Rank: S?*
*Length: 5.5 in (14 cm)*

**Global Range:** Reaches the northern limits of its range in southern Oregon and southwestern Wyoming. It is found on the West Coast as far south as the mountains of Baja California del Norte, and in the interior as far south as northern Sonora and west Texas.

**Habitat:** Plain Titmice use two main habitat types, oak and juniper woodland. They also may be found in chaparral or urban and suburban areas that retain some oaks.

**Reproduction:** The Plain Titmouse's breeding season begins around March in the warmer parts of its range. It usually nests in an abandoned woodpecker hole or natural cavity in a tree, but can dig its own hole in soft wood. The clutch of 6 to 8 (range 3-9) eggs is incubated for 14-16 days by the female. Young titmice fledge in about 21 days, and remain with the family group for another 3 or 4 weeks.

**Food Habits:** Plain Titmice are mainly insectivorous, gleaning insects (leafhoppers, aphids, beetles, caterpillars, ants, flies) and spiders from tree limbs, twigs, or the ground. They also eat oak and willow catkins, leaf buds, seeds, berries, and acorns.

**Ecology:** Pairs of Plain Titmice mate for life and defend a territory, usually between 2 and 10 acres, throughout the year. Unlike chickadees, these titmice do not form flocks after the breeding season.

**Comments:** The Warner Valley population is a different subspecies from the Plain Titmice west of the Cascade Mountains, and may someday be separated as Ridgway's Titmouse (*P. zaleptus*).

*References:* Dixon 1949, 1954, 1956; Gill and Slikas 1992, Gilligan et al. 1994, Hertz et al. 1976.

# Bushtit
## *(Psaltriparus minimus)*

*Order: Passeriformes*
*Family: Aegithalidae*
*State Status: None*
*Federal Status: None*
*Global Rank: G5*
*State Rank: S5*
*Length: 4 in (10 cm)*

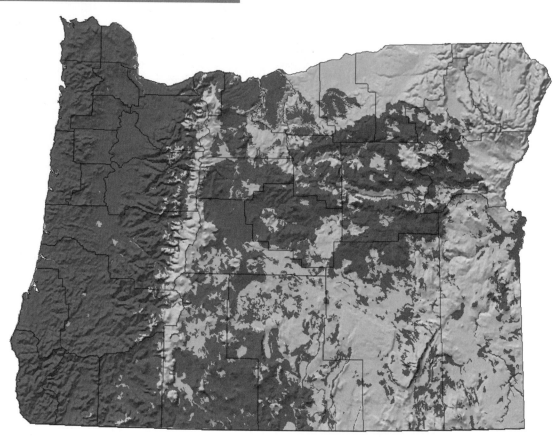

**Global Range:** Resident in western North America from southwestern British Columbia south and east to northern Baja California, Colorado, and central Texas, and south through the highlands of Mexico and Central America to Guatemala.

**Habitat:** Bushtits are associated with thick, brushy plants, but are not particular about where they find them. They occur in riparian woodlands, chaparral, juniper woodlands, aspen groves, mountain mahogany, brushy areas in coniferous forests, shrubs growing in residential areas, and, rarely, in sagebrush communities.

**Reproduction:** This species probably begins nesting in Oregon in April. It weaves a fairly large, pendulous nest of plant material and spider webs, usually suspended in a bush. The normal clutch of 5 to 7 eggs (range 4-15) is incubated for 12-13 days. Family flocks remain together over winter, often joining flocks of other small passerines.

**Food Habits:** This small bird gleans a variety of insects from foliage and stems of plants, including aphids, beetles, treehoppers, leafhoppers, caterpillars, wasps, and ants. It also eats some spiders, seeds, nectar, and berries.

**Ecology:** Bushtits defend a small territory, from a few tenths of an acre to a few acres, depending on habitat, during the breeding season. Winter flocks (sometimes interspecific) number 20-50 birds. They do not remain at high elevations during the winter.

**Comments:** Gilligan et al. (1994) report Bushtits have recently increased in abundance in the Portland metropolitan area.

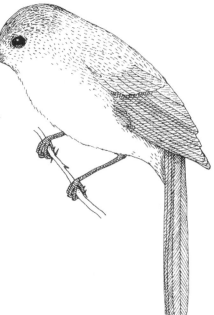

*References:* DeGraaf et al. 1991, Gilligan et al. 1994, Laudenslayer and Balda 1976.

## Red-breasted Nuthatch
### (Sitta canadensis)

*Order: Passeriformes*
*Family: Sittidae*
*State Status: None*
*Federal Status: None*
*Global Rank: G5*
*State Rank: S5*
*Length: 4.5 in (11 cm)*

*References:* Anderson 1976, Gilligan et al. 1994, Kilham 1973, 1974b, 1975.

**Global Range:** Breeds from the south coast of Alaska across Canada to the Atlantic, but is absent from much of the middle of the continent. In the West, it breeds in forested areas south to southern California and east to Colorado. In the East, it is found in New England, the upper Midwest, and in the Appalachian Mountains south to North Carolina.

**Habitat:** Red-breasted Nuthatches breed in virtually every coniferous forest habitat in Oregon, from sea level to timberline. They also use mixed coniferous-deciduous forests. In winter, they can be seen in an even wider variety of habitats at lower elevations.

**Reproduction:** The breeding season begins in late April at low elevations. Nuthatches are cavity nesters that use a natural hole in a tree or dig one. The normal clutch is 5 or 6 (range 4-7) eggs, and the female incubates for about 12 days. Young fledge in 3 weeks, but remain with the parents until they are nearly 2 months old.

**Food Habits:** Nuthatches forage by flying to a tree and working down the trunk, searching for insects in the bark. Despite this conspicuous behavior, their main food consists of seeds of coniferous trees (pines, spruces, firs, Douglas-firs), which are available in winter when insects are scarce. Animal food includes beetles, wasps, caterpillars, insect eggs, craneflies, moths, larvae of pine beetles, and spiders.

**Ecology:** This species is territorial throughout the year, and occurs at a density of a pair every 5 or 10 acres. Nest trees must be at least 12 inches in diameter. The Red-breasted Nuthatch smears an area around the entrance of the nest with pitch, even when using an artificial nest box; presumably, this practice deters predators.

**Comments:** Red-breasted Nuthatches are common in stands of conifers in suburbs.

237

# White-breasted Nuthatch
## (Sitta carolinensis)

*Order: Passeriformes*
*Family: Sittidae*
*State Status: None*
*Federal Status: None*
*Global Rank: G5*
*State Rank: S4*
*Length: 5.5 in (14 cm)*

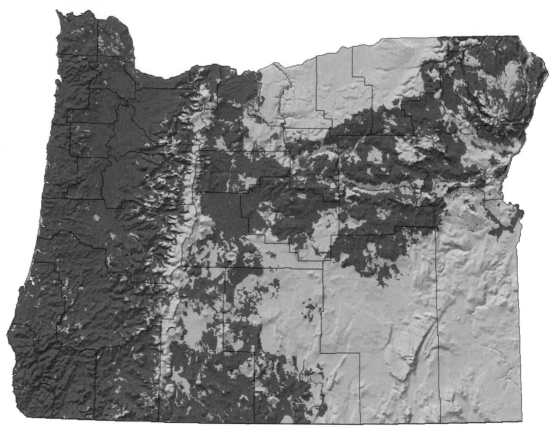

**Global Range:** Resident across North America, from southern Canada south to the highlands of Mexico. It is absent from the Great Plains, where there are no suitable woodlands.

**Habitat:** While the Red-breasted Nuthatch prefers denser, coniferous forests, the White-breasted species is found in deciduous forests, mixed coniferous-deciduous forests, or open woodlands, including oak, ponderosa pine, and juniper woodlands. It is absent from higher-elevation coniferous forests.

**Reproduction:** This species begins breeding in April. It is more of a secondary cavity nester than the Red-breasted Nuthatch. It nests in natural cavities in trees or abandoned woodpecker holes. The usual clutch has about 8 (range 5-10) eggs, and is incubated by the female for 12 days, with the male bringing food. Young fly in 2 weeks, and remain with the parents for another 2 weeks.

**Food Habits:** In spring and summer, this nuthatch eats mainly insects (moth and beetle larvae, beetles, caterpillars, ants) and spiders. In the nonbreeding season, it eats large seeds, such as acorns, as well as pine and other seeds.

**Ecology:** Pairs of White-breasted Nuthatches are solitary and remain mated in the nonbreeding season. They have a territory that varies with habitat from a few acres to nearly 100 acres.

**Comments:** This species avoids dense, humid coniferous forest habitats that are found in much of lower-elevation western Oregon. It is rare along the Oregon coast.

*References:* Gilligan et al. 1994, Kilham 1968, 1971, 1972.

238

# Pygmy Nuthatch
## (Sitta pygmaea)

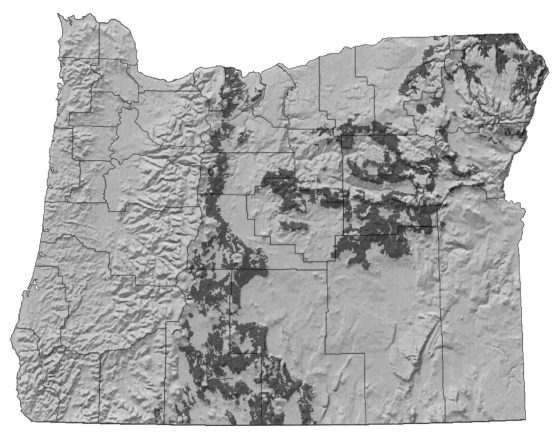

*Order: Passeriformes*
*Family: Sittidae*
*State Status: Sensitive*
*Federal Status: None*
*Global Rank: G5*
*State Rank: S4?*
*Length: 4 in (10 cm)*

**Global Range:** This small nuthatch of western North America is resident from southern British Columbia south to the mountains of northern Baja California. Inland, it is found in the Rocky Mountains of Idaho, south through the mountains of the Southwest, to the highlands of Mexico. It ranges as far east as western Texas.

**Habitat:** In other parts of its range, the species can be found in a variety of open coniferous woodland community types (Jeffrey pine, pinyon-juniper woodland, Monterey cypress), but in Oregon it is believed to be restricted to mature ponderosa pine woodlands with less than 70% canopy closure.

**Reproduction:** Its breeding season begins around late April. It digs a nest cavity in a decaying large (average 20-inch diameter) ponderosa pine snag. The clutch of 6 to 8 (range 4-10) eggs is incubated for 15 to 16 days by the female, who is fed by the male. Young leave the nest at 3 weeks of age, and are fed for an additional 4 weeks.

**Food Habits:** This nuthatch is mostly insectivorous, gleaning insects and spiders from cones and shoots high in pine trees. It takes wasps, ants, spittle bugs, beetles, moths, caterpillars, and grasshoppers. When insects are unavailable, it eats pine seeds. On occasion, it catches flying insects.

**Ecology:** Most territories are 2 to 4 acres and are defended only during breeding season. Following breeding season this nuthatch is very social. It digs large roost cavities that over 100 birds may use. This may be a heat conservation adaptation, given the cold winters of its preferred habitat and its small body size.

**Comments:** A nonbreeding adult male will often assist a breeding pair in rearing the nestlings.

*References:* Bock 1969, Gilligan et al. 1994, Knorr 1957, Marshall 1992, Norris 1958, Sydeman et al. 1988.

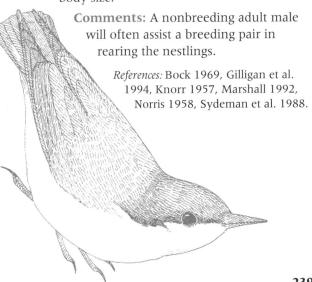

# Brown Creeper
*(Certhia americana)*

*Order: Passeriformes*
*Family: Certhiidae*
*State Status: None*
*Federal Status: None*
*Global Rank: G5*
*State Rank: S4*
*Length: 5.5 in (14 cm)*

**Global Range:** The Brown Creeper has a complicated distribution. It is resident from coastal southern Alaska south through the western United States, and into the highlands of Mexico and Central America, as far south as Guatemala. It breeds across southern Canada to the Atlantic Coast, and south in the Appalachian Mountains to North Carolina.

**Habitat:** The habitat varies with location, but in Oregon this is a species of both coniferous forests and deciduous woodlands, present, but rare, near the ocean, and found up to timberline. It is not found in the juniper woodlands of eastern Oregon, but uses riparian woodlands and trees bordering marshes.

**Reproduction:** This species begins breeding in late April. It nests in a cup of plant material concealed in the trunk of a tree, or in a crevice formed by peeling tree bark. The usual clutch is 6 (range 3-9) eggs. The female incubates for just over 2 weeks. Young leave the nest after about 2 weeks.

**Food Habits:** The Brown Creeper searches for food in reverse fashion to a nuthatch, flying to the base of a tree and foraging for insects in the bark while climbing up the tree, often in a spiral fashion. Food includes leafhoppers, flat bugs, plant lice, scale insects, insect eggs, flies, moths, caterpillars, cocoons, pupae, ants, spiders, and aphids. A small amount of plant material (nuts and seeds) is also eaten.

**Ecology:** Brown Creepers usually occur in low density (only a few pairs per 100 acres). They leave higher elevations during winter.

**Comments:** Brown Creepers use forest patches in the suburbs; however, studies have found them more abundant in old-growth forests.

*References:* Davis 1978, Gabrielson and Jewett 1940, Gilligan et al. 1994.

*Order: Passeriformes*
*Family: Troglodytidae*
*State Status: None*
*Federal Status: None*
*Global Rank: G5*
*State Rank: S5*
*Length: 5.5 in (14 cm)*

**Global Range:** Rock Wrens are birds of western North America. They breed from southern Canada south through the interior western United States. They are resident from northern California east to Texas and south through the highlands of Mexico and Central America to Costa Rica.

**Habitat:** Rock Wrens are specialists adapted for breeding in rock piles, talus slopes, and rimrock. They occur in these habitats in a variety of vegetation types, from arid deserts to rocky outcrops and canyons in forests. As a result their distribution tends to be discontinuous and uncorrelated with vegetation cover type. They are not found in humid, dense forests of northwestern Oregon.

**Reproduction:** In Oregon, the Rock Wren begins its breeding season by May. The nest is a cup of plant material hidden among rocks or in a crevice. The female incubates the clutch of 5 or 6 (range 4-10) eggs, while being fed by the male. The nestlings are fed by both sexes. There are apparently 2 clutches per year.

**Food Habits:** Rock Wrens forage for insects in rocky sites with no plant cover. They eat insects and their larvae (including beetles and grasshoppers), spiders, and other invertebrates, including earthworms.

**Ecology:** The Rock Wren's song is presumed to have a territorial function. Depending on habitat, various studies have found from 5 to 25 breeding males per 100 acres. This species is found in more arid situations than the Canyon Wren. Most Rock Wrens migrate south during winter.

**Comments:** Rocky areas in forest clear-cuts are often occupied by Rock Wrens.

*References:* DeGraaf et al. 1991, Gabrielson and Jewett 1940, Gilligan et al. 1994, Smyth and Bartholomew 1966, Wauer 1964.

# Canyon Wren
## (Catherpes mexicanus)

*Order: Passeriformes*
*Family: Troglodytidae*
*State Status: None*
*Federal Status: None*
*Global Rank: G5*
*State Rank: S4*
*Length: 5.5 in (14 cm)*

*References:* DeGraaf et al. 1994, Gilligan et al. 1994, Miller and Stebbins 1964.

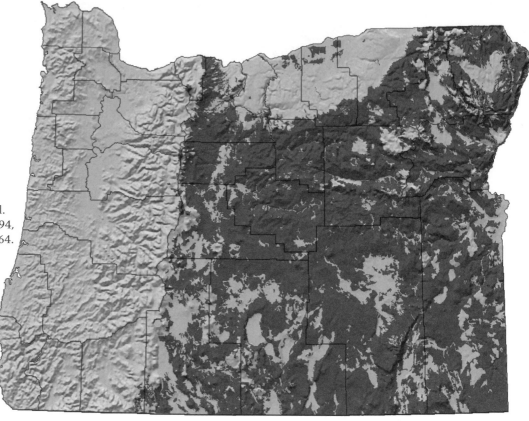

**Global Range:** Resident in western North America from southern British Columbia south to Baja California and the highlands of central Mexico. It is found as far east as western South Dakota and central Texas. It is absent west of the Cascade Mountains and Coast Ranges north of central California.

**Habitat:** The Canyon Wren shares many habitat affinities with the Rock Wren, but it frequents larger rocky features (steep canyon walls, gorges, rimrock cliffs) that offer some shade. It also is more likely to be found near streams, but it does not use rock outcrops in forest clear-cuts. It is not influenced by vegetation cover type, although most of its range in Oregon is within desert regions.

**Reproduction:** The Canyon Wren's breeding season is under way by May. The nest is a cup of twigs, lined with feathers or plant down, usually placed on a ledge in a cave or on a cliff face. The usual clutch is 5 or 6 (range 3-7) eggs, and 2 broods are raised each year. The male feeds the female during incubation, and both sexes feed the nestlings.

**Food Habits:** Canyon Wrens feed on small invertebrates they capture on or around rocks, as well as a variety of insects and spiders.

**Ecology:** Little is known about this species. Its distinctive song would appear to have a territorial function, and pairs are solitary. During the winter these wrens apparently leave higher-elevation canyons, and some probably migrate south, both sensible moves for a bird dependent on insects for food.

**Comments:** Canyon Wrens sometimes nest in buildings, and treat stone walls like canyon or rimrock walls.

## Bewick's Wren
### *(Thryomanes bewickii)*

*Order: Passeriformes*
*Family: Troglodytidae*
*State Status: None*
*Federal Status: None*
*Global Rank: G4*
*State Rank: S4*
*Length: 5.5 in (14 cm)*

*References:* DeGraaf et al. 1991, Gilligan et al. 1994, Kroodsma 1973, Root 1969a.

**Global Range:** Breeds along the west coast of North America, from southern British Columbia to Baja California. Inland, it occurs south to the highlands of central Mexico, and east from Nevada almost to the Atlantic Coast. It is absent from areas near the Gulf of Mexico east of Texas.

**Habitat:** This wren is found in tall brush and thickets, including the chaparral of southern Oregon, forest edges and clear-cuts, riparian woodland, thickets in canyon bottoms, and residential areas. It does not occur in arid deserts or dense, moist forests.

**Reproduction:** Bewick's Wrens begin breeding in April, and there are 2 or 3 broods each year. They nest in natural tree cavities or recesses in rock outcrops, buildings, or in old woodpecker holes. The clutch of 3 to 6 (range 3-11) eggs is incubated by the female, who is fed by the male. Young fledge in 2 weeks, but are fed by the parents their first month.

**Food Habits:** As with most wrens, small insects and spiders make up most (97%) of the Bewick's Wren's diet. Among other items, it eats beetles, leafhoppers, treehoppers, ants, wasps, caterpillars, moths, and grasshoppers. The other 3% consists mostly of seeds.

**Ecology:** In one Oregon study, Bewick's Wren territories averaged about 5 acres, which is the same as their home range. The territory is maintained throughout the year where the birds are resident. Some, breeding higher in the foothills, withdraw to the lower western Oregon valleys in winter.

**Comments:** Bewick's and House wrens have very similar nest site preferences and compete where they are sympatric. Bewick's Wren is expanding its range in eastern Oregon.

243

# House Wren
## (Troglodytes aedon)

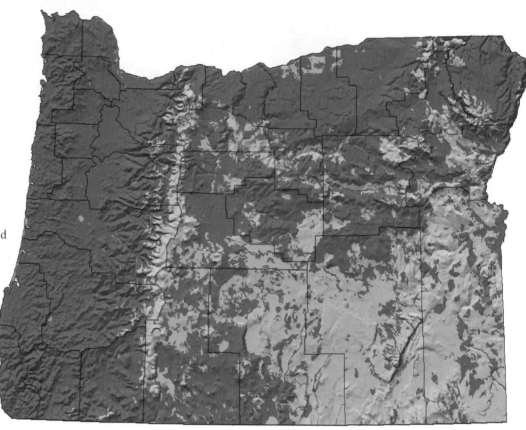

*Order: Passeriformes*
*Family: Troglodytidae*
*State Status: None*
*Federal Status: None*
*Global Rank: G5*
*State Rank: S4*
*Length: 5 in (13 cm)*

*References:* Brawn and Balda 1988, Drilling and Thompson 1988, Gilligan et al. 1994, Kroodsma 1973, Root 1969a, Zeiner et al. 1990a.

**Global Range:** There are several subspecies of the House Wren; these are treated as full species by some ornithologists. Taken as a whole, the House Wren breeds from central Canada and all but the southeastern United States south through Mexico and Central America into South America. It is resident from southern Mexico, south to its southern range limits in central Chile and Argentina.

**Habitat:** Although they are found throughout Oregon, House Wrens have a patchy distribution, being found only in relatively open habitats and brushy areas at forest edges. They use clear-cuts, riparian thickets in open coniferous forests, aspen groves, and riparian woodlands east of the Cascade Mountains. They also can be found in oak, oak-Douglas-fir, and oak-ponderosa pine woodlands, cottonwood riparian woodlands, coastal lodgepole pine woodlands, and residential areas, always near patches of brush.

**Reproduction:** The House Wren's breeding season in Oregon is under way by May, when it seeks out a natural cavity in a tree, an old woodpecker hole, or a nest box for its nest. The normal clutch of 6 to 8 (range 5-12) eggs is incubated by the female for about 2 weeks, while the male brings her food. The young leave the nest in about 2 weeks, and the male will feed them as the female incubates the second clutch of the year.

**Food Habits:** The diet of the House Wren is mainly insects (beetles, caterpillars, grasshoppers, crickets, butterflies, ants, bees, wasps, flies, ticks, plant lice), spiders, millipedes, and snails.

**Ecology:** In Oregon, House Wrens maintain territories of about 5 acres that are defended against other House Wrens and competing Bewick's Wrens. Most House Wrens migrate south to the southern tier of states and Mexico in winter.

**Comments:** House Wrens will destroy the eggs of other birds in their territory, and fill unused nest cavities with sticks to reduce competition.

*Order: Passeriformes*
*Family: Troglodytidae*
*State Status: None*
*Federal Status: None*
*Global Rank: G5*
*State Rank: S4*
*Length: 4 in (10 cm)*

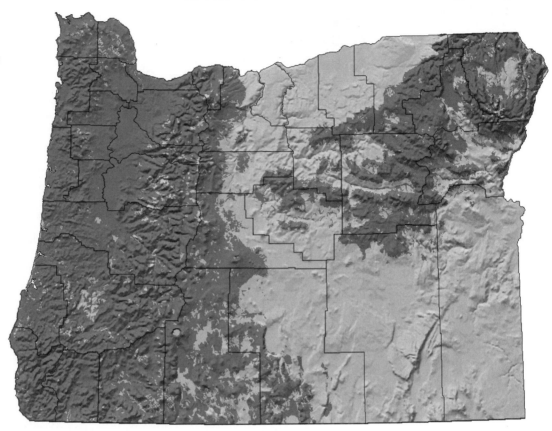

**Global Range:** The Winter Wren is a holarctic species with a complicated distribution in the middle latitudes of the Northern Hemisphere. In Eurasia, it breeds from Iceland to central Siberia, and south to North Africa and Japan. In North America, it breeds from the Alaskan coast south to California and east across the continent in central Canada to the Atlantic Coast. In the East, it descends in the Appalachian Mountains to northern Georgia.

**Habitat:** Winter Wrens live in dense, low thickets and undergrowth of dense forests (coniferous, mixed, or deciduous), and along riparian areas in woodlands. They are frequently restricted to areas near water. They use all dense, moist forest types, but are not found in ponderosa pine woodlands away from riparian zones.

**Reproduction:** At lower elevations, the breeding season can begin in late March, but it is undoubtedly later for those wrens breeding near timberline. The nest is a rather elaborate sphere made of plant material and lined with feathers. A normal clutch of 5 to 8 (range 4-16) eggs is incubated by the female for 14-17 days. Males mate with several females and help rear each brood in sequence.

**Food Habits:** The Winter Wren is an insectivorous bird that gleans prey from foliage or the ground under thickets. It eats larval and adult insects (aphids, beetles, lace bugs, moths, ants, sawflies, caterpillars), spiders, and an occasional seed.

**Ecology:** Winter Wrens have relatively small territories of an acre or 2. This wren moves to lower elevations or migrates south in winter.

**Comments:** Winter Wrens are more abundant in late-successional forests.

*References:* Armstrong 1956, Cody and Cody 1972, Gilligan et al. 1994, Zeiner et al. 1990a.

# Marsh Wren
## *(Cistothorus palustris)*

*Order: Passeriformes*
*Family: Troglodytidae*
*State Status: None*
*Federal Status: None*
*Global Rank: G5*
*State Rank: S5*
*Length: 5 in (13 cm)*

**Global Range:** Breeds coast to coast in North America, from southern Canada south to extreme northern Mexico in the West and to northern Florida in the East. It is generally absent from the southern tier of states, and its distribution is discontinuous away from coastal areas.

**Habitat:** Marsh Wrens are well named, since they occur in thick stands of emergent aquatic vegetation on the edges of large marshes, rivers, ponds, and coastal estuaries. They are nearly always in the cover of cattails, bulrushes, or sedges, but seem to prefer larger marshes with narrow-leaved rather than broad-leaved cattails.

**Reproduction:** At lower elevations in Oregon, the breeding season is under way in late April, but at higher elevations (up to 4,000 feet around lakes in the Cascade Mountains), it is likely to be later. The male Marsh Wren builds several spherical nests that are usually attached to vegetation and suspended over shallow water. The usual clutch of 4 or 5 eggs is incubated by the female for 12-14 days. Young fledge in about 2 weeks, and may be fed by the male while the female tends the second clutch.

**Food Habits:** The Marsh Wren gleans insects from aquatic vegetation, the surface of the water, or damp ground. The diet includes aquatic insects and their larvae (dragonflies, craneflies, mosquitoes), caterpillars, spiders, snails, and an occasional seed. Bird eggs are sometimes eaten as well.

**Ecology:** Marsh Wren territories are rather small, ranging from about a tenth to a half of an acre. Males may mate with more than one female. Marsh Wrens move to lower elevations or migrate south in winter.

**Comments:** Because of their habitat requirements, the species has a discontinuous distribution and is absent from large areas of forests, deserts, and upland vegetation types.

*References:* Gilligan et al. 1994, Leonard and Picman 1986, Picman 1977, Verner 1975, Verner and Engelsen 1970.

## American Dipper
### *(Cinclus mexicanus)*

*Order: Passeriformes*
*Family: Cinclidae*
*State Status: None*
*Federal Status: None*
*Global Rank: G5*
*State Rank: S4*
*Length: 8 in (20 cm)*

**Global Range:** Resident from coastal southern Alaska south through the Yukon and British Columbia to southern California and east to the Rocky Mountains of Colorado and New Mexico. Farther south, it reappears in the mountains of Mexico and Central America as far south as Panama.

**Habitat:** Dippers are closely associated with rapidly flowing rivers and streams, often in coniferous forests up to timberline, but in any habitat type down to sea level. They occasionally inhabit the shores of mountain ponds and lakes.

**Reproduction:** The breeding season is under way by April, at least at lower elevations. Dippers build their nest in a concealed crevice or depression in a stream bank, sometimes behind a waterfall. The usual clutch of 4 or 5 (range 3-6) eggs is incubated by the female for 16 days. The young fledge in about 4 weeks, and are fed by the parents for 2 additional weeks. Males may have several mates in their territory.

**Food Habits:** The dipper's main food items are aquatic insects and their larvae (caddisflies, beetles, midges, stoneflies, mayflies, mosquitoes). They also eat some small fish, worms, snails, and clams. Food is usually captured while the dipper walks submerged on the bottom of a rocky stream. Some flying insects are taken on the wing.

**Ecology:** Fitting their life style, dippers defend a linear territory along a stream. Territories are reported to be from 1,000 to 6,000 feet long. Most birds remain on the territory throughout the year, although some move downstream from higher elevations in winter.

**Comments:** American Dippers require clear water for foraging. Human activities that increase sediment in streams may destroy breeding habitat.

*References:* Bakus 1959, Gilligan et al. 1994, Price and Bock 1973, Thut 1970.

# Golden-crowned Kinglet
## (Regulus satrapa)

Order: Passeriformes
Family: Muscicapidae
State Status: None
Federal Status: None
Global Rank: G5
State Rank: S4
Length: 4 in (10 cm)

**Global Range:** The Golden-crowned Kinglet has a complicated distribution that follows the coniferous forests of North America. It is resident from southern Alaska south to central California on the Pacific Coast, and to the mountains of Arizona and New Mexico. It is also resident in the Great Lakes and northeastern United States. In between, it breeds in the forests across central Canada, but not in the prairie provinces.

**Habitat:** This kinglet is associated with dense coniferous forests, especially those with a spruce component. It breeds from sea level to timberline, but in eastern Oregon it is found mostly in higher-elevation forests above the ponderosa pine zone.

**Reproduction:** The breeding season for this species is under way by May. The nest is a cup hidden in the foliage of a conifer. The usual clutch is 8 or 9 (range 5-10) eggs, incubated by the female for about 2 weeks. Both sexes feed the young.

**Food Habits:** Golden-crowned Kinglets are insectivorous. They glean insects and insect eggs from foliage and also hawk flying insects. Bark beetles and scale insects are important food items, but they also eat aphids and other insects and

spiders. They also consume tree sap from woodpecker holes, some seeds, and berries.

**Ecology:** One study found that each pair of kinglets used about 5 acres of habitat. As one might expect from an insectivorous bird, most, but not all, kinglets breeding at higher elevations tend to move downslope in the winter.

**Comments:** Golden-crowned Kinglets are more abundant in large tracts of late-sucessional coniferous forests than in fragmented or younger forests.

*References:* DeGraaf et al. 1991, Galati 1991, Gilligan et al. 1994, Keast and Saunders 1991, Zeiner et al. 1990a.

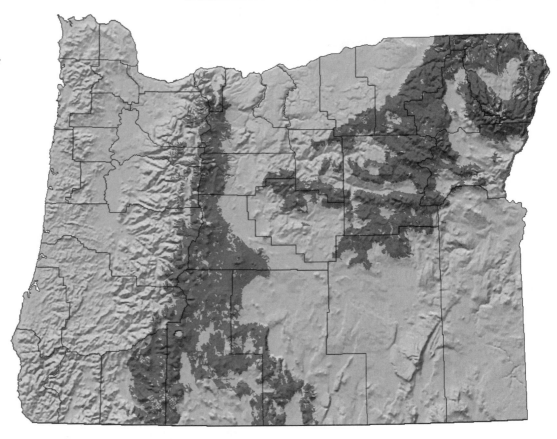

## Ruby-crowned Kinglet
### (Regulus calendula)

*Order: Passeriformes*
*Family: Muscicapidae*
*State Status: None*
*Federal Status: None*
*Global Rank: G5*
*State Rank: S4*
*Length: 4 in (10 cm)*

**Global Range:** Breeds from Alaska, across Canada, to the Atlantic Coast. In the West, they breed as far south as the mountains of southern California and New Mexico. In the East, they are found only as far south as the Great Lakes and northern New York. They are absent as breeders from the Great Plains and southern United States.

**Habitat:** Although the Ruby-crowned Kinglet uses both coniferous forests and mixed coniferous-deciduous woodlands in other parts of its range, in Oregon it breeds only in higher-elevation coniferous forests, such as lodgepole pine, mountain hemlock, Douglas-fir, Engelmann spruce, and true fir (*Abies*) forests. As a result, it is absent from our Coast Range Douglas-fir/western hemlock forests.

**Reproduction:** Ruby-crowned Kinglets are breeding by June. They build their deep cup nest of moss and plant fibers, bound with spider webs, in the fork of a conifer limb. The usual clutch is 8 or 9 eggs, incubated by the female for 14-17 days. Both sexes feed the young.

**Food Habits:** Insects form the main part of the Ruby-crowned Kinglet's diet, including true bugs, wasps, ants, beetles, butterflies, moths, flies, and caterpillars. They also feed on spiders, some seeds, tree sap, and berries.

**Ecology:** Two studies have found densities of 40 birds per 100 acres. In winter, these kinglets move to lower-elevation forests and woodlands, where they may join small mixed-species foraging flocks.

**Comments:** An Idaho study found this species associated with older hemlock-grand fir forests.

*References:* DeGraaf et al. 1991, Gabrielson and Jewett 1940, Gilligan et al. 1994, Keast and Saunders 1991, Morse 1970, Rea 1970.

249

# Blue-gray Gnatcatcher
### (Polioptila caerulea)

*Order: Passeriformes*
*Family: Muscicapidae*
*State Status: None*
*Federal Status: None*
*Global Rank: G5*
*State Rank: S3*
*Length: 4.5 in (11 cm)*

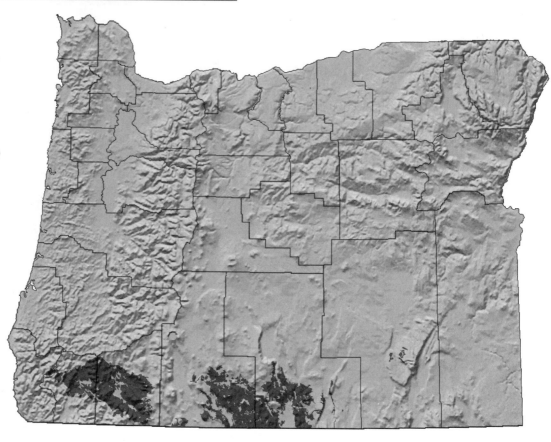

**Global Range:** Breeds across the United States, south from southern Oregon and southern Maine. It is resident from Florida and southern Texas south through Mexico to northern Guatemala.

**Habitat:** Throughout its extensive range, the Blue-gray Gnatcatcher is found in a variety of woodlands and brushy habitats. In Oregon, it is apparently restricted to the buckbrush chaparral of southwestern Oregon and mountain mahogany chaparral in the Klamath Basin and the mountains of southeastern Oregon.

**Reproduction:** The breeding season for this gnatcatcher gets under way in April, and most migrate south by August. This species builds its nest cup of plant material in a fork of a limb. The normal clutch of 4 to 5 (range 3-7) eggs is incubated for 15 days by both parents. The young fledge in about 2 weeks, and are tended by the parents for about 3 more weeks.

**Food Habits:** Blue-gray Gnatcatchers eat insects, insect eggs, and spiders. Insects taken include beetles, flies, ants, and gnats. Most food is gleaned from leaves and branches, but some insects are taken on the wing.

**Ecology:** Blue-gray Gnatcatchers maintain territories from about 2 to 7 acres in size. The female sometimes assists the male in territorial defense. This species is often parasitized by the Brown-headed Cowbird.

**Comments:** This species is peripheral to Oregon. It was not a known breeder in the state until 1962.

*References:* Gilligan et al. 1994, Root 1967, 1969b.

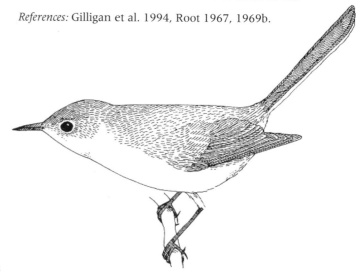

# Western Bluebird
## (Sialia mexicana)

*Order: Passeriformes*
*Family: Muscicapidae*
*State Status: Sensitive*
*Federal Status: None*
*Global Rank: G5*
*State Rank: S4*
*Length: 7 in (18 cm)*

**Global Range:** Breeds in western North America, from southern British Columbia east to the mountains of Colorado and New Mexico and south in the highlands to southern Mexico. It is absent from the most arid portions of the interior West.

**Habitat:** Western Bluebirds occupy a variety of habitat types, which change from western to eastern Oregon. In every habitat, they require nest holes or nest boxes. In western Oregon, they breed in forest clear-cuts with standing snags, around farms in agricultural lands, in riparian woodlands, and in open oak-ponderosa pine woodlands. In eastern Oregon, they use agricultural areas, open ponderosa pine or Douglas-fir woodlands (where stands of aspen are present), and juniper woodlands.

**Reproduction:** Western Bluebirds begin breeding by early May, and may rear 2 broods per year. They are secondary cavity nesters, and nest in abandoned woodpecker holes. They also make extensive use of nest boxes. The clutch of 4 to 6 (range 3-8) eggs is incubated about 2 weeks. Young leave the nest in 15-18 days, and are fed by the male while the female starts the second clutch.

**Food Habits:** This bluebird eats mostly insects, which it catches in the air or on the ground. It catches grasshoppers, caterpillars, ants, and beetles. Also included in the diet are spiders, earthworms, and snails. When available, it also eats a fair amount of fruit and berries.

**Ecology:** Western Bluebirds use an area of 1 to several acres, which is thought to be the same as their defended territory. During winter, they move from higher forests to lower valleys.

**Comments:** Competition for nest holes with introduced European Starlings is thought to be the main factor causing the decline of the Western Bluebird west of the Cascade Mountains during the last 4 or 5 decades. Only populations in western Oregon are state listed as Sensitive.

*References:* Anderson 1970, Gilligan et al. 1994, Marshall 1992.

**251**

# Mountain Bluebird
## *(Sialia currucoides)*

*Order: Passeriformes*
*Family: Muscicapidae*
*State Status: None*
*Federal Status: None*
*Global Rank: G5*
*State Rank: S4*
*Length: 7.5 in (19 cm)*

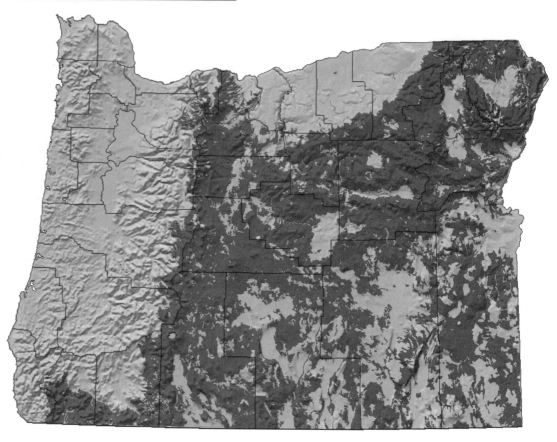

**Global Range:** Breeds from the southern Yukon to southern California east to North Dakota and Oklahoma. They are absent near the coast. In winter, some move south to Texas and northern Mexico.

**Habitat:** Mountain Bluebirds live in open coniferous forests, juniper woodlands, the edges of meadows, clear-cuts, and recently burned areas of higher elevation forests. They occupy true fir (*Abies*), lodgepole pine, subalpine fir, and mixed-coniferous woodlands.

**Reproduction:** In Oregon, their breeding season commences in late April (probably later at higher elevations). Mountain Bluebirds nest in natural cavities or woodpecker holes in trees at the edge of forest clearings. The clutch of 5 or 6 (range 4-8) eggs is incubated by both parents for about 2 weeks. Fledglings are tended by both sexes, and may remain to assist in the rearing of later broods.

**Food Habits:** About 90% of the Mountain Bluebird's diet consists of insects, including beetles, ants, bees, wasps, cicadas, caterpillars, grasshoppers, and crickets. The remaining 10% is mainly fruits and berries.

**Ecology:** Mountain Bluebird territories generally cover 5-15 acres around the nest, including flycatching perches.

**Comments:** Reflecting the range of habitats used by this species, the Mountain Bluebird can be found breeding east of the Cascade Mountains from about 2,000 feet to timberline.

*References:* DeGraaf et al. 1991, Gilligan et al. 1994, Power 1966.

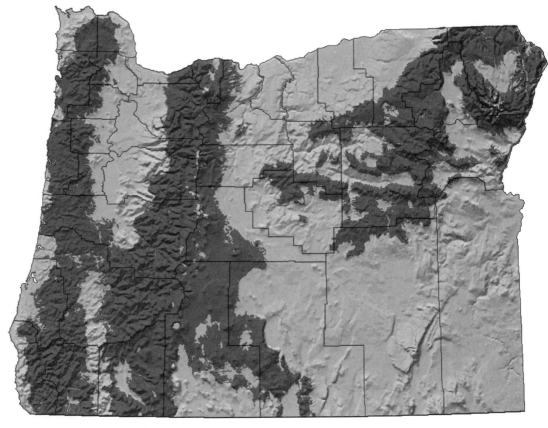

## Townsend's Solitaire
### *(Myadestes townsendi)*

*Order: Passeriformes*
*Family: Muscicapidae*
*State Status: None*
*Federal Status: None*
*Global Rank: G5*
*State Rank: S4*
*Length: 9 in (23 cm)*

**Global Range:** Breeds from south-central Alaska south in western North America to southern California, Arizona, and the highlands of central Mexico. It is found as far east as the Rocky Mountains, from Montana south to northern New Mexico.

**Habitat:** This species breeds in open forests and the forested edges of clear-cuts from the ponderosa pine zone up to timberline. It occupies a variety of montane and subalpine forest types, especially areas with some shrub understory.

**Reproduction:** Townsend's Solitaires are nesting by May in Oregon. They nest in rocks or other cover on or near the ground. The usual clutch is 4 eggs (range 3-4), and both sexes attend the young. The breeding biology is incompletely known.

**Food Habits:** The diet varies seasonally, consisting mostly of insects and spiders in summer, and fruits and berries in winter. Juniper berries are an important winter food. Insects eaten include beetles, moths, caterpillars, ants, termites, bees, wasps, and true bugs. Fruits, other than juniper berries, vary with location, but may include madrone, gooseberries, poison oak, mistletoe, honeysuckle, and wild cherries.

**Ecology:** Several studies have found Townsend's Solitaires occurring at a density of about 3 pairs per 100 acres of habitat in the breeding season. They apparently maintain winter territories as well, the size of which is related to the food supply. During winter, this species moves from its breeding habitat downslope, mostly into juniper woodlands.

**Comments:** Because of its preference for open habitats, clear-cuts in forests may benefit this species. Townsend's Solitaire is a local breeder in Oregon's Coast Range.

*References:* DeGraaf et al. 1991, Gilligan et al. 1994, Lederer 1977, Salomonson and Balda 1977, Zeiner et al. 1990a.

# Veery
## *(Catharus fuscescens)*

*Order: Passeriformes*
*Family: Muscicapidae*
*State Status: None*
*Federal Status: None*
*Global Rank: G5*
*State Rank: S4?*
*Length: 7 in (18 cm)*

*References:* DeGraaf
et al. 1991, Gilligan
et al. 1994,
Harrison 1978.

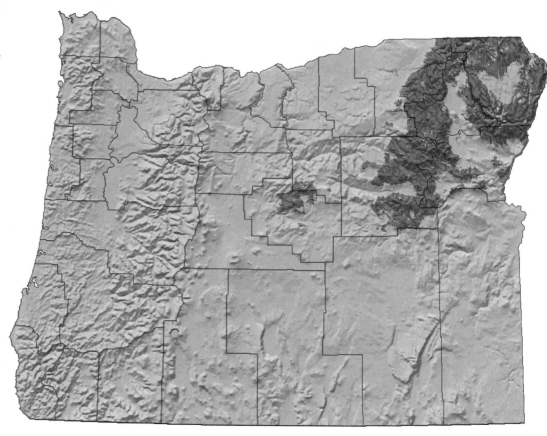

**Global Range:** Breeds from southern British Columbia east across southern Canada and the northern United States to the Atlantic Coast. Inland, it is found as far south as Colorado and the Appalachian Mountains of northern Georgia. The Oregon populations are at the western and southern limit of its range in the Pacific states.

**Habitat:** The Veery is a bird of damp deciduous forests and woodlands. In its Oregon range, it is found only in riparian woodlands and thickets along streams and rivers in both the mountains and valleys of the Blue Mountain region, including cottonwood riparian forests along the John Day River.

**Reproduction:** The breeding biology of the Veery has not been studied in Oregon, but elsewhere it lays eggs in May or June. The nest is a cup of plant materials on or near the ground, usually near the base of a tree or

shrub. The normal clutch size is 4 (range 3-6) eggs, which are incubated for 10-12 days. Young fledge in slightly less than 2 weeks.

**Food Habits:** About 60% of the diet is animal matter (insects and other invertebrates), mostly found on the forest floor, but occasionally gleaned from leaves. The remainder of its diet is plant material, especially small fruits and berries.

**Ecology:** The Veery requires a dense understory of low trees or shrubs in its riparian habitat. No information is available about the specifics of the natural history of this species in Oregon. An Idaho study found the occurrence of Veerys in cottonwood stands was positively correlated with increasing stand size.

**Comments:** The Veery winters in northern and central South America. Most leave Oregon by the middle of September.

*Order: Passeriformes*
*Family: Muscicapidae*
*State Status: None*
*Federal Status: None*
*Global Rank: G5*
*State Rank: S5*
*Length: 7 in (18 cm)*

*References:* DeGraaf
et al. 1991,
Gabrielson and
Jewett 1940,
Gilligan et al. 1994,
Zeiner et al. 1990a.

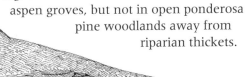

**Global Range:** Swainson's Thrush has an extensive breeding range across North America, from Alaska across the forests of Canada to the Atlantic Coast. It is found as far south as southern California and northern New Mexico in the West and Virginia in the East. It is absent from the Great Plains, much of the Midwest, and the Southeast. Winters are spent from Mexico south to South America.

**Habitat:** Swainson's Thrush is a bird of moist forest thickets. In western Oregon, it is found throughout dense, lowland Douglas-fir forests, as well as in riparian areas within the forest. In eastern Oregon, it occurs in higher-elevation coniferous forests and aspen groves, but not in open ponderosa pine woodlands away from riparian thickets.

**Reproduction:** The breeding season begins in late May, and most eggs have been laid by mid-July. The nest cup of plant material is usually on a lower limb of a tree, near the trunk. The female incubates the clutch of 3 or 4 (range 3-5) eggs for 10-13 days, and the young fledge within 2 weeks.

**Food Habits:** Swainson's Thrushes eat mainly animal matter, which they find on the forest floor or glean from leaves. They eat insects and their larvae (ants, bees, beetles, caterpillars, mosquitoes, craneflies), spiders, centipedes, millipedes, snails, and earthworms. Plant material eaten includes smaller fruits and berries (including wild cherries, blackberries, huckleberries, twinberry seeds, and elderberries).

**Ecology:** Although the conspicuous song of the Swainson's Thrush probably helps to establish territorial rights, the territory size is not reported. One study found 45-70 pairs per 100 acres, while another counted only 13 pairs in the same size area.

**Comments:** An Idaho study found this species more abundant in large tracts of undisturbed late-successional forests than in fragmented or harvested forests.

# Hermit Thrush
## (*Catharus guttatus*)

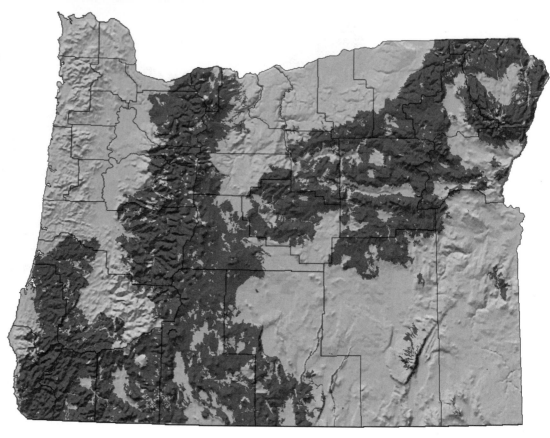

*Order: Passeriformes*
*Family: Muscicapidae*
*State Status: None*
*Federal Status: None*
*Global Rank: G5*
*State Rank: S4*
*Length: 7 in (18 cm)*

**Global Range:** Breeds from southern Alaska and most of Canada south to nearly the Mexican border in the western United States, and south to Virginia in the East. It is absent from the Great Plains and most of the southeastern United States.

**Habitat:** In Oregon, this thrush breeds in a wide variety of dense coniferous forests at all elevations. It is rarely present in ponderosa pine woodlands outside of riparian areas. It also makes limited use of denser juniper woodlands at the upper limit of their distribution.

**Reproduction:** Hermit Thrushes have begun their breeding season in Oregon by May. They build a nest of plant materials in a protected depression on the ground or on a low tree limb. The usual clutch is 3 or 4 (range 3-6) eggs, which the female incubates for 12-13 days. The young develop quickly, leaving the nest in about a week and a half.

**Food Habits:** The spring and summer diet is mostly animal, including crickets, ants, caterpillars, beetles, true bugs, grasshoppers, earthworms, spiders, and, occasionally, salamanders. Other foods include fruits, berries (madrone, snowberry, serviceberry, mistletoe, poison oak), and seeds, especially in fall and winter.

**Ecology:** One study in Arizona found that Hermit Thrushes defended territories of from 1 to 3 acres. They occur at densities of 1 to 10 pairs per 100 acres. Some Hermit Thrushes winter in western Oregon, but not necessarily the ones that breed there. Others fly south, the species wintering as far south as Guatemala.

**Comments:** An Idaho study found that Hermit Thrushes were more abundant in selectively harvested forests than in unharvested forest.

*References:* Gilligan et al. 1994, Johnston 1949, Morse 1972, Sealy 1974.

## American Robin
### (Turdus migratorius)

*Order: Passeriformes*
*Family: Muscicapidae*
*State Status: None*
*Federal Status: None*
*Global Rank: G5*
*State Rank: S5*
*Length: 10 in (25 cm)*

**Global Range:** Breeds throughout North America, from Alaska and Canada south through the United States in all but the extreme southwestern deserts, and again in the mountains to central Mexico.

**Habitat:** Robins breed in a wide range of habitats, but are restricted to areas with trees or, at the very least, large shrubs. These include forests of every sort, from sea level to timberline, most agricultural areas, urban and suburban areas, oak, pine, juniper, and riparian woodlands, and chaparral. While they find homes in riparian woodlands and towns of arid eastern Oregon, they are absent from treeless expanses of desert shrub communities.

**Reproduction:** The breeding season varies with elevation, starting in early spring in the lowlands of western Oregon, but later near timberline. The nest is a cup of plant material lined with mud and grass, situated in a tree or on a level ledge of a structure. The usual clutch is 4 (range 3-7) eggs, which the female incubates for 11-14 days. Young leave the nest in about 2 weeks, and the pair frequently renests.

**Food Habits:** The proportion of animal material in the diet doubles from 40% to 80% with the coming of spring and summer. Prey items include earthworms, beetles, grasshoppers, ants, termites, cutworms, caterpillars, butterflies, flies, millipedes, centipedes, spiders, and snails. Plant items are more important in fall and winter, and include many fruits and berries (cherries, serviceberries, mistletoe, orchard fruits, and grapes).

**Ecology:** American Robins defend a territory of about a half to several acres. After the breeding season and in migration, they are quite gregarious, and flocks forage on lawns, meadows, and pastures.

**Comments:** This is one of Oregon's most abundant birds. While present throughout the year, they leave higher-elevation breeding grounds for lower valleys or migrate south in winter.

*References:* Aldrich and James 1991, Gilligan et al. 1994, James and Shugart 1974, Wheelwright 1986.

# Varied Thrush
## (Ixoreus naevius)

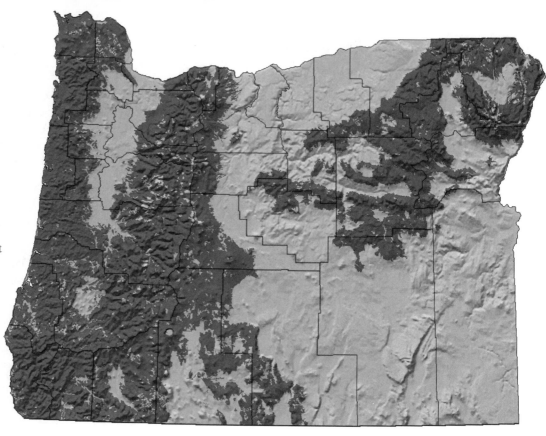

*Order:* Passeriformes
*Family:* Muscicapidae
*State Status:* None
*Federal Status:* None
*Global Rank:* G5
*State Rank:* S4
*Length:* 9.5 in (24 cm)

*References:* DeGraaf et al. 1991, Gabrielson and Jewett 1940, Gilligan et al. 1994, Martin 1970, Zeiner et al. 1990a.

**Global Range:** This is a breeding bird of the coniferous forests of western North America. It breeds in most of Alaska, the Yukon, British Columbia, and in the mountains of Oregon, Washington, and in the Coast Ranges of northern California.

**Habitat:** The Varied Thrush is most common in dense, older coniferous forests. Although it breeds in low-elevation and ponderosa pine forests, it is more typical of higher-elevation forest types, above the ponderosa pine zone. It is found in western hemlock and Sitka spruce forests along the northern Oregon coast, but does not use juniper woodlands.

**Reproduction:** Varied Thrushes begin their breeding season by early April in Oregon. The nest cup of plant material is usually placed on a limb of a conifer, 10-15 feet off the ground. The clutch of 3 (range 2-5) eggs is incubated by the female for about 2 weeks.

**Food Habits:** Only 20% of the annual diet of the Varied Thrush consists of animals;

however, more animals are eaten in the breeding season. Plant material eaten includes acorns, berries, fruit (madrone, manzanita, blackberry, snowberry, poison oak, honeysuckle, orchard fruits), and seeds. Animal foods include beetles, ants, bees, wasps, caterpillars, millipedes, centipedes, crickets, flies, grasshoppers, sow bugs, and earthworms.

**Ecology:** A study in northern Idaho found this species was associated with late-successional forests. No information is available about its territory size or density. In winter, it is often forced to lower-elevation valleys by storms. Birds from higher-elevation habitats in Oregon may migrate south to California in winter.

**Comments:** During winter, this thrush occurs in towns, orchards, juniper woodlands, and, locally, in the riparian thickets of eastern Oregon.

# Wrentit
## *(Chamaea fasciata)*

*Order: Passeriformes*
*Family: Muscicapidae*
*State Status: None*
*Federal Status: None*
*Global Rank: G5*
*State Rank: S5*
*Length: 6 in (15 cm)*

*References:*
Erickson 1938,
Geupel and
DeSante 1990,
Gilligan et al.
1994, Williams
and Koenig 1980.

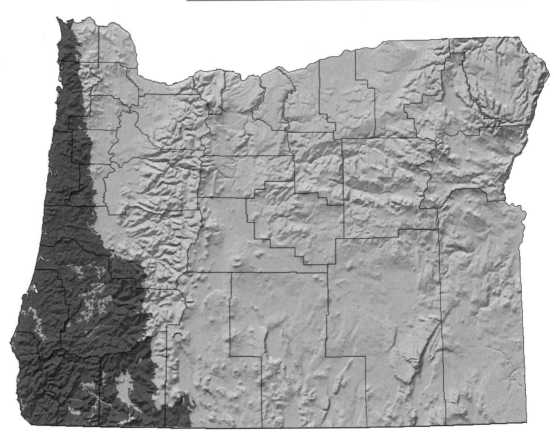

**Global Range:** Reaches the northern limit of its distribution on the Oregon side of the mouth of the Columbia River. It is found south along the coast, in the interior valleys of southern Oregon and western California, and along the west slope of the Sierra Nevada, to northern Baja California.

**Habitat:** The Wrentit is closely associated with thick brush and chaparral. It will use brushy areas in open oak or pine forests. The species composition of brushy communities is less important than vegetation structure: along the Oregon coast it will use salal and huckleberry thickets, in the San Francisco Bay area it is found in coyote bush, and in southern California it uses chamise chaparral.

**Reproduction:** The breeding season is under way by March. The pair builds a cup nest of plant material and spider webs, placed

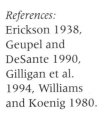

in a bush, near an opening in the chaparral. The usual clutch is 4 (range 3-5) eggs, which are incubated by both parents (the male takes the day shift). Young fledge in just over 2 weeks.

**Food Habits:** The overall diet is split half-and-half between plant and animal items, but in the breeding season it eats mostly, if not entirely, insects (ants, wasps, beetles, true bugs, flies, caterpillars) and spiders. It eats a variety of fruits and berries, including salal, poison oak, blackberry, sumac, elderberry, snowberry, and grape.

**Ecology:** Wrentit territories are small (1-3 acres) and are defended throughout the year. In favorable habitat, nearly every acre will be occupied by a resident pair of Wrentits. Following the breeding season, the young of the year move upslope.

**Comments:** Wrentits occupy brushfields growing in regenerating clear-cuts, and appear to be expanding their range northward along the east slopes of the inner Coast Range in response to timber harvesting.

# Gray Catbird
## (Dumetella carolinensis)

*Order: Passeriformes*
*Family: Mimidae*
*State Status: None*
*Federal Status: None*
*Global Rank: G5*
*State Rank: S4?*
*Length: 9 in (23 cm)*

*References:* DeGraaf
et al. 1991,
Gabrielson and
Jewett 1940,
Gilligan et al. 1994,
Littlefield 1990,
Peterson 1990.

**Global Range:** Breeds coast to coast, from southern Canada to the southeastern United States. In the West, it occurs along the coast only in British Columbia, and its western limit moves inland, generally following the Rocky Mountains and nearby ranges, south and east to New Mexico. It winters from the Gulf Coast south to Panama.

**Habitat:** In the East, the Gray Catbird occupies a variety of disturbed sites (roadsides, suburban gardens), but in Oregon it is closely associated with riparian vegetation along rivers and streams in the valleys and slopes of the Blue Mountain region.

**Reproduction:** In Oregon, this species begins breeding when it arrives in late May. Its short breeding season is over quickly, and it flies south in August. The cup nest of plant material is placed fairly low (3-10 feet high) in a thicket. The usual clutch of 4 (range 3-6) eggs is incubated by the female for 12-13 days. Young leave within 2 weeks. A second clutch is possible.

**Food Habits:** Gray Catbirds have a diet that is balanced between animal (mainly insects and spiders) and plant material and varies seasonally. More insects are eaten in spring and summer, and a variety of fruits and berries are added to the diet as they ripen later in the year.

**Ecology:** There has been little study of the natural history of the Gray Catbird in Oregon. Gabrielson and Jewett (1940) were unable even to confirm its breeding in the state. Like the Wrentit, it tends to confine its activities to thickets, being more easily heard than seen.

**Comments:** The limits of the range for this species appear to be expanding westward, generally and in Oregon. The Gray Catbird has been observed in summer at Malheur National Wildlife Refuge and may breed in the vicinity.

## Sage Thrasher
### (Oreoscoptes montanus)

*Order: Passeriformes*
*Family: Mimidae*
*State Status: None*
*Federal Status: None*
*Global Rank: G5*
*State Rank: S4*
*Length: 8.5 in (22 cm)*

**Global Range:** Breeds from southern British Columbia south to the southwestern edge of the Great Basin in eastern California and east to western South Dakota and northern New Mexico.

**Habitat:** As its name implies, the Sage Thrasher is associated with arid desert shrub communities dominated by sagebrush (*Artemisia*). It is also found in other arid desert communities (greasewood, saltbush, rabbitbrush). There are often occasional junipers in its sagebrush habitat, but it avoids areas where junipers are dominant.

**Reproduction:** This species begins breeding in late April and leaves the state for its wintering grounds in Mexico and the southwestern United States in September. The cup nest of plant material is usually located in sagebrush. The normal clutch is 4 or 5 (range 3-7) eggs, which both sexes incubate for 15 days. The young remain in the nest at least 11 days.

**Food Habits:** The Sage Thrasher feeds mostly on insects (grasshoppers, beetles, Mormon crickets, ants, bees) in spring and summer. It also eats fruits and berries as they ripen.

**Ecology:** Sage Thrashers have a territory of about 2- 3 acres. There may be as many as 30 pairs per square kilometer. They prefer larger bushes with dense foliage for nesting, frequently with an eastern exposure.

**Comments:** In some parts of its range, the Sage Thrasher also breeds in mountain mahogany thickets where they grow adjacent to stands of sagebrush.

*References:* DeGraaf et al. 1991, Gilligan et al. 1994, Medin 1990, Petersen and Best 1991, Reynolds and Rich 1978, Rotenberry and Wiens 1989.

# American Pipit
## (*Anthus rubescens*)

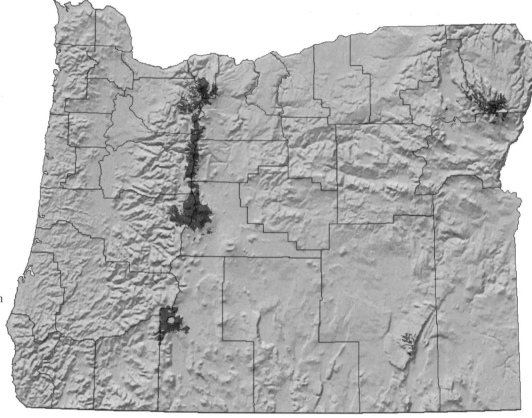

*Order: Passeriformes*
*Family: Motacillidae*
*State Status: None*
*Federal Status: None*
*Global Rank: G5*
*State Rank: SU*
*Length: 6.5 in (17 cm)*

*References:* Gabrielson
and Jewett 1940,
Gibb 1956, Gilligan
et al. 1994, Verbeek
1970, Zeiner et al.
1990a.

**Global Range:** The American Ornithologists' Union (1989) elevated the eastern Asian and North American populations of Water Pipits to full species status, known as the Water Pipit (*Anthus spinoletta*) and the American Pipit (*Anthus rubescens*), respectively. American Pipits breed from Alaska, northern Canada, and Greenland, south in the Rocky Mountains, to New Mexico, and west to eastern Oregon and the Sierra Nevada of California.

**Habitat:** In the northern part of its range, the American Pipit breeds in tundra. In Oregon, this is a bird of alpine communities. It prefers to nest on the ground of mossy south-facing slopes, which are free of snow earlier in the year. It also nests on level areas near streams.

**Reproduction:** The breeding season in Oregon gets under way in mid-June, appropriately late, given the high elevation of its nesting habitat. The nest cup is made of plant material and lined with grass or hair, frequently under the cover of an overhanging rock or a bush. The usual clutch is 4 or 5 (range 3-7) eggs, which are incubated by the female for about 2 weeks. Young are independent in about a month.

**Food Habits:** The American Pipit feeds on insects (including grasshoppers, crickets, beetles, ants, flies) and spiders during the breeding season, usually taken on the ground or from the surface of a snow field. It will forage in shallow water. In winter, along the coast, it adds molluscs and crustaceans to its diet. It also eats some seeds and berries.

**Ecology:** Breeding territories of about half an acre are defended. During winter, it moves to lower elevations, where it forms large flocks. Some American Pipits winter in Oregon, but not necessarily the ones that breed here. It mainly winters in the southern United States, Mexico, and Central America.

**Comments:** Other than rosy finches, this is the only Oregon bird that breeds exclusively in alpine habitat.

## Cedar Waxwing
*(Bombycilla cedrorum)*

*Order: Passeriformes*
*Family: Bombycillidae*
*State Status: None*
*Federal Status: None*
*Global Rank: G5*
*State Rank: S5*
*Length: 7 in (18 cm)*
Pg. 484

**Global Range:** Breeds from coast to coast in North America, from central Canada south to northern California, Kansas, and the Appalachian Mountains of northern Georgia.

**Habitat:** In Oregon, Cedar Waxwings are forest birds, using the edges of coniferous forests, mixed coniferous-deciduous forests, and deciduous riparian forests and woodlands in coniferous forests. They rarely use quaking aspen groves in eastern Oregon, and are found in mesic mountain forests, rather than riparian woodlands of hot desert valleys.

**Reproduction:** The Cedar Waxwing begins breeding rather late, usually waiting until June to lay eggs. The nest cup is placed near the end of a branch high in a tree. The usual clutch is 4 or 5 (range 3-6) eggs, which the female incubates for about 2 weeks. The young fledge in about 2-1/2 weeks.

**Food Habits:** This bird is primarily a frugivore, but during the breeding season up to one-quarter of its diet consists of insects (including beetles, ants, flies, caterpillars, grasshoppers, crickets, mayflies, and scale insects). It eats a wide variety of fruits, berries, and blossoms (including orchard fruits like cherries, and native species such as mistletoe, madrone, huckleberry, and rose hips). Tree sap is an important winter food.

**Ecology:** Cedar Waxwings often breed in small colonies, many birds nesting in the same tree, as close as 7 feet to one another. They defend only a small area around the nest. In winter, they congregate in large, nomadic flocks, seeking out concentrations of berries.

**Comments:** In winter, they often visit ornamental shrubs in residential areas, sometimes in the company of Bohemian Waxwings.

*References:* Gabrielson and Jewett 1940, Gilligan et al. 1994, Jewett et al. 1953, Putnam 1949, Rothstein 1971.

# Loggerhead Shrike
## (Lanius ludovicianus)

*Order: Passeriformes*
*Family: Laniidae*
*State Status: Sensitive*
*Federal Status: None*
*Global Rank: G4G5*
*State Rank: S4*
*Length: 9 in (23 cm)*

*References: Gilligan et al. 1994, Marshall et al. 1996, Reynolds 1979, Smith 1973a, 1973b, Woods and Cade 1996.*

**Global Range:** Breeds in central Canada and from eastern Washington and Oregon east to New York and south throughout the United States to southern Mexico.

**Habitat:** Loggerhead Shrikes occur in almost any fairly open vegetation type where there are occasional tall shrubs or trees for perching and nesting. This includes parts of sagebrush, bitterbrush, greasewood, and other desert communities, juniper woodlands, very open pine or oak woodlands, and mountain shrub communities.

**Reproduction:** The nesting season is under way by April. The nests is a lined cup of plant material. Clutches average 4 or 5 eggs, which the female incubates for 14-16 days, while being fed by the male. The young fledge in about 3 weeks, and are independent by 7 weeks. Two broods are raised each year.

**Food Habits:** Loggerhead Shrikes are passerines that have the foraging strategy and diet of small raptors. They fly out from a perch to attack prey items, which include many types of large insects (beetles, grasshoppers, dragonflies, crickets, butterflies, and moths). They also eat just about any other animal they can catch, including small birds, mammals, reptiles, amphibians, and fish. They eat more invertebrates in spring and summer, and more warm-blooded prey in winter. They also eat some carrion.

**Ecology:** The Loggerhead Shrike defends a breeding territory of 20-40 acres. Where they winter in Oregon, they remain territorial, but most depart the state by October. This species is declining in Idaho, and may also be declining throughout its range.

**Comments:** Where food is abundant, Loggerhead Shrikes will store kills by impaling them around their territory on twigs or thorns. Only populations in the high lava plains and Columbia Basin ecoregions are considered sensitive by the Department of Fish and Wildlife.

*Order: Passeriformes*
*Family: Sturnidae*
*State Status: None*
*Federal Status: None*
*Global Rank: G5*
*State Rank: SE*
*Length: 8 in (20 cm)*

*References:* Bray et al. 1975, Feare 1984, Gilligan et al. 1994, Ingold 1989, Weitzel 1988, Zeiner et al. 1990a.

**Global Range:** The European Starling is native to Europe and western Asia from Iceland east to Lake Baikal and south to Italy, Iran, and Pakistan. It has been widely introduced in North America, from Alaska east to Newfoundland and south to Baja California and Florida. It has also been introduced to South Africa, Australia, New Zealand, Hawaii, and most larger islands in the Caribbean Sea.

**Habitat:** European Starlings occupy urban and agricultural areas, pastures, open woodlands, riparian forests, and edges of meadows and clear-cuts. They are nearly always found near human development. Since they are cavity nesters, they are not found in treeless grasslands and deserts, but may be common around towns and ranches in eastern Oregon. They are not found in dense forests.

**Reproduction:** European Starlings are secondary cavity nesters that often evict other primary and secondary cavity nesters, including woodpeckers and American Kestrels, from their nest holes. They also nest in buildings. Starlings begin breeding in April and may raise 2 broods of 4 to 6 (range 3-8) young. The eggs are incubated by both parents for about 2 weeks. The young fledge at 3 weeks of age.

**Food Habits:** Insects (including beetles, ants, and grasshoppers) make up about half the European Starling's diet. They also eat other invertebrates (earthworms, snails, spiders), grain, seeds, nuts, and fruit. Most food is found on the ground or just below the surface of moist soil.

**Ecology:** European Starlings defend a small territory of 1-2 feet around the nest. They frequently nest in proximity to one another. During winter, they aggregate into large flocks, which may travel many miles from roosts to feeding areas. They may join flocks of blackbirds or other species.

**Comments:** Starlings are a pest species that are in part responsible for the decline of many native cavity-nesting birds. In addition, they destroy grain and orchard crops, resulting in considerable economic losses. Starlings were first introduced to the Portland area in the late 19th century, but soon disappeared. They were next noted in the Willamette Valley in the 1940s, and have since colonized the entire state.

# Solitary Vireo
## (*Vireo solitarius*)

*Order: Passeriformes*
*Family: Vireonidae*
*State Status: None*
*Federal Status: None*
*Global Rank: G5*
*State Rank: S4?*
*Length: 5.5 in (14 cm)*

**Global Range:** The Solitary Vireo has a complicated distribution in North America, breeding coast to coast across Canada and in most of the western part of the continent south to Honduras. It is absent from the plains east of the Rocky Mountains and from most of the southeastern United States, descending from the Northeast only in the Appalachian Mountains as far south as Georgia.

**Habitat:** This vireo breeds in a variety of open forest habitats throughout its range. In Oregon, it is found in open, usually mixed, coniferous-deciduous woodlands, quaking aspen groves, ponderosa pine and other coniferous woodlands, riparian woodlands, and forest edges.

**Reproduction:** The Solitary Vireo begins its breeding season by early May. It often places its nest cup of plant fibers on the branch of a small conifer. There are usually 4 (range 3-5) eggs to a clutch, which the female incubates for about 2 weeks.

**Food Habits:** Solitary Vireos are primarily insectivorous, eating caterpillars, moths, beetles, bees, wasps, ants, stoneflies, dragonflies, crickets, and grasshoppers. They also eat spiders, seeds, small fruits, berries, and leaf galls.

**Ecology:** This species defends a territory of several acres around the nest. It migrates south by October, journeying to its wintering grounds in the southern United States, Mexico, and Central America.

**Comments:** Nests of the Solitary Vireo are subject to parasitism by Brown-headed Cowbirds. Recent genetic studies (Johnson 1995) suggest that the Pacific Coast subspecies (*V. s. cassinii*) may warrant specific recognition as Cassin's Vireo (*Vireo cassinii*).

*References:* Gabrielson and Jewett 1940, Gilligan et al. 1994, Hamilton 1962, Johnson 1995, Johnson et al. 1988, Marvil and Cruz 1989.

# Hutton's Vireo
## (Vireo huttoni)

Order: Passeriformes
Family: Vireonidae
State Status: None
Federal Status: None
Global Rank: G5
State Rank: S4
Length: 4.5 in (11 cm)

References: DeGraaf et al. 1991, Gilligan et al. 1994, Hamilton 1962, Zeiner et al. 1990a.

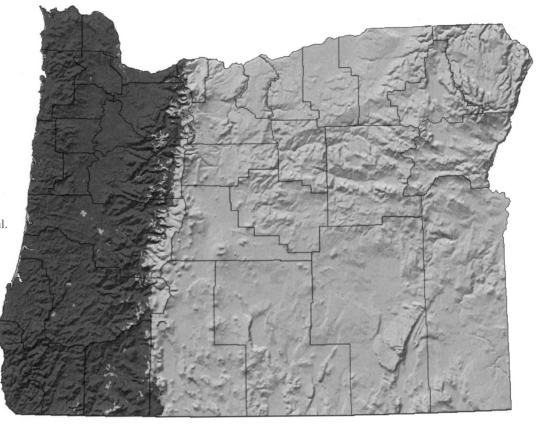

**Global Range:** Resident west of the Cascade Mountains, from southern British Columbia south, in the Coast Ranges and western foothills of the Sierra Nevada, to the mountains of northern Baja California. A few are found east of the Cascade Mountains in southwestern Klamath County. There is a disjunct population in southern Baja California, and a second major resident range from southeastern Arizona south through Mexico to Guatemala.

**Habitat:** Throughout its range, this species is associated with oak or pine-oak woodlands and riparian woodlands in canyons. In Oregon, it is found in a variety of mixed coniferous-deciduous forests and woodlands, including areas where oak, madrone, and bigleaf maple mix with western hemlock and Douglas-fir.

**Reproduction:** At lower elevations, the breeding season is under way by May. The cup nest is placed at a fork in a tree limb, usually 7-25 feet high. The clutch of 4 (range 2-5) eggs is incubated by both sexes for just over 2 weeks. Young fledge in another 2 weeks.

**Food Habits:** Hutton's Vireos glean insects (beetles, moths, bees, wasps, ants) and spiders from foliage. They also eat some seeds, fruits, berries, and plant galls. Some insects are caught in the air by flying out from a perch.

**Ecology:** Pairs nest alone, usually each using 2 or 3 acres of habitat. There may be some post-breeding movement upslope. During winter, residents at higher elevations descend to the lowlands, but few migrate south out of the state. They are subject to nest parasitism by the Brown-headed Cowbird.

**Comments:** The Hutton's Vireo's bill is more robust than that of the Ruby-crowned Kinglet, from which it is otherwise difficult to distinguish, suggesting it may take larger or harder-bodied prey items.

# Warbling Vireo
## *(Vireo gilvus)*

*Order: Passeriformes*
*Family: Vireonidae*
*State Status: None*
*Federal Status: None*
*Global Rank: G5*
*State Rank: S5*
*Length: 5 in (13 cm)*

*References:* Douglas et al. 1992, Gabrielson and Jewett 1940, Gilligan et al. 1994, Johnson et al. 1988, Zeiner et al. 1990a.

**Global Range:** Breeds from southern Alaska and British Columbia east across Canada to the Great Lakes, and farther east to the Atlantic Coast of New England, as well as south to southern California, Arizona, the highlands of central Mexico, and the Gulf Coast of Louisiana. It is absent from the extreme southeastern United States.

**Habitat:** The Warbling Vireo is another bird that breeds in stands of deciduous trees, such as red alder, bigleaf maple, or quaking aspen. Although these trees can occur within a mosaic of coniferous forest or along riparian corridors, Warbling Vireos do not use interiors of dense coniferous forests. They also breed in oak and cottonwood woodlands and in suburbs. They are absent from vast expanses of treeless desert in eastern Oregon.

**Reproduction:** This vireo returns to Oregon in April and is nesting by mid-May. The nest cup is built at the end of a branch high (20-90 feet) in a tree. The clutch of 4 (range 3-5) eggs is incubated by both parents for about 12 days. The young fledge in about 2 weeks.

**Food Habits:** Warbling Vireos are mostly insectivorous, gleaning prey from leaves high in the canopy. Their food includes caterpillars, moths, true bugs, beetles, spiders, lepidopteran eggs, aphids, grasshoppers, ants, scale insects, flies, and dragonflies. A small amount (3-10%) of the diet consists of seeds and berries (e.g., elderberry, snowberry, poison oak).

**Ecology:** The Warbling Vireo has a conspicuous song, presumably used in territorial defense. The home range is about a 120-foot radius around the nest. This species leaves Oregon for its winter home in Mexico and Central America by October. It is frequently the victim of nest parasitism by the Brown-headed Cowbird.

**Comments:** This is the most abundant vireo in western Oregon.

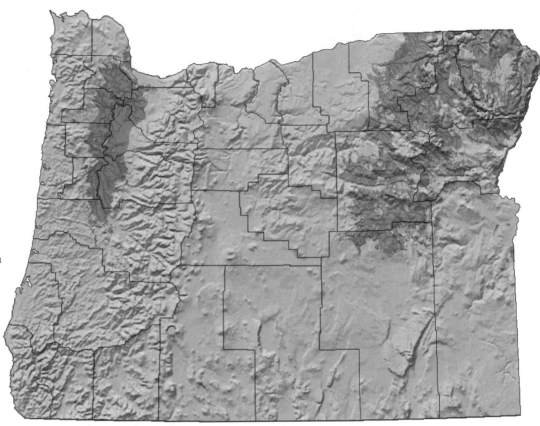

# Red-eyed Vireo
## (Vireo olivaceus)

*Order:* Passeriformes
*Family:* Vireonidae
*State Status:* None
*Federal Status:* None
*Global Rank:* G5
*State Rank:* S4
*Length:* 6 in (15 cm)

*References:* Gabrielson and Jewett 1940, Gilligan et al. 1994, Johnson and Zink 1985, Williamson 1971.

**Global Range:** As currently recognized, this species has an extensive breeding range in the Western Hemisphere. It is absent from most of the United States west of the Rockies. The Oregon populations are on the western limit of its distribution. It breeds from northern British Columbia, eastern Washington, and northern Idaho east across Canada, and throughout all of the eastern United States. It also breeds along the coastal areas of Mexico and Central America, into northern and central South America.

**Habitat:** In Oregon, the Red-eyed Vireo is closely associated with cottonwood, alder, and willow riparian woodland and gallery forest. It also uses quaking aspen and Oregon ash groves. Larger stands with mature trees are preferred.

**Reproduction:** In Oregon, the breeding season of the Red-eyed Vireo is under way by June. The nest cup is built fairly low (5-10 feet) in a deciduous tree. The usual clutch of 4 (range 3-5) eggs is incubated by both parents for about 2 weeks. The young fledge in another 2 weeks.

**Food Habits:** Insects make up most (about 85%) of the Red-eyed Vireo's diet, like that of most vireos, during the breeding season. It consumes mayflies, dragonflies, caterpillars, moths, beetles, wasps, bees, ants, scale insects, spiders, and snails. The remainder of the diet includes wild fruits and berries (such as elderberries and blackberries).

**Ecology:** There is little information about this species in Oregon. On the East Coast, one study found the territory was between 1 and 2 acres. The male is said to forage higher in the tree canopy than the female, subdividing the food supply. Our Red-eyed Vireos winter in the Amazon Basin.

**Comments:** Although primarily found in the Blue Mountains region, these vireos also breed locally in the Willamette Valley and along the Columbia River in Multnomah County. There are also scattered breeding records throughout the northern part of eastern Oregon.

# Orange-crowned Warbler
## (Vermivora celata)

*Order: Passeriformes*
*Family: Emberizidae*
*State Status: None*
*Federal Status: None*
*Global Rank: G5*
*State Rank: S5*
*Length: 4.5 in (11 cm)*

**Global Range:** Breeds from Alaska across Canada to Labrador. It is absent from the United States east of the Rocky Mountains, but in the West is found breeding south to southern California and the mountains of Arizona, New Mexico, and west Texas.

**Habitat:** Orange-crowned Warblers breed in thickets of shrubs, which can occur as the undergrowth of forests or woodlands, along riparian corridors, in chaparral and mountain mahogany, at the edges of forest clearings, and in oak woodland. In eastern Oregon, they use thick stands of mixed deciduous trees (aspen, alder, willow) and ponderosa pine growing around subalpine meadows.

**Reproduction:** At lower elevations, the breeding season is under way by May. The nest cup is often on the ground or low in a thicket. The normal clutch is 5 (range 4-6) eggs. The nestlings fledge in 8-10 days.

**Food Habits:** Orange-crowned Warblers eat a variety of insects (caterpillars, beetles, wasps, ants, flies, true bugs, plant lice, bees, leafhoppers, scale insects, and aphids) and spiders. A small quantity of seeds, leaf galls, and fruit is also eaten. Most food is gleaned from leaves and twigs, but they also eat sap flowing from woodpecker holes.

**Ecology:** The song of the Orange-crowned Warbler likely serves to mark territorial limits. The home range in one California study was 5 acres, although densities in some areas are as high as one singing male per 2-1/2 acres. Although a few Orange-crowned Warblers remain in western Oregon in the winter, the majority travel to the southern United States, Mexico, and South America.

**Comments:** Forest harvest probably increases the amount of habitat available for this species.

*References:* DeGraaf et al. 1991, Gilligan et al. 1994, Zeiner et al. 1990a.

# Nashville Warbler
## *(Vermivora ruficapilla)*

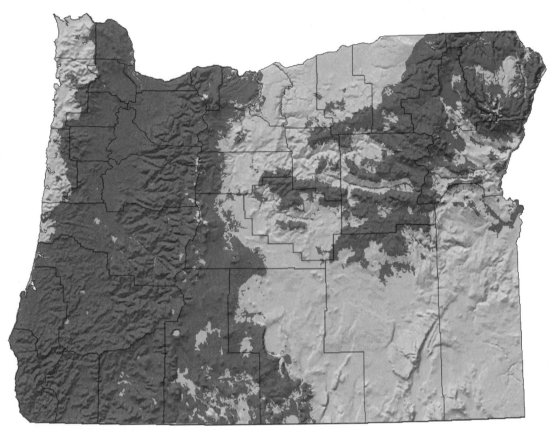

*Order: Passeriformes*
*Family: Emberizidae*
*State Status: None*
*Federal Status: None*
*Global Rank: G5*
*State Rank: S4?*
*Length: 4.5 in (11 cm)*

**Global Range:** The Nashville Warbler has an unusual distribution, breeding in the West from southern British Columbia south to southern California and east to the Rocky Mountains of Montana and Wyoming. In the East, it breeds in southern Canada east of Saskatchewan, and in the northeastern United States south to Maryland.

**Habitat:** Nashville Warblers are birds of brushy areas from the foothills to high elevations of mountains. Chaparral and buckbrush (genus *Ceanothus*), regenerating clear-cuts and burns, forest edges, the undergrowth of open, mixed forests, and shrubs bordering wet meadows or bogs are all used.

**Reproduction:** Depending on location and elevation, the Nashville Warbler's breeding season begins from April to May. The nest is usually placed in a depression on the ground, under the cover of some vegetation. There are usually 4 (range 3-5) eggs, which are incubated for 11-12 days by the female. The young are ready to leave the nest in another 11-12 days.

**Food Habits:** This warbler gleans food from the trunks and foliage of trees and shrubs. It mainly eats insects (caterpillars, beetles, wasps, ants, flies, true bugs, plant lice, bees, leafhoppers, aphids, and grasshoppers).

**Ecology:** In an Ontario study, the territory was only about half an acre. This species is often the subject of nest parasitism by Brown-headed Cowbirds. Most Nashville Warblers have left for their wintering grounds in Mexico and Central America by the end of September.

**Comments:** After the breeding season and prior to fall migration, this species may disperse upslope.

*References:* DeGraaf et al. 1991, Gilligan et al. 1994, Johnson 1976, Lawrence 1948, Zeiner et al. 1990a.

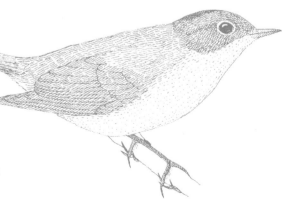

# Yellow Warbler
*(Dendroica petechia)*

*Order: Passeriformes*
*Family: Emberizidae*
*State Status: None*
*Federal Status: None*
*Global Rank: G5*
*State Rank: S4*
*Length: 4.5 in (11 cm)*

*References:* Ficken and Ficken 1965, 1966, Gabrielson and Jewett 1940, Gilligan et al. 1994, Weatherhead 1989.

**Global Range:** Breeds throughout most of North America, and is resident along the Florida Keys and the coasts of Mexico, Central America, and northern South America. It is absent from most of the southeastern United States, from central Texas east to Florida.

**Habitat:** This warbler is closely associated with various types of riparian vegetation, including willows and cottonwoods. It occupies riparian thickets in valleys, and follows them upward to mid-elevations in mountains. It makes use of willow thickets in mountain meadows and moist quaking aspen groves. It is absent from the expanses of eastern Oregon's desert valleys, except where there are riparian thickets or, rarely, trees around human settlements.

**Reproduction:** In Oregon, the Yellow Warbler is breeding by late May. The nest is usually fairly low (3-8 feet high) in a bush or tree. The normal clutch is 4 or 5 (range 3-6) eggs, which are incubated for 11-12 days by the female. The young leave the nest in another 9-12 days.

**Food Habits:** Like most warblers, this species feeds mostly on insects and spiders, gleaning them from leaves in the upper canopy of deciduous trees. Insects eaten include caterpillars, bark beetles, aphids, grasshoppers, flies, plant lice, bees, and wasps.

**Ecology:** The territory size of the Yellow Warbler is small, usually less than an acre. The territory and home range sizes vary considerably with location and habitat quality. Most Yellow Warblers migrate south by September, heading for their winter range along the coasts of Mexico, Central America, and eastern South America.

**Comments:** This species has declined in some regions due to extensive nest parasitism by Brown-headed Cowbirds.

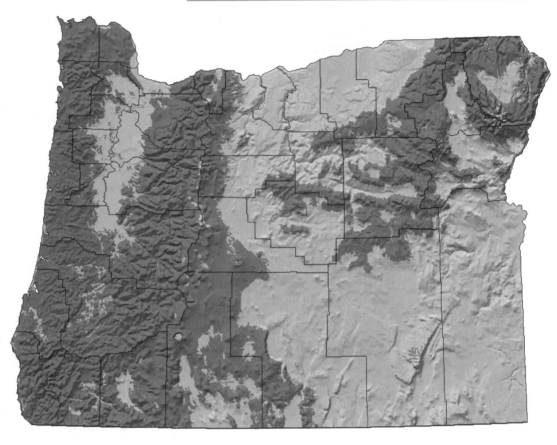

*Order: Passeriformes*
*Family: Emberizidae*
*State Status: None*
*Federal Status: None*
*Global Rank: G5*
*State Rank: S5*
*Length: 5 in (13 cm)*

*References:* DeGraaf et al. 1991, Gabrielson and Jewett 1940, Hubbard 1969, Zeiner et al. 1990a.

**Global Range:** Breeds from Alaska across Canada, and in the northeastern U.S., south in the Appalachian Mountains as far as Virginia. In the western U.S., it breeds south to central California, and in the mountains of southern California and the Great Basin, in the Rocky Mountains, and south, in the mountains of Mexico and Central America, to Guatemala. It is absent from much of the central and southeastern United States.

**Habitat:** In Oregon, the Yellow-rumped Warbler is found in all types of coniferous and mixed coniferous-deciduous forests. It is more numerous in the mountains. It prefers open forests and forest edges, especially around meadows or lakes. In the Rocky Mountains, it uses quaking aspen groves and can be expected in aspen groves in the mountains of eastern Oregon.

**Reproduction:** Yellow-rumped Warblers begin breeding by May. The nest cup is placed out on the limb of a conifer, often 15-20 feet high. The clutch of 4 or 5 eggs is incubated by the female for 12-13 days. Both sexes feed the young for another 2 weeks, when they fledge.

**Food Habits:** During the breeding season this warbler is mostly insectivorous, both gleaning insects from foliage and catching some in the air. It eats flies, beetles, ants, bees, wasps, craneflies, gnats, scale insects, aphids, plant lice, and caterpillars. It will eat nectar from flowers and sap from sapsucker holes. In winter, it turns to fruits, berries, and seeds for food.

**Ecology:** Studies in the East indicate a territory size of about 2 acres. Yellow-rumped Warblers occur at densities of about one pair per 10 acres. Following the breeding season, these warblers move upslope, where they may associate in mixed-species flocks prior to migration. Most winter south to Mexico and Guatemala, but some remain in Oregon, mostly in the Columbia Basin and low valleys west of the Cascade Mountains. In mild winters, hundreds of birds can be found along the Oregon coast.

**Comments:** The western race of this warbler was formerly known as Audubon's Warbler (*Dendroica auduboni*).

273

# Black-throated Gray Warbler
### (Dendroica nigrescens)

*Order: Passeriformes*
*Family: Emberizidae*
*State Status: None*
*Federal Status: None*
*Global Rank: G5*
*State Rank: S5*
*Length: 5 in (13 cm)*

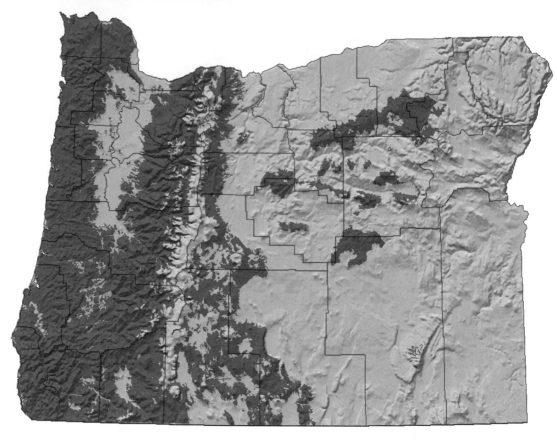

**Global Range:** The breeding range of the Black-throated Gray Warbler extends from southern British Columbia south to the mountains of northern Baja California and southeast to the Rocky Mountains of Wyoming. It occurs further south through New Mexico, and just enters northern Mexico.

**Habitat:** This warbler uses a wide range of forests, woodlands, and brushy areas at forest edges, including the brushy regeneration in recent clear-cuts. It can be found in deciduous and mixed deciduous-coniferous forests, as well as in oak and juniper woodlands. Dense, moist coniferous forests (e.g., coastal western hemlock-Sitka spruce forests) are avoided.

**Reproduction:** The breeding season for this species gets under way in May. The cup nest of plant material is placed low (3-10 feet high) in a tree. A normal clutch contains 4 (range 3-5) eggs, which the female takes charge of incubating. Both sexes attend to the needs of the nestlings.

**Food Habits:** As is typical for wood warblers, the Black-throated Gray Warbler gleans insects from foliage and catches a few in the air. It forages at low to mid-levels in the tree canopy. It has been

suggested that males forage higher in the canopy than females.

**Ecology:** Oregon studies have found about 15 singing males per 100 acres in Douglas-fir-alder forests in the western Cascades, and 4-12 birds per 100 acres in Willamette Valley oak woodlands. Most Black-throated Gray Warblers leave Oregon for their wintering grounds in the southwestern United States and Mexico by October.

**Comments:** Timber harvest may help create habitat for this species in western Oregon.

*References: DeGraaf et al. 1991, Gabrielson and Jewett 1940, Gilligan et al. 1994, Morrison 1982, Zeiner et al. 1990a.*

# Townsend's Warbler
## (Dendroica townsendi)

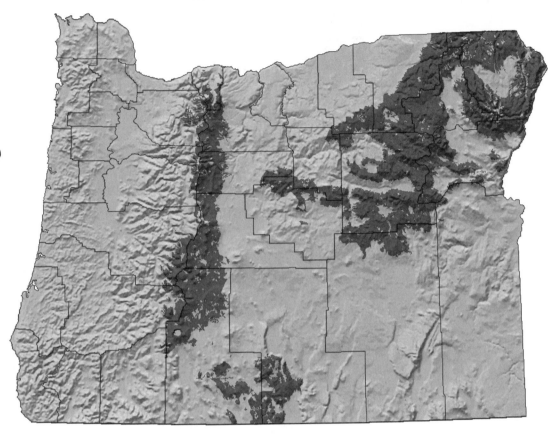

*Order: Passeriformes*
*Family: Emberizidae*
*State Status: None*
*Federal Status: None*
*Global Rank: G5*
*State Rank: S4*
*Length: 4.5 in (11 cm)*

**Global Range:** Has a relatively small breeding distribution, from southeastern Alaska south through British Columbia to the Rocky Mountains of Wyoming, and the Cascade and Blue Mountains of the Pacific Northwest.

**Habitat:** Most Townsend's Warblers in Oregon breed in higher-elevation coniferous forests (true fir, Engelmann spruce, lodgepole pine, Douglas-fir, mixed conifer) and woodlands. There are a few unusual records of nesting in lower-elevation coniferous forests near the coast.

**Reproduction:** Given its higher-elevation habitats, the breeding season does not fully begin until June. The nest cup is hidden in the limb of a conifer, often near the crown of a tall tree. The clutch of 3 to 5 eggs is incubated by the female.

**Food Habits:** Townsend's Warblers forage in foliage at the very tops of tall conifers, gleaning insects such as wasps, ants, beetles, caterpillars, flies, leafhoppers, bees, and scale insects. They also eat spiders and small amounts of seeds or leaf galls.

**Ecology:** The tree-top habits of this warbler may help to explain why so little is known about the specifics of its natural history. A study in Idaho and Montana found that it was more abundant in late-successional than in rotation-aged Douglas-fir forests. In winter and during migration, it uses a wide variety of woodland and brushy thicket habitats. During migration, it will join mixed-species flocks of other passerines.

**Comments:** A few Townsend's Warblers winter in western Oregon, but not necessarily the same ones that breed in the state. Most winter in Mexico and Central America.

*References:* DeGraaf et al. 1991, Gabrielson and Jewett 1940, Gilligan et al. 1994, Zeiner et al. 1990a.

275

# Hermit Warbler
## *(Dendroica occidentalis)*

*Order: Passeriformes*
*Family: Emberizidae*
*State Status: None*
*Federal Status: None*
*Global Rank: G5*
*State Rank: S4*
*Length: 4.5 in (11 cm)*

**Global Range:** The Hermit Warbler has an even smaller breeding range than Townsend's Warbler. It breeds from central Washington south to central California in forested areas from the Cascade Mountains and Sierra Nevada to the coast, and in the San Gabriel and San Bernardino mountains of southern California.

**Habitat:** The Hermit Warbler primarily uses Douglas-fir, western hemlock, Sitka spruce, lodgepole pine, mountain hemlock, and true fir forests. It is found in a variety of successional stages, but often selects areas where very tall trees tower above the canopy.

**Reproduction:** The Hermit Warbler's breeding season is under way by May. The nest cup is placed on a branch in the middle of a tall conifer. Clutches usually have 4 (range 3 to 5) eggs.

**Food Habits:** Hermit Warblers hunt in the tops of coniferous forests, both gleaning insects from foliage and darting out to catch flying insects on the wing. They eat caterpillars, beetles, true bugs, flies, wasps, and spiders.

**Ecology:** An Oregon study found as many as 143 Hermit Warblers per 100 acres, while a California

study in the Sierra Nevada found just over 6 per 100 acres. Most Hermit Warblers migrate south to wintering grounds in coastal southern California, Mexico, and Central America, leaving by October.

**Comments:** Like Townsend's Warbler, the preference of this species for life high in coniferous trees has made study of its natural history difficult.

*References:* Gabrielson and Jewett, 1940, Gilligan et al. 1994, Wiens and Nussbaum 1975.

# American Redstart
## *(Setophaga ruticilla)*

*Order: Passeriformes*
*Family: Emberizidae*
*State Status: None*
*Federal Status: None*
*Global Rank: G5*
*State Rank: SU*
*Length: 5 in (13 cm)*

*References:* DeGraaf et al. 1991, Gabrielson and Jewett 1940, Holmes et al. 1989, Sherry and Holmes 1988.

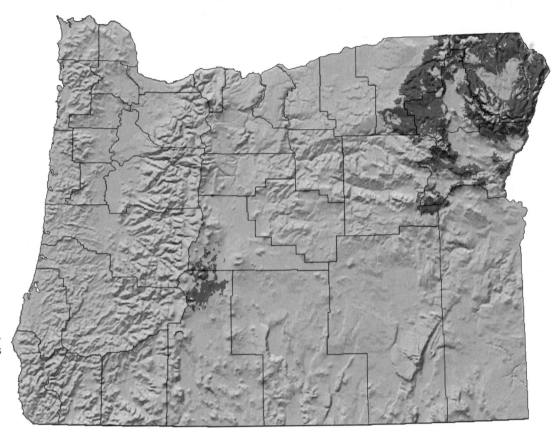

**Global Range:** Breeds from southeastern Alaska across Canada south to northern Florida. In the West, it occurs in northern Washington and northeastern Oregon, then its range moves inland, following the Rocky Mountains to Colorado, becoming continuous in the southeastern United States east of Texas.

**Habitat:** Throughout much of its range, the American Redstart is found in open deciduous forests with a brushy understory. In the mountains of Oregon, it is more specialized in its habitat use, breeding primarily in riparian thickets along streams, which are frequently within a matrix of a variety of coniferous forests and woodlands. It also uses similar deciduous vegetation around the margins of mountain lakes.

**Reproduction:** The breeding season of the American Redstart begins around June. The nest cup is placed on the limb of a shrub or tree, usually only 5-10 feet high. A normal clutch will have 4 (range 3-5) eggs, which are incubated for 12 days by the female. Young fledge quickly, leaving the nest in about a week and a half.

**Food Habits:** The American Redstart gleans insects from foliage and branches, and catches some flying insects on the wing. It also eats a few spiders. When available, a small amount of fruit will be included in the diet.

**Ecology:** In the eastern United States, breeding densities of American Redstarts have ranged from 10 to 51 per hectare. This species is frequently parasitized by Brown-headed Cowbirds.

**Comments:** The American Redstart was not confirmed as a breeder in Oregon until after 1940, and populations remain small and scattered. It has bred at Davis Lake in the central Cascade Mountains.

# Northern Waterthrush
## (Seiurus noveboracensis)

*Order: Passeriformes*
*Family: Emberizidae*
*State Status: None*
*Federal Status: None*
*Global Rank: G5*
*State Rank: S2?*
*Length: 6 in (15 cm)*

**Global Range:** Breeds throughout most of Alaska and Canada, but only enters the northern conterminous United States in the Pacific Northwest and in the Great Lakes Region, east to New England.

**Habitat:** The typical habitat of the Northern Waterthrush is riparian thickets in forests, near rapidly flowing water. On occasion, it will also use dense vegetation at the edges of lakes.

**Reproduction:** In Oregon, the breeding season for this warbler begins in late May. The nest is a cup of vegetation, built on the ground or in a hole in a stream bank. There are usually 4 or 5 (range 3-6) eggs, incubated by the female.

**Food Habits:** As one might expect from a species associated with water, this species differs from most other warblers by including aquatic species in its diet. Among other food items, it eats aquatic and terrestrial insects, spiders, molluscs, small fishes, and snails.

**Ecology:** Little information is available about the natural history of this species in Oregon. Northern Waterthrushes winter primarily in Mexico, Central America, and northern South America.

**Comments:** The Northern Waterthrush has a very limited distribution in Oregon, with known breeding populations along 3 creeks in the central Cascade Range.

*References:* Contreras 1988, DeGraaf et al. 1991, Gilligan et al. 1994.

278

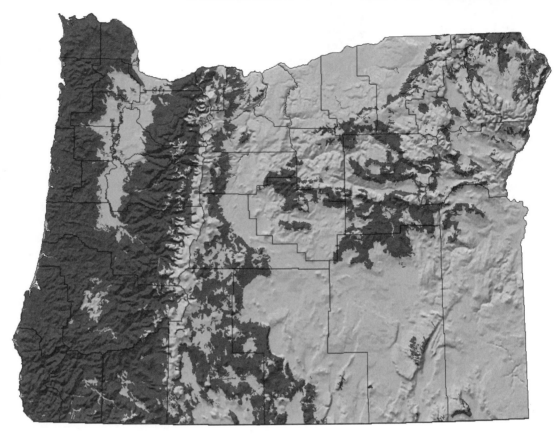

## Macgillivray's Warbler
### (Oporornis tolmiei)

*Order: Passeriformes*
*Family: Emberizidae*
*State Status: None*
*Federal Status: None*
*Global Rank: G5*
*State Rank: S4*
*Length: 5 in (13 cm)*

**Global Range:** Breeds from southern Alaska and the Yukon south to southern California and east to western South Dakota and the mountains of central New Mexico.

**Habitat:** This is another warbler that nests in thickets, which means it is found in local situations in a variety of vegetation types from lowlands to treeline, but typically in riparian vegetation, in or outside of forests, patches of forest undergrowth, forest edges, regenerating clear-cuts, young quaking aspen stands, chaparral, and mountain brush communities. It is absent from the broad, arid desert valleys of eastern Oregon.

**Reproduction:** Its breeding season is under way by late May. A nest cup is placed in a small tree or shrub, usually no more than a few feet off the ground. The usual clutch is 4 (range 3-6) eggs, which the female incubates for about 2 weeks. The young fledge in about a week and a half.

**Food Habits:** This warbler forages close to the ground in dense shrubs, mostly catching insects (beetles, wasps, ants, flies, true bugs, bees) and spiders.

**Ecology:** Densities vary widely, depending on habitat quality. In Idaho, this species was found to be closely associated with willow riparian habitats. Most leave Oregon for wintering grounds in Mexico and Central America by October.

**Comments:** Forestry practices such as clear-cutting probably increase habitat for this species.

*References:* DeGraaf et al. 1991, Douglas et al. 1992, Gabrielson and Jewett 1940, Gilligan et al. 1994, Pitocchelli 1990.

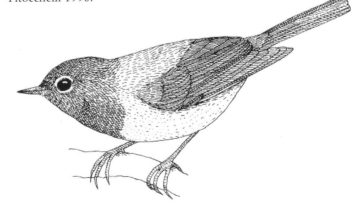

# Common Yellowthroat
## (Geothlypis trichas)

Order: Passeriformes
Family: Emberizidae
State Status: None
Federal Status: None
Global Rank: G5
State Rank: S5
Length: 5 in (13 cm)

**Global Range:** The Common Yellowthroat breeds coast to coast in North America, from south-coastal Alaska and the southern Yukon south to central Mexico. It is absent from some very arid regions, such as most of Baja California and parts of the Great Basin, Mojave, and Sonoran deserts.

**Habitat:** Common Yellowthroats are most closely associated with cattail marshes, coastal marshes, and streams flowing through wet meadows with borders lined with willows or other brushy vegetation. They seem to prefer areas with dense grasses or forbs interspersed with shrubs, especially if water is nearby. In deserts, they are restricted to riparian areas.

**Reproduction:** The breeding season for the Common Yellowthroat is under way by late April. The nest can be built in low bushes or attached to emergent vegetation over water. The normal clutch is 4 (range 3-6) eggs, which the female incubates for about 12 days. Young fledge in about 9-10 days. There are usually 2 broods a year.

**Food Habits:** A variety of insects (dragonflies, mayflies, damselflies, beetles, moths, butterflies, grasshoppers, flies, ants, aphids, leafhoppers) and

spiders are gleaned from foliage. Some seeds and fruit are also eaten.

**Ecology:** Common Yellowthroats defend a small territory, usually between 1 and 2 acres in size. They tend to be evenly spaced in marsh and riparian habitat. By October, most leave Oregon for wintering grounds in the southern United States, Mexico, and Central America.

**Comments:** Common Yellowthroat nests are frequently parasitized by Brown-headed Cowbirds.

*References:* DeGraaf et al. 1991, Douglas et al. 1992, Gabrielson and Jewett 1940, Gilligan et al. 1994, Zeiner et al. 1990a.

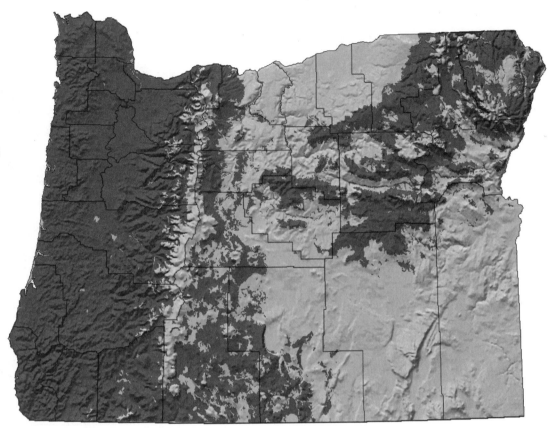

# Wilson's Warbler
## (Wilsonia pusilla)

*Order: Passeriformes*
*Family: Emberizidae*
*State Status: None*
*Federal Status: None*
*Global Rank: G5*
*State Rank: S5*
*Length: 4.5 in (11 cm)*

**Global Range:** Breeds from Alaska and the Yukon across most of Canada to New England and northern New York. In the West, it breeds south to southern California and the mountains of Nevada, Utah, and northern New Mexico. It is absent from most of the United States east of the Rocky Mountains.

**Habitat:** Like many warblers, Wilson's Warbler breeds in brushy areas within forests (coniferous, deciduous, or mixed), especially in willow or alder thickets at forest edges, near water, or in riparian vegetation along streams. In the open woodlands of eastern Oregon mountains, it is closely associated with riparian woodlands. It is found from low-elevation forests near the coast to treeline.

**Reproduction:** The breeding season for Wilson's Warbler gets under way in May. A nest cup of plant material is usually placed on the ground, hidden by grass, or in branches at the base of a tree. The normal clutch of 5 (range 3-6) eggs is incubated for 11-13 days by the female. The nestlings mature quickly, and fledge in about a week and a half.

**Food Habits:** Over 90% of the Wilson Warbler's diet is insects (including caterpillars, beetles, wasps, ants, flies, bees, and true bugs), which are gleaned from the ground or vegetation. It also eats some spiders, fruit pulp, and seeds. Some insects are taken in flight.

**Ecology:** Wilson's Warbler usually defends a territory of 1-3 acres during the breeding season. In the Sierra Nevada of California, some males are polygynous. By October, most Wilson's Warblers leave Oregon for wintering grounds in coastal Mexico and Central America.

**Comments:** Wilson's Warbler has been extirpated from parts of its range by heavy nest parasitism by Brown-headed Cowbirds.

*References:* DeGraaf et al. 1991, Gilligan et al. 1994, Stewart 1973, Stewart et al. 1978, Zeiner et al. 1990a.

# Yellow-breasted Chat
## (Icteria virens)

*Order: Passeriformes*
*Family: Emberizidae*
*State Status: None*
*Federal Status: None*
*Global Rank: G5*
*State Rank: S4?*
*Length: 7 in (18 cm)*

**Global Range:** Breeds from southern-most Canada and southern New England throughout most of the United States, south to central Mexico. It is absent from high mountains, dense forests, and extremely arid areas.

**Habitat:** This warbler breeds in brushy areas (such as blackberry or willow thickets) and in riparian woodlands along streams. It will use tangles of brush in the open or occurring as understory in deciduous or mixed deciduous-coniferous woodlands.

**Reproduction:** Yellow-breasted Chats are recorded nesting in Oregon as early as May 6th. They build nest cups of plant material placed in dense bushes. The usual clutch of 5 (range 3-6) eggs is incubated for about 2 weeks. The young leave the nest about 8-11 days after hatching.

**Food Habits:** Like most warblers, this species has a diet mostly composed of insects (including grasshoppers, beetles, true bugs, ants, bees, wasps, moths, mayflies, and caterpillars), which are gleaned from vegetation. When they are available, it also eats some fruit and berries (blackberries, strawberries, elderberries, wild grapes).

**Ecology:** There is some disagreement as to the territorial nature of the Yellow-breasted Chat. It may defend breeding territories of 1 to a few acres or, in some places, may form loose breeding colonies with only a small area about the nest defended. This may reflect limited habitat availability. The species winters from south Texas and Florida south to Panama.

**Comments:** Loss of riparian habitat has resulted in a decline of the Yellow-breasted Chat in many regions. It is listed as a "species of special concern" in California.

*References:* Gabrielson and Jewett 1940, Gilligan et al. 1994, Hunter et al. 1988, Thompson and Nolan 1973, Zeiner et al. 1990a.

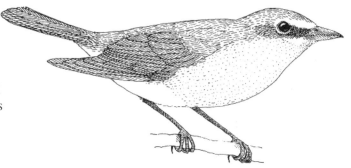

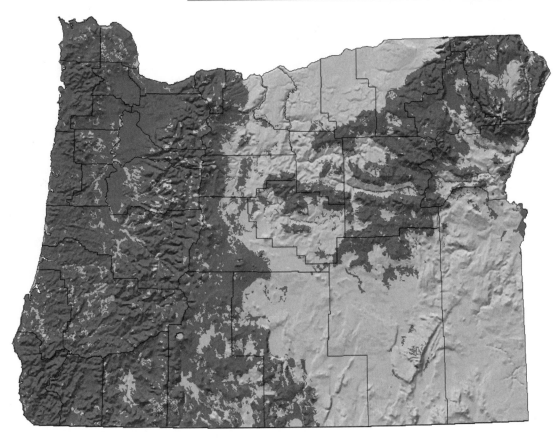

*Order: Passeriformes*
*Family: Emberizidae*
*State Status: None*
*Federal Status: None*
*Global Rank: G5*
*State Rank: S4*
*Length: 7 in (18 cm)*

**Global Range:** Breeds from southeastern Alaska and the Northwest Territories south to the mountains of extreme northern Mexico. It also breeds as far east as western South Dakota and west Texas. Its range is discontinuous in the interior West, being restricted to mountainous areas.

**Habitat:** Western Tanagers are forest birds, breeding in both coniferous and mixed coniferous-deciduous forests and woodlands. They breed in oak woodlands elsewhere, and may use them in Oregon as well. They use all forest types from the coast to high elevations, but are not typical of riparian woodlands in the arid valleys of eastern Oregon.

**Reproduction:** This tanager begins nesting when it arrives in Oregon in April. Eggs have been recorded by early May. The nest is usually a lined cup of plant material placed on the limb of a conifer. The usual clutch is 3 or 4 (range 3-5) eggs, incubated by the female for about 2 weeks. Both parents see to the needs of the nestlings.

**Food Habits:** The Western Tanager eats about 80% insects and 20% fruits, varying by season. Insects eaten include beetles, wasps, bees, ants, caterpillars, scale insects, grasshoppers, and termites. Among various fruits and berries, it eats cherries, elderberries, serviceberries, and raspberries. It will also eat buds.

**Ecology:** In Oregon, breeding densities of Western Tanagers ranged from about 20 to 40 per 100 acres. A study in Idaho and Montana found this species was more typical of late-successional Douglas-fir/ponderosa pine forests than younger stands. Most leave Oregon in October for wintering grounds along the coasts of Mexico and Central America.

**Comments:** During migration, Western Tanagers may join mixed-species flocks of other passerines.

*References:* Gabrielson and Jewett 1940, Gilligan et al. 1994, Jewett et al. 1953, Wiens and Nussbaum 1975, Zeiner et al. 1990a.

# Black-headed Grosbeak
## (Pheucticus melanocephalus)

Order: Passeriformes
Family: Emberizidae
State Status: None
Federal Status: None
Global Rank: G5
State Rank: S5
Length: 7 in (18 cm)

References: DeGraaf
et al. 1991,
Gabrielson and
Jewett 1940,
Gilligan et al. 1994,
Jewett et al. 1953,
Weston 1947,
Zeiner et al. 1990a.

**Global Range:** Breeds in western North America, from southern British Columbia and western North Dakota south through most of the western United States into the central Mexican highlands. It only breeds in the mountains of northern Baja California, not in the remainder of the peninsula (although some winter in southern Baja California).

**Habitat:** The Black-headed Grosbeak is found primarily in stands of deciduous trees with a shrub understory, although it also uses brushy areas in mixed deciduous-coniferous forests and woodlands. It uses oak, aspen, cottonwood, willow, maple, and alder woodlands, especially along stream courses and in canyons. It frequents forest edges and shores of lakes and rivers.

**Reproduction:** This grosbeak arrives in Oregon in April, and is nesting by May. The cup nest of plant material is placed 6-12 feet high in a tree or shrub. The normal clutch is 3 or 4 (range 2-5) eggs, incubated by both male and female for about 2 weeks. The young fledge in about 12 days.

**Food Habits:** Black-headed Grosbeaks are fairly omnivorous, and their diet varies seasonally. They glean spiders and insects from the ground or vegetation, including beetles, grasshoppers, caterpillars, bees, wasps, flies, ants, scale insects, and true bugs. They also eat a few snails. About 40% of the diet is plant material, including tree buds, fruit (elderberries, cherries, blackberries, and mistletoe berries), and seeds.

**Ecology:** The prominent song of the Black-headed Grosbeak is part of its territorial behavior. Both members of the pair defend the territory, but the male may defend the area around the female more than a fixed area of land. This species winters in central and southern Mexico.

**Comments:** Black-headed Grosbeaks are absent from the vast expanses of desert scrub in the valleys of southeastern Oregon, but breed in quaking aspen stands on desert mountain ranges and in riparian forests.

# Lazuli Bunting
## *(Passerina amoena)*

*Order: Passeriformes*
*Family: Emberizidae*
*State Status: None*
*Federal Status: None*
*Global Rank: G5*
*State Rank: S4*
*Length: 5 in (13 cm)*

*References:* Baker
and Baker 1990,
DeGraaf et al.
1991, Emlen et al.
1975, Gabrielson
and Jewett 1940,
Whitmore 1975,
Zeiner et al. 1990a.

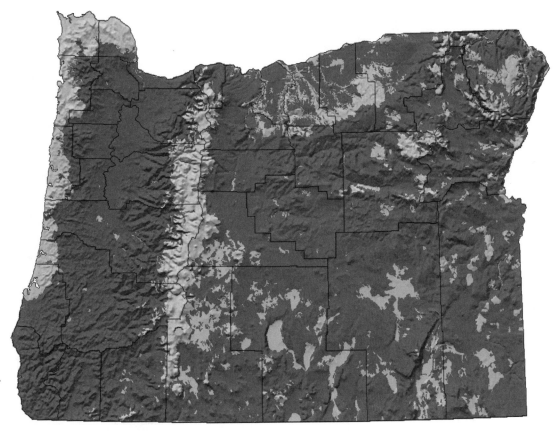

**Global Range:** Primarily a breeding bird of the western United States, though its breeding range extends into southern Canada and Baja California. Absent from northwestern Washington and arid deserts from eastern California to Texas, but otherwise found from the Pacific Coast east to the Great Plains.

**Habitat:** Lazuli Buntings breed in shrubs in or at the edges of open habitats. They are not found in dense forests of any type, but just about anything else will do, including oak, juniper, and riparian woodlands, forest edges along clear-cuts, willows of mountain meadows, and chaparral. In sagebrush-covered valleys, they are most likely to occur in willow, cottonwood, or rose thickets along streams. They use subalpine scrub to at least 6,000 feet high in the Mount Jefferson area.

**Reproduction:** Arrives in Oregon in May and is nesting by June. The nests are cups of vegetation, commonly placed a few feet off the ground in a shrub. The clutch of 4 (range 3-5) eggs is incubated by the female for about 12 days. The young fledge in about 2 weeks.

**Food Habits:** The diet varies seasonally, consisting of over 50% insects in spring and summer, turning to seeds later in the year. Insects eaten include caterpillars, grasshoppers, beetles, ants, bees, and true bugs. Lazuli Buntings mainly eat grass seeds, including needlegrass, wild oat, canary grass, and bluegrass.

**Ecology:** Singing males have poorly defined territories and are spaced several hundred feet apart. Following the breeding season, Lazuli Buntings may form flocks and move to higher elevations. They leave for wintering grounds in southern Arizona and Mexico by October.

**Comments:** The Lazuli Bunting hybridizes with the Indigo Bunting in the Great Plains. The Indigo Bunting seems to be more aggressive and is expanding its range west at the expense of the Lazuli Bunting. Some authors consider them the same species.

285

# Green-tailed Towhee
## *(Pipilo chlorurus)*

*Order: Passeriformes*
*Family: Emberizidae*
*State Status: None*
*Federal Status: None*
*Global Rank: G5*
*State Rank: S4*
*Length: 6.5 in (17 cm)*

*References:* DeGraaf et al. 1991, Gabrielson and Jewett 1940, Gilligan et al. 1994, Hayward et al. 1976, Hering 1948, Ryser 1985, Zeiner et al. 1990a.

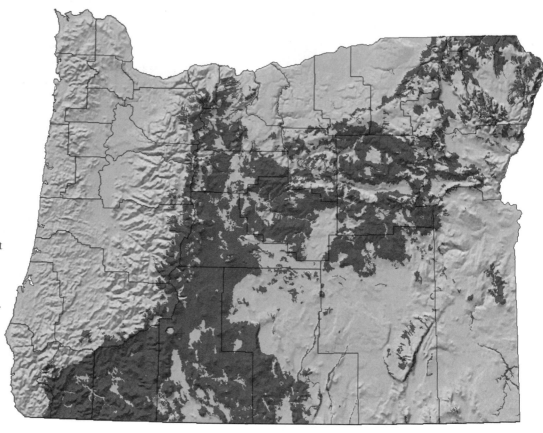

**Global Range:** The Green-tailed Towhee is a bird of the western portion of the lower 48 United States. Its northern limit is in the Blue Mountains of Washington, just over the Oregon state line. From there it breeds south to southern California and east to Wyoming and eastern New Mexico.

**Habitat:** During breeding season, this species is generally found in brushy areas on mountain slopes, sometimes as high as 6,000-7,000 feet. It occupies the undergrowth of sagebrush, bitterbrush, manzanita, mountain mahogany, and buckbrush in open ponderosa pine woodlands (Gilligan et al. 1994). Although it is usually associated with ponderosa pine or juniper woodlands, it breeds in higher-elevation coniferous woodlands in the Klamath Region, west as far as eastern Curry County. In arid eastern Oregon, it occurs only in riparian woodlands and sagebrush thickets on mountain slopes, but not on sagebrush or salt-desert shrub flats.

**Reproduction:** This towhee arrives in Oregon around May and is breeding by June. The nest is a cup of plant materials, built on the ground or low in a bush. The normal clutch has 4 (range 2-5) eggs. Both sexes feed the young.

**Food Habits:** The food habits of this species are poorly studied. Like other towhees, it scratches litter on the ground, presumably looking for seeds and insects. It probably takes a variety of forb and shrub seeds, with relatively more insects added to the diet in the breeding season. It will take some berries from shrubs when they are available.

**Ecology:** A study of territoriality in the Green-tailed Towhee was carried out in Colorado, where very small areas (less than half an acre) were defended. In Arizona, this species occupied open areas in a spruce-fir forest.

**Comments:** The Green-tailed Towhee is migratory, spending winters in the southwestern United States and Mexico.

## Spotted Towhee
*(Pipilo maculatus)*

*Order: Passeriformes*
*Family: Emberizidae*
*State Status: None*
*Federal Status: None*
*Global Rank: G5*
*State Rank: S5*
*Length: 8 in (20 cm)*

*References:* Davis 1957, 1960, DeGraaf et al. 1991, Gilligan et al. 1994, Williams and Koenig 1980, Zeiner et al. 1990a.

**Global Range:** Widely distributed in western North America, from the southern fringes of Canada south through the highlands of Mexico and Central America to Guatemala. It occurs as far east as the western Great Plains. An isolated population is found on Socorro Island, off the western coast of Mexico.

**Habitat:** Throughout its range, the Spotted Towhee is found in a variety of habitat types. It prefers thick brush near open areas, which typically occurs in forest openings, brushy ravines, and chaparral on mountain slopes, and in suburban landscaping. In eastern Oregon, this towhee is found in the mountains and in willow and other riparian growth near water, but not in arid desert valleys.

**Reproduction:** The Spotted Towhee begins breeding activities in April, and there are eggs in the nest in May and June. Its nest cup of plant material is usually built on the ground or low in a shrub. The normal clutch is 3 or 4 (range 2-6) eggs, which the female incubates for a little less than 2 weeks. The young leave the nest in about a week and a half.

**Food Habits:** About 70% of this towhee's diet is plant material, including a wide variety of grass, forb, and shrub seeds, acorns (where available), fruits, and berries. The diet varies seasonally, with more animal material included in the spring and summer (beetles, moths, caterpillars, true bugs, sow bugs, millipedes, spiders, snails). Occasionally, a small reptile or amphibian may be eaten.

**Ecology:** Territory sizes for the Spotted Towhee vary geographically, probably related to habitat quality and structure, and range in size from about half an acre to several acres. This species is a permanent resident at low elevations west of the Cascade Mountains, but most breeders in eastern Oregon migrate south for the winter.

**Comments:** After the breeding season, but prior to winter, this towhee tends to move to higher elevations in mountains. This species was formerly considered conspecific with the Rufous-sided Towhee (*Pipilo erythrophthalmus*) of eastern North America (American Ornithologists' Union 1995).

# California Towhee
## (Pipilo crissalis)

*Order: Passeriformes*
*Family: Emberizidae*
*State Status: None*
*Federal Status: None*
*Global Rank: G4G5*
*State Rank: S4?*
*Length: 9 in (23 cm)*

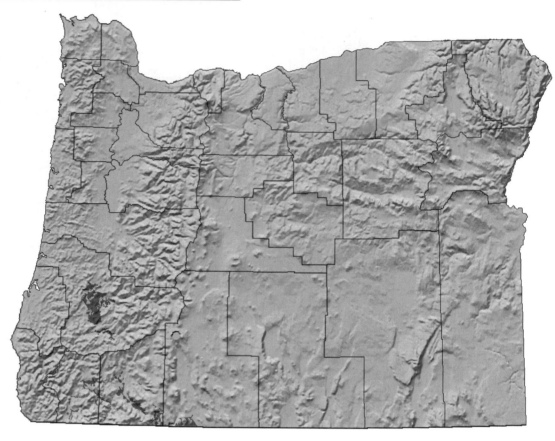

**Global Range:** Two races of the Brown Towhee (*Pipilo fuscus*) are now considered distinct species (American Ornithologists' Union 1989). The populations resident in Oregon are assigned to the California Towhee (*Pipilo crissalis*), which occurs from the Rogue River and Klamath regions of southern Oregon south to the tip of Baja California. It is found from the west slope of the Sierra Nevada to the Pacific Coast.

**Habitat:** California Towhees occupy a variety of thick, shrubby chaparral and brush communities. In California, they seek the cover of coyote bush in the San Francisco Bay region and chamise in the Los Angeles area. In Oregon, they are generally associated with buckbrush (*Ceanothus*) patches, often within a matrix of oak woodland.

**Reproduction:** The breeding season for the California Towhee is under way by May. The nest is built in a shrub or tree, and can be as high as 25-35 feet off the ground, but is usually lower. A normal clutch has 4 (range 2-6) eggs, which the female incubates for about 11 days. There may be 2 or more clutches per year.

**Food Habits:** The California Towhee forages for food items by scratching aside leaf litter on the surface of the ground. It eats seeds and insects (sometimes it flies out to catch insects on the wing). The diet varies seasonally, more insects (about a third of the diet) are eaten in spring and summer, and more seeds in fall and winter. Some fruit is also eaten when available. Nestlings are fed insects.

**Ecology:** California Towhees are thought to be monogamous, establishing a long-term pair bond. They defend a territory of 2 to 5 acres, but defense is strongest in the breeding season.

**Comments:** While the Spotted Towhee usually forages under the cover of a bush, the California Towhee searches for food in the open, usually within a few feet of cover.

*References:* Davis 1951, 1957; DeGraaf et al. 1991, Gilligan et al. 1994, Zeiner et al. 1990a, Zink 1988, Zink and Dittman 1991.

## Chipping Sparrow
*(Spizella passerina)*

*Order: Passeriformes*
*Family: Emberizidae*
*State Status: None*
*Federal Status: None*
*Global Rank: G5*
*State Rank: S4*
*Length: 5 in (13 cm)*

*References:* DeGraaf et
al. 1991, Gabrielson
and Jewett 1940,
Jewett et al. 1953,
Linsdale 1936, Ryser
1985, Zeiner et al.
1990a.

**Global Range:** Breeds from southeastern Alaska across most of Canada to the Atlantic Coast, and south through most of the United States into the highlands of Mexico and as far south as Nicaragua.

**Habitat:** Chipping Sparrows are found in open coniferous woodlands and at the edges of meadows and forest clearings, including clear-cuts. They will also use arid mountain slopes with only a few scattered trees. In Nevada, they use mountain mahogany, ponderosa pine, limber pine, quaking aspen, and sagebrush communities. Only Jewett et al. (1953) mention them breeding in sagebrush communities in Washington. Chipping Sparrows forage in grassy, open areas and adapt to lawns and pastures. Littlefield (1990) mentions these sparrows breeding in juniper woodlands on Steens Mountain and in ponderosa pine woodlands of the Blue Mountains, but not in the sagebrush flats of Malheur National Wildlife Refuge. Most references emphasize their presence in mountain woodlands rather than in desert valleys. In Arizona, Laudenslayer and Balda (1976) found them in the ponderosa pine-pinyon-juniper ecotone. Small (1994) assigns them to a variety of woodland and forest habitats up to timberline in California.

**Reproduction:** Chipping Sparrows are breeding by late April. Their nest cup of plant materials is usually placed on a tree limb, but also in bushes, where trees are unavailable. The clutch usually contains 4 (range 3-5) eggs, and is incubated by the female for about 2 weeks. The young fledge in another 2 weeks.

**Food Habits:** This sparrow eats both insects and a variety of grass and forb seeds, usually gleaned from bare ground or areas with short grass cover. Insects (including beetles, leafhoppers, caterpillars, grasshoppers, true bugs, ants, and wasps) are more important in spring and summer.

**Ecology:** The territory size is reported to vary from about 3 to 7 acres. Most winter in the southwestern United States and Mexico. Family groups tend to move upslope following the breeding season.

**Comments:** This species often lines its nest with horsehair, and is colloquially known as the "hair bird." What materials were used prior to the reintroduction of the horse to North America is open to conjecture.

**289**

# Brewer's Sparrow
## (Spizella breweri)

*Order:* Passeriformes
*Family:* Emberizidae
*State Status:* None
*Federal Status:* None
*Global Rank:* G4
*State Rank:* S4
*Length:* 5 in (13 cm)

**Global Range:** A disjunct population of Brewer's Sparrow breeds in the southern Yukon and northern British Columbia. Otherwise, this is mostly a species of the interior West, breeding from southern British Columbia south to southern California and east to western South Dakota and New Mexico.

**Habitat:** Although Brewer's Sparrows are most closely associated with sagebrush, including sagebrush clearings in open coniferous forests (including clear-cuts), they also breed in bitterbrush, shadscale, greasewood, and grasslands with occasional shrubs.

**Reproduction:** This sparrow arrives in Oregon in April and is breeding by the end of the month. The nest cup of plant material is placed in a low shrub. There are 3 to 4 (sometimes 5) eggs in a normal clutch. Following the breeding season, adults and young often congregate into large wandering flocks.

**Food Habits:** The diet varies seasonally, with insects (including leafhoppers, beetles, aphids, and caterpillars) eaten in spring, and a variety of seeds consumed later in the year.

**Ecology:** Brewer's Sparrow can be quite abundant in favorable habitat, with densities as high as 150-500 per square kilometer. The territory size is usually about an acre. This species winters in the southwestern United States and Mexico, leaving Oregon by the first part of October.

**Comments:** Brewer's Sparrow is thought to be declining in Idaho, perhaps due to loss of habitat. A few breed in a stand of dwarf pine in the Mt. Jefferson Wilderness. This habitat is similar to that used by the Yukon subspecies.

*References:* Gabrielson and Jewett 1940, Gilligan et al. 1994, Jewett et al. 1953, Medin 1990, Rotenberry and Wiens 1989, 1991, Ryser 1985, Wiens et al. 1985, Wiens and Van Horne 1990, Zeiner et al 1990a.

*Order: Passeriformes*
*Family: Emberizidae*
*State Status: None*
*Federal Status: None*
*Global Rank: G5*
*State Rank: S2?*
*Length: 5 in (13 cm)*

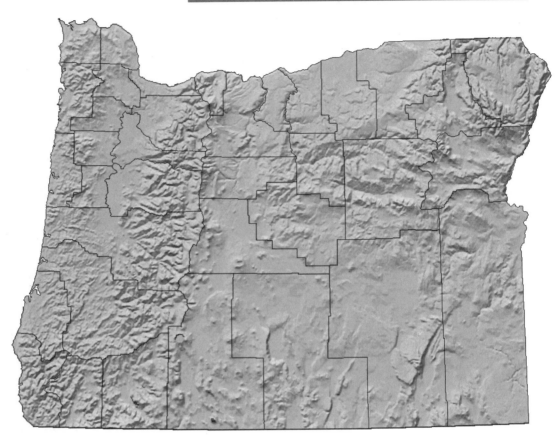

**Global Range:** Five Oregon records from Jackson and Klamath counties, including the sighting of a pair accompanied by 4 young, represent the northern limit of the breeding range of this sparrow. It otherwise breeds from northern California south to Baja California, southeast through southern Nevada to west Texas, and farther south to the highlands of Mexico.

**Habitat:** In other parts of its range, it occupies a variety of brushy habitats, including chaparral, sagebrush, and the shrub understory of pinyon-juniper woodlands. In Oregon, it has been associated with stands of buckbrush (*Ceanothus*).

**Reproduction:** This species has been seen in Oregon from May to July. The cup nest is placed in a low shrub. The normal clutch ranges from 3 to 4 (rarely 5) eggs.

**Food Habits:** This species has not been well studied, and its food habits are assumed to be similar to those of its relatives: insects in the spring, some fruit when available, and seeds at other times.

**Ecology:** Aside from the fact that this species prefers arid shrublands and migrates south to Arizona, New Mexico, and Mexico in winter, little information is available about its natural history.

**Comments:** Given the extent of its habitat in southern Oregon, this sparrow may be more common than currently known.

*References:* DeGraaf et al. 1991, Gilligan et al. 1994, Ryser 1985, Zeiner et al. 1990a.

# Vesper Sparrow
## (Pooecetes gramineus)

*Order: Passeriformes*
*Family: Emberizidae*
*State Status: Sensitive*
*Federal Status: None*
*Global Rank: G5*
*State Rank: S4*
*Length: 6 in (15 cm)*

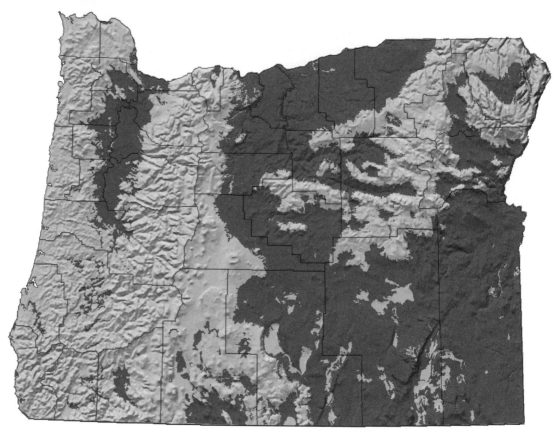

**Global Range:** Breeds from the southern Northwest Territories south across southern Canada to the Atlantic Coast, and throughout most of the northern three-quarters of the United States. It is absent as a breeder from the Southeast, Southwest, and most of California.

**Habitat:** Vesper Sparrows are birds of open habitats, including grasslands, pastures, prairies, juniper woodlands, sagebrush-covered valleys, mountain meadows, and agricultural areas. They are not found in forests.

**Reproduction:** The Vesper Sparrow generally returns to Oregon in April and is breeding by May. Its cup nest is located in a depression in the ground. The normal clutch has 3 to 5 (range 3-6) eggs, which the female incubates for just under 2 weeks. The young remain with the pair for about 5 weeks after hatching, being fed by the male while the female incubates the second brood.

**Food Habits:** Vesper Sparrows consume both insects and seeds, presumably more insects during the breeding season than at other times. They eat beetles, caterpillars, grasshoppers, and moths, among other insects. They tend to prefer seeds of broad-leafed plants, but consume waste grain in agricultural areas.

**Ecology:** Vesper Sparrows defend territories of about 1 to 3 acres. Most Vesper Sparrows winter in central California, the southwestern United States, along the Gulf of Mexico, and in Mexico, leaving Oregon by October.

**Comments:** This species is declining in Idaho, and has nearly vanished from western Oregon, where it was common early this century. Only the populations in the western interior valleys are considered Sensitive by the Department of Fish and Wildlife.

*References: DeGraaf et al. 1991, Gilligan et al. 1994, Marshall et al. 1996, Pylypec 1991, Rodenhouse and Best 1983, Zeiner et al. 1990a.*

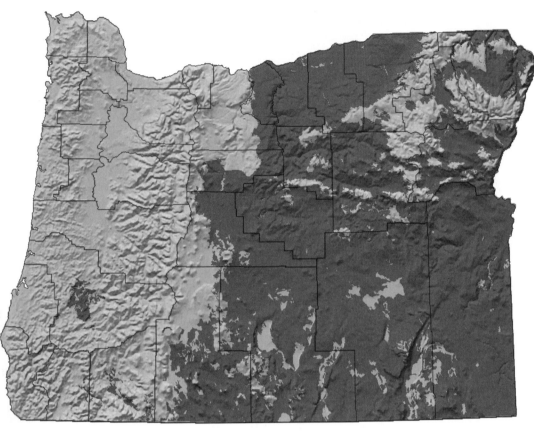

## Lark Sparrow
*(Chondestes grammacus)*

*Order: Passeriformes*
*Family: Emberizidae*
*State Status: None*
*Federal Status: None*
*Global Rank: G5*
*State Rank: S4?*
*Length: 6 in (15 cm)*

*References: DeGraaf et al. 1991, Gabrielson and Jewett 1940, Gilligan et al. 1994, Hayward et al. 1976, Ryser 1985, Zeiner et al. 1990a.*

**Global Range:** Breeds from southern Canada south throughout most of the United States, except along most of the Atlantic and Gulf coasts. It is resident from the southwestern United States south through the highlands of Mexico.

**Habitat:** Lark Sparrows breed in two major habitat types that often intergrade with one another in eastern Oregon: grasslands with scattered bushes (often sagebrush) and sagebrush communities with a grass understory. They also make use of the sagebrush understory in open pine or juniper woodlands. Their range extends into the Rogue River region, where they are found on open grassy hillsides and in valleys. They may also be found in many agricultural areas.

**Reproduction:** In Oregon, the breeding season for the Lark Sparrow is well under way by mid-May. It builds a nest cup of plant materials, placed low in a bush or in a slight depression in the ground. There are ordinarily 4 or 5 (range 3-6) eggs in a clutch, which the female incubates for just under 2 weeks. The young develop rapidly and leave the nest in a week and a half.

**Food Habits:** The Lark Sparrow eats both seeds and soft-bodied insects. Grass and forb seeds make up about three-fourths of the diet. Grasshoppers are a preferred animal food, composing half of the animal part of the diet. Other insects taken include caterpillars and weevils. The diet varies seasonally, becoming half insect during the breeding season.

**Ecology:** Territories are actively defended only early in the breeding season, when they may be anywhere from 3 to 9 acres in size. Most Lark Sparrows migrate south to their wintering grounds in the southwestern United States and Mexico by October; however, a fair number winter in the Rogue River Valley.

**Comments:** Western Oregon populations north of the Rogue River Valley have suffered a dramatic decline since the early part of the century.

293

# Black-throated Sparrow
## (Amphispiza bilineata)

*Order: Passeriformes*
*Family: Emberizidae*
*State Status: Sensitive*
*Federal Status: None*
*Global Rank: G5*
*State Rank: S2?*
*Length: 5 in (13 cm)*

*References:* DeGraaf et al.
1991, Gilligan et al.
1994, Marks et al. 1980,
Marshall et al. 1996,
Smyth and
Bartholomew 1966,
Zeiner et al. 1990a.

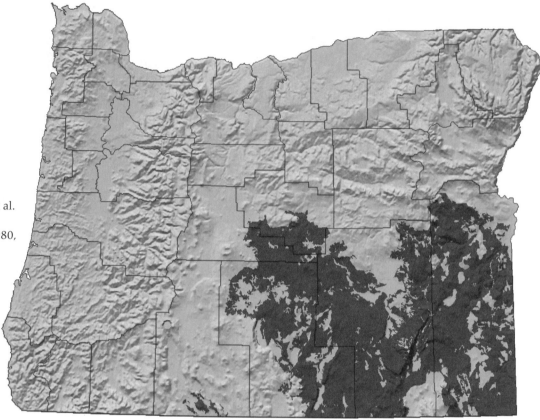

**Global Range:** Breeds from southeastern Oregon east to southwestern Wyoming and south through the desert Southwest to south-central Mexico.

**Habitat:** Black-throated Sparrows are birds of the desert, but they prefer to nest at the interface of valleys and hills, with scattered desert shrubs and a grass understory, often near rock piles. They are less frequent in open, flat valleys with sagebrush, juniper, or salt-desert shrub communities. Just across the state line, Idaho populations are found in sagebrush, spiny hopsage, or horsebrush communities in which plants are more than half a meter high.

**Reproduction:** By May, Black-throated Sparrows have returned to Oregon for their breeding season. Their nest is a cup of plant material, lined with plant down or animal fur, and placed low in a bush. A typical clutch consists of 3 or 4 eggs. Details of their breeding biology are poorly known.

**Food Habits:** During the breeding season, many insects are eaten. The Black-throated Sparrow can survive on water obtained from its food, although it sometimes eats green vegetation (shoots of grasses and forbs) as well. Soft-bodied insects and spiders are preferred foods. Seeds of desert plants probably form the bulk of the diet after breeding season.

**Ecology:** Various studies have found that the territory of the Black-throated Sparrow ranges from about 1 to 4 acres in size. This species is apparently excluded from stands of dense sagebrush by Sage and Brewer's sparrows. This sparrow winters in the southwestern United States and Mexico, departing Oregon by late September. One aspect of the Black-throated Sparrow that has been well-studied is its water balance. In cooler periods, it can survive on seeds without drinking water. While feeding on insects, it does not need access to free water. Its kidneys are more efficient than those of most birds, allowing it to exploit arid habitats.

**Comments:** This species is at the northern edge of its range in Oregon, and it may not breed in many areas every year. There are isolated records of possible breeding by the Black-throated Sparrow from Deshutes, Klamath, and Wheeler counties.

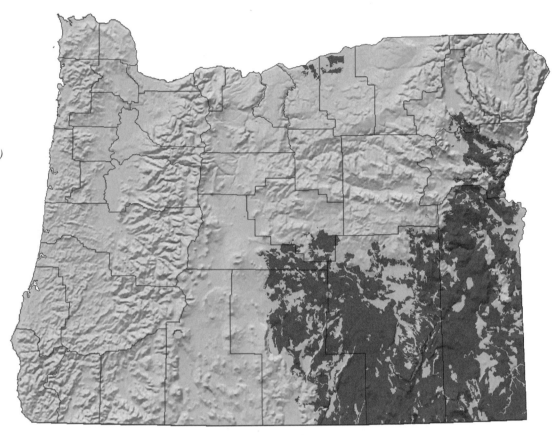

# Sage Sparrow
*(Amphispiza belli)*

*Order: Passeriformes*
*Family: Emberizidae*
*State Status: None*
*Federal Status: None*
*Global Rank: G5*
*State Rank: S4*
*Length: 5.5 in (14 cm)*

**Global Range:** Breeds in the interior West, from eastern Oregon and Washington, east to Wyoming, and south to New Mexico, most of desert and coastal California, and northern Baja California.

**Habitat:** In most of its range, including Oregon, the Sage Sparrow occupies sagebrush-covered valleys. It uses other desert shrub communities (greasewood, shadscale) to a lesser extent. It is far less likely to breed in grasslands if sagebrush or other shrubs are too scattered, or in sagebrush steppe where junipers are too frequent. In coastal California, it switches its preference to chamise chaparral and coastal sage scrub.

**Reproduction:** By April, the Sage Sparrow is breeding in Oregon. Its nest cup of plant materials is lined with grass, fur, or other fine material, and placed about a foot off the ground in a bush. The normal clutch is 3 or 4 (sometimes 5) eggs, which are incubated for about 2 weeks.

**Food Habits:** In the breeding season, many soft-bodied insects, ants, and spiders are eaten. The Sage Sparrow finds its food on the ground or by gleaning foliage. Seeds and some green foliage are also eaten, and seeds become a more important diet item following the breeding season.

**Ecology:** The Sage Sparrow's territory is about 4 to 8 acres, and in some parts of the Great Basin it can be quite dense (up to 200 per square kilometer). Nesting success is higher in years with more rainfall. By October, most Sage Sparrows leave Oregon for winter ranges in California, the desert Southwest, and Mexico.

**Comments:** The Oregon breeding distribution of Sage Sparrows varies from year to year.

*References:* DeGraaf et al. 1991, Gabrielson and Jewett 1940, Gilligan et al. 1994, Johnson and Marten 1992, Petersen and Best 1985, Rotenberry and Wiens 1989, 1991, Wiens et al. 1985, Wiens and Van Horn 1990, Zeiner et al. 1990a.

# Savannah Sparrow
### *(Passerculus sandwichensis)*

*Order: Passeriformes*
*Family: Emberizidae*
*State Status: None*
*Federal Status: None*
*Global Rank: G5*
*State Rank: S5*
*Length: 5 in (13 cm)*

**Global Range:** The Savannah Sparrow has an extensive breeding distribution in North America, ranging from Alaska across Canada to the Atlantic Coast, and south to central California in the West and Maryland in the East. It is also resident from the highlands of Mexico south to Guatemala.

**Habitat:** The Savannah Sparrow usually lives up to its name, breeding in grasslands, fields, pastures, mountain meadows, low, wet prairies, and grassy areas around the edges of lakes, ponds, and rivers.

**Reproduction:** The breeding season for this sparrow is well under way by May. The nest cup of plant material is placed in a well-hidden depression in the ground. The typical clutch is 4 or 5 (range 3-6) eggs, which both parents incubate for about 12 days. The young depart the nest about 2 weeks after hatching. There are 2 clutches a year.

**Food Habits:** Like most sparrows, the Savannah Sparrow becomes primarily carnivorous during the breeding season, eating insects (mostly beetles and grasshoppers, but also caterpillars, ants, aphids, butterflies, moths, dragonflies, and true bugs), spiders, snails, and worms. At other times of the year, the diet is mainly seeds from a variety of grasses and forbs.

**Ecology:** Savannah Sparrows have relatively small breeding territories, usually less than an acre. In some circumstances they breed in loose colonies. Most Savannah Sparrows depart Oregon by October for wintering grounds in the southwestern United States south to Mexico and Central America, although a few winter in western Oregon, especially along the coast.

**Comments:** This species breeds in emergent vegetation along coastal estuaries.

*References:* DeGraaf et al. 1991, Gabrielson and Jewett 1940, Gilligan et al. 1994, Jewett et al. 1953, Williams 1987, Zeiner et al. 1990a.

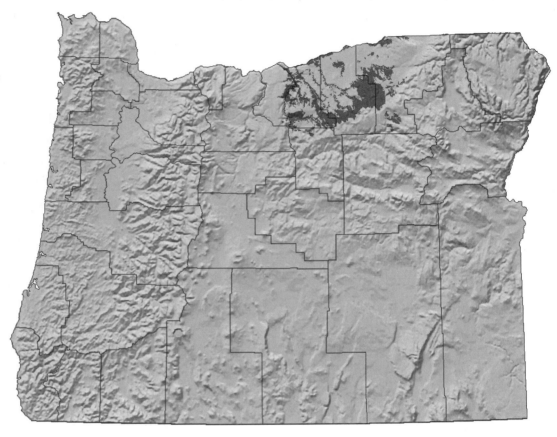

# Grasshopper Sparrow
## *(Ammodramus savannarum)*

*Order: Passeriformes*
*Family: Emberizidae*
*State Status: Sensitive*
*Federal Status: None*
*Global Rank: G4*
*State Rank: S2?*
*Length: 5 in (13 cm)*

**Global Range:** This is mostly a bird of the central and eastern United States, but it breeds from the southern fringe of Canada south to California, east to northern New Mexico, and northeast to Virginia. There are disjunct populations in Florida and southern Arizona-northern Sonora.

**Habitat:** Grasshopper Sparrows inhabit short grasslands in which occasional shrubs may grow. If the shrub cover exceeds 35%, the area becomes unacceptable. In some places, they are found in the grass understory of woodlands, but in eastern Oregon they prefer native bunchgrass grasslands on the north slopes of hills with scattered shrubs. In other parts of their range, they use cultivated grasslands and pastures.

**Reproduction:** Breeding is under way by June, when the nest cup of plant materials is built in a hollow in the ground. Because of limited habitat, many pairs may nest in proximity to one another. The clutch usually has 4 or 5 (range 3-6) eggs, which the female incubates for about 12 days. The young leave the nest in about a week and a half, before they have completely fledged.

**Food Habits:** During the breeding season, about three-fifths of the Grasshopper Sparrow's diet is composed of various insects. It also eats spiders and snails. The vegetable part of its diet is mostly seeds of grasses, sedges, forbs, and waste grain. It forages on the ground.

**Ecology:** Grasshopper Sparrow territories have been found to vary from about 1 to 4 acres. In Oregon, densities have ranged from about 7 to 50 per square mile. This sparrow leaves Oregon in the fall for its wintering grounds in the southwestern United States, Mexico, and Central America.

**Comments:** The breeding locations of Grasshopper Sparrow colonies in Oregon are unstable. Populations move about from year to year, depending on the location of suitable habitat.

*References:* DeGraaf et al. 1991, Gilligan et al. 1994, Janes 1983, Marshall 1992, Zeiner et al. 1990a, Zink and Avise 1990.

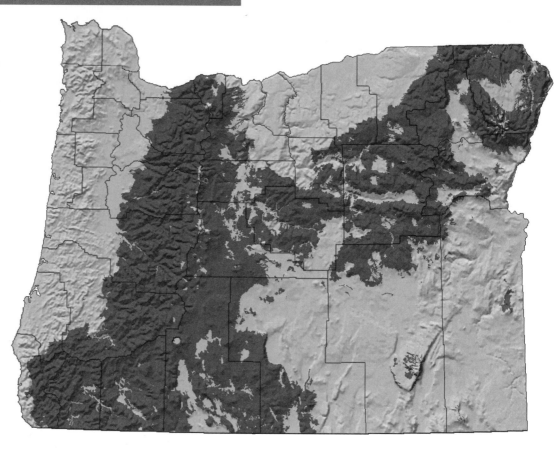

# Fox Sparrow
## *(Passerella iliaca)*

*Order: Passeriformes*
*Family: Emberizidae*
*State Status: None*
*Federal Status: None*
*Global Rank: G5*
*State Rank: S4*
*Length: 7 in (18 cm)*

**Global Range:** Breeds from Alaska across most of Canada, except for the southern portions of provinces east of British Columbia, to the Atlantic Coast. In the western United States, it breeds in the Cascade, Sierra Nevada, Rocky, and other interior mountains to southern California and Colorado.

**Habitat:** The primary habitat requirement of the Fox Sparrow is thick, shrubby vegetation, which can occur along rivers, in forests, at forest edges and clearings, or in chaparral or other shrubby vegetation types. In Oregon, it frequents willow riparian habitats in desert mountain ranges and patches of manzanita and buckbrush in open ponderosa pine woodlands.

**Reproduction:** Depending on elevation, the Fox Sparrow will begin breeding in late May or June. The nest is usually a cup of plant material, placed on the ground or in a low bush. A typical clutch has 3 to 5 eggs, which are incubated for 12-14 days.

**Food Habits:** In typical sparrow pattern, insects (including beetles and craneflies), millipedes, and spiders are the main food in the breeding season. At other times, the diet turns to seeds

and berries. Fox Sparrows forage by kicking away ground litter to uncover food items.

**Ecology:** The Fox Sparrow is territorial, but little is known about the size of the defended area. In a Grant County, Oregon, study, male Fox Sparrows used about 20 acres. Fox Sparrows winter in Oregon, but not the same individuals that breed here, which are replaced by birds from populations breeding as far north as Alaska. Oregon populations may winter in the southern United States and northern Mexico.

**Comments:** Although Fox Sparrows breed in riparian thickets in arid country, they are absent from vast expanses of desert vegetation.

*References:* Burns and Zink 1990, DeGraaf et al. 1991, Gabrielson and Jewett 1940, Gilligan et al. 1994, Hayward et al. 1976, Zeiner et al. 1990a.

# Song Sparrow
## (Melospiza melodia)

*Order: Passeriformes*
*Family: Emberizidae*
*State Status: None*
*Federal Status: None*
*Global Rank: G5*
*State Rank: S5*
*Length: 6 in (15 cm)*

*References:* DeGraaf et al. 1991, Gabrielson and Jewett 1940, Gilligan et al. 1994, Tompa 1962, Zeiner et al. 1990a.

**Global Range:** Breeds from southern Alaska across central and southern Canada to the Atlantic Ocean. They occur south to Baja California in the West and to Georgia in the East, but do not breed in much of Texas or the southeastern United States. Disjunct populations breed in the highlands of Mexico.

**Habitat:** The key habitat element for the Song Sparrow seems to be thickets of deciduous shrubs, commonly willows along streams, marshes, lakes, or ponds. In western Oregon they make use of blackberry thickets as well as thickets in forests and forest edges. In the eastern part of the state, they are more restricted to riparian corridors. They are usually found near water, and make use of emergent vegetation. Song Sparrows are found in thickets in suburbs and towns as well.

**Reproduction:** Depending on location and altitude, Song Sparrows may begin breeding as early as mid-April in Oregon. They build a nest cup on the ground or in a shrub or tree. Normal clutches have 3 to 5 (range 2-6) eggs, which the female incubates for about 2 weeks. The young are independent in about 5 weeks. Song Sparrows have 2 or 3 broods per year.

**Food Habits:** During the breeding season, about half of the Song Sparrow's diet consists of insects (beetles, grasshoppers, ants, wasps, flies, true bugs, leafhoppers, and termites). Other food includes seeds of grasses and forbs, and many fruits and berries (blackberries, elderberries, wild cherries, grapes).

**Ecology:** Song Sparrow territories are generally around an acre or less. Many Song Sparrow nests are parasitized by Brown-headed Cowbirds. Birds breeding at high elevations move to lower valleys in winter.

**Comments:** Song Sparrows are absent from most of the arid desert valleys of eastern Oregon.

# Lincoln's Sparrow
*(Melospiza lincolnii)*

*Order: Passeriformes*
*Family: Emberizidae*
*State Status: None*
*Federal Status: None*
*Global Rank: G5*
*State Rank: S4*
*Length: 5.5 in (14 cm)*

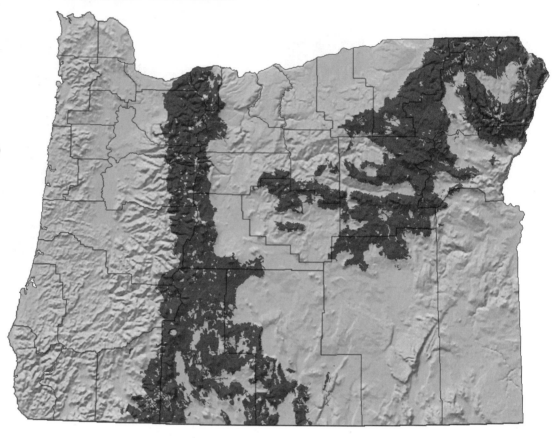

**Global Range:** Breeds from southern Alaska across central Canada to the Atlantic Ocean. In the eastern United States, it is found only as far south as northern New York, but in the West it descends in the mountains to central California, northern New Mexico, and eastern Arizona.

**Habitat:** Lincoln's Sparrow is restricted to thickets of deciduous (especially willow and alder) vegetation, growing along streams or in and around wet meadows, marshes, lakes, or ponds, in the forests of the Cascade and Blue mountains of Oregon.

**Reproduction:** Depending on elevation, the breeding season for Lincoln's Sparrow gets under way from late May to July. This species is a ground nester, placing its nest cup of plant material in a hidden depression. The normal clutch of 4 or 5 (range 3-6) eggs is incubated for about 2 weeks by the female. The young are ready to leave the nest in just under 2 weeks.

**Food Habits:** Like most sparrows, Lincoln's Sparrow eats more animal material (about 70% of the diet) in the breeding season. By fall, over 90% of the diet is made up of plant material (mostly seeds of grasses and forbs). Insects (beetles, ants, flies, grasshoppers,

true bugs, millipedes) and spiders are the main animal food items.

**Ecology:** Like the very similar Song Sparrow, most Lincoln's Sparrow territories are small (an acre or less). During winter, these sparrows leave the mountains for lower elevations. They winter as far south as Panama, and those present in Oregon in winter may not be the same ones that breed here.

**Comments:** In areas of sympatry, Lincoln's Sparrows and Song Sparrows may have overlapping territories, but the Song Sparrow is the dominant competitor.

*References:* DeGraaf et al. 1991, Gabrielson and Jewett 1940, Gilligan et al. 1994, Miller 1939, Zeiner et al. 1990a.

*Order: Passeriformes*
*Family: Emberizidae*
*State Status: None*
*Federal Status: None*
*Global Rank: G5*
*State Rank: S5*
*Length: 6 in (15 cm)*

*References:* DeGraaf et
al. 1991, Gabrielson
and Jewett 1940,
Gilligan et al. 1994,
Lewis 1975,
Petrinovich and
Patterson 1983, Ralph
and Pearson 1971,
Zeiner et al. 1990a,
Zink et al. 1991.

**Global Range:** Breeds from Alaska across northern Canada to the Atlantic Coast. Absent from the eastern United States and from southern Canada, east of British Columbia. In the West, however, it breeds south to coastal southern California, and to the Rocky Mountains of northern New Mexico.

**Habitat:** White-crowned Sparrows, like so many of their relatives, breed in brushy thickets, without much regard for dominant vegetation type. They occur in willow riparian corridors and quaking aspen stands around the streams and lakes of the mountains of eastern Oregon. From the Cascade Mountains west, they can be found at forest edges, and in blackberry and other thickets in forests or valleys, suburban gardens, and riparian woodlands. Trees are not an essential habitat component, and these sparrows are not characteristic of dense forest interiors.

**Reproduction:** This sparrow is breeding in the lowlands of Oregon by early May. Its nest cup of plant material may be located on the ground or in a low bush. There are usually 3 or 4 (range 2-6)

eggs, which the female incubates for about 2 weeks. Young are often dependent on the male for as long as 6 weeks after hatching. The female may begin another clutch during this time.

**Food Habits:** White-crowned Sparrows feed on spiders and insects (beetles, caterpillars, mosquitoes, ants, true bugs, flies), taking more animal food in spring and summer. They also eat a wide variety of plant foods, especially in fall and winter, including a variety of grass and forb seeds, blossoms, leaves, green shoots, berries, fruits, and catkins.

**Ecology:** Depending on habitat, territories can range in size from a tenth of an acre on a vacant city lot to 2 acres near timberline. They leave higher elevations for more moderate valleys in winter, and many undoubtedly migrate south. Birds from Oregon may winter as far south as central Mexico. Most of the birds that winter east of the Cascade Mountains come from more northerly populations.

**Comments:** This species is thought to be declining in Idaho.

301

# Dark-eyed Junco
## (Junco hyemalis)

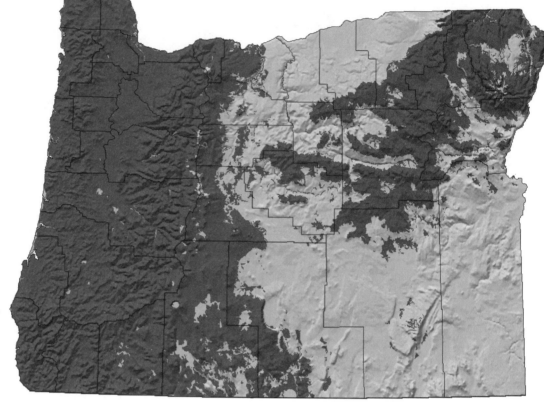

Order: Passeriformes
Family: Emberizidae
State Status: None
Federal Status: None
Global Rank: G5
State Rank: S5
Length: 6 in (15 cm)

References: DeGraaf
et al. 1991,
Gabrielson and
Jewett 1940,
Gilligan et al. 1994,
Hayward et al.
1976, Ryser 1985,
Smith and
Andersen 1982,
Zeiner et al. 1990a.

**Global Range:** Breeds from Alaska across most of Canada to the Atlantic Coast. In the eastern United States, it is found as a breeding bird as far south as the Great Lakes and New York, and farther south, in the Appalachian Mountains, to Georgia. In the West, it occurs south, mainly in moister habitats, to northern Baja California and western Texas. It is absent from most of the Great Plains, Midwest, and South.

**Habitat:** Dark-eyed Juncos breed in deciduous, mixed, or coniferous forests and woodlands, including riparian corridors along forest streams, and quaking aspen stands. They also use higher-elevation riparian vegetation in the arid mountains of eastern Oregon. They nest in brushy areas within forests and along edges of forest openings and meadows. Interiors of dense forests are avoided. They are found in all forest types, from sea level to high elevation, including forested suburbs.

**Reproduction:** Depending on elevation, Dark-eyed Juncos may begin breeding by late April. Dark-eyed Juncos are ground nesters, placing their nest cup in a well-hidden depression of their own making. The normal clutch of 4 or 5 (range 3-6) eggs is incubated by the female for about 11-12 days. Young are independent in about 5 weeks. There may be several broods in a season.

**Food Habits:** Juncos forage for seeds and insects on the ground, often moving aside leaf litter to search for food items. The diet varies seasonally. Seeds are eaten in all seasons, but more animal material is eaten during the breeding season.

**Ecology:** Dark-eyed Juncos have a small territory of less than several acres. They retreat to lower elevations from the mountains in winter, and several subspecies from the north winter in Oregon. They are thought to be declining in neighboring Idaho. During winter, Dark-eyed Juncos form flocks.

**Comments:** Two sources, apparently based on the work of Gashwiler (1977), indicate that this junco breeds in the juniper woodlands of central Oregon.

# Bobolink
## *(Dolichonyx oryzivorus)*

*Order: Passeriformes*
*Family: Emberizidae*
*State Status: Sensitive*
*Federal Status: None*
*Global Rank: G5*
*State Rank: S2*
*Length: 7 in (18 cm)*

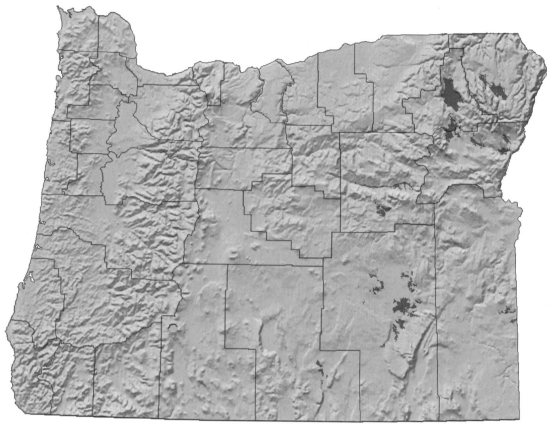

**Global Range:** Breeds nearly coast to coast, from southern Canada south across the northern United States. It reaches the Atlantic Coast in the East, but remains east of the Cascade Mountains in Oregon and Washington. In the East, it breeds as far south as North Carolina.

**Habitat:** The Bobolink is a bird of open prairies, grasslands, wet meadows, pastures, and grain crops. In Oregon, there are only a few disjunct populations that breed in irrigated hay meadows fringed with willows, or in wet, grassy meadows with local growths of forbs and sedges. Many of these areas are mowed and/or grazed, which facilitates nesting of Bobolinks.

**Reproduction:** Bobolinks migrate to Oregon and begin their breeding season in May. They build their cup nest of plant materials and locate it in a hollow on the ground. The usual clutch has 5 or 6 (range 4-7) eggs, which are incubated for a week and a half to 2 weeks. The young leave the nest at 2 weeks of age, before they are able to fly.

**Food Habits:** Bobolinks eat grass and forb seeds as well as insects. During the breeding season, more insects are included in the diet, especially caterpillars, which can be extremely abundant at this time. They will also forage in cultivated fields and eat waste grain.

**Ecology:** Bobolinks breed in small colonies, although it is likely that an area around the nest is defended. Bobolinks migrate to northeastern South America in winter, leaving Oregon by October.

**Comments:** Estimates place the total number of Bobolinks breeding in Oregon at less than 1,000 individuals.

*References:* Bollinger and Gavin 1989, DeGraaf et al. 1991, Gilligan et al. 1994, Marshall 1992, Marshall et al. 1996, Wittenberger 1978.

# Red-winged Blackbird
## (Agelaius phoeniceus)

*Order: Passeriformes*
*Family: Emberizidae*
*State Status: None*
*Federal Status: None*
*Global Rank: G5*
*State Rank: S5*
*Length: 8 in (20 cm)*

*References:* DeGraaf et al. 1991, Gabrielson and Jewett 1940, Gilligan et al. 1994, Holm 1973, Orians 1961, Orians and Horn 1969, Payne 1969, Zeiner et al. 1990a.

**Global Range:** Breeds coast to coast in North America, from southeastern Alaska across Canada and south through the United States, most of Mexico, and coastal Central America, to Costa Rica.

**Habitat:** The most typical habitat for the Red-winged Blackbird is emergent vegetation growing in and around freshwater, alkaline, and brackish water marshes and wetlands. They also use the vegetation lining streams, canals, or virtually any body of water. Sometimes a colony is established in an upland situation, usually a grassy hillside or pasture with clumps of bushes or low trees for nesting. They breed from the coast to high elevation.

**Reproduction:** The breeding season is well under way by late April. Nest cups of plant material are attached to emergent vegetation or placed in bushes or trees, often overhanging water. The normal clutch has 4 eggs (range 3-5), which are incubated by the female for 10-12 days. Males are polygynous, and may have as many as 15 mates. Young fledge in a week and a half, but are tolerated in the nest area for another week and a half.

**Food Habits:** Red-winged Blackbirds eat a very broad range of foods. During breeding season,

animal matter is more important, including insects (caterpillars, grubs, grasshoppers, damselflies, mayflies), spiders, snails, and other molluscs. In winter, a variety of seeds and waste grain from cultivated fields is eaten. They eat fruits and berries when available.

**Ecology:** Male Red-winged Blackbirds aggressively defend a relatively small territory, usually only a few tenths of an acre, in which females of the harem nest. Following breeding season, large flocks form and forage widely, especially in agricultural areas, sometimes 50 miles away from roosts. They leave higher elevations during winter, and often form large flocks in western interior valleys.

**Comments:** In some parts of its range, the Red-winged Blackbird is considered an agricultural pest. This is one of the most-studied birds in North America.

304

## Tricolored Blackbird
### (Agelaius tricolor)

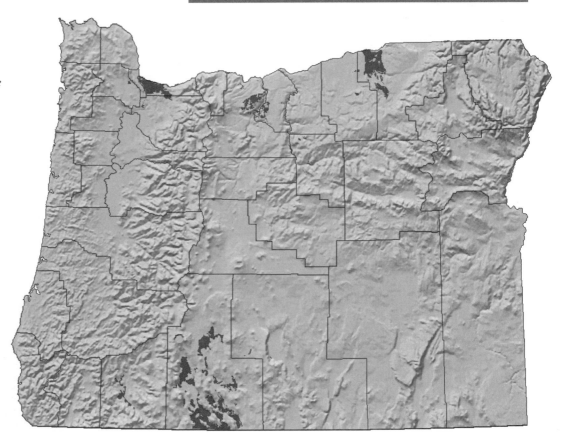

*Order: Passeriformes*
*Family: Emberizidae*
*State Status: Sensitive*
*Federal Status: SC*
*Global Rank: G3*
*State Rank: S2*
*Length: 8 in (20 cm)*

**Global Range:** Has a restricted breeding distribution from southern Oregon south through cismontane California to northern Baja California.

**Habitat:** This blackbird generally prefers to breed in freshwater marshes with emergent vegetation (cattails) or in thickets of willows or other shrubs. In Oregon, it has bred in tangles of Himalayan blackberry growing in and around wetlands. It often is found breeding in the company of Red-winged Blackbirds.

**Reproduction:** Tricolored Blackbirds are breeding by early April. They are resident throughout most of their range, but migrate to Oregon breeding grounds. The nest of plant fibers is attached to emergent vegetation or secured in a thicket of shrubs. Clutches normally have 4 (range 2-6) eggs. The female incubates the clutch for about 11 days, and the young leave the nest 2 weeks after hatching.

**Food Habits:** Like the related Red-winged Blackbird, the Tricolored Blackbird eats mostly animal food in the breeding season (grasshoppers, crickets, damselflies, beetles, spiders, snails, small tadpoles). It feeds on a variety of seeds and waste grain following breeding. Tricolored Blackbirds may fly as far as 4 miles from their nest while foraging.

**Ecology:** This blackbird is colonial rather than territorial, defending only an area a few feet away from the nest. Following the breeding season, it forms large flocks. Most of Oregon's Tricolored Blackbirds winter in California.

**Comments:** Although some colonies number over 1,000 birds, the location of colonies is unpredictable from year to year, making monitoring and conservation difficult. Small colonies have been found at several locations in north-central Oregon, including one colony in the Portland area.

*References:* DeGraaf et al. 1991, Gilligan et al. 1994, Marshall 1992, Orians 1961, Payne 1969.

305

# Western Meadowlark
## (Sturnella neglecta)

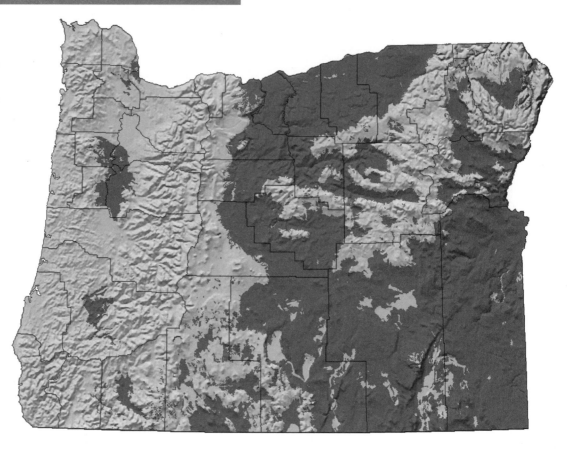

*Order: Passeriformes*
*Family: Emberizidae*
*State Status: None*
*Federal Status: None*
*Global Rank: G5*
*State Rank: S4*
*Length: 9 in (23 cm)*

**Global Range:** Despite its name, the Western Meadowlark has an extensive breeding range, from British Columbia east to western New York and Tennessee and south to Texas and Baja California. It is also found breeding in the central Mexican highlands. It is absent from the East Coast and the southeastern United States.

**Habitat:** Western Meadowlarks are most typical of open grasslands, grassy hillsides, pastures, meadows, and sagebrush plateaus with a grass understory. They readily adapt to agricultural operations, often perching on fences and posts at the edges of alfalfa or grain fields. Sometimes they are found in open woodlands.

**Reproduction:** In Oregon, Western Meadowlarks may begin breeding as early as April, with a peak in May. The nest is a dome of grass placed in a depression in the ground. The normal clutch of 5 (range 3-7) eggs is incubated by the female for about 2 weeks. The young are fed by both parents for about a month.

**Food Habits:** A majority of the Western Meadowlark's diet is animal material, more so in spring and summer. It eats beetles, crickets,

grasshoppers, caterpillars, craneflies, sow bugs, spiders, snails, a few bird eggs, and some carrion. In winter, it consumes more plant material, including a variety of grass and forb seeds and waste grain.

**Ecology:** Territory size is reported to range from a few to over 10 acres. The Western Meadowlark is sometimes polygynous. Following the breeding season, it may form small flocks. It leaves higher elevations in winter, and many migrate south.

**Comments:** The Western Meadowlark is Oregon's state bird.

*References:* DeGraaf et al. 1991, Gabrielson and Jewett 1940, Gilligan et al 1994, Jewett et al. 1953, Johnson and Temple 1990, Ryser 1985.

# Yellow-Headed Blackbird
## (Xanthocephalus xanthocephalus)

*Order: Passeriformes*
*Family: Emberizidae*
*State Status: None*
*Federal Status: None*
*Global Rank: G5*
*State Rank: S5*
*Length: 9.5 in (24 cm)*

**Global Range:** Breeds from British Columbia across central Canada to the Great Lakes, and south to southern California, northern New Mexico, and Kansas.

**Habitat:** Yellow-headed Blackbirds are more strictly restricted to freshwater and alkaline marshes than Red-winged Blackbirds. They breed in emergent cattails, reeds, and bulrushes in large marshes of eastern Oregon's valleys and in smaller marshes fringing mountain lakes. They use marshes locally in western Oregon.

**Reproduction:** This blackbird arrives in Oregon in April and is breeding by May. The nest of aquatic vegetation is suspended over water in emergent vegetation or, sometimes, placed in willows at the edge of a marsh or lake. The clutch normally numbers 4 (range 3-5) eggs, which the female incubates for about 2 weeks. Young fledge in about 3 weeks.

**Food Habits:** This blackbird finds its food on marsh vegetation or on moist ground near water. It also catches some insects in flight. In the breeding season, it eats many insects (dragonflies, damselflies, caterpillars, grasshoppers, ants, wasps), spiders, and snails. It also eats a variety of seeds and grain, as well as the leaves of some forbs and grasses.

**Ecology:** Each male may have several mates in a territory of several hundred square meters. Following the breeding season, Yellow-headed Blackbirds may join large flocks of other blackbirds. They tend to defend their territory interspecifically against Marsh Wrens, hawks, crows, and other large birds. They leave higher elevation marshes in winter, and most migrate south by October for wintering grounds in the southwestern United States and Mexico.

**Comments:** The young leave the nest before they can fly, often perching on tules close to the water. In such circumstances, they are sometimes caught by large fish.

*References:* Bump 1986, DeGraaf et al. 1991, Gabrielson and Jewett 1940, Gilligan et al. 1994, Leonard and Picman 1986, Verner 1975, Willson 1966, Zeiner et al. 1990a.

**307**

# Brewer's Blackbird
## *(Euphagus cyanocephalus)*

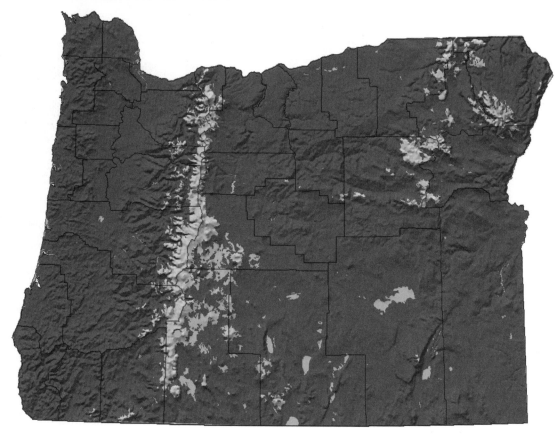

*Order: Passeriformes*
*Family: Emberizidae*
*State Status: None*
*Federal Status: None*
*Global Rank: G5*
*State Rank: S5*
*Length: 9 in (23 cm)*

*References:*
DeGraaf et al. 1991, Gabrielson and Jewett 1940, Gilligan et al. 1994, Horn 1968, 1970; Jewett et al. 1953, Ryser 1985, Williams 1952, Zeiner et al. 1990a.

**Global Range:** Breeds from British Columbia across southern Canada and the northern United States to the Great Lakes. In the West, it is resident from western Oregon and Washington south to Baja California and east through the Great Basin to eastern Colorado. It breeds, but does not winter, east of the Cascade Mountains to northern Idaho and Montana.

**Habitat:** This blackbird uses a remarkably wide range of habitats. Like other blackbirds, it may nest in marshes or willow riparian woodlands and prefers to be at water's edge. However, it also breeds in arid salt-desert shrub, sagebrush, and grasslands, long distances from the nearest water. Cultivated fields, farmsteads, rural towns, suburbs, and even urban centers provide additional habitat, as do mountain meadows, quaking aspen groves, juniper woodlands, and cleared or burned forests. It breeds from the coast to at least 6,000 feet, and avoids only the interior of dense forests.

**Reproduction:** This species begins nesting in small colonies by late April. The nest cup, usually plant material held together with mud, is placed low in bushes or high in trees. The usual clutch is 5 or 6 (range 3-7) eggs, which the female incubates for 12-13 days. Young are independent in about 4 weeks.

**Food Habits:** Over the year, the diet is about two-thirds plant material, but animal matter is more important during the breeding season. It eats beetles, caterpillars, grasshoppers, crickets, spiders, snails, termites, and aphids. Waste grain is an especially important food in fall and winter, and it also eats grass and forb seeds, fruits, and berries.

**Ecology:** Only the immediate area of the nest is defended, and the male more often defends the female than the nest. Males frequently have more than one mate. Following breeding season, they congregate in large flocks, which are sometimes interspecific. They leave higher elevations in winter, and some probably migrate south.

**Comments:** This species can cause damage to agricultural crops, but it also assists in keeping insect pests under control.

308

## Brown-headed Cowbird
*(Molothrus ater)*

*Order: Passeriformes*
*Family: Emberizidae*
*State Status: None*
*Federal Status: None*
*Global Rank: G5*
*State Rank: S5*
*Length: 6.5 in (17 cm)*

*References: DeGraaf et al. 1991, Gabrielson and Jewett 1940, Gilligan et al. 1994, Ryser 1985, Scott and Ankney 1983, Weatherhead 1989, Woodward 1983, Zeiner et al. 1990a.*

**Global Range:** Breeds coast to coast in North America, from southeastern Alaska east across Canada and south to northern Baja California, central Mexico, and the Gulf Coast.

**Habitat:** Once a bird of grasslands and prairies in most of its range in the West, the Brown-headed Cowbird occupies sagebrush-covered valleys, juniper woodlands, grasslands, and marshy areas. Following human activity, this species has expanded its range into urban and agricultural areas; open coniferous, deciduous, and riparian woodlands; mountain meadows and other forest clearings; and forest edges. Rural towns, ranches, and any type of livestock operation will be frequented by this species.

**Reproduction:** The Brown-headed Cowbird is a brood parasite, females laying their eggs in the nests of other birds, which then incubate the egg and rear the young. They are engaged in their unorthodox form of breeding by May. Each female usually lays only one egg per nest, and may lay over 2 dozen eggs in a season. Cowbird eggs often hatch in 10 days, prior to their hosts' eggs, the cowbird nestlings subsequently out-competing their fellow nestlings for food. Host birds may feed young cowbirds for up to 7 weeks.

**Food Habits:** About three-quarters of the diet is animal matter, especially in the spring. They eat grasshoppers, beetles, true bugs, flies, wasps, ants, and caterpillars. A variety of grass and forb seeds are also eaten, as are fruits and berries when they are available.

**Ecology:** Although they occupy semi-exclusive home ranges, they are thought to defend an area around their mate rather than a specific territory. They are known to parasitize over 200 species of birds, but some groups (warblers, vireos, flycatchers, thrushes, finches) are affected more than others. They will join flocks of other blackbirds following the breeding season, and many migrate south for the winter.

**Comments:** This species was less common in Oregon early in the century, but has experienced rapid population growth in recent decades.

**309**

# Bullock's Oriole
## (Icterus bullockii)

*Order: Passeriformes*
*Family: Emberizidae*
*State Status: None*
*Federal Status: None*
*Global Rank: G5*
*State Rank: S4*
*Length: 7 in (18 cm)*

**Global Range:** Breeds from southern Canada south through the western United States to the highlands of northern Mexico. It is found as far east as South Dakota and central Texas.

**Habitat:** Bullock's Orioles are mostly found in deciduous and broad-leafed woodlands, such as maple, alder, oak, or madrone woodlands and forests, often at the edges of clearings. They nest in cottonwood or willow riparian woodlands, city parks, and cottonwoods or other trees surrounding farmsteads, ponds, or irrigation ditches. They often nest near water, but sometimes can be found in arid juniper woodlands.

**Reproduction:** Bullock's Orioles migrate to Oregon in April, and are building their pendulous nests of plant fiber, suspended 20-30 feet high from the branch of a tree, by May. Their normal clutch has 4 or 5 (range 3-6) eggs, which the female incubates for 2 weeks. The young fledge in another 2 weeks.

**Food Habits:** About 80% of the diet consists of animal material, mostly gleaned from leaves and branches of trees. They eat caterpillars, beetles, ants, leafhoppers, treehoppers, aphids, and grasshoppers, as well as a variety of fruits and berries (blackberries, elderberries, orchard crops) and some nectar.

**Ecology:** Breeding territories are defended up to several hundred feet from the nest, resulting in pairs being spaced throughout the available habitat. By October, most Bullock's Orioles have flown south to their wintering grounds in Mexico and Central America.

**Comments:** Bullock's Orioles are monogamous, and often a pair returns in subsequent years to nest in the same tree. Bullock's Oriole was formerly considered conspecific with the Baltimore Oriole (*Icterus galbula*) of the eastern United States and known as the Northern Oriole (American Ornithologists' Union 1995).

*References:* DeGraaf et al. 1991, Gabrielson and Jewett 1940, Gilligan et al. 1994, Rohwer and Manning 1990, Zeiner et al. 1990a.

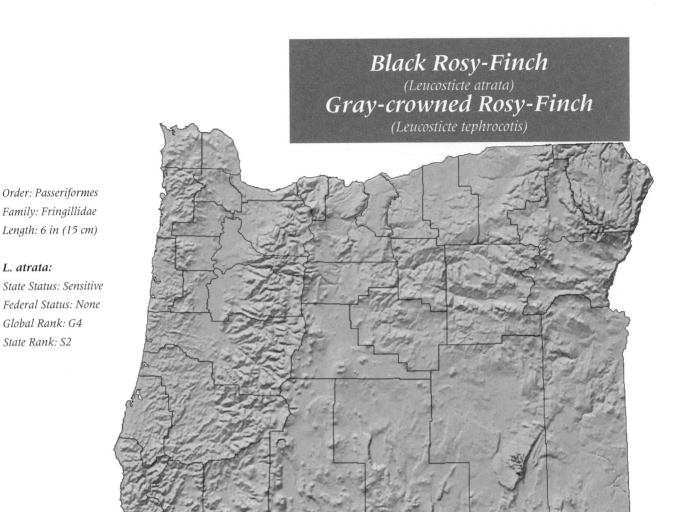

## Black Rosy-Finch
### (Leucosticte atrata)
## Gray-crowned Rosy-Finch
### (Leucosticte tephrocotis)

Order: *Passeriformes*
Family: *Fringillidae*
Length: *6 in (15 cm)*

**L. atrata:**
State Status: *Sensitive*
Federal Status: *None*
Global Rank: *G4*
State Rank: *S2*

*Distribution map for Black Rosy-Finch. Map for Gray-crowned Rosy-Finch is on following page.*

**Global Range:** Rosy-Finches are birds of colder environments. They breed in Asia, from the Altai Mountains south to the Lake Baikal region and east to the Pacific Ocean, including the Commander Islands. In North America, they are found breeding from most of Alaska south through British Columbia to the Sierra Nevada of California and the Rocky Mountains of northern New Mexico.

**Habitat:** These finches breed on open, rocky ground. In the Arctic, such treeless barrens may occur at sea level, but in Oregon, Rosy-Finches find their habitat in alpine environments above timberline. They usually breed among patches of snow.

**Reproduction:** At higher elevations, breeding is postponed until June. The nest cup of moss or grass is placed on a rock ledge or concealed in a crevice. The usual clutch is 4 or 5 (range 2-6) eggs, which are incubated for 2 weeks by the female. The young are fed by the parents for about 5 weeks and remain with the family group until fall.

**Food Habits:** During the breeding season, they take what insects they can find, often ones frozen and exposed by melting snow. They catch some insects on the wing. Seeds and green parts of

**L. tephrocotis:**

*State Status: None*

*Federal Status: None*

*Global Rank: G5*

*State Rank: S2S3*

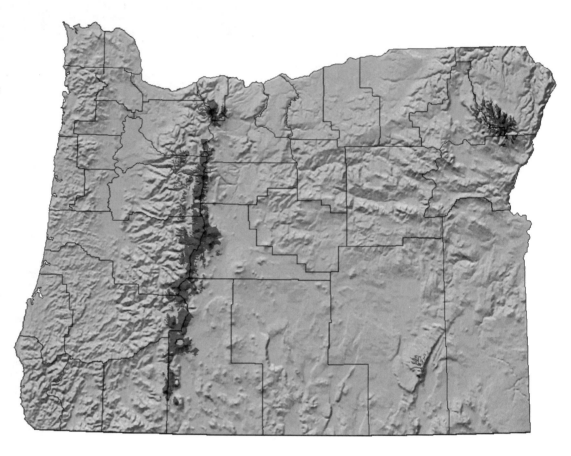

*Distribution map for Gray-crowned Rosy-Finch.*

alpine plants are an important part of the diet, especially after the breeding season.

**Ecology:** Males do not defend a specific area, rather a radius of about 100 feet around the female. They forage at least as far as half a mile from the nest. In winter, they retreat to lower elevations, being joined by migrants from the north. The ecological information in this account applies to both species of Rosy-Finch that breed in Oregon.

**Comments:** The American Ornithologists' Union (1993) again recognizes three species of North American rosy-finches, two of which, the Black Rosy-Finch (*L. atrata*) and the Gray-crowned Rosy-Finch (*L. tephrocotis*), occur in Oregon. Only the Black Rosy-Finch population on Steens Mountain is considered "sensitive" by the Department of Fish and Wildlife. Although there are no confirmed records, it is possible that the Black Rosy-Finch also breeds locally in the Wallowa Mountains. The Black Rosy-Finch is ranked "G4" by the Nature Conservancy. *Leucosticte arctoa* is the Latin name for the Rosy-Finch of Eurasia.

*References:* DeGraaf et al. 1991, French 1959, Gilligan et al. 1994, Johnson 1975, Marshall 1992, Udvardy 1977, Zeiner et al. 1990a.

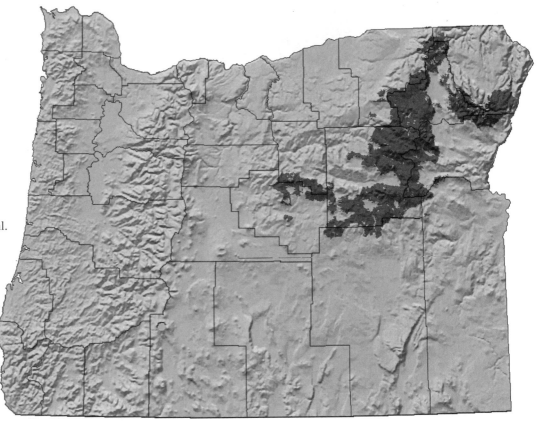

## Pine Grosbeak
### *(Pinicola enucleator)*

Order: *Passeriformes*
Family: *Fringillidae*
State Status: *None*
Federal Status: *None*
Global Rank: *G5*
State Rank: *S2?*
Length: *8 in (20 cm)*

References: DeGraaf et al. 1991, French 1954, Gilligan et al. 1994, Zeiner et al. 1990a.

**Global Range:** A holarctic species that breeds in Eurasia, from Scandinavia east to Siberia and south to Japan. In North America, it breeds from Alaska across northern Canada, south to the Great Lakes and New England. From the Rocky Mountains west, it is resident farther south, in the Sierra Nevada of central California and the mountains of northern New Mexico and eastern Arizona.

**Habitat:** Pine Grosbeaks occupy high-elevation forest communities. They do not use lower ponderosa pine woodlands, instead preferring lodgepole pine, subalpine fir, true fir, Douglas-fir, Engelmann spruce, and other mixed coniferous forest types. They are usually encountered around wet meadows, lakes, or streams.

**Reproduction:** The Pine Grosbeak breeds fairly late, in late May or June, as might be expected at high elevations. The nest cup of twigs is located in the branch of a conifer. The typical clutch has 4 (range 2-5) eggs, which the female incubates for 2 weeks, while the male feeds her. The young leave the nest at about 2 weeks of age.

**Food Habits:** This species eats some spiders and insects (grasshoppers, beetles, caterpillars, flies), especially in the breeding season, but feeds mostly on plant material, including buds, seeds, fruits, and berries. It will eat pine, fir, and larch seeds, but also eats seeds from deciduous trees, and will eat berries when they are available. The Pine Grosbeak forages in both coniferous trees and riparian thickets.

**Ecology:** The only information on territoriality comes from a Utah spruce-fir forest, where a 26-acre territory was recorded. This species forms small flocks after breeding season, some of which move to lower elevations. At least some individuals may remain on their breeding grounds through the winter.

**Comments:** Pine Grosbeaks favor open, rather than dense, forest stands, and frequently forage near water. They may take baths in soft snow. The breeding range in Oregon is poorly known.

# Purple Finch
*(Carpodacus purpureus)*

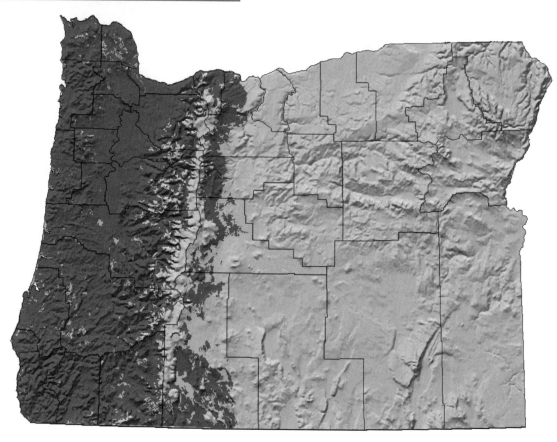

*Order: Passeriformes*
*Family: Fringillidae*
*State Status: None*
*Federal Status: None*
*Global Rank: G5*
*State Rank: S4*
*Length: 6 in (15 cm)*

**Global Range:** Breeds from the southern Yukon and British Columbia east across southern Canada to the Atlantic provinces, New England, and the Great Lakes. A southern extension of resident birds occurs in the Pacific states, south to the mountains of Baja California del Norte.

**Habitat:** Purple Finches prefer open areas or edges of low to mid-elevation coniferous forests, but use similar edges in mixed coniferous-deciduous forests. They also occupy forested residential areas and riparian thickets in forests. They are absent from the interior of dense forests.

**Reproduction:** The Purple Finch is breeding in Oregon by May. The nest cup of plant material is usually located high in a coniferous tree. A typical clutch will have 4 or 5 (range 3-6) eggs, which the female incubates for about 2 weeks, while the male brings food. The young leave the nest in another 2 weeks.

**Food Habits:** During the nonbreeding season, Purple Finches primarily feed on seeds of forbs, grasses, maple, ash, and other hardwoods. They also eat some buds and berries when they are available. In spring, they add insects, such as caterpillars and

beetles, to their diet. Most of their food items are gleaned from branches.

**Ecology:** Singing males are presumably engaged in territorial display. One study found that males used an area of about 5 acres. Following breeding, this species may join mixed species flocks of other finches.

**Comments:** Purple Finches frequent bird feeders in winter. Ash seeds are an important item in the diet in western Oregon.

*References:* DeGraaf et al. 1991, Gilligan et al. 1994, Jewett et al. 1953, Pulliam and Enders 1971, Salt 1952, Zeiner et al. 1990a.

# Cassin's Finch
*(Carpodacus cassinii)*

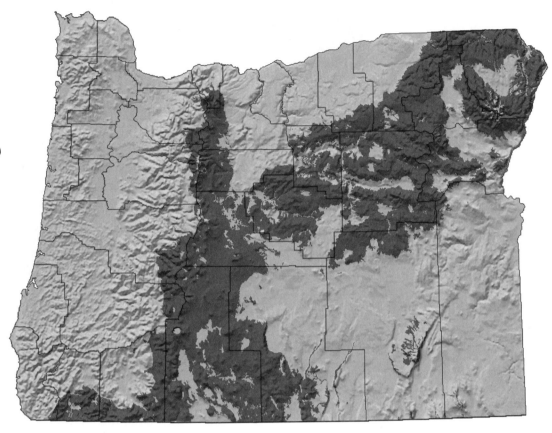

Order: *Passeriformes*
Family: *Fringillidae*
State Status: *None*
Federal Status: *None*
Global Rank: *G5*
State Rank: *S4*
Length: *6.5 in (17 cm)*

**Global Range:** Cassin's Finch is a species of western North America. It is resident from southern British Columbia south to the mountains of northern Baja California and east to the Rocky Mountains, from Montana to New Mexico.

**Habitat:** Cassin's Finch inhabits edges and open areas of middle-elevation coniferous forests. It is typical of lodgepole pine, ponderosa pine, subalpine fir, and other mixed coniferous woodlands. It is most often seen near open areas, rather than in dense forest interiors, and will use quaking aspen stands within a coniferous forest matrix.

**Reproduction:** The Cassin's Finch is breeding in Oregon by May. The nest is a cup of plant material, located fairly high on the branch of a coniferous tree. There are normally 4 or 5 (range 3-6) eggs in a clutch. The female incubates the eggs for just under 2 weeks, while being fed by the male.

**Food Habits:** This finch either forages on the ground or gleans food from the foliage of trees. It eats mostly plant material, especially seeds of grasses and forbs, but also seeds and buds of coniferous trees. A few insects and berries are included in the diet.

**Ecology:** A Utah study found that males defended an area around the female, rather than a fixed territory. This finch leaves higher-elevation forests for more moderate climates in winter.

**Comments:** During winter, Cassin's Finch often joins mixed species flocks that may include Evening Grosbeaks or Red Crossbills.

*References:* DeGraaf et al. 1991, Gabrielson and Jewett 1940, Gilligan et al. 1994, Mewaldt and King 1985, Salt 1952, Samson 1976, Zeiner et al. 1990a.

**315**

# House Finch
## (Carpodacus mexicanus)

*Order: Passeriformes*
*Family: Fringillidae*
*State Status: None*
*Federal Status: None*
*Global Rank: G5*
*State Rank: S5*
*Length: 5.5 in (14 cm)*

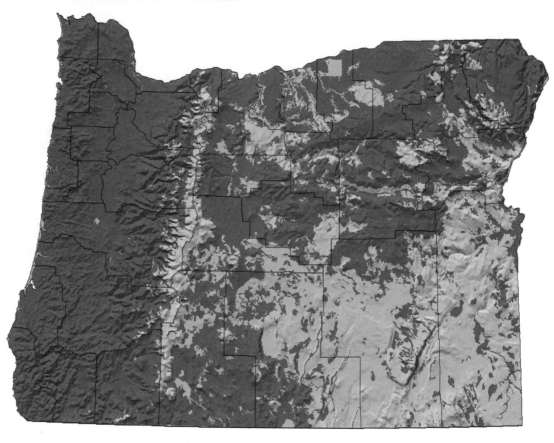

**Global Range:** Native to western North America, where it is resident from central British Columbia south to southern Mexico and east to the Rocky Mountains and western Texas. It has been introduced in Hawaii and the eastern United States, where it now breeds from New England to Georgia, and west to the Great Lakes and Missouri.

**Habitat:** Throughout its wide range, it occupies a variety of brushy scrub and woodland habitats, as well as making use of cities, towns, farmsteads, and agricultural areas. In Oregon, it occurs in grasslands, shrub thickets, and woodlands at lower elevations, especially where humans have settled. Aside from areas with buildings and houses, it is not found in arid sagebrush or salt-desert shrub-covered valleys or in dense forests.

**Reproduction:** These finches have a relatively short breeding season from April to June. They nest in natural tree cavities or shrubs, as well as on ledges and in crevices of buildings. There are usually 4 or 5 (range 2-6) eggs, which are incubated by the female for about 2 weeks, while the male brings food. The young are fed for 3 weeks after hatching. There may be more than one brood per year.

**Food Habits:** About 86% of the diet consists of the seeds of grasses and forbs (including dandelion, thistle, other weeds, and cultivated plants). They will eat fruit and berries when they are available. A small amount of insects is also eaten.

**Ecology:** Only very small areas (6-30 feet) around the nest or female are defended as territories, but not vigorously. House Finches drink water daily, and will often fly considerable distances to find it. After the breeding season, they form small wandering flocks.

**Comments:** House Finches can be responsible for damage to agricultural crops and gardens. They have expanded their range in western Oregon considerably in recent decades.

*References:* DeGraaf et al. 1991, Gabrielson and Jewett 1940, Gilligan et al. 1994, Salt 1952, Zeiner et al. 1990a.

## Red Crossbill
*(Loxia curvirostra)*

*Order: Passeriformes*
*Family: Fringillidae*
*State Status: None*
*Federal Status: None*
*Global Rank: G5*
*State Rank: S4*
*Length: 6 in (15 cm)*

*References:* Benkman 1990,
DeGraaf et al. 1991,
Gabrielson and Jewett
1940, Gilligan et al. 1994,
Knox 1990, Zeiner et al.
1990a.

**Global Range:** Widely distributed holarctic species. Breeds from the British Isles and Scandinavia east to Siberia and south to northern Africa and the Philippines. In North America, occurs from southeastern Alaska across southern Canada to the Atlantic Coast, and south, from the Pacific Ocean to the Rockies, through the highlands of Mexico to Nicaragua. It is absent from most of the eastern U.S. south of the Great Lakes and New York, but extends down the Appalachian Mountains to North Carolina.

**Habitat:** Red Crossbills breed in every type of coniferous forest in Oregon, although they are most common in lower-elevation forests west of the Cascade Mountains and in the ponderosa pine zone east of the Cascade crest.

**Reproduction:** This species breeds throughout the year, whenever and wherever it happens upon a good crop of seeds of coniferous trees. The nest cup of plant materials is placed on the branch of a conifer. There are usually 3 or 4 (range 2-5) eggs,

which the female incubates for about 2 weeks. The first week after hatching, the female tends the young at the nest while being fed by the male. Young stay with their parents for about 7 weeks.

**Food Habits:** In winter, Red Crossbills feed almost entirely on conifer seeds extracted with their crossed scissor-like bills from the bracts of cones. The diet is more varied in summer, including seeds and buds of deciduous trees (maple, alder, willow) and about 30% insects (beetles, ants, caterpillars, aphids).

**Ecology:** Perhaps because of their dependence on locally abundant seed crops, Red Crossbills are not strongly territorial, often nesting within 100 feet of one another. They usually move to lower-elevation forests in winter, but wander widely, even searching for food in towns and suburbs.

**Comments:** A study in Idaho and Montana found Red Crossbills favored rotation-aged, rather than late-successional, Douglas-fir forests. Several unrecognized sibling species may occur among nomadic populations of Red Crossbills in western North America (Groth 1993).

# Pine Siskin
## *(Carduelis pinus)*

*Order: Passeriformes*
*Family: Fringillidae*
*State Status: None*
*Federal Status: None*
*Global Rank: G5*
*State Rank: S5*
*Length: 4.5 in (11 cm)*

*References:* DeGraaf et al. 1991, Gabrielson and Jewett 1940, Gashwiler and Ward 1966, Gilligan et al. 1994, Zeiner et al. 1990a.

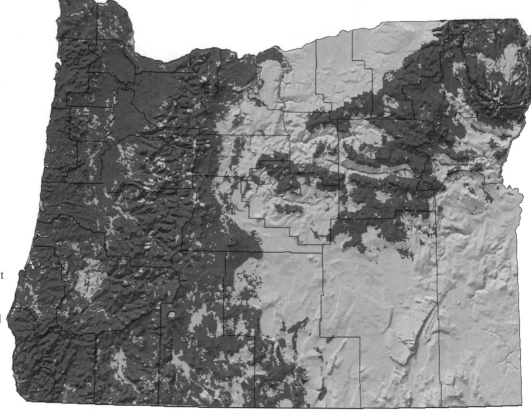

**Global Range:** The Pine Siskin has a range much like that of the Red Crossbill. It breeds from southeastern Alaska across southern Canada to the Atlantic Ocean. In the West, it occurs south through the United States from the Rocky Mountains west to the Pacific Ocean, and breeds farther south in the highland areas of northern and central Mexico. It does not breed in most of the eastern United States south of the Great Lakes or New Jersey, with the possible exception of the Appalachian Mountains south to North Carolina.

**Habitat:** This species is usually associated with coniferous forests of every type, but also will use deciduous trees at the edges of coniferous forests. Unlike the Red Crossbill, this species prefers older forests. It is less common in open ponderosa pine forests. A study in Idaho and Montana found it preferred late-successional stands to rotation-aged forests. Douglas-fir, true fir, Engelmann spruce, western hemlock, mountain hemlock, and mixed conifer forests are its most typical habitat.

**Reproduction:** In Oregon, the nesting season begins by the middle of April. The nest cup of plant material is usually placed in the branch of a conifer, 15-35 feet off the ground. In the Willamette Valley, it has also nested in oaks and maples, most likely in the proximity of conifers. The clutch of 3 or 4 (range 2-6) eggs is incubated by the female for about 2 weeks. The young leave the nest 2 weeks after hatching.

**Food Habits:** The diet of the Pine Siskin varies with the season, and includes more insects (caterpillars, aphids, scale insects, grasshoppers) in summer. At other times, the seeds, sap, and buds of both coniferous and deciduous trees (alder, birch, maple) become more important foods. It also eats forb seeds and some fruit and berries when they are available.

**Ecology:** This species nests in loose colonies. Nests are often within 3 to 6 feet of one another. In winter, it leaves higher-elevation forests for warmer lowlands, although some probably migrate south.

**Comments:** The Pine Siskin is often a member of large mixed-species flocks that frequent towns and suburbs in winter.

318

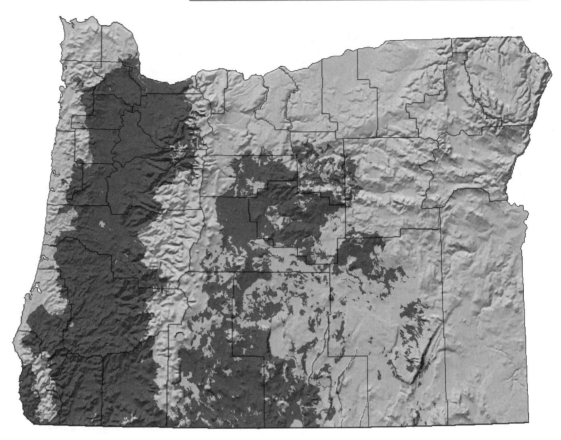

## Lesser Goldfinch
### (Carduelis psaltria)

*Order: Passeriformes*
*Family: Fringillidae*
*State Status: None*
*Federal Status: None*
*Global Rank: G5*
*State Rank: S4*
*Length: 4 in (10 cm)*

**Global Range:** Vancouver, Washington, marks the northern limit of this bird's breeding range. It is rare in Idaho, but breeds farther south in the interior through Nevada, Utah, Colorado, and the Southwest, as well as in all the Pacific states. South of the United States, it is resident throughout Mexico, Central America, and northern South America.

**Habitat:** Throughout its wide distribution it is found in many habitat types, usually containing scattered trees or bushes, but also in fields, pastures, and around human settlements. It is usually near water, which tends to restrict it to riparian areas or other moist habitats in eastern Oregon. It nests in oak and juniper woodlands, but does not use dense coniferous forests, and is rare or absent at higher elevations.

**Reproduction:** The Lesser Goldfinch's breeding season is under way by early May. The nest cup of plant material is placed in a tree or shrub. The clutch of 4 or 5 (range 3-6) eggs is incubated for about 12 days by the female, who is fed by the male.

**Food Habits:** Lesser Goldfinches are unusual in that they eat very little animal matter, even during the breeding season. Seeds make up about 96% of their

diet. They eat a few insects, as well as some buds, berries, flowers, and leaves. They live in arid environments and, not eating food with a high moisture content, must drink water frequently.

**Ecology:** Although they may defend an area of up to 100 feet around their nests, they sometimes nest in small colonies. Following breeding season, they may wander upslope in large flocks. Lesser Goldfinches breeding in eastern Oregon probably migrate south for the winter.

**Comments:** During breeding season, these birds will travel as far as half a mile from the nest for water.

*References:* Coutlee 1968, DeGraaf et al. 1991, Gabrielson and Jewett 1940, Gilligan et al. 1994, Linsdale 1957, Ryser 1985, Zeiner et al. 1990a.

# American Goldfinch
### *(Carduelis tristis)*

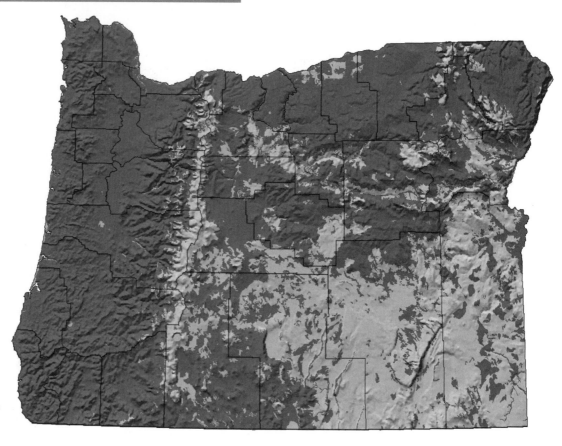

*Order:* Passeriformes
*Family:* Fringillidae
*State Status:* None
*Federal Status:* None
*Global Rank:* G5
*State Rank:* S4
*Length:* 5 in (13 cm)

*References:* DeGraaf et al. 1991, Gabrielson and Jewett 1940, Gilligan et al. 1994, Jewett et al. 1953, Watt and Dimberio 1990, Wiseman 1975, Zeiner et al. 1990a.

**Global Range:** The American Goldfinch has a much smaller range than the Lesser Goldfinch, breeding from southern Canada south to northern Baja California in the West, and south to Georgia in the East. It is absent as a breeder from much of the Southwest and the Gulf Coast region, although it winters in these areas, as well as in northern Mexico.

**Habitat:** American Goldfinches tend to nest in areas where thickets of shrubs or low trees grow in or adjacent to weedy fields, pastures, croplands, edges of marshes, and other open areas, most commonly in lower-elevation valleys. They are restricted to riparian areas and towns in the desert valleys of eastern Oregon. They are not birds of dense forests, and seem to prefer willow riparian woodland.

**Reproduction:** This species waits until June to commence breeding. The nest is a tightly woven cup of plant material, placed in a tree or shrub, usually 5-15 feet off the ground. The usual clutch is 5 (range 4-6) eggs, which the female incubates for about 2 weeks, while being fed by the male. The young leave the nest in about 2 to 2-1/2 weeks.

**Food Habits:** Much of the biology of the American Goldfinch revolves around its favored food, the seeds of thistles, dandelions, goldenrod, or other composites. These plants grow in open, often disturbed, environments. Some tree seeds (birch, alder, conifer) contribute to the diet as well. Breeding is put off until these seeds are ripe. During breeding season, insects (caterpillars and aphids) make up about half the diet. Buds and green plant material are also eaten.

**Ecology:** This species is a solitary nester that defends a territory about 100 feet in diameter. Following the breeding season, there is an upslope movement of flocks of goldfinches. They are so closely associated with thistles that one of the older vernacular names for the species is "thistle bird."

**Comments:** Ryser (1985) reports the nest is so well woven that it will temporarily hold water, which has resulted in the young drowning during rainstorms.

## Evening Grosbeak
### (Coccothraustes vespertinus)

*Order: Passeriformes*
*Family: Fringillidae*
*State Status: None*
*Federal Status: None*
*Global Rank: G5*
*State Rank: S5*
*Length: 7 in (18 cm)*

*References:* Bekoff et
al. 1987, DeGraaf et
al. 1991, Gabrielson
and Jewett 1940,
Gilligan et al. 1994,
Jewett et al. 1953,
Prescott 1991, Scott
and Bekoff 1991,
Wiens and
Nussbaum 1975,
Zeiner et al. 1990a.

**Global Range:** Breeds from British Columbia east to the Atlantic Coast in a relatively narrow band across central and southern Canada. In the East, breeds no farther south than New York, but is resident in the West through the forested lowlands and mountains of the Pacific Northwest, east to the Rocky Mountains, and south to the highlands of Mexico.

**Habitat:** Evening Grosbeaks breed in middle- to high-elevation coniferous forests, especially Engelmann spruce and true fir (*Abies*) forests, but also in a variety of mixed-conifer forests. They can be found at forest edges around meadows and fairly open coniferous woodlands. Elsewhere in their range, these grosbeaks will use mixed coniferous-deciduous forests, although there are certainly patches of deciduous trees (aspen, willow, maple) within their breeding habitat in Oregon.

**Reproduction:** The breeding season is under way by June. The nest cup of plant materials is placed high in a coniferous tree. A typical clutch has 3 or 4 (range 2-5) eggs, which are incubated for about 12-14 days by the female, while the male brings her food. The young are ready to leave the nest in 2 weeks.

**Food Habits:** During the breeding season, some insects (caterpillars, ants, bees, wasps, true bugs) are included in the diet. At other times, conifer seeds (fir, pine, spruce, western red cedar) and seeds and buds of a variety of broadleaf trees (vine maple, bigleaf maple, dogwood, willow, oak) are eaten. Some fruits (e.g., mountain ash, chokecherry) are included in the diet when they are available.

**Ecology:** In an Oregon mountain hemlock-silver fir forest, 53 individuals per 100 acres were found. No information about territorial defense was found for this species. Some individuals may remain on their high-elevation breeding grounds through the winter, but there is a general downslope movement toward warmer valleys.

**Comments:** In winter and spring, small flocks of Evening Grosbeaks can be seen in cities and towns, often feeding on bigleaf maple samaras. Numbers vary from year to year, and it is likely that some migrate south for the winter.

# House Sparrow
## *(Passer domesticus)*

*Order:* Passeriformes
*Family:* Passeridae
*State Status:* None
*Federal Status:* None
*Global Rank:* G5
*State Rank:* SE
*Length:* 6 in (15 cm)

*References:* Bennett 1990, Gabrielson and Jewett 1940, Gilligan et al. 1994, Grinnell and Miller 1944, Zeiner et al. 1990a.

**Global Range:** So closely is this species associated with humans that its original range is difficult to determine. It is native to the Old World, from the United Kingdom east to China and from northern Russia south to northern Africa and Burma. It is widely introduced throughout the New World, South Africa, Australia, New Zealand, and Hawaii.

**Habitat:** House Sparrows are mainly found in urban and agricultural areas. They thrive around livestock on remote ranches and in small towns or settlements in arid regions. They are most likely to occur where there are both nest sites and a reliable supply of food. They are seldom found away from human development.

**Reproduction:** House Sparrows typically have 2 or 3 broods per year and, although breeding is concentrated in spring and summer, they may breed at any time. They are usually cavity nesters, but will also nest in crannies of buildings. The clutch of 3 to 5 (range 2-8) eggs is incubated for about 2 weeks by the female. Young fledge in another 2 weeks. House Sparrows often nest in loose colonies.

**Food Habits:** The young are fed both insects and seeds. While the adult diet consists mostly of seeds and grain, they, too, eat some insects. Around human habitations they forage for whatever edible scraps are available.

**Ecology:** Usually only a small area (a few meters) around the nest is defended. Nonbreeding birds often congregate in flocks, which may forage up to a mile or two from their night roost. They are nonmigratory.

**Comments:** House Sparrows compete with native cavity nesters for nest sites. They also destroy eggs and nestlings of native species. House Sparrows were introduced into the eastern United States about 1850, and arrived in California about 1871. Sometimes this species is called the English Sparrow. Some authorities place it in the family Ploceidae.

# Mammals

# Virginia Opossum
## (Didelphis virginiana)

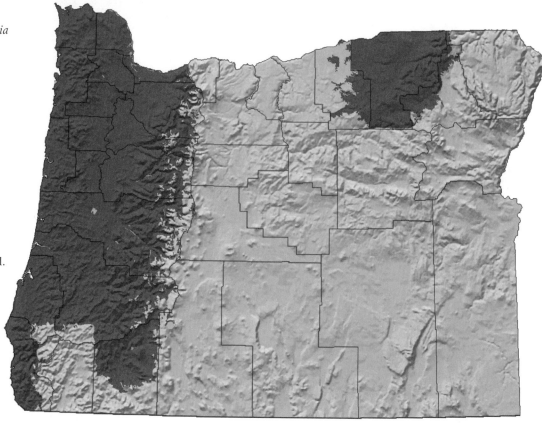

*Order: Didelphimorphia*
*Family: Didelphidae*
*State Status: None*
*Federal Status: None*
*Global Rank: G5*
*State Rank: SE*
*Length: 31 in (78 cm)*

*References:* Bailey 1936, Fitzgerald et al. 1994, Hopkins and Forbes 1979, 1980; Maser et al. 1981, McManus 1974.

**Global Range:** Native to eastern North America from South Dakota east to New York and south to the highlands of Central America. It has been introduced along the Pacific Coast from southern British Columbia to northern Baja California.

**Habitat:** The Virginia opossum can exploit a variety of habitats, but is usually found in moist areas, often near water. This is especially true where it occurs in arid parts of the Columbia Basin. It can be found in marshes, riparian forests, croplands, suburbs, towns, cities, forests, and wooded canyons.

**Reproduction:** Opossums breed from January through October and are capable of producing 2 litters of 9 (range 5-13) young per litter. They are marsupials, which means the young are born in a near-embryonic state after a gestation period of 12 to 13 days. The young continue development attached to a teat inside their mother's pouch, where they remain for the next 2 months. Family groups remain together for about 4 months. Females breed at 1 year of age.

**Food Habits:** Virginia opossums are opportunistic omnivores that eat just about any animal or plant material they come across and can catch, gather, or subdue. Insects, other invertebrates, and carrion form the bulk of the diet, but many small vertebrates are eaten, including rattlesnakes. They also eat fruit, grain, and bird eggs and, around human habitation, steal pet food.

**Ecology:** Opossums are solitary except for mothers with young. Their home range is 10 to 50 acres. They are nocturnal and active throughout the year. They are killed by a variety of larger predators, including coyotes and Great Horned Owls. Most do not survive more than a year or two.

**Comments:** Virginia Opossums were released into the wild in Oregon around 1910-1912. If confronted, they often assume aggressive postures, and their bites can be dangerous. They may become catatonic ("playing 'possum") if harassed. Many are run over by motor vehicles, a major cause of mortality. They were formerly placed in the Order Marsupialia.

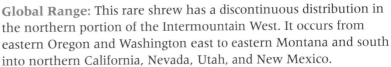

# Preble's Shrew
*(Sorex preblei)*

*Order: Insectivora*
*Family: Soricidae*
*State Status: None*
*Federal Status: SC*
*Global Rank: G4*
*State Rank: S3*
*Length: 3.5 in (9 cm)*

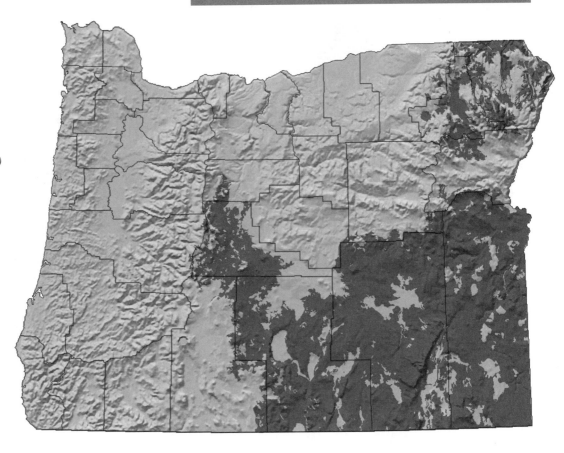

*References:* Cornely et al.
1992, Ports and George
1990, Verts 1975, Williams
1984.

**Global Range:** This rare shrew has a discontinuous distribution in the northern portion of the Intermountain West. It occurs from eastern Oregon and Washington east to eastern Montana and south into northern California, Nevada, Utah, and New Mexico.

**Habitat:** Preble's shrew usually occurs near permanent or intermittent streams in arid to semi-arid shrub/grass associations, as well as within dense high-elevation coniferous forests. Although it has been found in openings in coniferous forests grown to grass or sagebrush, it also frequents sagebrush thickets and stands of willow or aspen in moister parts of the Great Basin.

**Reproduction:** There is no information on the reproductive biology of this shrew, although it is unlikely to differ greatly from that of other shrews.

**Food Habits:** Although there have been no studies of the food habits of Preble's shrew, it probably feeds on small invertebrates. Based on skull and tooth morphology, some biologists have suggested it may eat soft-bodied prey.

**Ecology:** This shrew has been found in areas occupied by 3 other species of shrews. Because it is so rarely captured, little is known about its natural history.

**Comments:** This shrew is difficult to trap using standard snap traps. Its apparent rarity may be an artifact of inadequate sampling.

# Vagrant Shrew
## *(Sorex vagrans)*

*Order: Insectivora*
*Family: Soricidae*
*State Status: None*
*Federal Status: None*
*Global Rank: G5*
*State Rank: S4*
*Length: 4 in (10 cm)*

**Global Range:** Opinions differ about the taxonomy of the "*Sorex vagrans* species complex" (see next entry), and consequently the distribution of this species depends on which revision is followed. At a minimum, the vagrant shrew occurs from southern British Columbia south to the mountains of central California and east to the Rocky Mountains. Hall (1981) includes populations from Alaska south to northern Mexico, but Hennings and Hoffmann (1977) refer many of these populations to *S. monticolus.*

**Habitat:** Vagrant shrews may be found in a variety of habitats, but are almost always near wet areas such as riparian woodlands, marshes, or meadows at the edges of forests. They may be abundant in meadows in the Willamette Valley.

**Reproduction:** Breeding takes place from January to early November, but is concentrated during the spring and summer. The average litter size is 5 or 6 (range 2-9). The gestation period is 20 days. In some regions, there may be 2 or 3 litters per year.

**Food Habits:** Vagrant shrews have an unspecialized diet. They eat insects, insect larvae, spiders, earthworms, snails, slugs, some green plant material,

and, occasionally, small vertebrates. They consume more than their weight each day.

**Ecology:** Vagrant shrews forage under the surface litter of moist ground and beneath logs, stumps, and rocks. Along the Oregon coast, where their average home range size is 372 square meters, they make use of the runway systems constructed by voles.

**Comments:** Like all shrews, they are active both day and night. Their surface area to body weight ratio is so high that they must eat frequently or succumb to hypothermia.

*References:* Eisenberg 1964, Hawes 1977, Hooven et al. 1975, Maser et al. 1981.

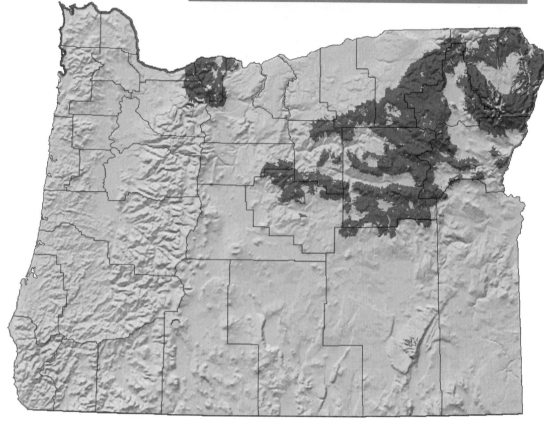

**Montane Shrew**
*(Sorex monticolus)*
**Baird's Shrew**
*(Sorex bairdi)*
**Fog Shrew**
*(Sorex sonomae)*

*Order: Insectivora*
*Family: Soricidae*
*State Status: None*
*Federal Status: None*
*Length: 4 in (10 cm)*

*Distribution map for montane shrew.*
*Global Rank: G4G5*
*State Rank: SU*

**Global Range:** The montane shrew occurs in higher mountains from Alaska and western Canada south to northern Mexico. It is not found east of the Rocky Mountains or south of the northern Cascade Mountains in Oregon. Baird's shrew is endemic to northwestern Oregon, and the fog shrew occurs west of the crest of the Cascade Mountains from central Oregon south, in coastal mountains, to San Francisco Bay.

**Habitat:** These shrews are found in cool, moist areas, usually within coniferous or deciduous forests. They frequent damp meadows, mossy banks of small streams, sphagnum bogs, and marshes, and often take refuge under downed logs or decaying ground litter. In the Cascade Mountains, they inhabit riparian areas and uplands in Douglas-fir/western hemlock forests, often with a mixture of deciduous trees.

**Reproduction:** These species begin breeding in January or February and continue through summer and early fall. Litters average 4 (range 4-6). The gestation period is 3 weeks, with young weaned in another 3 weeks. There may be 2 litters per year, but shrews rarely survive more than one breeding season.

**Food Habits:** As is typical of shrews, these species feed on a variety of small invertebrates, including beetles, worms, sowbugs, snails, earthworms, centipedes, and some vegetable matter (such as underground fungi).

**Ecology:** Previous inclusion of *S. monticolus, S. bairdi, S. sonomae,* and *S. pacificus* in the species *S. vagrans* makes assignment of ecological information to any of the currently recognized species difficult.

327

**Comments:** The taxonomy and natural history of several closely related shrews that occur in Oregon, known as the "*Sorex vagrans* species complex," have long been subjects of research, leading to differing interpretations of the specific affinity of specimens from western Oregon. In this atlas, we follow the taxonomy suggested by Carraway (1990) and Jones et al. (1992). However, Findley (1955), Hennings and Hoffmann (1977), George (1988), and Smith and Belk (1996) present other opinions. Carraway (1990) considers most western Oregon specimens of *S. monticolus* to be *S. bairdi*. [Alexander (1996) notes that the correct spelling of the Latin name for Baird's shrew is *Sorex bairdi*, not *Sorex bairdii*, as used by many authors.]

*References:* Alexander 1996, Carraway 1990, Hawes 1977, Maser et al. 1981.

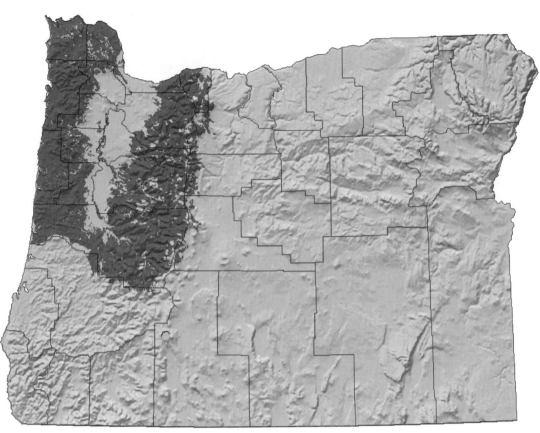

*Distribution map for Baird's shrew.*
*Global Rank: G4*
*State Rank: SU*

*Distribution map for fog shrew.*
*Global Rank: G5*
*State Rank: SU*

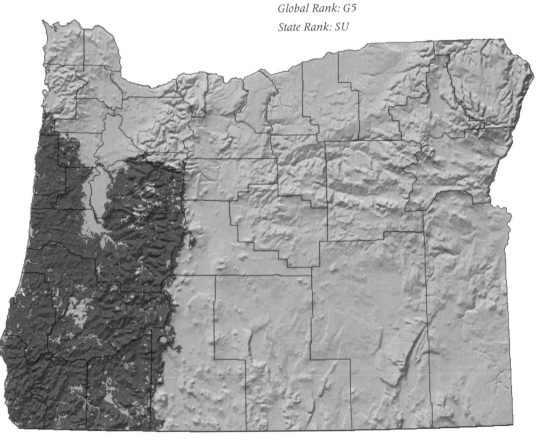

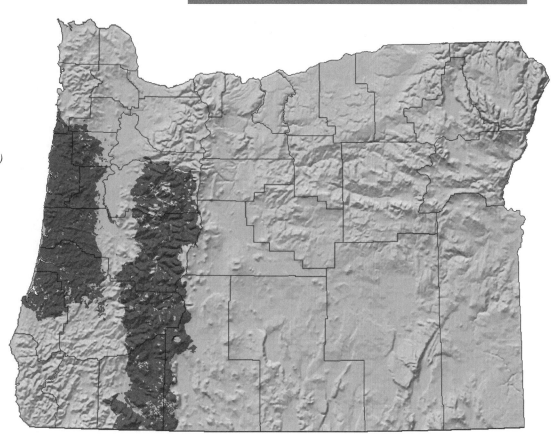

## Pacific Shrew
### (Sorex pacificus)

*Order: Insectivora*
*Family: Soricidae*
*State Status: None*
*Federal Status: None*
*Global Rank: G3G4*
*State Rank: S3S4*
*Length: 5.5 in (14 cm)*

**Global Range:** Western Oregon in the Coast, Cascade, and Siskiyou mountains south to the California border. Some authors think the species is present near the coast south to San Francisco Bay, although Carraway (1990) places these populations in *S. sonomae*. Following Carraway (1990), this species is endemic to Oregon.

**Habitat:** The Pacific shrew is found in humid forests, marshes, and thickets, often near riparian vegetation; it has been found in early successional stages of forests, but less often in mature coniferous forests. It requires downed logs, brushy thickets, or ground debris for cover and feeding.

**Reproduction:** This species breeds from February to September, with an apparent peak of reproduction during the summer. The average litter size is 4 or 5 (range 2-6), and there may be more than 1 litter per year.

**Food Habits:** The Pacific shrew feeds on snails, slugs, centipedes, insect larvae, earthworms, and some vegetable matter. Shrews must feed frequently, and will cache food items near their nest to feed on during periods of reduced activity (i.e., during daylight hours).

**Ecology:** Pacific shrews are the largest members of the "*Sorex vagrans* complex." They are primarily nocturnal and hunt by smell, although they can catch flying insects out of the air, apparently using visual cues. Because some populations formerly considered *S. vagrans* or *S. pacificus* are now considered *S. bairdi* and *S. sonomae*, it is difficult to assign previous accounts of ecology to currently recognized species.

**Comments:** Populations north of Douglas County have been considered by some (Maser et al. 1981) to represent a separate species, the Yaquina shrew (*S. yaquinae*).

*References:* Carraway 1985, 1990, Maser and Hooven 1974, Maser et al. 1981.

# Water Shrew
*(Sorex palustris)*

*Order: Insectivora*
*Family: Soricidae*
*State Status: None*
*Federal Status: None*
*Global Rank: G5*
*State Rank: S4*
*Length: 6 in (15 cm)*

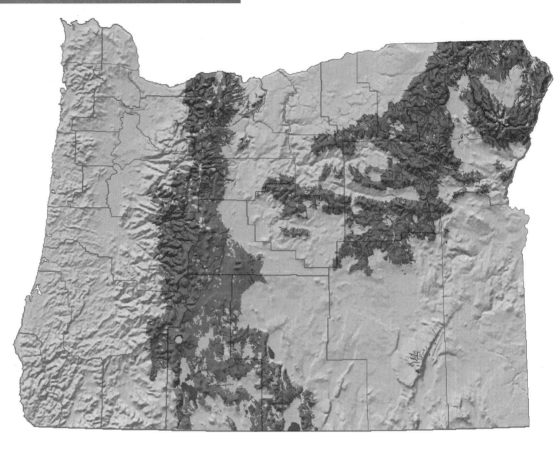

**Global Range:** Found from Alaska and central Canada across the northern United States and south in mountainous areas almost to the Mexican border. Absent from the Plains states, Texas, and the southeastern United States.

**Habitat:** As their name implies, water shrews are closely associated with streams, marshes, lakes, and ponds. They are found in aquatic microhabitats in forests, and also in riparian and other wet areas with sufficiently thick vegetation.

**Reproduction:** These shrews breed from February through August, with most litters born from March through July. There are 2 or 3 litters per year, and litter size is about 6 (range 3-10). Gestation lasts about 21 days.

**Food Habits:** As one might expect, aquatic invertebrates make up the bulk of the water shrew's diet. Occasionally, water shrews eat small aquatic and terrestrial vertebrates, such as tadpoles, fish, and fish eggs.

**Ecology:** Water shrews use water for both foraging and escape behavior. They have a fringe of stiff hairs on their feet

to assist swimming. The shrew must swim to stay submerged, due to the buoyancy of a layer of air trapped in the fur, which serves as insulation.

**Comments:** There can be hazards to a semi-aquatic existence. The most common predators of the water shrew include other stream-dwellers, such as garter snakes, weasels, salmonids, and large frogs.

*References:* Beneski and Stinson 1987, Conaway 1952, Sorenson 1962.

*Order: Insectivora*
*Family: Soricidae*
*State Status: None*
*Federal Status: None*
*Global Rank: G5*
*State Rank: S4*
*Length: 6.5 in (17 cm)*

**Global Range:** Coastal and Cascade mountains from southwestern British Columbia south to Gualala on the Mendocino coast of northern California. Apparently absent from the Klamath Mountains.

**Habitat:** This shrew is found in moist forests, swamps, marshes, and riparian areas.

**Reproduction:** Breeding season begins as early as February in the warmer parts of its range. Litters, usually of 3 or 4 young, are born from March through June.

**Food Habits:** The Pacific marsh shrew has a mixed diet of aquatic and terrestrial invertebrates, including earthworms, snails, slugs, beetles, and spiders as well as stonefly nymphs, mayflies, alderflies, and phantom crane flies.

**Ecology:** These shrews are usually most active from dusk to dawn, but sometimes venture out in the daytime. They are at home on both land and water, and are good swimmers.

**Comments:** The ecology of this species is poorly known.

*References:* George 1988, Maser et al. 1981, Pattie 1973, Whitaker and Maser 1976.

# Trowbridge's Shrew
## (Sorex trowbridgii)

*Order: Insectivora*
*Family: Soricidae*
*State Status: None*
*Federal Status: None*
*Global Rank: G5*
*State Rank: S4*
*Length: 5 in (13 cm)*

**Global Range:** Southern British Columbia, south through Oregon and Washington, from the coast to the Cascade Mountains. In California, the range of this species is limited to the Coast Ranges and the Sierra Nevada, south to Santa Barbara County and Kern County, respectively.

**Habitat:** Trowbridge's shrew generally occurs in forests (both deciduous and coniferous), and seems to prefer somewhat dryer areas than other shrews. It can be found near riparian areas and in ravines, but is more abundant several meters away from streams.

**Reproduction:** This shrew breeds from February to June. Timing of the breeding season depends on local climate; populations in warmer areas reach sexual maturity earlier in the year. In Oregon, the average litter size is 4. There may be 2 or 3 litters per year.

**Food Habits:** This shrew has a relatively unspecialized diet. Most frequently, it eats beetles, snails, spiders, and larval insects. It is known to eat Douglas-fir seeds and some other plant material.

**Ecology:** Trowbridge's shrews dig their own burrows in the layer of organic debris on the forest floor. In areas where they occur with montane shrews, the latter forage on the forest floor, and Trowbridge's shrews forage under the surface.

**Comments:** Research in Oregon has detected the highest population densities of Trowbridge's shrew in mature forests. This shrew can be found in clearcuts, but it takes about 6 years for population densities to approach those in mature forests.

*References:* Gashwiler 1976a, George 1989, Maser et al. 1981, Whitaker and Maser 1976.

## Merriam's Shrew
### (Sorex merriami)

*Order: Insectivora*
*Family: Soricidae*
*State Status: None*
*Federal Status: None*
*Global Rank: G5*
*State Rank: S3*
*Length: 4 in (10 cm)*

**Global Range:** Western United States, from the east slope of the Cascade and Sierra Nevada mountains east to the Dakotas and south to the mountains of east-central Arizona.

**Habitat:** Unlike most shrews, *Sorex merriami* seems to occur in arid environments such as sagebrush steppe and juniper woodland. It has also been found in mountain mahogany stands and in wet areas such as stream banks or willows adjacent to its typical habitat.

**Reproduction:** In Washington, pregnant females were found as early as mid-March. Based on a sample of 3 females, the litter size seems to be about 6 (range 5-7).

**Food Habits:** In warmer months, caterpillars seem to dominate the diet. Other foods include spiders, larval and adult beetles, cave crickets, and wasps.

**Ecology:** While foraging, this shrew is thought to use the burrows and runways of other small mammals, especially the sagebrush vole (*Lemmiscus curtatus*). This is likely a response to an arid habitat lacking loose soil and ground litter. Owls are the only known predator.

**Comments:** Like our other arid-land shrew, *Sorex preblei*, the ecology of Merriam's shrew is poorly known.

*References:* Armstrong and Jones 1971, George 1988, Johnson and Clanton 1954.

333

# Shrew-mole
## (Neurotrichus gibbsii)

Order: Insectivora
Family: Talpidae
State Status: None
Federal Status: None
Global Rank: G5
State Rank: S4
Length: 4 in (10 cm)

**Global Range:** Occurs from southern British Columbia south to Monterey County, California. It is found from the coast inland to the Cascade Mountains in Washington and Oregon, and in the Siskiyou Mountains, northern Sierra Nevada, and Coast Ranges in California.

**Habitat:** Shrew-moles occupy a variety of habitats from sea level to high mountains. They frequent areas with a cover of thick vegetation and, usually, some water or moist soil. At low elevations in the Northwest, they are often found in the bottom of ravines shaded by bigleaf maples. They also occur in wetter parts of coniferous forests, wet meadows, and along the edges of marshes.

**Reproduction:** The breeding season can last from February to September, but is concentrated in the spring. The average litter size is 3 (range 1-4).

**Food Habits:** The food habits of the shrew-mole have been studied in western Oregon; the most important items in its diet are earthworms, centipedes, snails, slugs, and insects. It also eats some vegetable material.

**Ecology:** Shrew-moles may be somewhat social, and may travel in small groups. They are good swimmers

and are active both night and day. Although shrew-moles burrow, they do not cause as much damage as their larger cousins.

**Comments:** Shrew-moles can be recognized by their small size and fringe of bristles along the tail. They are blind, and are the only vertebrate that has a pigmented layer covering the front of the lens of the eye.

*References:* Carraway and Verts 1991a, Dalquest and Orcutt 1942, Maser et al. 1981.

**334**

*Order: Insectivora*
*Family: Talpidae*
*State Status: None*
*Federal Status: None*
*Global Rank: G5*
*State Rank: S4*
*Length: 8.5 in (22 cm)*

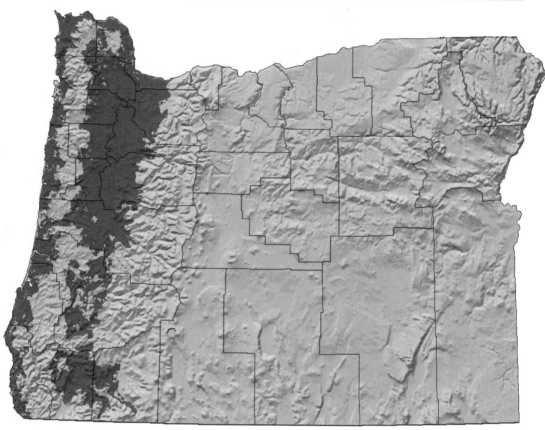

**Global Range:** Ranges from extreme southwestern British Columbia south to Cape Mendocino, California. In Oregon and Washington, it ranges from the coast to the west slope of the Cascade Mountains (highest elevation about 1,700 meters), but in California it is restricted to a narrow band along the coast.

**Habitat:** This mole is found in moist areas with deep friable soil. It occurs in pastures, grasslands, meadows, suburban lawns, and open forests.

**Reproduction:** Mating takes place early in the year (February), and litters of 2 or 3 (range 1-4) young are born in late March or early April. Young are ready to breed the year following their birth.

**Food Habits:** Earthworms form the bulk of this mole's diet, but it also takes insect larvae, centipedes, slugs, and some adult insects. It may consume considerable amounts of grass roots.

**Ecology:** Townsend's moles construct elaborate tunnel systems just under the surface of the ground while foraging, but these are used only once. Densities of 12 moles per hectare have been recorded in Tillamook County.

**Comments:** This mole may cause considerable damage to pastures and crops.

*References:* Carraway et al. 1993, Giger 1973, Kuhn et al. 1966, Maser et al. 1981.

# Coast Mole
## *(Scapanus orarius)*

*Order: Insectivora*
*Family: Talpidae*
*State Status: None*
*Federal Status: None*
*Global Rank: G5*
*State Rank: S5?*
*Length: 6.5 in (17 cm)*

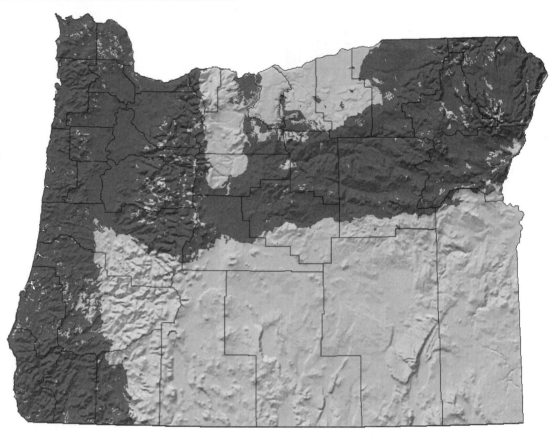

**Global Range:** A Pacific Northwest species found from southwestern British Columbia to northwestern California, ranging east just into Idaho. It is absent from the arid areas of southeastern Oregon.

**Habitat:** Coast moles are found far inland as well as near the coast, occurring in loose soil in a variety of habitat types, including meadows, deciduous riparian woodland, sagebrush scrub, and coniferous forests. They are more likely to occur in forests than are Townsend's moles.

**Reproduction:** One litter of 3 (range 2-4) young is produced each year, generally in March or April.

**Food Habits:** A study of food habits in western Oregon showed that coast moles primarily eat earthworms (about 70% of their diet), but also take insect larvae, beetles, centipedes, and a small amount of vegetable matter.

**Ecology:** Coast moles are solitary except during the breeding season. They dig elaborate tunnel systems, as expected of moles, and are primarily active at night.

**Comments:** Because of its preference for forested areas, the coast mole is less of an agricultural or economic pest than its relative, Townsend's mole.

*References:* Hartman and Yates 1985, Maser et al. 1981, Whitaker et al. 1979.

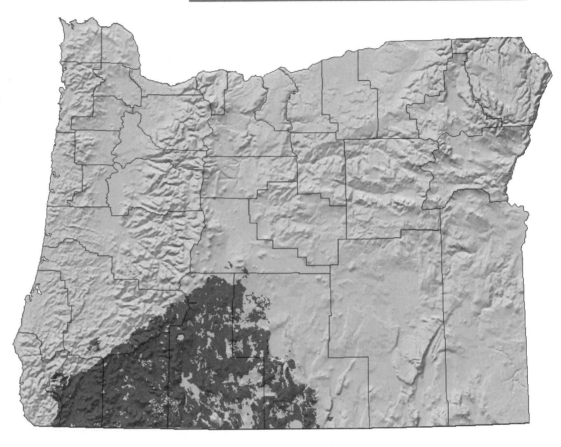

*Order: Insectivora*
*Family: Talpidae*
*State Status: None*
*Federal Status: None*
*Global Rank: G5*
*State Rank: S4?*
*Length: 7 in (18 cm)*

**Global Range:** Occurs from the valleys of southwestern Oregon south through most of non-desert California into northern Baja California. Distribution is discontinuous, especially in the arid parts of its range.

**Habitat:** The broad-footed mole is generally found in meadows, grasslands, or edges of streams and marshes where the soil is moist. It is most common in low valleys, but one has been captured at the rim of Crater Lake.

**Reproduction:** This species breeds from March through late June, giving birth to only one litter of 4 (range 2-5) young a year.

**Food Habits:** This mole eats earthworms, crickets and other small insects, insect larvae, crustaceans, and, occasionally, small vertebrates.

**Ecology:** As one would expect from a mole, this species forages by digging burrows under the surface of the ground, to almost 1 meter deep. It is active throughout the year, and more active in the daytime. It constantly expands its tunnel system, which is larger when food is scarce.

**Comments:** The broad-footed mole reaches the northern limits of its range in southern Oregon. This species can be a garden pest in suburban areas.

*References:* Bailey 1936, Grim 1958, Whitaker et al. 1979.

337

# California Myotis
## *(Myotis californicus)*

*Order: Chiroptera*
*Family: Verspertilionidae*
*State Status: None*
*Federal Status: None*
*Global Rank: G5*
*State Rank: S4*
*Length: 3 in (8 cm)*

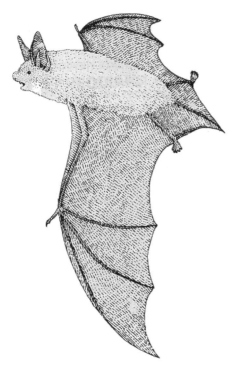

**Global Range:** Ranges from the Alaskan panhandle south through British Columbia and the western United States, generally west of the Rocky Mountains, and farther south through most of Mexico.

**Habitat:** Generally, this bat forages around the edges of clumps of trees or over or near open water. It flies well above ground in open country. It roosts in cliff faces, crevices in trees, or caves. It is a crevice-dweller that frequently uses structures (under tarpaper, between shingles) for roosting.

**Reproduction:** California myotis mate in the fall, but delay fertilization until spring. A single young is born in May or June. Females form maternity colonies during the breeding season.

**Food Habits:** In eastern Oregon, this bat feeds primarily on butterflies and small flies. It also eats some beetles and moths.

**Ecology:** In the Northwest, this species hibernates during the winter. It is active longer and emerges from hibernation earlier west of the Cascades.

**Comments:** This species does not have fixed long-term roosts, but seeks shelter after nocturnal foraging. This may indicate that individual home ranges are large.

*References:* Barbour and Davis 1969, Maser et al. 1981, Simpson 1993, Whitaker et al. 1981.

# Western Small-footed Myotis
## (Myotis ciliolabrum)

Order: Chiroptera
Family: Vespertilionidae
State Status: Sensitive
Federal Status: SC
Global Rank: G5
State Rank: S4
Length: 3 in (8 cm)

**Global Range:** Western North America, from southern Alberta, Saskatchewan, and British Columbia south through the western United States to central Mexico.

**Habitat:** While primarily associated with cliffs and rocky canyons in arid grasslands and desert scrub, this bat also is found in ponderosa pine and mixed conifer forests. It finds night roosts and day retreats in rock crevices, under boulders or, sometimes, beneath bark, and hibernates in caves and mines.

**Reproduction:** Little is known about this bat's reproductive biology. Litters of 1 or perhaps 2 young are born from May to July, and females form small maternity colonies. High rainfall or low temperatures are thought to delay births.

**Food Habits:** The western small-footed myotis feeds on a variety of small insects taken in flight, including moths, flies, beetles, wasps, true bugs, leafhoppers, and butterflies.

**Ecology:** This species coexists with the California myotis, but avoids competition by foraging over rocks rather than over water. It flies along cliffs and rocky slopes at heights of 1 to 3 meters.

**Comments:** Until recently, western populations were considered part of the small-footed myotis, *Myotis leibii*. Electrophoretic studies support the separation of the species into eastern and western species.

*References:* Barbour and Davis 1969, Herd 1987, Marshall et al. 1996, van Zyll de Jong 1984.

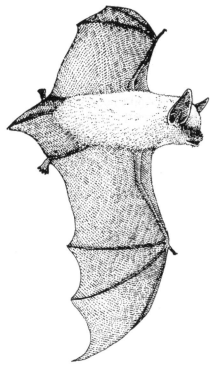

**339**

# *Yuma Myotis*
## (Myotis yumanensis)

*Order: Chiroptera*
*Family: Vespertilionidae*
*State Status: Sensitive*
*Federal Status: SC*
*Global Rank: G5*
*State Rank: S3*
*Length: 3.5 in (9 cm)*

**Global Range:** Western North America from central British Columbia south to southern Mexico. The range is difficult to plot due to frequent misidentification of specimens.

**Habitat:** The Yuma myotis is more closely associated with water than most other North American bats. It is found in a wide variety of habitats, including riparian, desert scrub, moist woodlands, and open forests. In western Oregon, it frequents older Douglas-fir forests, Sitka spruce forests, and oak and ponderosa pine woodlands.

**Reproduction:** This bat probably breeds in the fall. Females form maternity colonies of up to 5,000 individuals in April, and a single young is born in late May to July.

**Food Habits:** Moths, midges, flies, and termites are the most important food items. Bees, wasps, beetles, leafhoppers, and lacewings are also eaten. Like other members of the genus, Yuma myotis requires free water for drinking.

**Ecology:** This species forages a few centimeters above the surface of open water. It establishes large colonies in buildings, mines, caves, or under bridges, and may be locally abundant, but is sensitive to disturbance.

**Comments:** The location of winter roost sites is unknown. Because of its use of buildings for roost sites, populations of this bat may have increased since European settlement of North America. All recent Oregon records of this species are from the Klamath region or western Oregon.

*References:* Barbour and Davis 1969, Dalquest 1947, Marshall et al. 1996, Maser et al. 1981.

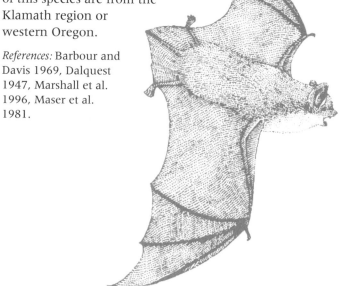

## Little Brown Myotis
### *(Myotis lucifugus)*

*Order: Chiroptera*
*Family: Vespertilionidae*
*State Status: None*
*Federal Status: None*
*Global Rank: G5*
*State Rank: S4*
*Length: 3.5 in (9 cm)*

**Global Range:** Occurs throughout most of North America north of Mexico, except for northern Canada and Alaska. There is a disjunct record from central Mexico.

**Habitat:** This bat is closely associated with water. As a result, it is found in moist forests or other areas with trees, such as riparian woodlands in the arid parts of the state.

**Reproduction:** Maternity colonies are often located in structures, caves, or hollow trees. Mating is in the fall, and fertilization is delayed. One (occasionally 2) young is born from May to August, but usually in late June or July. The gestation period is 50-60 days.

**Food Habits:** Little brown myotis depend heavily on flying insects for food. Along the Oregon coast, they eat midges, mosquitoes, and flies. Generally, they hunt for insects that are one-eighth to one-half inch long. In other areas they are known to eat small moths, caddisflies, beetles, ants, and craneflies.

**Ecology:** These bats hibernate in caves with high humidity. They are most active 2 to 3 hours after sunset, with a secondary peak of activity after midnight. They tend to congregate in night roosts after feeding bouts.

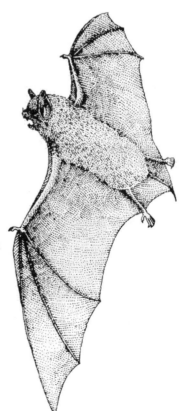

**Comments:** This bat commonly uses structures for roosts, and can become a pest in attics of houses and other buildings. Control activities have drastically reduced populations in parts of its range. It is known to have a life-span of over 20 years.

*References:* Fenton and Barclay 1980, Fenton and Bell 1979, Maser et al. 1981, O'Farrell and Studier 1975.

# Long-Legged Myotis
## *(Myotis volans)*

*Order: Chiroptera*
*Family: Vespertilionidae*
*State Status: Sensitive*
*Federal Status: SC*
*Global Rank: G5*
*State Rank: S3*
*Length: 3.75 in (9.5 cm)*

**Global Range:** Extends from northern British Columbia south to central Mexico and east to the western Great Plains.

**Habitat:** The long-legged myotis is associated with coniferous forests, including Douglas-fir, true fir, Sitka spruce, lodgepole pine, and ponderosa pine forests. It also frequents oak and mixed evergreen woodlands. However, in the more arid parts of its range, it frequents riparian forests. It seeks shelter in a variety of roosting sites, including crevices in cliff faces, abandoned buildings, caves, and mines.

**Reproduction:** This bat breeds in the fall and delays fertilization until spring. Pregnant females have been found from mid-April to mid-August in various parts of the range. The litter size is 1. Large maternity colonies are formed in the spring and summer.

**Food Habits:** Moths make up the bulk of the diet, but other insects, such as flies, termites, lacewings, wasps, true bugs, leafhoppers, and small beetles are also taken.

**Ecology:** This species is active throughout the night, but concentrates its activity in the early part of the evening. It flies quickly through the forest canopy in search of prey. Some long-legged myotis overwinter in Oregon in caves and mines.

**Comments:** This is one of the more common species of *Myotis*. Based on recovery of banded individuals, the species can live to be at least 21 years old.

*References:* Barbour and Davis 1969, Marshall et al. 1996, Maser et al. 1981, Warner and Czaplewski 1984.

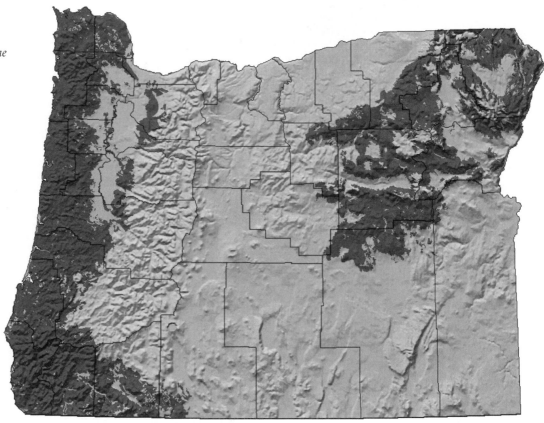

## Fringed Myotis
### (Myotis thysanodes)

*Order: Chiroptera*
*Family: Vespertilionidae*
*State Status: Sensitive*
*Federal Status: SC*
*Global Rank: G5*
*State Rank: S3*
*Length: 3.5 in (9 cm)*

**Global Range:** Western North America, from south-central British Columbia south through the western United States to southern Mexico. Generally west of the Rocky Mountains, but a disjunct population occurs in the Black Hills of South Dakota and Wyoming.

**Habitat:** Although the species is found in a wide variety of habitats throughout its range, it seems to prefer forested or riparian areas. There are scattered records from waterholes in deserts, but these tend to be within flying distance of forested areas. Most Oregon records are west of the Cascade Mountains.

**Reproduction:** The reproductive biology of Oregon's populations has not been studied well. The fringed myotis mates in the fall and delays fertilization until spring. Litter size is 1 young, born in late June to mid-July after a gestation period of 50-60 days. Females form maternity colonies of up to several hundred individuals.

**Food Habits:** It is thought to forage by picking up food items from shrubs or the ground. It consumes beetles, moths, harvestmen, crickets, craneflies, and spiders.

**Ecology:** This species is migratory. There are only 2 winter records from Oregon. Its nursery colonies are established in caves, mines, and buildings. It is very sensitive to human disturbance.

**Comments:** This bat is rare in Oregon. It is most common in southwestern Oregon, and breeds regularly in the cave at Oregon Caves National Monument.

*References:* Manning and Jones 1988, Marshall 1992, Maser et al. 1981, O'Farrell and Studier 1980.

343

# Long-eared Myotis
*(Myotis evotis)*

*Order: Chiroptera*
*Family: Vespertilionidae*
*State Status: Sensitive*
*Federal Status: SC*
*Global Rank: G5*
*State Rank: S3*
*Length: 3.5 in (9 cm)*

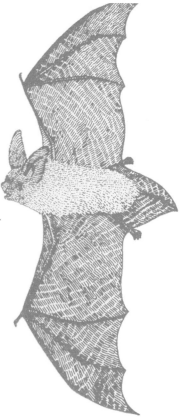

**Global Range:** Western North America from central British Columbia south to Baja California. Found east as far as western North Dakota, eastern Arizona, and central New Mexico.

**Habitat:** This species is associated primarily with forested habitats and forested edges, including juniper woodlands, open areas in ponderosa pine woodlands, Douglas-fir, spruce, true fir, and subalpine forests, as well as willow and alder forests along streams. It also occurs in arid shrublands if suitable roosting sites are available.

**Reproduction:** Information about its reproduction is lacking, but the long-eared myotis appears to give birth to a single young in late June or July. Small maternity colonies are formed in late spring or early summer.

**Food Habits:** Moths seem to constitute the major food of the long-eared myotis, although it has also been reported to eat beetles, spiders, small flies, bees, ants, termites, and dragonflies.

**Ecology:** The long-eared myotis emerges late in the evening, and feeds by picking prey items off the surface of foliage. Although most probably migrate out of state during the coldest part of the year, a few have been found in caves in western Oregon during winter.

**Comments:** In Oregon, the long-eared myotis is distributed widely, but it is not abundant. It is probably more common in forested areas than in the arid plains of southeast Oregon.

*References:* Cross et al. 1990, Manning and Jones 1989, Marshall et al. 1996, Maser et al. 1981, Whitaker et al. 1981.

*Order: Chiroptera*
*Family: Vespertilionidae*
*State Status: Sensitive*
*Federal Status: None*
*Global Rank: G5*
*State Rank: S4?*
*Length: 4 in (10 cm)*

**Global Range:** Found throughout most of the United States and southern Canada from coastal Alaska to Nova Scotia and south to southern Texas. Absent from southern California and along the coast of the Gulf of Mexico, including all of Florida.

**Habitat:** This is a bat of forested areas and is reported to be most abundant in older Douglas-fir/ western hemlock forests, although it also occupies ponderosa pine forests. It forages over ponds and streams in the woods, and typically finds a day roost under a flap of loose bark.

**Reproduction:** The silver-haired bat mates in autumn, and delays fertilization until spring. After a 50-60 day gestation period, females give birth to 1 or, more frequently, 2 young in late June or July.

**Food Habits:** A study along the Oregon coast found that these bats preferred soft-bodied prey, with moths, termites, and flies being the most important food items. It also eats ants, beetles, true bugs, spittlebugs, and planthoppers.

**Ecology:** This species is most active about 3 hours after sunset and again 7 or 8 hours after sunset. It has been suggested that each bat maintains an individual foraging route some 46 to 91 meters in diameter. When hunting, their flight is slow and erratic.

**Comments:** On a continent-wide basis, populations of this bat may have declined due to deforestation over the last several centuries. In Oregon, loss of older Douglas-fir forests and associated snags may threaten it.

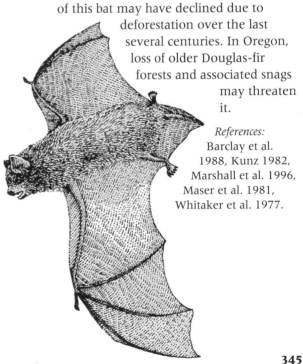

*References:*
Barclay et al. 1988, Kunz 1982, Marshall et al. 1996, Maser et al. 1981, Whitaker et al. 1977.

345

# Western Pipistrelle
## (Pipistrellus hesperus)

*Order: Chiroptera*
*Family: Vespertilionidae*
*State Status: None*
*Federal Status: None*
*Global Rank: G5*
*State Rank: S4*
*Length: 3 in (8 cm)*

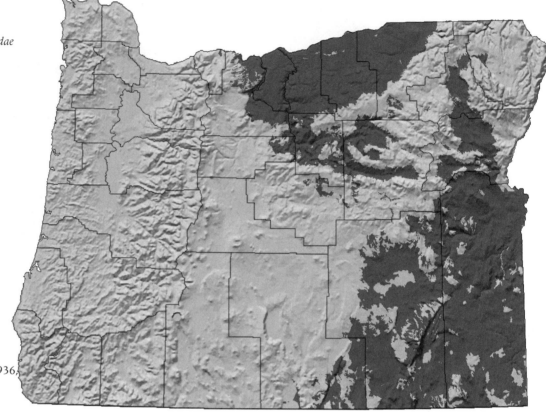

*References:* Bailey 1936, Barbour and Davis 1969, Koford and Koford 1948.

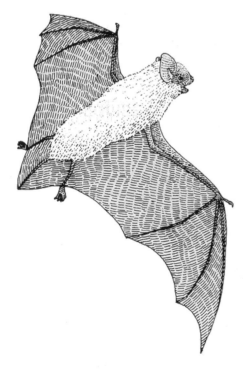

**Global Range:** Reaches the northern limits of its range in southeastern Washington; occurs as far east as Oklahoma and as far south as southern Mexico.

**Habitat:** This is a small bat of deserts and lowlands. It is found both over desert flats and in rocky canyons. While it is frequently found in desert scrub communities, such as greasewood and sagebrush, it has been found in both juniper woodlands and sedge.

**Reproduction:** The western pipistrelle breeds in the fall and delays fertilization until spring. The litter size is 1 or (usually) 2, and young are born in June after a gestation period of about 40 days.

**Food Habits:** The western pipistrelle feeds on swarms of small flying insects, such as flying ants, mosquitoes, small flies, caddisflies, stoneflies, and some beetles, moths, and leafhoppers.

**Ecology:** This species emerges from day roosts well before dark, and remains out after dawn; its activity peaks are just after dusk and before dawn. It is a very slow flyer. Consequently, it is difficult to capture in mist nets, which it detects and avoids.

**Comments:** Barbour and Davis (1969) report seeing western pipistrelles emerge in the open desert, miles from any possible day roosts. This led them to speculate that the western pipistrelle, which would dry out rapidly in hot, arid habitats, may find shelter in rodent burrows. This is the smallest bat in the United States.

346

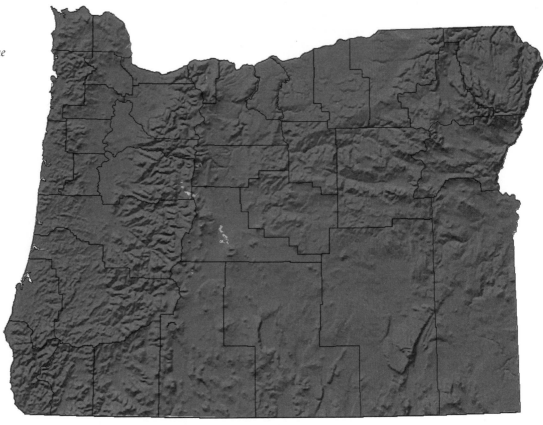

*Order: Chiroptera*
*Family: Vespertilionidae*
*State Status: None*
*Federal Status: None*
*Global Rank: G5*
*State Rank: S4*
*Length: 4.5 in (11 cm)*

**Global Range:** Found from southern Canada south to northern South America and from coast to coast. They also inhabit islands in the Caribbean Sea.

**Habitat:** This bat is more common in deciduous than in coniferous forests. It has adapted to human development and is also known by the common name, "house bat." It readily takes up residence in a variety of artificial structures. It forages over open areas such as meadows and pastures or even city streets. It has been known to take up residence in hollow trees or crevices in cliffs and may be more common now than before European settlement.

**Reproduction:** Mating occurs in the fall, but fertilization is delayed until spring. Females form maternity colonies of several hundred individuals. In the Northwest, the usual litter is 1 young, which is born in late June or early July.

**Food Habits:** This bat feeds primarily on beetles. Termites are also taken. It rarely eats moths, so common in the diets of other bats.

**Ecology:** Generally, big brown bats forage in meadows or above the forest canopy. They do not migrate, rarely traveling more than 50 miles between summer and winter roosts.

**Comments:** In mountainous areas, males tend to occur at higher elevations than females. This bat has a life span of at least 19 years. This is one of our most studied bats.

*References:* Brigham 1991, Kurta and Baker 1990, Maser et al. 1981, Philips 1966.

# Hoary Bat
*(Lasiurus cinereus)*

*Order: Chiroptera*
*Family: Vespertilionidae*
*State Status: None*
*Federal Status: None*
*Global Rank: G5*
*State Rank: S4?*
*Length: 5.5 in (14 cm)*

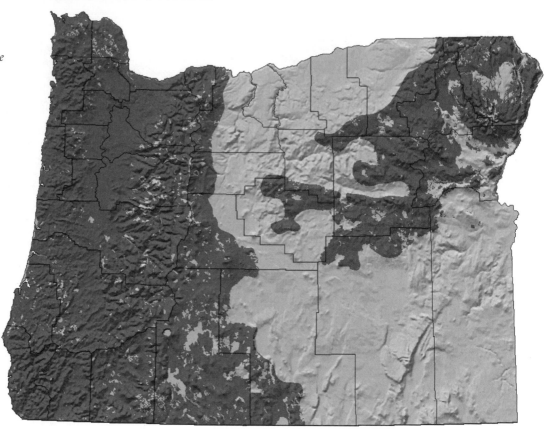

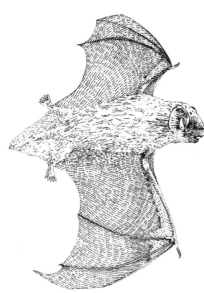

**Global Range:** Found throughout most of North America except for Alaska and the Northwest Territories. It is absent from Central America, but also occurs in northwestern and central South America. An isolated subspecies, now endangered, is native to Hawaii.

**Habitat:** This is a solitary, forest-dwelling species that roosts in trees during the day. Both coniferous and deciduous forests are used, and it forages along riparian corridors and brushy areas in forests.

**Reproduction:** The hoary bat mates in the fall and delays implantation until spring. Some sources say the litter size is always 2, others say 2 is the average (range 1-4). Young are born from early June to early July.

**Food Habits:** Generally, hoary bats specialize in feeding on moths, but they are also known to eat beetles, flies, grasshoppers, termites, dragonflies, mosquitoes, and wasps.

**Ecology:** This species is migratory. Little is known about its winter distribution. Males and females are seldom encountered together in the spring and summer. Some speculate that they are segregated by elevation. In California, females use lowlands and coastal valleys, and males use foothills and mountains.

**Comments:** These bats are fast (21.3 kilometers per hour), direct fliers that forage about 10 to 15 feet above the ground. Females may carry young less than a week old along when foraging. Hall (1981) argues that the correct generic name for this bat is *Nycteris*.

*References:* Findley and Jones 1964, Maser et al. 1981, Shump and Shump 1982.

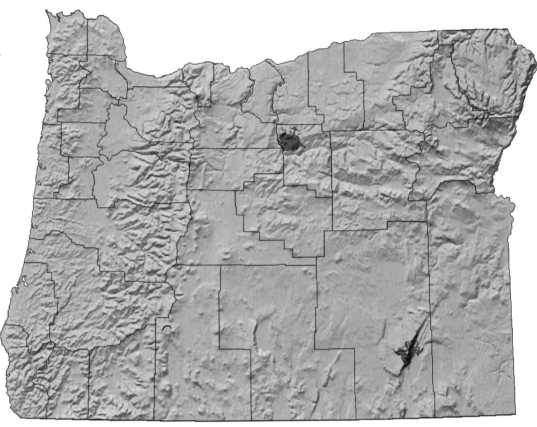

## Spotted Bat
*(Euderma maculatum)*

*Order: Chiroptera*
*Family: Vespertilionidae*
*State Status: None*
*Federal Status: SC*
*Global Rank: G4*
*State Rank: S1*
*Length: 4.5 in (11 cm)*

*References:* Barbour and Davis 1969, Fenton et al. 1983, 1987, Keller 1987, Poche 1981, Poche and Ruffner 1975, Wai-Ping and Fenton 1989, Watkins 1977.

**Global Range:** The range limits of the spotted bat are poorly known. It was first detected in southern British Columbia in 1983, and is known to occur as far east as Montana and Texas and as far south as central Mexico.

**Habitat:** The spotted bat has been observed in a wide variety of habitat types, from ponderosa pine forests to desert water holes. It is known to nest in crevices in cliffs, which may be more important in determining its distribution than any particular vegetation type.

**Reproduction:** As might be expected, little is known about the reproductive biology of the spotted bat. Most likely a single young is born in June or July. There is evidence to suggest that the species mates in the spring and does not delay fertilization. Juvenile bats have been collected in early September.

**Food Habits:** The spotted bat primarily feeds on moths. In one study, moths comprised 65% of all stomach contents.

**Ecology:** Spotted bats are solitary foragers. They emerge from day roosts after sunset, and are most active between midnight and 3:00 a.m.

**Comments:** The spotted bat is considered one of the rarest mammals in North America. Recently, it was discovered that it emits audible echolocation calls, and surveys are discovering that it is more common and widespread than previously thought. It was discovered in canyons in Owyhee County, Idaho, in 1987. There are two records from eastern Oregon, where it is probably a rare but widely distributed species; however, more surveys are required to determine its distribution and status.

**349**

# Townsend's Big-eared Bat
## *(Plecotus townsendii)*

*Order: Chiroptera*
*Family: Vespertilionidae*
*State Status: Sensitive*
*Federal Status: SC*
*Global Rank: G4*
*State Rank: S4*
*Length: 4 in (10 cm)*

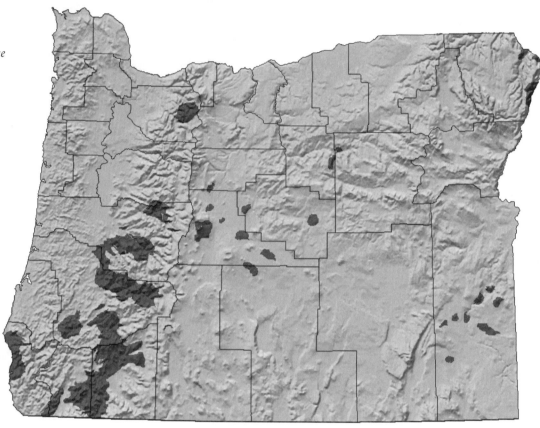

**Global Range:** Southern British Columbia south to southern Mexico. Found from the Pacific Coast east to the Great Plains, with isolated populations in the Ozarks and the Blue Ridge Mountains of West Virginia and Virginia.

**Habitat:** Two subspecies occur in Oregon, one in forested regions west of the Cascades, the other in arid eastern Oregon. The presence of suitable roost sites is more important than the vegetation type in determining the distribution of this bat. It roosts in buildings, caves, mines, and bridges.

**Reproduction:** This bat mates in late fall and during winter hibernation, and delays fertilization until spring. A single young is born around June. The length of the gestation period varies from 56 to 100 days, depending on temperature.

**Food Habits:** Townsend's big-eared bat feeds primarily on moths, but will also ingest beetles, true bugs, and flies. In addition to foraging on the wing, it will take insects from foliage.

**Ecology:** During the winter, this bat hibernates in relatively cold areas of caves and mines. Its body temperature can drop to as low as 1° C.

**Comments:** This bat is very intolerant of human disturbance at both winter hibernacula and summer roosts. It is one of our better-studied bats despite the fact that it is relatively rare and populations are declining. This taxon was formerly known as *Corynorhinus rafinesquii*. Recent morphological and karyological research may result in a change in generic name back to *Corynorhinus* (Frost and Timm 1992).

*References:* Barbour and Davis 1969, Kunz and Martin 1982, Marshall 1992, Pearson et al. 1952.

## Pallid Bat
*(Antrozous pallidus)*

*Order: Chiroptera*
*Family: Vespertilionidae*
*State Status: Sensitive*
*Federal Status: None*
*Global Rank: G5*
*State Rank: S3S4*
*Length: 5 in (13 cm)*

**Global Range:** Southern British Columbia south to central Mexico and east to Texas and Oklahoma. Distribution in the Northwest extends east to about the Idaho border.

**Habitat:** The pallid bat is found in arid regions, but can be found also in open forest types (ponderosa pine, oaks). It occurs in a variety of desert vegetation types (sagebrush, juniper, salt-desert shrub), and typically uses cliff-faces, caves, mines, or buildings for roosts.

**Reproduction:** The pallid bat mates during late fall and winter, but fertilization is delayed until spring. The length of the gestation period varies with temperature, but usually is about 9 weeks. A litter of 1 or 2 (2 is more common) young is born from May to June.

**Food Habits:** A variety of flightless arthropods is taken, including Jerusalem crickets, beetles, grasshoppers, and scorpions. Some moths are also taken, and this bat is known to eat some small vertebrates such as lizards and pocket mice.

**Ecology:** The pallid bat is unusual in that it forages on the ground, where it is adept at walking around in search of prey. This trait also has disadvantages, for the bat is subject to predation by snakes and owls. It is active relatively late in the evening.

**Comments:** This bat is intolerant of disturbance and readily abandons roosts. Populations have declined since Bailey (1936) wrote about the species in Oregon.

*References:* Hermanson and O'Shea 1983, Marshall 1992, O'Shea and Vaughan 1977.

351

# Brazilian Free-tailed Bat
*(Tadarida brasiliensis)*

*Order: Chiroptera*
*Family: Molossidae*
*State Status: None*
*Federal Status: None*
*Global Rank: G5*
*State Rank: S2*
*Length: 4 in (10 cm)*

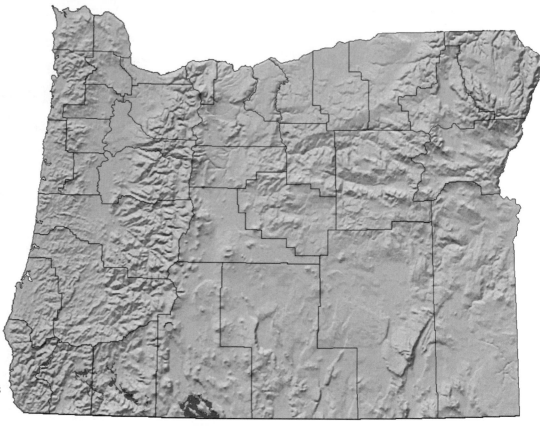

*References:* Barbour and Davis 1969, Jewett 1955, Wilkins 1989.

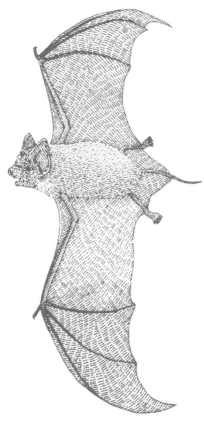

**Global Range:** The Brazilian free-tailed bat has an extensive distribution in the Western Hemisphere. Its range reaches its northern limit in southwestern Oregon, and extends across the continent to South Carolina. It occurs throughout Mexico and Central and South America, being absent only from the tropical northern edge of the continent, the Amazon Basin, and southern Chile and Argentina.

**Habitat:** This is a colonial species that frequents caves, hollow trees, and, more recently, buildings. It can occur in great numbers; several million of these bats occupy Carlsbad Caverns, New Mexico. It is usually found in warmer areas at low elevations. Habitat does not seem to play as big a role as climate in determining its presence.

**Food Habits:** These bats capture insects with their interfemoral membrane. Their diet consists mainly of moths, which they pursue over the forest canopy or over open meadows, grasslands, or pastures.

**Ecology:** Although migratory in other parts of its range, Oregon populations seem to be permanent residents. When females return to maternity colonies to nurse, they do not appear to seek out their own young. These bats have long, narrow wings, and are rapid flyers. Top speed is estimated to be at least 60 miles per hour.

**Comments:** These bats are strong flyers that can travel as far as 30 miles from colonies to foraging areas.

352

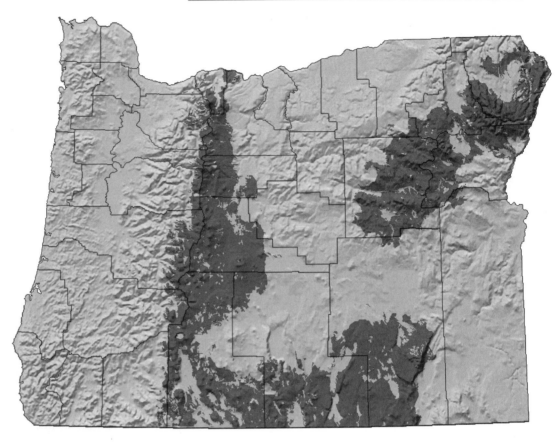

*Order: Lagomorpha*
*Family: Ochotonidae*
*State Status: None*
*Federal Status: None*
*Global Rank: G5*
*State Rank: S4?*
*Length: 7 in (18 cm)*

**Global Range:** Found discontinuously in mountains from central British Columbia and Alberta to the southern Sierra Nevada in California and the Rocky Mountains of northern New Mexico.

**Habitat:** The American pika is generally found in rocky talus slopes, often at the interface with a meadow, or other rocky areas within forests. Occasionally it colonizes other rocky areas, including mine tailings.

**Reproduction:** American pikas have 2 litters per year. Litter size varies from 2 to 5 young. They are able to breed early in the year due to stored food and access to meadows via snow tunnels. At lower elevations, the first litter may appear in March, while at higher elevations it may be as late as May. By 4 weeks of age, the young are independent of their mothers.

**Food Habits:** American pikas feed primarily on grasses and sedges. Lichens may be an important food in winter. They also consume some flowering plants and shoots of woody vegetation. The species composition of the diet varies with the local flora.

**Ecology:** American pikas set up territories in talus slopes and rock piles. They consume some food directly, and harvest some plant material, mostly tall grasses, which they dry and then store in "haypiles." They do not hibernate, and this accumulation of food is important for survival through the winter.

**Comments:** Neighboring American pikas tend to be of the opposite sex, and there is some overlap in home ranges. Haypiles alone are not sufficient to provide food during the winter, and some foraging occurs via snow tunnels.

*References:* Huntly et al. 1986, Smith 1974, Smith and Weston 1990.

353

# Pygmy Rabbit
## (Brachylagus idahoensis)

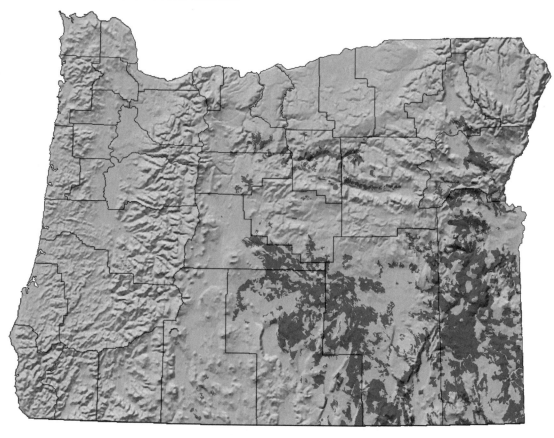

Order: Lagomorpha
Family: Leporidae
State Status: Sensitive
Federal Status: SC
Global Rank: G5
State Rank: S2?
Length: 11 in (28 cm)

**Global Range:** Restricted to the northern parts of the Great Basin. Generally, its range largely coincides with that of Great Basin or big sagebrush (*Artemisia tridentata*). An isolated population is found in southeastern Washington.

**Habitat:** Pygmy rabbits are most closely associated with areas supporting tall, dense clumps of Great Basin sagebrush, although they are also reported to frequent areas dominated by greasewood. Soil also plays a role in their distribution, since they require deep, friable soil to dig their burrows.

**Reproduction:** The beginning of breeding season varies with the quality of local habitat, but can be as early as February or as late as March. The pygmy rabbit can have up to 3 litters of about 6 young (range 4 to 8) per year. Gestation period is not known but is probably slightly under a month.

**Food Habits:** During most of the year, the pygmy rabbit feeds almost exclusively on the leaves of Great Basin sagebrush. However, during summer, grass may account for up to 30-40% of the diet.

**Ecology:** This rabbit is unique in that it digs its own extensive burrow system. It is active year round and throughout the day, but is most active around dawn and dusk. Its primary predators include weasels, coyotes, and owls.

**Comments:** This species has a discontinuous distribution. Loss of favorable habitat to agriculture, over-grazing, and conversion of sagebrush to exotic grasslands presents a threat to this species. Roads and cleared areas seem to be barriers to dispersal. Some authors consider this species to be in the genus *Sylvilagus*.

*References:* Green and Flinders 1980, Marshall 1992, Weiss and Verts 1984, White et al. 1982.

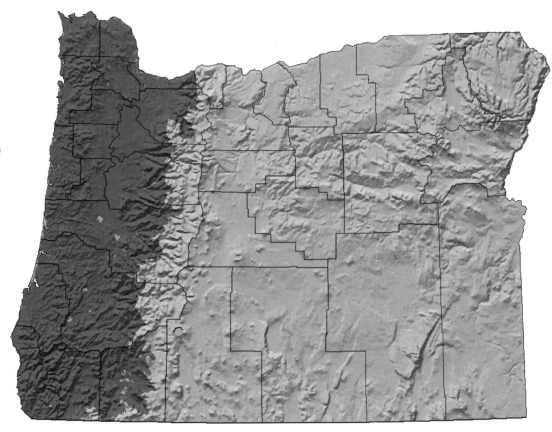

## Brush Rabbit
*(Sylvilagus bachmani)*

*Order: Lagomorpha*
*Family: Leporidae*
*State Status: None*
*Federal Status: None*
*Global Rank: G5*
*State Rank: S5*
*Length: 13 in (33 cm)*

**Global Range:** The Columbia River forms the northern limit of the range of the brush rabbit. From there, its range extends south in coastal mountains and valleys to the tip of Baja California. It also can be found in the foothills of the Sierra Nevada.

**Habitat:** Although the brush rabbit may feed at the edge of grassy meadows, it is seldom far from dense brushy cover. It occurs in a variety of habitat types, including brushy areas in forests, chaparral, and in thickets bordering sand dunes.

**Reproduction:** In Oregon, this species breeds from February through August. Females have 3 (or possibly 4) litters each year. About 3 (range 1-6) young are born after a gestation period of 27 days. The young first breed in the spring following their birth.

**Food Habits:** Accounts of the foods of brush rabbits differ. Some say that grasses are the primary food item, while Maser et al. (1981) suggest these rabbits eat forbs (such as false dandelion and plantain) and berries, as well as grasses. During winter, when forbs are in short supply, they turn to leaves and twigs of woody plants like salal, and also eat needles from Douglas-fir branches that have blown down.

**Ecology:** This species is active during the first half of the night and again in the morning. Generally, its home range coincides with the size of the thickets it inhabits. Usually, thickets smaller than 460 square meters are not permanently occupied. It is prey to a wide variety of mammalian and avian predators.

**Comments:** While not a serious pest of Douglas-fir forests, this rabbit can cause damage to home gardens.

*References:*
Chapman 1971, 1974, Maser et al. 1981.

**355**

# Eastern Cottontail
## *(Sylvilagus floridanus)*

*Order: Lagomorpha*
*Family: Leporidae*
*State Status: None*
*Federal Status: None*
*Global Rank: G5*
*State Rank: SE*
*Length:: 17 in (43 cm)*

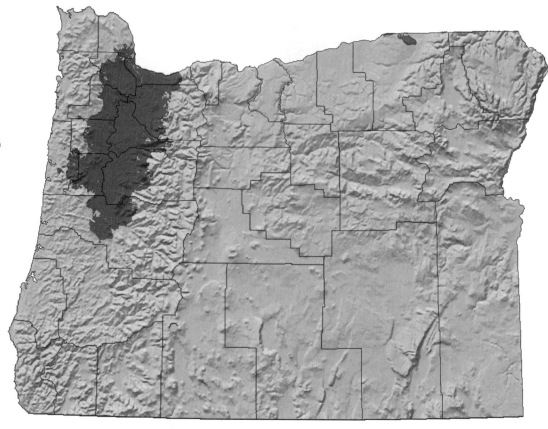

**Global Range:** Native to North America from southern Alberta east to Quebec and south to Panama. It is also found in northern South America. In the Southwest, it ranges as far west as Arizona and northern Mexico. It is introduced and established in Oregon and Washington.

**Habitat:** The eastern cottontail occupies a wide variety of habitats throughout its range. It requires bushes or other cover nearby, but otherwise occurs in pastures, grasslands, riparian forests, open woodlands, suburbs, croplands, and marshes. It is absent from dense forests.

**Reproduction:** Breeding biology varies with location. In Oregon, this species is thought to breed from January to August. The nest is a depression under a bush, lined with grass and fur. There may be as many as 7 litters of 5 (range 2-7) young per year. Young themselves are capable of breeding when they are 2-1/2 months old. The gestation period is about 4 weeks. Females are capable of producing from 2 to 3 dozen young per year.

**Food Habits:** Eastern cottontails eat a wide variety of plant material. Grasses are the most important food item, followed by forbs, clover, alfalfa, plantain, and the bark of trees, shrubs, and vines. In winter, the diet shifts toward bark, buds, and shoots of woody plants.

**Ecology:** Male eastern cottontails form a dominance hierarchy, and the dominant male mates with many females. They are not thought to be territorial, and have a home range of 3 to 20 acres. Densities may be as high as 4 per acre. Eastern cottontails are most active near dawn and dusk. They do not hibernate.

**Comments:** Eastern cottontails were introduced to Linn and Benton counties in 1939 and 1941. They have spread north to other parts of the Willamette Valley. They may cause damage to fruit trees.

*References:* Chapman et al. 1980, Fitzgerald et al. 1994, Graf 1955, Trethewey and Verts 1971, Verts and Carraway 1981.

*Order: Lagomorpha*
*Family: Leporidae*
*State Status: None*
*Federal Status: None*
*Global Rank: G5*
*State Rank: S4*
*Length: 14 in (36 cm)*

**Global Range:** This is a species of the Intermountain West, ranging from southern Canada south to central Arizona and New Mexico and from east of the Cascade and Sierra Nevada mountains to the western edge of the Great Plains.

**Habitat:** This species occurs in rocky ravines or riparian thickets in sagebrush-covered deserts but not in broad expanses of sagebrush proper. It also is found in juniper woodland and around the edges of ponderosa pine forests but not at high elevations in mountains.

**Reproduction:** This prolific species breeds from March to July and may produce 4 or 5 litters, each numbering 4 to 5 young, per year. The gestation period is less than 1 month. There is one report of a 3-month old female breeding, but usually females do not breed until the spring following their birth.

**Food Habits:** The mountain cottontail feeds primarily on grass when it is available in spring and summer. However, it will eat sagebrush and western juniper when necessary.

**Ecology:** Usually, this species feeds near cover and is active mainly at night. Over-winter mortality can be as high as 85-90%. Predators include bobcats, coyotes, owls, and large snakes, including rattlesnakes.

**Comments:** Although generally found east of the Cascades, the mountain cottontail occurs as far west as Ashland in southwestern Oregon, occupying areas of rabbitbrush and ponderosa pine. This species is also known as Nuttall's cottontail.

*References:* Chapman 1975, McKay and Verts 1978, Powers and Verts 1971, Verts et al. 1984

# Snowshoe Hare
### (Lepus americanus)

*Order: Lagomorpha*
*Family: Leporidae*
*State Status: None*
*Federal Status: None*
*Global Rank: G5*
*State Rank: S4*
*Length: 15 in (38 cm)*

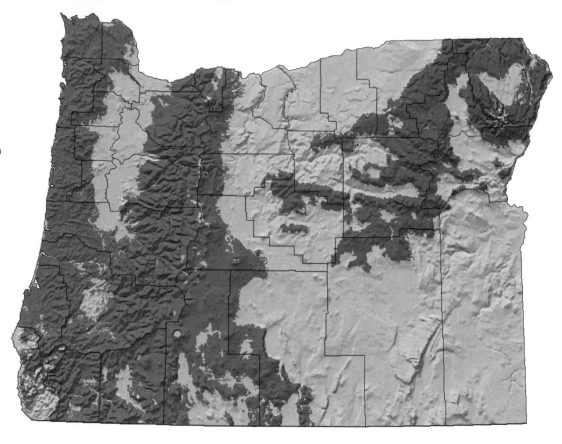

**Global Range:** This is a boreal species that ranges from Alaska and northern Canada south to the mid-Atlantic states and the Pacific Northwest, and, in mountainous areas of the West, to central California and northern New Mexico.

**Habitat:** The snowshoe hare occurs in coniferous forests where there is appropriate brushy cover. This includes riparian vegetation along streams and thickets of salal.

**Reproduction:** The breeding season begins about March and continues through the summer. Litters of about 3 young (range 1-6) are born as early as the middle of April. There can be 2 or possibly 3 litters per year.

**Food Habits:** In summer, forbs, grasses, bracken fern, and leaves of plants such as salal and lupine are eaten. In winter, the diet turns more to needles, bark, and twigs of conifers such as Douglas-fir and western hemlock.

**Ecology:** These hares are primarily nocturnal and solitary. They can be quite abundant in some years. One study found 1.6 to 3.0 hares per hectare in a western Oregon clear-cut. They can impede reforestation efforts.

**Comments:** Through most of its range the species molts into a white winter coat, but the subspecies found along the Oregon coast remains brown all year.

*References:* Keith and Windberg 1978, Maser et al. 1981, Nagorsen 1985.

# White-tailed Jackrabbit
## *(Lepus townsendii)*

*Order: Lagomorpha*
*Family: Leporidae*
*State Status: Sensitive*
*Federal Status: None*
*Global Rank: G5*
*State Rank: S4?*
*Length: 24 in (61 cm)*

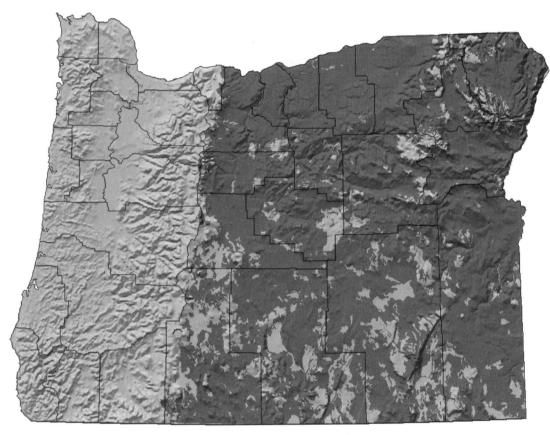

**Global Range:** Found from central Canada south to northern New Mexico. The Cascade and Sierra Nevada mountains form the western limit of its range, and it occurs as far east as western Illinois.

**Habitat:** This species has an unusual mix of habitat preferences. It inhabits open regions such as sagebrush deserts and grasslands, but also can be found in open areas in coniferous forests and even in alpine meadows.

**Reproduction:** The breeding season varies with climate. Generally, the white-tailed jackrabbit mates in the spring. Litters of 4 or 5 (range 1-11) young are born after a gestation period of 36-43 days. In cold areas, there is only 1 litter per year, but in milder climates there may be 4 litters per year.

**Food Habits:** This jackrabbit feeds primarily on grasses and forbs in spring and summer and on leaves and stems of woody plants in winter. It may also eat cultivated crops.

**Ecology:** This species is nocturnal and crepuscular. In areas it shares with black-tailed jackrabbits, it tends to be found in more open grasslands rather than in sagebrush or salt-desert shrub communities.

**Comments:** Once considered common, this species now seems to be distributed discontinuously in a few mountain ranges and higher valleys of eastern Oregon. The disappearance of native grasses may have contributed to its decline.

*References:* Bailey 1936, Lim 1987, Marshall 1992.

# Black-tailed Jackrabbit
## *(Lepus californicus)*

*Order: Lagomorpha*
*Family: Leporidae*
*State Status: None*
*Federal Status: None*
*Global Rank: G5*
*State Rank: S4*
*Length: 22 in (56 cm)*

*References:* Best 1996, Gross et al. 1974, Haskell and Reynolds 1947, Smith 1990.

**Global Range:** This is a species of the semi-arid and arid West. Its range extends east into the Great Plains and south to central Mexico. There are introduced populations along the Atlantic Coast.

**Habitat:** The black-tailed jackrabbit is found in open habitats, from lower coastal valleys, pastures, and fields to desert areas with scattered shrubs (such as sagebrush, shadscale, and greasewood). It can be found in thickets of chaparral and around the edges of forests.

**Reproduction:** This species has an extended breeding season from late winter to late summer. The gestation period is 41-47 days. One to 4 litters of 2 to 4 (range 1-8) young are produced each year. Sexual maturity is attained at 1 year of age.

**Food Habits:** The black-tailed jackrabbit eats grasses and forbs in summer. It will also consume crops and hay. In winter, it eats the buds, bark, and leaves of woody plants.

**Ecology:** Active throughout the year, it is primarily crepuscular and nocturnal. It has been known to reach densities of at least 260 individuals per square mile. This adaptable species is active during the day in areas where disturbance is unlikely, such as grass lawns separating airport runways.

**Comments:** This species fluctuates in abundance. Bailey (1936) estimated an overall density of 1 per acre in eastern Oregon, or 20,000,000 jackrabbits!

## Mountain Beaver
*(Aplodontia rufa)*

*Order: Rodentia*
*Family: Aplodontidae*
*State Status: None*
*Federal Status: None*
*Global Rank: G5*
*State Rank: S4*
*Length: 14 in (36 cm)*

**Global Range:** Found from the Cascade Mountains to the coast from southern British Columbia to Cape Mendocino, California. Another race is found on the east slope of the Sierra Nevada, and two isolated subspecies occur in small areas along the northern California coast.

**Habitat:** This is a forest dweller, or more exactly, a denizen of brushy thickets within early to mid-successional deciduous and coniferous forests, but it is more abundant in the former. Since the thickets favored by mountain beavers are often found growing on moist, deep soils near streams, there is a relationship between the mountain beaver and water.

**Reproduction:** A single litter of 2 or 3 (sometimes 4) young is born each year after a gestation period of 28-30 days. The timing of the breeding season depends on climate, and litters may be born as late as May (in Jefferson County).

**Food Habits:** Ferns are the most important component of the mountain beaver's diet, but it varies seasonally, apparently shifting to foods with the highest protein content. In spring, nursing females feed on the new growth of coniferous trees, while in the fall the leaves of red alder are preferred. Grasses, mosses, and forbs are also eaten.

**Ecology:** The mountain beaver digs extensive burrow systems in soft, deep soil. It can climb trees to harvest food, and uses runways on the surface. Mountain beavers are active year round, and tend to be somewhat nocturnal.

**Comments:** The mountain beaver is considered an economic pest because it feeds on young tree seedlings in reforestation projects.

*References:* Carraway and Verts 1993, Crouch 1968, Maser et al. 1981.

361

# Least Chipmunk
## (Tamias minimus)

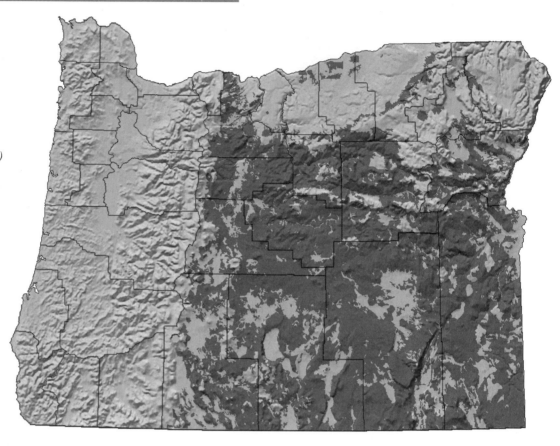

*Order: Rodentia*
*Family: Sciuridae*
*State Status: None*
*Federal Status: None*
*Global Rank: G5*
*State Rank: S4*
*Length: 7.5 in (19 cm)*

**Global Range:** Occurs throughout most of central and southern Canada and extends south through the Rocky and Sierra Nevada mountains and northern Great Basin. There are isolated populations in the mountains of eastern Arizona and central New Mexico.

**Habitat:** This chipmunk is associated with sagebrush, which may include sagebrush-covered desert valleys or sagebrush understories of open coniferous forests at higher elevations. It also has been recorded from alpine tundra.

**Reproduction:** After winter hibernation, which may be short, the breeding season begins in March. A single litter of 5 or 6 young (range 2-7) is produced each year. Gestation period is 20 to 30 days.

**Food Habits:** The least chipmunk has a varied diet which includes seeds, nuts, fruits, berries, green vegetation, roots, bulbs, and fungi. Conifer seeds are a preferred food item.

**Ecology:** Like other chipmunks, the least chipmunk is active during daylight hours. Its home range is from 1 to 4 acres, but, in favorable habitat, several home ranges may overlap, leading to densities as high as 30 per acre.

**Comments:** This species is a good climber and will sometimes nest in trees.

*References:* Jones et al. 1985, Sullivan 1985, Zeveloff 1988.

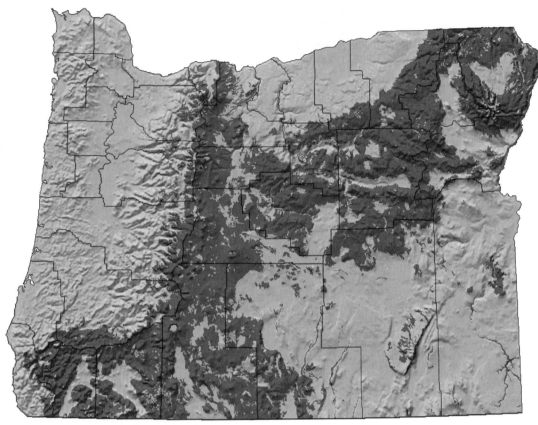

*Order: Rodentia*

*Family: Sciuridae*

*State Status: None*

*Federal Status: None*

*Global Rank: G5*

*State Rank: S4*

*Length: 8.5 in (22 cm)*

**Global Range:** Interior mountains from central British Columbia south through the Rocky Mountains to northern Utah and through the Cascade, Siskiyou, and Sierra Nevada mountains to central California.

**Habitat:** This chipmunk is found in brushy areas within coniferous forests. Shrubs such as snowberry, mountain mahogany, antelope brush, and buckbrush are typical of these areas, which occur in open ponderosa pine and Douglas-fir forests and woodlands.

**Reproduction:** Breeding season begins in late April or early May. A litter of 4 or 5 (range 3-8) young is born from late May to early June. Young breed the following spring.

**Food Habits:** This omnivorous chipmunk eats seeds, fruits, bird eggs, berries, flowers, bulbs, roots, fungi, and the buds of woody plants.

**Ecology:** Yellow-pine chipmunks are inactive most of the winter, but do not accumulate body fat in fall, suggesting they rely on stored food to survive cold periods. They are active about 5 months of the year, between November and March. They maintain dens in or under stumps or logs, and have a home range of 2 to 2.4 acres. They interact with other chipmunks, being dominated by Townsend's chipmunk and dominant over the least chipmunk.

**Comments:** This species may benefit from open brushy forests and ground litter resulting from timber harvest.

*References: Broadbrooks 1970, Kenagy and Barnes 1988, Maser and Maser 1988, Sutton 1992.*

# Townsend's Chipmunk
## *(Tamias townsendii)*

*Order: Rodentia*
*Family: Sciuridae*
*State Status: None*
*Federal Status: None*
*Global Rank: G5*
*State Rank: S4*
*Length: 10 in (25 cm)*

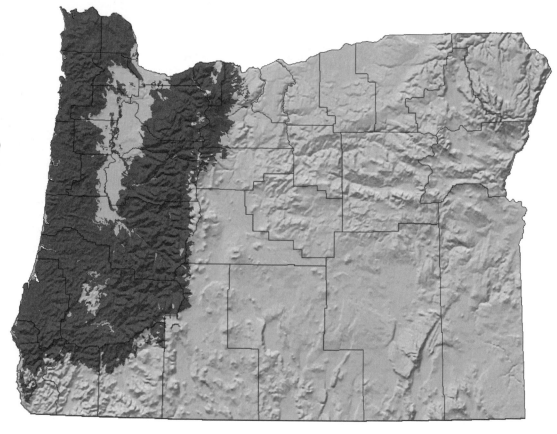

**Global Range:** This species complex extends from southern British Columbia south through Oregon and Washington, from the Cascade Mountains to the coast. It occurs further south in California in the Coast Ranges to Marin County and in the Sierra Nevada to Yosemite National Park. *Tamias siskiyou* occurs in the Siskiyou Mountains and the west slope of the southern Cascades in Oregon. *T. senex* is found along most of the east slope of the Cascades in Oregon and west and south to the California coast and Sierra Nevada. The coastal populations from about Cape Mendocino, California, to Marin County are *T. ochrogenys*, the yellow-cheeked chipmunk. Because most information on the biology of these chipmunks dates from the time when all were considered *T. townsendii*, this account applies to all populations in the species complex.

**Habitat:** These are species of coniferous forests, usually found in patches of chaparral or riparian zones within the forest. Populations in the Siskiyou Mountains prefer more open and less humid forests, and are most abundant in pine forests along the crests of ridges.

**Reproduction:** These chipmunks breed in the spring. A single litter of about 4 young (range 3-6) is born in May, but remains underground in the nest until July.

**Food Habits:** The diet varies seasonally. Subterranean fungi and huckleberries are eaten in winter. They also eat seeds, insects (primarily beetles), nuts, grasses, and roots.

**Ecology:** These chipmunks hibernate in areas where snow accumulates. The home range is about 2 acres. Rarely do home ranges of more than 2 individuals overlap. The life span is at least 7 years.

**Comments:** These chipmunks eat conifer seeds, including those of Douglas-fir, and are considered pests by the forest products industry. They may be abundant in clear-cuts. State and Global Ranks of the recently recognized species remain unclear.

*References:* Gannon and Forbes 1995, Gannon and Lawlor 1989, Gashwiler 1976b, Levenson and Hoffmann 1984, Maser et al. 1981, Sutton 1987, 1993, 1995.

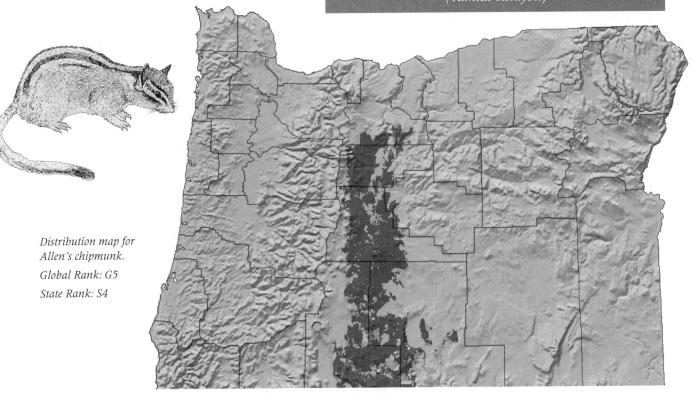

*Distribution map for
Allen's chipmunk.*

*Global Rank: G5*

*State Rank: S4*

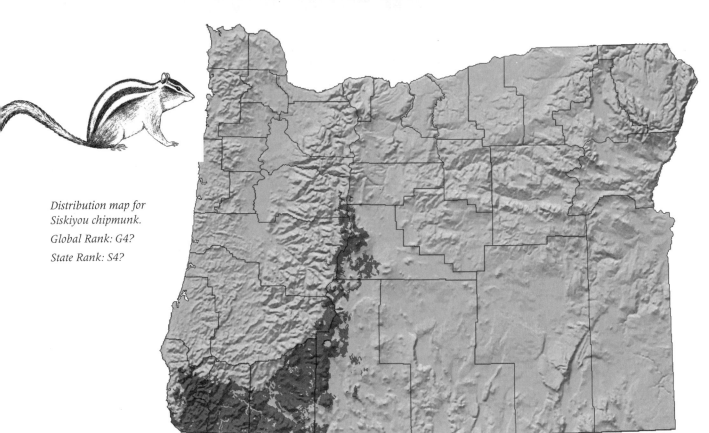

*Distribution map for
Siskiyou chipmunk.*

*Global Rank: G4?*

*State Rank: S4?*

365

# Yellow-bellied Marmot
### (Marmota flaviventris)

*Order: Rodentia*
*Family: Sciuridae*
*State Status: None*
*Federal Status: None*
*Global Rank: G5*
*State Rank: S4*
*Length: 24 in (61 cm)*

*References:* Armitage and Downhower 1974, Barash 1989, Frase and Hoffmann 1980.

**Global Range:** Found throughout interior western North America from southern Canada south through the Sierra Nevada of California and in the Rocky Mountains to northern New Mexico.

**Habitat:** Yellow-bellied marmots are strongly associated with talus or piles of rocks. They may occur on mountain slopes, in meadows in forests, or in alpine and higher semi-desert areas, providing sufficient rock accumulations are present.

**Reproduction:** There is a single litter of 4 or 5 (range 3-8) young per year. The gestation period is about 30 days. Marmots breed shortly after they emerge from hibernation, which varies from late February to May depending on local climate.

**Food Habits:** Marmots eat a variety of herbaceous foods, including grasses and forbs in spring and flowers and ripening seeds in summer. Where marmots are abundant, they may invade hay or alfalfa fields.

**Ecology:** Marmots live both singly and in harems. Because young marmots must gain sufficient weight to survive the winter, those born earlier in the year are more likely to survive. Winter mortality can be significant.

**Comments:** Yellow-bellied marmots have a scattered distribution in eastern Oregon, being absent from broad valleys but present along rimrock canyons.

*Order: Rodentia*
*Family: Sciuridae*
*State Status: Sensitive*
*Federal Status: None*
*Global Rank: G5*
*State Rank: S4?*
*Length: 9 in (23 cm)*

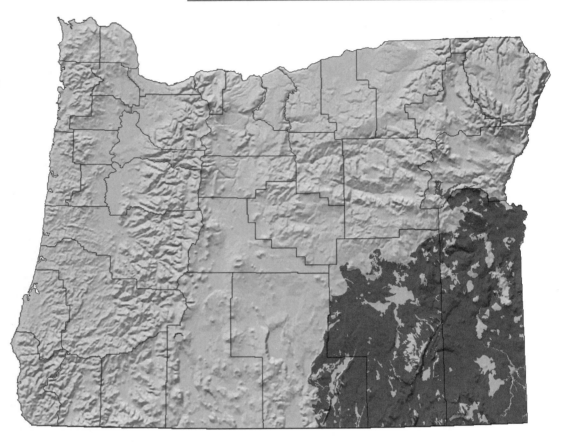

**Global Range:** This is a species of southwestern and Great Basin deserts that reaches the northern limit of its range in southeastern Oregon. It is found eastward to western Utah and New Mexico and south through the Mojave Desert to the tip of Baja California.

**Habitat:** This species can be found in low, open valleys of arid eastern Oregon where sagebrush, greasewood, shadscale, and other low desert shrubs furnish cover.

**Reproduction:** The breeding biology of Oregon's populations has not been studied well, but elsewhere mating occurs in March, followed by a 30-35 day gestation period. The average litter is 8 or 9 young (range 5-14), but this varies with food availability. There is likely only 1 litter per year in the northern Great Basin.

**Food Habits:** This omnivorous squirrel feeds on green vegetation, seeds, insects, and carrion. It will venture into cultivated areas to feed on ripening grain or alfalfa. The diet varies seasonally, tending towards green vegetation in spring and seeds in autumn.

**Ecology:** This species of ground squirrel is active throughout the year, and is not known to hibernate or aestivate. In summer, it is active in the early morning and late afternoon, avoiding the hottest temperatures of the day. It is active throughout the day in winter.

**Comments:** This species occurs only at scattered localities and at low densities in southeastern Oregon. There is anecdotal evidence that it is less common now than in the early 1970s. Bailey (1936) considered it scarce.

*References:* Belk and Smith 1991, Kenagy and Bartholomew 1985, Marshall 1992.

# Townsend's Ground Squirrel
## (*Spermophilus townsendii*)

*Order: Rodentia*
*Family: Sciuridae*
*State Status: None*
*Federal Status: None*
*Global Rank: G5*
*State Rank: S4*
*Length: 9 in (23 cm)*

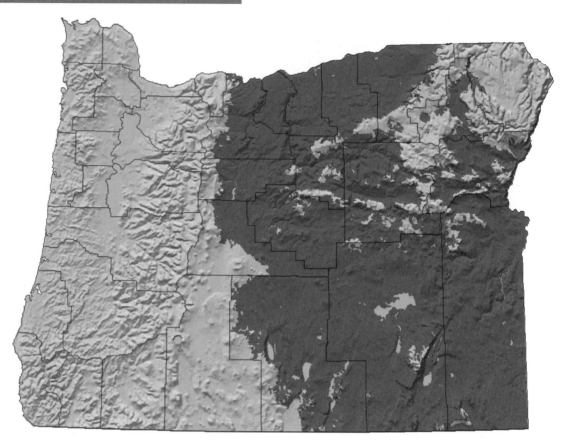

**Global Range:** A species of the Intermountain West, found from southeastern Washington south to southern Nevada and east to western Utah.

**Habitat:** This squirrel is typical of arid desert shrub communities dominated by sagebrush, shadscale, and greasewood. This species can become abundant around agricultural areas in the low, hot river valleys in eastern Oregon, where it may be a serious agricultural pest.

**Reproduction:** Depending on locality, this species breeds from January to March. A single litter of 7-10 young (range 4-12) is born after a 24-day gestation period. Breeding is influenced by climate, and the species may not breed during drought years.

**Food Habits:** This species feeds primarily on green vegetation (grasses and forbs) and seeds. It also may eat some woody vegetation of shrubs, which it will climb while foraging.

**Ecology:** This is a diurnal species that hibernates in winter, and also undergoes summer dormancy after the grass has dried. The adults enter winter hibernation earlier than the young of the year. These squirrels are most active in the early morning.

**Comments:** Chromosomal evidence suggests that two of the subspecies found in Oregon (*mollis* and *canus*) may be distinct species. *Spermophilus townsendii mollis* occurs only in the southeastern corner of the state.

*References:* Bailey 1936, Rickart 1987, 1988.

# Washington Ground Squirrel
## (Spermophilus washingtoni)

*Order: Rodentia*
*Family: Sciuridae*
*State Status: Sensitive*
*Federal Status: SC*
*Global Rank: G2*
*State Rank: S2*
*Length: 9 in (23 cm)*

*References:* Betts 1990,
Marshall 1992,
Rickart and Yensen
1991.

**Global Range:** The original range of this species is a relatively small area of southeastern Washington and an even smaller area of adjacent northeastern Oregon. Much of this area has been converted to agriculture, and no longer supports populations of this ground squirrel.

**Habitat:** The Washington ground squirrel is found in arid deserts and grasslands, most frequently in sagebrush or grasslands that are associated with river banks, hillsides, or ravines.

**Reproduction:** These squirrels emerge from hibernation in late January or early February and begin breeding a few weeks later. One litter of about 8 (range 5-11) young is produced around March. The length of gestation is unknown for this species, but most ground squirrels have gestation periods of 23-30 days.

**Food Habits:** This species has a diet typical of omnivorous ground squirrels. It eats green vegetation (forbs, grasses), flowers, bulbs, roots, seeds, seed pods, and insects.

**Ecology:** Frequently, Washington ground squirrels are found in colonies of up to 250 individuals. These squirrels are diurnal, and have protracted periods of summer and winter dormancy.

**Comments:** The Washington ground squirrel is absent from many areas where it was once considered common. There are about 35 confirmed colonies in Oregon, the largest located on the Research Natural Area of the Boardman Bombing Range, Morrow County.

# Belding's Ground Squirrel
## (Spermophilus beldingi)

*Order:* Rodentia
*Family:* Sciuridae
*State Status:* None
*Federal Status:* None
*Global Rank:* G5
*State Rank:* S5
*Length:* 10 in (25 cm)

*References:* Jenkins and Eshelman 1984, Zeveloff 1988.

**Global Range:** Found in the northern Great Basin and surrounding mountains, reaching its northern limit in northeastern Oregon. It is found as far east as Utah, as far south as the central Sierra Nevada, and as far west as Fort Klamath, Klamath County, Oregon.

**Habitat:** This is a squirrel of open habitats, but it has little regard for where they are located. It is found in alpine meadows, sagebrush flats, grasslands, croplands, and pastures, but avoids rocky or heavily forested areas.

**Reproduction:** This species breeds shortly after emerging from hibernation, which varies according to climate. A single litter of about 6 young is born after a gestation period of 25-28 days.

**Food Habits:** These ground squirrels feed primarily on green plant material such as grass seeds and the leaves and stems of forbs and grasses. Occasionally they will eat insects and other animal matter. Dandelion leaves and flowers are favorite foods.

**Ecology:** This is a colonial species that spends much of the year underground. It may hibernate from late July to middle March. It is diurnal and some members of the colony are always on watch for predators.

**Comments:** This species occurs at high elevations where it may have only 3 months a year to reproduce, grow, and gain weight for hibernation.

# Columbian Ground Squirrel
### (Spermophilus columbianus)

*Order: Rodentia*
*Family: Sciuridae*
*State Status: None*
*Federal Status: None*
*Global Rank: G5*
*State Rank: S4?*
*Length: 15 in (38 cm)*

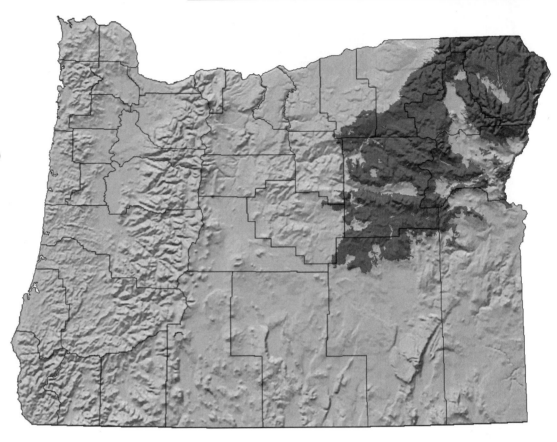

**Global Range:** Found from southeastern British Columbia and southwestern Alberta south through western Montana, northern Idaho, and the mountainous areas of northeastern Oregon and eastern Washington.

**Habitat:** These ground squirrels are typical of high grassy plateaus and open areas in forests. They also are found around the edges of mountain meadows or in mounds in meadows that are subject to flooding. Occasionally, they enter grain fields in valleys of the Blue Mountains. They readily use recent clear-cuts.

**Reproduction:** Mating occurs soon after these squirrels emerge from hibernation (which varies with climate). A single litter of 3 or 4 (range 2-7) young is born after a 24-day gestation period.

**Food Habits:** Their diet varies seasonally, according to availability. In spring, grasses, roots, bulbs, leaves, and some insects are eaten. Ripening seeds are eaten in summer, and in fall, berries, ripe seeds, and grains are consumed.

**Ecology:** The Columbian ground squirrel is a colonial species that is adapted to moist conditions. These squirrels spend about 70% of the year underground. Females exhibit strong site fidelity, remaining close to their natal nests and sometimes inheriting the burrow system from their mothers.

**Comments:** On April 13, 1919, a Columbian ground squirrel emerged through 6 inches of snow at Wallowa Lake.

*References:* Bailey 1936, Elliot and Flinders 1991, Wiggett and Boag 1989, Zammuto and Millar 1985.

371

# California Ground Squirrel
## *(Spermophilus beecheyi)*

*Order: Rodentia*
*Family: Sciuridae*
*State Status: None*
*Federal Status: None*
*Global Rank: G5*
*State Rank: S5*
*Length: 18 in (46 cm)*

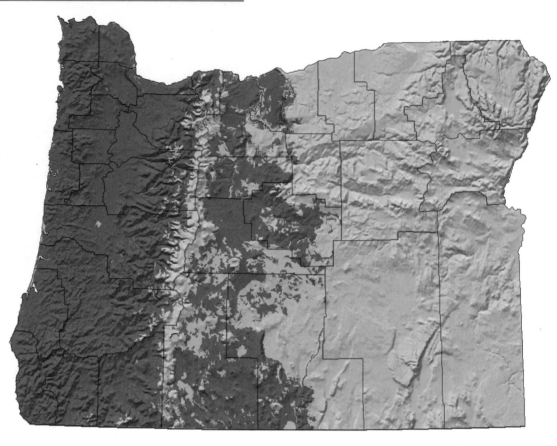

**Global Range:** This species of the west coast of North America is found from southern Washington (east of the Cascades) south through western Oregon and all of non-desert California into northern Baja California.

**Habitat:** The California ground squirrel frequents open, grassy areas within a mosaic of many other habitat types. It is found in pastures, along roadsides, in oak woodlands, chaparral, and even in disturbed areas in forests. In western Oregon, it is most common in low valleys.

**Reproduction:** This species breeds in late March and early April. One litter of 5 or 6 young (range 3-15) is born each year after a 25-30 day gestation period. The young emerge from the burrow about 2 months later.

**Food Habits:** This ground squirrel is widely omnivorous. It eats fruits, leaves, green seeds, flowers, bulbs, roots, insects, berries, grains, nuts, and acorns. It is known to eat tanoak acorns and some carrion.

**Ecology:** California ground squirrels accumulate fat during the summer and become inactive in underground burrows during cold weather. Sometimes a few will emerge on warm winter days. Usually, these squirrels can be found living in loosely organized colonies, some members of which serve as lookouts for possible danger.

**Comments:** The California ground squirrel can be a serious agricultural pest. It both consumes crops and digs extensive burrow systems that can be a hazard to livestock.

*References:* Dobson and Davis 1986, Linsdale 1946, Maser et al. 1981.

# Golden-mantled Ground Squirrel
*(Spermophilus lateralis)*

*Order: Rodentia*
*Family: Sciuridae*
*State Status: None*
*Federal Status: None*
*Global Rank: G5*
*State Rank: S4*
*Length: 11 in (28 cm)*

**Global Range:** This is a squirrel of the mountainous West. It occurs from the Rocky Mountains of central British Columbia south to the White Mountains of Arizona, east to the Rocky Mountain front ranges in Colorado and west to the Siskiyou Mountains of Oregon and California. It does not reach the Pacific Coast.

**Habitat:** This ground squirrel is found in open grassy or rocky areas in ponderosa pine, Douglas-fir, lodgepole pine, and true fir forests. It can be found from the upper edge of the juniper belt to above timberline. Given rocks for shelter, it is even found in meadows surrounded by sagebrush.

**Reproduction:** This species breeds shortly after emerging from hibernation. A single litter of 4 to 6 (range 2-8) young is born from July to August after a 26- to 33-day gestation period.

**Food Habits:** The diet of this omnivorous species includes fungi, seeds of conifers, leaves of shrubs and forbs, flowers, insects, bird eggs, and even some small lizards.

**Ecology:** Hibernation can begin in early September and last until mid-April. This squirrel is territorial, and usually is distributed throughout the available habitat. Young born late in the year do not have time to accumulate sufficient fat for both hibernation and reproduction and delay breeding until their second year.

**Comments:** Where abundant, the golden-mantled ground squirrel can do damage to reforestation projects by consuming emerging seedlings.

*References:* Bartels and Thompson 1993, Maser and Maser 1988, McKeever 1964.

# Eastern Fox Squirrel
## (Sciurus niger)

*Order: Rodentia*
*Family: Sciuridae*
*State Status: None*
*Federal Status: None*
*Global Rank: G5*
*State Rank: SE*
*Length: 20 in (51 cm)*

*References:*
Fitzgerald et al.
1994, Koprowski
1994b.

**Global Range:** Native to the United States east of the Rocky Mountains. Just enters southern Canada and northern Mexico and is absent from New England.

**Habitat:** This is the common squirrel of eastern deciduous and mixed coniferous-deciduous forests. It frequents riparian woodlands, suburbs, and urban parks. In Oregon, it is most common in metropolitan areas or smaller towns. Elsewhere on the West Coast, introduced populations have moved into open forests and woodlands adjacent to developed regions. It is most abundant in forests with oaks, where acorns may be stored for winter food.

**Reproduction:** Eastern fox squirrels build nests lined with leaves and grass in hollows or cavities in trees. A sphere of leaves may be used for nesting in the absence of tree cavities. Under favorable conditions, there may be 2 litters of 2 or 3 (range 1-6) young per year. The gestation period is just over 6 weeks, and young leave the nest at 3 months of age. Young are sexually mature the following spring.

**Food Habits:** Eastern fox squirrels feed mostly on plant material, such as acorns, nuts, tree buds, berries, leaves, and twigs. They also eat grain, insects, bird eggs and nestlings, and tree bark. They may cause damage in orchards.

**Ecology:** Male eastern fox squirrels may establish a dominance hierarchy. The home range varies from a few acres to 40 acres, and the home range of several individuals may overlap where densities are high. During breeding season, females defend an area around the nest. Like most squirrels, eastern fox squirrels are diurnal. They remain active during winter, and several may share a den during very cold weather. Young may disperse as far as 40 miles from their natal den.

**Comments:** Eastern fox squirrels make use of more open woodlands and forests than native western gray squirrels. They are an important game species in the eastern United States.

374

# Western Gray Squirrel
## (Sciurus griseus)

*Order:* Rodentia
*Family:* Sciuridae
*State Status:* Sensitive
*Federal Status:* None
*Global Rank:* G5
*State Rank:* S4?
*Length:* 22 in (56 cm)

*References:* Carraway and Verts 1994, Marshall et al. 1996, Maser and Maser 1988, Maser et al. 1981.

**Global Range:** Found from central Washington south through western Oregon into most of California, with the exception of the Central Valley, Mojave Desert, and coastal southern California. Also on the east side of the Cascade and Sierra Nevada mountains. They may be confused with the eastern gray squirrel, introduced into some urban areas in the Willamette Valley and locally elsewhere in Oregon and Washington (Koprowski 1994a).

**Habitat:** The primary habitat of this species seems to be deciduous or broadleaf evergreen woodlands dominated by oaks, sometimes mixed with pines. They also are present in riparian areas and in mixed forests of tanoak, maple, madrone, and conifers such as Douglas-fir, ponderosa pine, white fir, sugar pine, or Jeffrey pine. Their habitat usually has a broad-leafed component. They sometimes occupy urban parks or orchards where natural habitat is nearby.

**Reproduction:** This is primarily a squirrel of low elevations, allowing breeding to commence early in the year. A single litter of 2 or 3 young (range 1-5) is born from February to May, after a gestation period of about 44 days.

**Food Habits:** Western gray squirrels eat a variety of foods. Fungi, acorns, the seeds of conifers such as Douglas-fir, Sitka spruce, sugar pine, and true firs seem to be the primary foods. Some fruit, berries, green vegetation, and insects are also eaten. They eat fungi in spring and acorns, which they store, in fall and winter.

**Ecology:** This is an arboreal species that remains active throughout the year. Large stick nests are built in trees as high as 50 feet above the ground. These squirrels often forage on the ground. Their home range varies from a fraction of an acre to over 40 acres. Although not colonial, the squirrels of a local habitat patch display a social hierarchy.

**Comments:** This species is a game animal in Oregon. It may invade orchards of nut trees and become an agricultural pest. Competition with introduced eastern fox and gray squirrels, the loss of older trees, and conversion of oak woodlands may contribute to its decline in Oregon.

375

# Red Squirrel
## (Tamiasciurus hudsonicus)

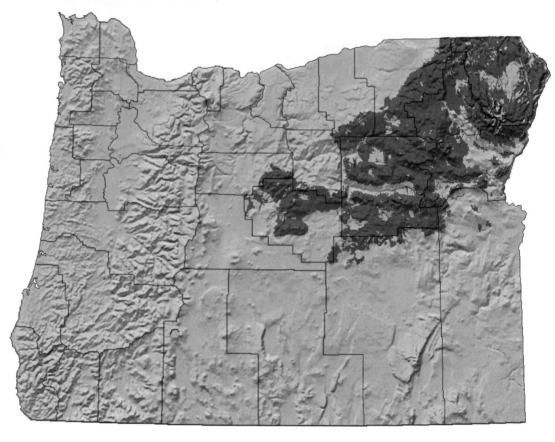

*Order: Rodentia*
*Family: Sciuridae*
*State Status: None*
*Federal Status: None*
*Global Rank: G5*
*State Rank: S4?*
*Length: 15 in (38 cm)*

**Global Range:** The red squirrel is a species of northern forests, from Alaska east to Newfoundland and south to the Smoky Mountains. In the West, it follows the Rocky and Blue mountains south to isolated mountain ranges of southern Arizona and New Mexico.

**Habitat:** Although it usually occurs in higher-elevation coniferous forests, such as lodgepole pine or spruce-fir forests, this squirrel also frequents mixed forests and even deciduous woodlands. It will follow willow-cottonwood riparian zones down into valleys if the seed crop higher in the mountains fails.

**Reproduction:** There is an extended breeding season from about March to July or even later in the summer. After a 31-35 day gestation period a litter of 4 or 5 (range 1-10) young is born in a nest located about 15-20 feet up in a tree. There is only 1 litter per year in the northern part of its range.

**Food Habits:** The red squirrel is an opportunistic omnivore. In spring and summer, it eats buds, flowers, fruits, fungi, insects, birds and their eggs, and, occasionally, small mammals. In winter, it feeds on seeds of coniferous trees, which it stores in caches.

**Ecology:** This diurnal squirrel is active throughout the year. The home range is as large as 6 acres, and individuals defend territories centered on seed caches.

**Comments:** The red squirrel may benefit coniferous forests by spreading tree seeds and by spreading spores of mycorrhizal fungi.

*References:* Lair 1985, Maser and Maser 1988, Sullivan 1990, Zeveloff 1988.

*Order: Rodentia*

*Family: Sciuridae*

*State Status: None*

*Federal Status: None*

*Global Rank: G5*

*State Rank: S5*

*Length: 14 in (36 cm)*

*References:* Buchanan et al. 1990, Maser and Maser 1988, Maser et al. 1981.

**Global Range:** A West Coast species found from western British Columbia south to San Francisco Bay. Inland, the species occurs as far south as the southern Sierra Nevada. There is an isolated population in the Sierra San Pedro Mártir of northeastern Baja California.

**Habitat:** This squirrel is a forest-dweller, typically found in coniferous forests from the coast to high mountains, but it will also use areas of deciduous trees within those forests and wooded suburbs in western Oregon.

**Reproduction:** Breeding season begins about March, and a litter of 4 to 6 (range 2-8) young is born in May or June, after a 35-40 day gestation period. One litter per year is the norm in Oregon.

**Food Habits:** Douglas' squirrels, also known as chickarees, have a broad diet. In spring, they will eat developing shoots and pollen cones of Douglas-fir. Later in the year, they will eat green vegetation, fruits, insects, fungi, berries, and seeds of both conifers and deciduous trees, such as bigleaf maple.

**Ecology:** These squirrels are active throughout the year. They defend a territory as large as 2 acres. Summer nests are built on tree limbs, but winters usually are spent in holes in trees.

**Comments:** The subspecies of chickaree from eastern Oregon is found in higher and more arid forests than coastal chickarees.

# Northern Flying Squirrel
## *(Glaucomys sabrinus)*

*Order: Rodentia*
*Family: Sciuridae*
*State Status: None*
*Federal Status: None*
*Global Rank: G5*
*State Rank: S4*
*Length: 11 in (28 cm)*

**Global Range:** This forest-dwelling squirrel is found from Alaska to Newfoundland, south to the Blue Ridge Mountains in the East and south in the Rocky Mountains to southern Utah. Along the Pacific Coast it occurs south to Mendocino County, and in the Cascade, Sierra Nevada, and San Bernardino mountains south to southern California.

**Habitat:** Flying squirrels inhabit forests with fairly tall trees. They use coniferous, mixed, and deciduous forests, but the latter are used more for feeding than for nesting. They frequent forests of Douglas-fir, redwood, ponderosa pine, lodgepole pine, true fir, and spruce. They can be found feeding on the ground along streams and at forest-meadow transitions.

**Reproduction:** In Oregon, a single litter of 2 to 4 (range 1-6) young is produced each year. Breeding season starts in March, and most young are born in May or June, after a 37-40 day gestation period.

**Food Habits:** Studies in Oregon have found that during summer this squirrel feeds almost exclusively on underground fungi. It also includes nuts, seeds, fruits, insects, and green vegetation in its diet.

**Ecology:** This squirrel is active throughout the year. Unlike most squirrels, it is nocturnal and, therefore, seldom seen. The northern flying squirrel is solitary and lives in holes in trees. Where these are unavailable, it will build a round nest of moss, lichen, and shredded bark in a fork formed by tree limbs. In the Northwest, its chief predator is the Spotted Owl.

**Comments:** These squirrels may play an important role in forest ecosystems by spreading the spores of hypogeous fungi. The northern flying squirrel does not really fly, but glides from one tree to another on a flap of skin that extends between its front and rear limbs. From a height of 18 meters it can glide to a point 50 meters away.

*References:* Maser and Maser 1988, Maser et al. 1981, Wells-Gosling and Heaney 1984.

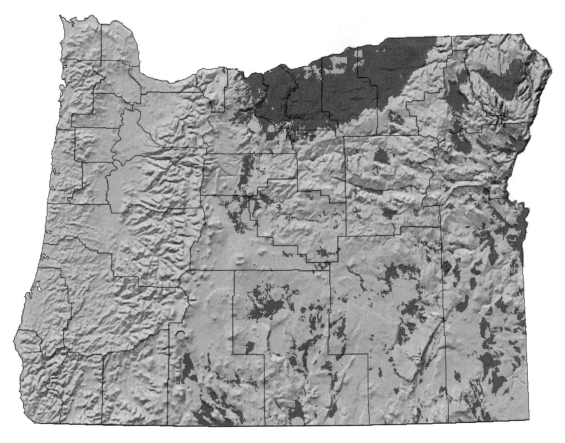

*Order: Rodentia*
*Family: Geomyidae*
*State Status: None*
*Federal Status: None*
*Global Rank: G5*
*State Rank: S4*
*Length: 8 in (20 cm)*

**Global Range:** Occurs from the Great Plains of southern Canada east to Wisconsin, south to New Mexico, and west to the Cascade and Sierra Nevada mountains.

**Habitat:** This species prefers deep soils found in meadows and along streams. It also is found in rocky or clay soils, brushy areas along roadsides, cultivated fields, and alpine meadows.

**Reproduction:** The breeding season lasts from March through August. A single litter of 4 to 7 young is born after a 28-day gestation period.

**Food Habits:** These gophers are generalized herbivores that eat grass stems and roots, bulbs, tubers, and forbs. They also feed on the leaves and stems of plants close to their burrow entrances.

**Ecology:** Their extensive tunnel systems may be 500 feet in length. They are solitary, and defend their burrow systems against others of their species. They are prey to horned owls, badgers, and coyotes. In south-central Oregon, densities have been found to range from 1.6 to 20.6 gophers per acre.

**Comments:** Bailey (1936) refers to populations of the northern pocket gopher under the names *T. fuscus*, *T. quadratus*, and *T. columbianus*.

*References:* Bailey 1936, Hansen and Remmenga 1961, Zeveloff 1988.

# Western Pocket Gopher
## *(Thomomys mazama)*

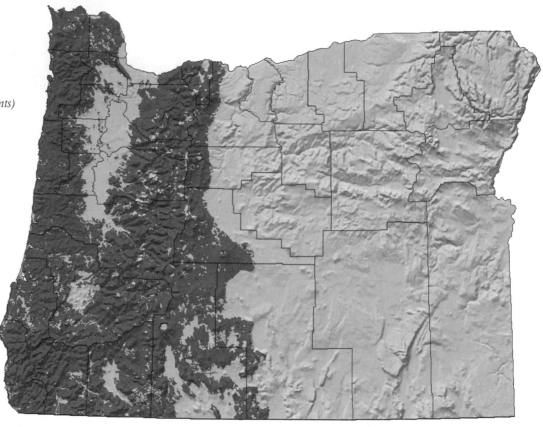

*Order: Rodentia*

*Family: Geomyidae*

*State Status: None*

*Federal Status: SC (one population—see comments)*

*Global Rank: G4G5*

*State Rank: SU*

*Length: 8 in (20 cm)*

**Global Range:** This Northwestern species is found from the Olympic Peninsula and Puget Sound region of Washington south through the Oregon Coast Range. It also occurs in the Cascade Mountains of Oregon and northern California.

**Habitat:** Western pocket gophers occupy open, grassy meadows and wet pastures found in mountain forests. Usually, they do not inhabit forest interiors.

**Reproduction:** There is one litter of 4 to 6 young each year. Breeding season begins early, and young are born from March to June, after a gestation period of about 28 days.

**Food Habits:** Western pocket gophers consume both the above- and below-ground portions of a variety of plants. Their diet includes forbs (false dandelion, clover, lupine), grasses, bulbs (wild onion and garlic are especially favored), plant roots, and the bark of trees.

**Ecology:** Like most gophers, this species forages in underground tunnel systems. It may be active on the surface at night or on overcast days. It collects food in external, fur-lined cheek pouches, and may store considerable amounts in underground chambers.

**Comments:** Sometimes this species is found on the floor of open ponderosa pine forests on the east slope of the Cascade Mountains. The subspecies, *T. m. helleri*, occurs only at the mouth of the Rogue River, Gold Beach, Curry County, and is a federal "species of concern."

*References:* Bailey 1936, Burton and Black 1978, Lidicker 1971, Maser et al. 1981.

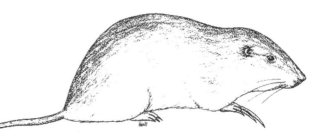

*Order: Rodentia*
*Family: Geomyidae*
*State Status: None*
*Federal Status: None*
*Global Rank: G4?*
*State Rank: S4*
*Length: 11.5 in (29 cm)*

**Global Range:** Until fairly recently, this gopher, whose range is restricted to the Willamette Valley, was the only mammal endemic to Oregon. Since Carraway's (1990) revision of *Sorex bairdi* and *Sorex pacificus*, those two shrews now join this gopher as state endemics.

**Habitat:** Camas pocket gophers live in grassy areas, in both lowlands and hills. Originally, they probably occurred in wet grasslands throughout the Willamette Valley. In 1914, they were abundant in lawns of the Council Crest neighborhood of Portland, at about 1,000 feet of elevation. They now occur in pastures, roadsides, and agricultural lands.

**Reproduction:** For a "pest" species, this pocket gopher is not very prolific. There is one litter of 3 to 5 young each year. Its gestation period is inferred to be 18-19 days. The breeding season lasts from late April to early July.

**Food Habits:** In natural habitats, it eats and stores roots of false dandelion, vetch, grasses, and wild onion. In agricultural areas, it eats vegetables (e.g., carrots, parsnips, potatoes) and the roots of fruit and nut trees.

**Ecology:** This is a large pocket gopher that digs extensive tunnel systems up to 127 millimeters in diameter and 240 meters in length. It is solitary, and active throughout the year.

**Comments:** This species has apparently adapted to the conversion of most of its original habitat to agriculture.

*References:* Bailey 1936, Carraway and Kennedy 1993, Verts and Carraway 1987a.

# Botta's Pocket Gopher
## (Thomomys bottae)

*Order: Rodentia*

*Family: Geomyidae*

*State Status: None*

*Federal Status: SC (one population—see comments)*

*Global Rank: G5*

*State Rank: S4?*

*Length: 9 in (23 cm)*

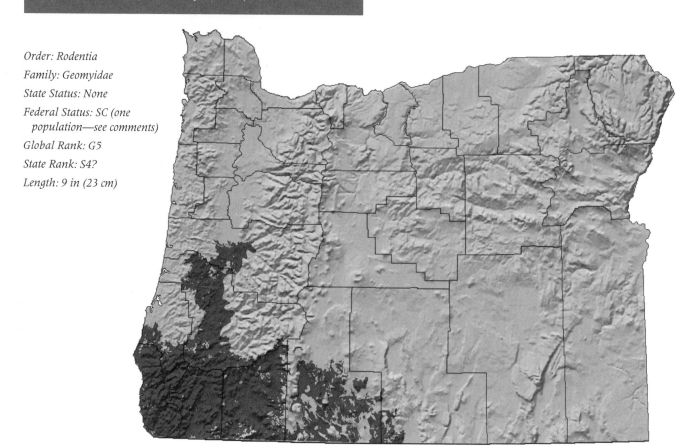

**Global Range:** Reaches its northern limit in Oregon, and extends south to Baja California and northern Mexico and east to Colorado, New Mexico, and western Texas. In arid regions, its distribution is discontinuous.

**Habitat:** Generally, this gopher is found in moist meadows, pastures, grasslands, and riparian areas. In Oregon, it is found in two low coastal valleys and the interior valleys of the southwestern part of the state. It requires deep soils. In parts of its range it is an agricultural and suburban pest.

**Reproduction:** This species can produce 1 to 3 litters of 4 or 5 (range 3-12) young per year. In California, the species breeds year round, with most litters born in spring, late summer, and early winter. The gestation period is 18-19 days.

**Food Habits:** This gopher feeds exclusively on vegetable matter, including roots, tubers, bulbs, grass stems, forbs, seedlings, seeds, and acorns.

**Ecology:** Botta's pocket gophers are active burrowers that gather food as they tunnel. They are active both day and night and are territorial, actively defending their burrow systems.

**Comments:** An isolated subspecies (*T. b. detumidis*) found along the Pistol River in Curry County is a federal "species of concern." Because of suspected interbreeding, some taxonomists consider *T. bottae* conspecific with the southern pocket gopher, *Thomomys umbrinus*.

*References:* Daly and Patton 1986, 1990; Gettinger 1984.

## Townsend's Pocket Gopher
*(Thomomys townsendii)*

*Order: Rodentia*
*Family: Geomyidae*
*State Status: None*
*Federal Status: None*
*Global Rank: G4G5*
*State Rank: S4*
*Length: 11 in (28 cm)*

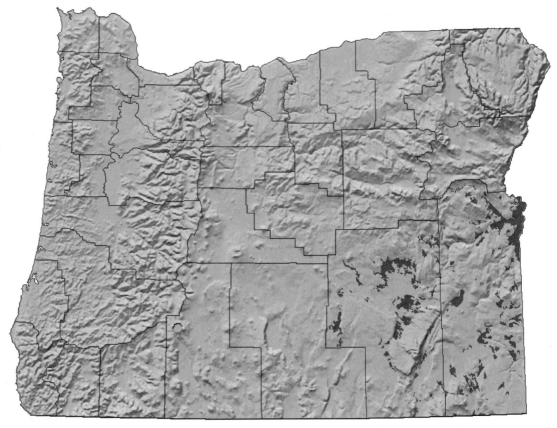

**Global Range:** Restricted to the northern part of the Great Basin. There are isolated populations in eastern California, northern Nevada and southern Oregon, and the western and eastern (but not central) parts of the Snake River plain in Idaho.

**Habitat:** This gopher has a strong association with deep, moist soils surrounding lakes or in bottomlands along rivers. It also will occupy irrigated farmland.

**Reproduction:** Townsend's pocket gopher begins breeding early in the spring and may produce 2 litters of 8 young (range 3-10) per year. The gestation period is about 19 days.

**Food Habits:** This species eats roots and stems of grasses and tubers, and leaves of surface vegetation. It also will eat alfalfa, grains, and other crops. In Idaho, it was observed to feed primarily on the runners of saltgrass.

**Ecology:** This gopher is solitary except during the breeding season. It is active mostly at night. It may store food in underground chambers. Townsend's pocket gophers are rarely found in communities with other pocket gopher species.

**Comments:** The drying of the Great Basin over the last 10,000 years may have reduced and fragmented the distribution of this species.

*References:* Bailey 1936, Hall 1946, Zeveloff 1988.

383

# Great Basin Pocket Mouse
## (Perognathus parvus)

*Order: Rodentia*
*Family: Heteromyidae*
*State Status: None*
*Federal Status: None*
*Global Rank: G5*
*State Rank: SU*
*Length: 7 in (18 cm)*

**Global Range:** As its name implies, this pocket mouse occurs throughout the Great Basin. It is also found as far north as southern British Columbia, following the arid lands east of the Cascade Mountains through Oregon and Washington.

**Habitat:** In the northern part of its range, the Great Basin pocket mouse occurs in sagebrush and shadscale-greasewood communities, as well as in grasslands, although it is limited to grasslands with lighter, sandy soil. It has been found in pinyon-juniper woodlands in Nevada, and occurs in juniper woodlands in Oregon where there is a sagebrush understory. It rarely has been found in the sagebrush understory on the edges of ponderosa pine woodlands.

**Reproduction:** Breeding season begins in April and May and continues at least through August. The average litter size is 5 (range 2-8), and the gestation period is about 25 days. There are probably 2 and sometimes 3 litters per year, but reproduction may be abandoned in years with poor rainfall.

**Food Habits:** Although the species feeds primarily on the seeds of grasses and desert shrubs, studies

from British Columbia indicate that green vegetable material and insects are also eaten.

**Ecology:** Following winters with good rainfall, the Great Basin pocket mouse can reach densities of 80 per hectare. It is nocturnal, and spends most of the winter in a state of torpor in its burrow, though the duration of torpor varies from year to year.

**Comments:** There is considerable geographic variation in the ecology of this species. Length of activity above ground, habitat associations, number of litters per year, and even food habits appear to change from the southern to northern parts of its range.

*References:* Genoways and Brown 1993, Kritzman 1974, Verts and Kirkland 1988.

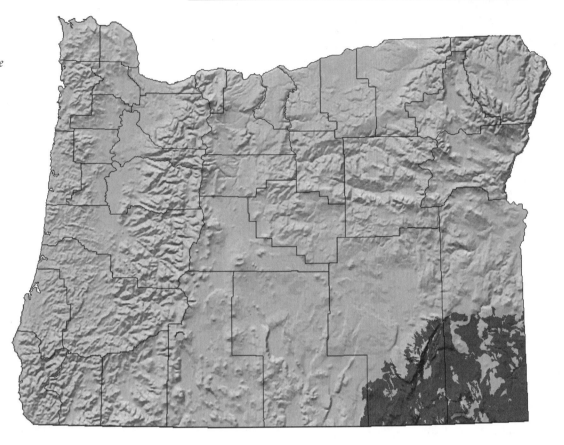

## Little Pocket Mouse
### (Perognathus longimembris)

*Order: Rodentia*
*Family: Heteromyidae*
*State Status: None*
*Federal Status: None*
*Global Rank: G5*
*State Rank: S4?*
*Length: 5 in (13 cm)*

**Global Range:** Found in the Great Basin and in arid regions of central and southern California, south into northern Mexico and east to central Arizona.

**Habitat:** This diminutive pocket mouse is found in a variety of desert shrub communities, from alkali flats grown to greasewood and saltbush to big sagebrush communities. In the southern part of its range, it also occurs in creosote bush and blackbrush communities.

**Reproduction:** This pocket mouse breeds from April to July. The average litter of 3 or 4 young (range 2-7) is born after a gestation period of 25 days. It may not attempt to breed in years with poor rainfall. There is probably only one litter per year.

**Food Habits:** Like most heteromyids, this pocket mouse is primarily granivorous. Seeds of grasses and other desert plants are gathered and usually carried back to the burrow in fur-lined external cheek pouches, where they are eaten in relative safety or cached. It has been known to collect flowers, and may eat insects or green vegetation during the breeding season.

**Ecology:** These are solitary, nocturnal desert rodents. They become torpid at low temperatures, and spend the winter underground. They can be

quite abundant locally (e.g., 400 per acre), but densities are know to fluctuate widely.

**Comments:** When foraging, pocket mice tend to stay under or near bushes, whereas coexisting kangaroo rats forage in the open.

*References:* Genoways and Brown 1993, Hall 1946, Kenagy and Bartholomew 1985, Zeveloff 1988.

**385**

# Dark Kangaroo Mouse
## (Microdipodops megacephalus)

*Order: Rodentia*
*Family: Heteromyidae*
*State Status: None*
*Federal Status: None*
*Global Rank: G5*
*State Rank: S4?*
*Length: 6 in (15 cm)*

**Global Range:** Of the two kangaroo mice, this species has the more northern distribution. It is found mainly in central and northern Nevada, but extends northward into southeastern Oregon. There are isolated subspecies in western Utah and eastern California.

**Habitat:** This rodent is found in sagebrush and shadscale communities with sandy or gravelly soil. It may occur around the margins of vegetated sand dunes.

**Reproduction:** The prolonged breeding season extends at least from April to September, but the majority of young are born in May and June. There are thought to be 2 litters per year, and the average litter size is 4 (range 2-7).

**Food Habits:** Kangaroo mice seem to be more omnivorous than other heteromyids. While seeds of grasses and desert shrubs are staples, they also consume insects, spiders, scorpions, and green vegetation. It is likely that more insects and vegetation are consumed in spring than during the rest of the year.

**Ecology:** These kangaroo mice spend winter in their burrows. They defend small territories around their burrows, which are usually located at the bases of bushes. They are able to hop on their hind legs, but are not as accomplished at bipedal locomotion as kangaroo rats.

**Comments:** Except at its base, the tail of the kangaroo mouse is thickened and may act as a fat storage organ. This species probably has a discontinuous distribution in the valleys of southeastern Oregon.

*References:* Genoways and Brown 1993, Hafner 1985, Hall 1946, O'Farrell and Blaustein 1974

## Ord's Kangaroo Rat
*(Dipodomys ordii)*

*Order: Rodentia*
*Family: Heteromyidae*
*State Status: None*
*Federal Status: None*
*Global Rank: G5*
*State Rank: S4*
*Length: 10 in (25 cm)*

**Global Range:** This is the most widely distributed kangaroo rat, ranging from southern Canada to central Mexico. The Cascade and Sierra Nevada mountains form the western limit of its range, and it occurs as far east as Oklahoma and Texas.

**Habitat:** In Oregon, this is a species of sagebrush and shadscale deserts, juniper woodlands, and grasslands.

**Reproduction:** There may be 1 or 2 litters of 3 (range 1-6) young per year. The gestation period is 29 days. In Oregon, breeding begins in the spring.

**Food Habits:** Ord's kangaroo rat has a diet that emphasizes seeds of grasses, forbs, and desert shrubs. It also eats some green vegetation and, rarely, insects.

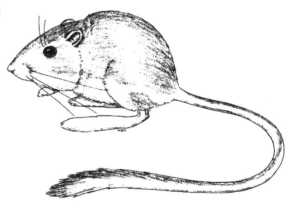

**Ecology:** Like other members of the genus, Ord's kangaroo rat is nocturnal and solitary. Despite the harsh climate experienced by northern populations, it does not hibernate. Like other kangaroo rats, it takes frequent dust baths, which prevent the accumulation of skin oils on the fur and may reduce numbers of ectoparasites. The home range size is about an acre, but home ranges can overlap. A density of 15.6 per hectare was observed in Texas.

**Comments:** Bailey (1936) comments that these are "the common kangaroo rats of eastern Oregon."

*References:* Garrison and Best 1990, Genoways and Brown 1993, Rogers and Hedlund 1980.

# Chisel-toothed Kangaroo Rat
## (Dipodomys microps)

Order: Rodentia
Family: Heteromyidae
State Status: None
Federal Status: None
Global Rank: G5
State Rank: S4?
Length: 10 in (25 cm)

**Global Range:** Found in the Intermountain West from southeastern Oregon and southwestern Idaho south through Nevada and Utah to the Mojave Desert in California and extreme northern Arizona.

**Habitat:** Chisel-toothed kangaroo rats are typically found in salt-desert shrub communities dominated by shadscale (*Atriplex confertifolia*) and greasewood, but occasionally they occur in sagebrush shrub. They frequent both sandy soils and alkaline playas of desert valleys. On the southern edge of their range, they are associated with blackbrush (*Coleogyne ramosissima*).

**Reproduction:** A single litter of 1 to 4 young (average litter size = 2.3) is born in the spring or early summer. The breeding season can last from March to July but varies with climate. Young of the year are not thought to breed until the following spring. The gestation period is about 31-32 days.

**Food Habits:** *Dipodomys microps* is unique among kangaroo rats in its ability to survive on a diet of shadscale leaves, which it harvests by climbing bushes. It shaves the salt-laden epithelium off with its lower incisors, and eats the protein-rich interior. It also eats seeds in other parts of its range.

**Ecology:** This can be the most abundant small mammal in some desert communities. The average life span is 1 year. The home range is just over 1 hectare. It is nocturnal and solitary except during the breeding season and is active throughout the year.

**Comments:** The chisel-toothed kangaroo rat often constructs an elaborate burrow system with many entrances, usually centered around the base of a shrub. In parts of its range, it may avoid competition with other kangaroo rats by eating leaves.

*References:* Csuti 1979, Genoways and Brown 1993, Hayssen 1991, Kenagy 1973.

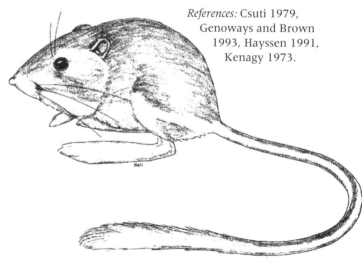

*Order: Rodentia*
*Family: Heteromyidae*
*State Status: None*
*Federal Status: None*
*Global Rank: G5*
*State Rank: S4?*
*Length: 12 in (30 cm)*

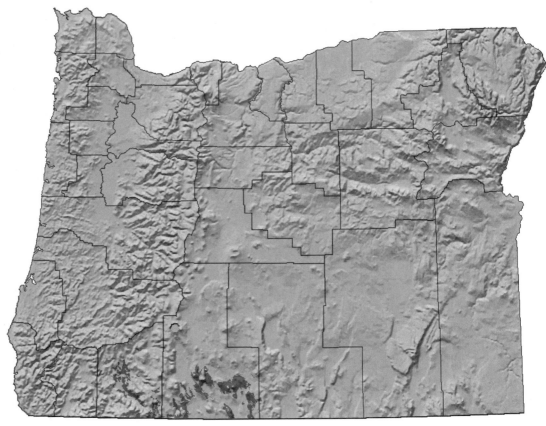

**Global Range:** A relatively small range in the foothills of the Coast Range, Siskiyou, and northern Sierra Nevada mountains of northern California and southern Oregon. An isolated subspecies once occupied Sutter Buttes in the middle of California's Central Valley.

**Habitat:** The California kangaroo rat prefers open, grassy areas which may be intermixed with oak woodland or patches of buckbrush (*Ceanothus*) or chaparral. It can burrow into hard packed clay or gravelly soils. The burrow entrance is usually at the base of a bush.

**Reproduction:** Breeding has been noted from February to September, but peaks in April and May. There may be 2 litters per year, but the spring litter averages 3 young, while fall litters average 2. The gestation period is 28-32 days.

**Food Habits:** During spring and summer, this kangaroo rat includes considerable green vegetation in its diet; at other times, seeds of grasses and bushes are more important. Seeds of buckbrush, manzanita, lupine, rabbitbrush, and bur-clover have been found in its cheek pouches.

**Ecology:** Kangaroo rats are nocturnal, and do not hibernate. They are solitary except during the breeding season. Unlike most kangaroo rats, the California kangaroo rat does not maintain large caches of food.

**Comments:** Oregon populations of this kangaroo rat formerly were assigned to *Dipodomys heermanni*, Heermann's kangaroo rat. Bailey (1936) described them as scarce.

*References:* Bailey 1936, Genoways and Brown 1993, Kelt 1988, Patton et al. 1976.

# American Beaver
## *(Castor canadensis)*

*Order: Rodentia*
*Family: Castoridae*
*State Status: None*
*Federal Status: None*
*Global Rank: G5*
*State Rank: S5*
*Length: 40 in (102 cm)*

**Global Range:** Occurs throughout most of the United States and Canada, except for the arctic northern fringe of the continent and extremely dry deserts without permanent streams. It is absent from southern California, but present in most of Arizona and New Mexico.

**Habitat:** The American beaver is an aquatic rodent that lives nearly anywhere there is permanent water. Although thought of as an inhabitant of mountains or coastal forests, it occurs throughout Oregon, including along natural and artificial watercourses and lakes in arid eastern Oregon.

**Reproduction:** American beavers have 1 litter per year. The breeding season begins early, in January or February, and litters of 3 to 5 (range 1-9) young are born after a 107-day gestation period. The age of first reproduction is 3 years (range 2-4).

**Food Habits:** Beavers are strictly herbivorous. They eat leaves, buds, branch ends, and bark of woody plants, as well as aquatic plants and herbaceous plants that grow near water. Willow and aspen are their favorite foods.

**Ecology:** Beavers are known for modifying their environment by building dams across streams to create ponds. They live in small colonies of related individuals, and usually only one female in the colony is reproductively active.

**Comments:** In Oregon, American beavers are considered furbearers. They were severely depleted by early fur trappers, but have recovered in most of the state. About 10,000 pelts a year are taken in Oregon, and over half a million pelts are taken each year from all of North America.

*References:* Bailey 1936, Jenkins and Busher 1979, Maser et al. 1981.

# Western Harvest Mouse
## *(Reithrodontomys megalotis)*

*Order: Rodentia*
*Family: Muridae*
*State Status: None*
*Federal Status: None*
*Global Rank: G5*
*State Rank: S4*
*Length: 6 in (15 cm)*

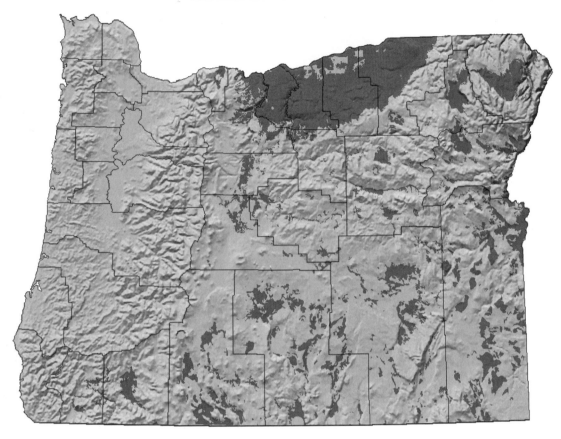

**Global Range:** Just enters Canada to the north, and is found as far south as southern Mexico. It extends from the Pacific Coast east to the Great Lakes, western Texas, and Oklahoma.

**Habitat:** This species lives in grassy areas, which can be along riverbanks, in pastures or meadows, at the edges of agriculture, or in early successional stages of forests. It has a scattered distribution in southeastern Oregon.

**Reproduction:** Western harvest mice breed throughout the year in western and central Oregon, but not in winter in the northeastern part of the state. The average litter size is 4 (range 1-6), and there can be several litters per year. Gestation takes about 23-24 days.

**Food Habits:** These mice eat the seeds of grasses, forbs, and shrubs. They also take some insects, leaves, grasses, and forb shoots.

**Ecology:** Harvest mice construct spherical nests out of plant fiber. Usually, nests are located on the ground under a bush, but sometimes are found several feet up in a shrub. Home ranges cover several thousand square meters and, in favorable habitat, these harvest micevary in abundance from 4 to 60 per hectare.

**Comments:** Western harvest mice are nocturnal. They are reported to be more active on moonless nights.

*References:* Bailey 1936, Kaufman et al. 1988, Webster and Jones 1982.

## Deer Mouse
### (Peromyscus maniculatus)

*Order: Rodentia*
*Family: Muridae*
*State Status: None*
*Federal Status: None*
*Global Rank: G5*
*State Rank: S5*
*Length: 7 in (18 cm)*

**Global Range:** Found coast to coast from central Canada to southern Mexico. It is absent only from the southeastern United States and from most of coastal Mexico.

**Habitat:** About the only habitat that deer mice don't use is open water. While they are found throughout Oregon in every habitat type, deer mice tend to be more abundant in early successional stages, such as forest clear-cuts.

**Reproduction:** Deer mice are capable of breeding throughout the year. However, populations along the Oregon coast may have fewer litters from October to February. The average litter of 5 or 6 young is born after a 23-day gestation period. The young may begin breeding at 2 months of age.

**Food Habits:** Deer mice are opportunistic omnivores. They eat seeds, green vegetation, insects, berries, and fungi.

**Ecology:** Because they are widespread and abundant, deer mice have been well studied. Their home range is 1 to 2 hectares, and population densities can exceed 60 per hectare. Active throughout the year, they are primarily nocturnal.

**Comments:** Deer mice are a staple prey item for almost everything carnivorous.

*References:* King 1968, Maser et al. 1981, Millar and Innes 1985.

# Canyon Mouse
*(Peromyscus crinitus)*

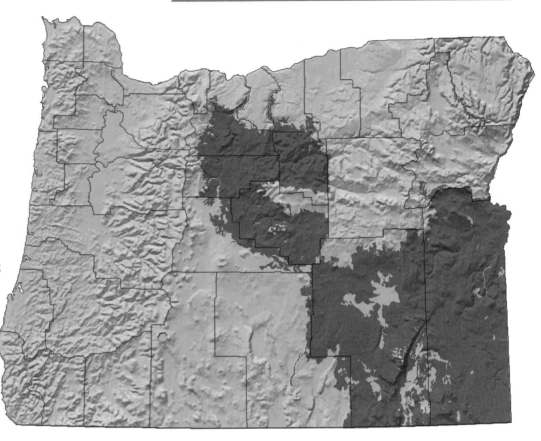

*Order: Rodentia*
*Family: Muridae*
*State Status: None*
*Federal Status: None*
*Global Rank: G5*
*State Rank: S4*
*Length: 7 in (18 cm)*

*References:* Bailey 1936, Johnson and Armstrong 1987, King 1968, Zeveloff 1988.

**Global Range:** Inhabits the arid Intermountain West. The northernmost record for this species is Maupin, Wasco County, Oregon. It occurs eastward to western Colorado and northwestern New Mexico, and southward to northern Mexico.

**Habitat:** Canyon mice are well named, since they are found exclusively in steep, rocky areas such as cliffs, ravines, talus slopes, or rimrock. The particular plant community around rocky areas does not influence their distribution. They have been found in rocky situations from sagebrush deserts to ponderosa pine or fir forests.

**Reproduction:** There may be many litters of 2 to 4 young each year. The gestation period is 23-28 days, and young are sexually mature 4 to 6 months after birth.

**Food Habits:** This mouse is an opportunistic omnivore. It eats seeds, green plant material, and insects.

**Ecology:** In areas where there are no other species of *Peromyscus*, canyon mice will use habitats other than those listed above, suggesting competitive exclusion. Their home range is about an acre, and they may reach densities of 15-20 per acre.

**Comments:** In some populations living in the lava beds of southeastern Oregon, most of the individuals have black fur, presumably an adaptation to match the color of their environment.

# Piñon Mouse
## *(Peromyscus truei)*

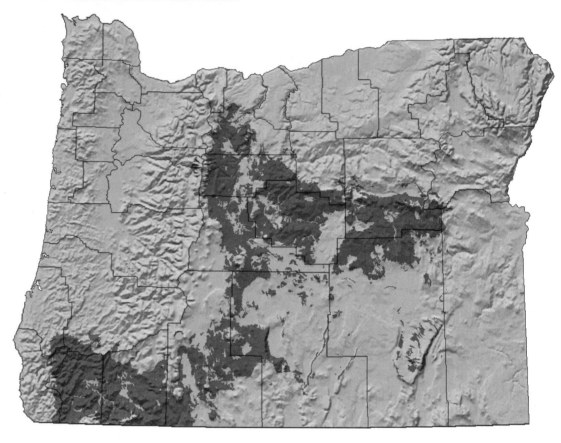

*Order: Rodentia*
*Family: Muridae*
*State Status: None*
*Federal Status: None*
*Global Rank: G5*
*State Rank: S4?*
*Length: 8 in (20 cm)*

**Global Range:** This species reaches its northern limit in central Oregon, extending south, at lower elevations, to northern Baja California, and east to Colorado and New Mexico. It also occurs throughout most of the interior of Mexico.

**Habitat:** Although it inhabits a narrower range of habitats than the deer mouse, the piñon mouse is found in a great variety of habitats, including brushy areas in redwood and Douglas-fir forests near the coast, oak woodland, and rocky outcrops in juniper woodlands (especially where large junipers dominate). In other parts of its range, it occupies sagebrush and shadscale desert shrub communities. It does not occur in higher-elevation forests.

**Reproduction:** Reproduction is concentrated in spring and summer but can occur throughout the year. The average litter size is 3 or 4 (range 1-6), and the gestation period is 25-27 days.

**Food Habits:** Like most *Peromyscus*, the piñon mouse has a diet of seeds, nuts, berries, fungi, insects, and green vegetation which varies by season and location. In California, piñon mice have been found to eat mostly insects in summer, but in other areas acorns are a primary food.

**Ecology:** This nocturnal mouse is active throughout the year. Only about 20% of the population survives for 1 year.

**Comments:** Two subspecies occur in Oregon. Near the coast, *P. t. gilberti* lives in brush and timber, while, east of the Cascade Mountains, *P. t. preblei* prefers rocky areas in canyons.

*References:* Carraway et al. 1993, Douglas 1969, Hoffmeister 1981, King 1968.

# Northern Grasshopper Mouse
### (Onychomys leucogaster)

*Order: Rodentia*
*Family: Muridae*
*State Status: None*
*Federal Status: None*
*Global Rank: G5*
*State Rank: S4?*
*Length: 6 in (15 cm)*

**Global Range:** Found in mid-continent from southern Canada to northern Mexico. It occurs as far west as eastern Oregon and northeastern California, and as far east as western Minnesota and the Gulf Coast of Texas.

**Habitat:** This is a species of arid grasslands and deserts that prefers areas with fine, sandy or silty soil. In eastern Oregon, it is found in sagebrush, juniper, grassland, and shadscale communities.

**Reproduction:** The northern grasshopper mouse breeds from February to October, but most litters are born in late spring and summer. The average litter size is about 4 (range 1-6), and the gestation period varies from 26-38 days. There can be multiple litters per year. Gestation is longer in lactating females.

**Food Habits:** The grasshopper mouse is unique among North American rodents in its largely carnivorous diet. While insects (grasshoppers, beetles, spiders,

scorpions, insect larvae) form the bulk of the diet, it also takes a variety of small vertebrates, especially other small mammals. Some green plant material and seeds are taken, especially in winter when insects are scarce.

**Ecology:** Grasshopper mice usually occur at low densities. Their home ranges are large (2.3 hectare), and they emit shrill whistles during breeding season. This may represent territorial defense. They kill small mammals larger than themselves, usually with a bite at the base of the skull.

**Comments:** These mice were never abundant in eastern Oregon, and may have suffered from loss of habitat to agriculture.

*References:* Bailey 1936, McCarty 1978, Riddle and Choate 1986.

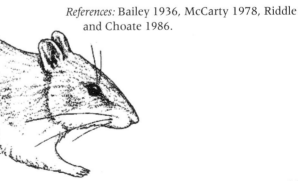

# Desert Woodrat
### (Neotoma lepida)

*Order: Rodentia*
*Family: Muridae*
*State Status: None*
*Federal Status: None*
*Global Rank: G5*
*State Rank: S4*
*Length: 12 in (30 cm)*

*References:* Bailey
1936, Hall 1946,
Hoffmeister 1986,
Mascarello 1978,
Matson 1976,
Zeveloff 1988.

**Global Range:** This species reaches the northern limit of its range in southeastern Oregon. It is found south to the tip of Baja California and east to western Colorado. Populations east of the Colorado River in western and central Arizona are considered a distinct species (*N. devia*) by most mammalogists.

**Habitat:** As its name implies, this is a species of arid desert valleys and rimrock. It occupies sagebrush, shadscale, and other arid shrub communities. It may occur in the sagebrush understory of open juniper woodlands. While often found in rocky areas, desert woodrats will also burrow under bushes in the open desert.

**Reproduction:** Desert woodrats have a long breeding season, from February to June or July, and have several litters of 2 or 3 (range 1-5) young per year. The gestation period is 30-36 days, and young are weaned in 3 to 5 weeks (varying with litter size). Young reach sexual maturity 2 to 3 months after birth.

**Food Habits:** The desert woodrat eats a variety of leaves, seeds, shoots, and fruit. The diet reflects the composition of the local plant community, and can include juniper berries, grass seeds, Mormon tea, and leaves of succulent plants.

**Ecology:** Desert woodrats often build an elaborate fortification of sticks and other vegetation around their burrow entrances. Populations can be dense (approaching 40 per hectare) in habitats with good food supplies. This species is nocturnal and active throughout the year.

**Comments:** Desert woodrats commonly occupy abandoned burrows of ground squirrels or kangaroo rats, adding their own defensive thickets of plant material around the entrance.

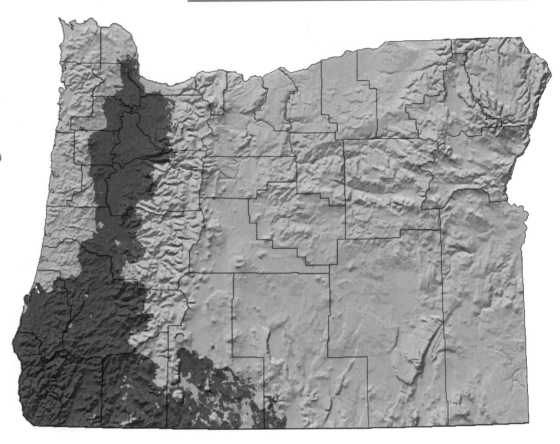

## Dusky-footed Woodrat
*(Neotoma fuscipes)*

*Order: Rodentia*
*Family: Muridae*
*State Status: None*
*Federal Status: None*
*Global Rank: G5*
*State Rank: S4*
*Length: 16 in (41 cm)*

**Global Range:** This is a West Coast species that reaches the northern limit of its range at the Columbia River in western Oregon. It extends south through most of non-desert California and into northern Baja California.

**Habitat:** This species is found in brushy undergrowth of forests, woodlands, and chaparral. It inhabits riparian areas, and typically builds large nests of twigs and leaves.

**Reproduction:** Usually, there is only one litter of 2 or 3 (range 1-4) young per year. The gestation period is about 33 days. The breeding season extends from February to July, but most young are born in March.

**Food Habits:** This species is vegetarian, feeding on forbs, leaves, seeds, nuts, acorns, fruits, fungi, and the inner bark of trees.

**Ecology:** Dusky-footed woodrats build large stick nests that are used as cover by a variety of other species. They are loosely colonial, and are most active at night.

**Comments:** In southern Oregon, dusky-footed woodrats are a primary food item of the Spotted Owl.

*References:* Carraway and Verts 1991b, Linsdale and Tevis 1951, Maser et al. 1981.

# *Bushy-tailed Woodrat*
## *(Neotoma cinerea)*

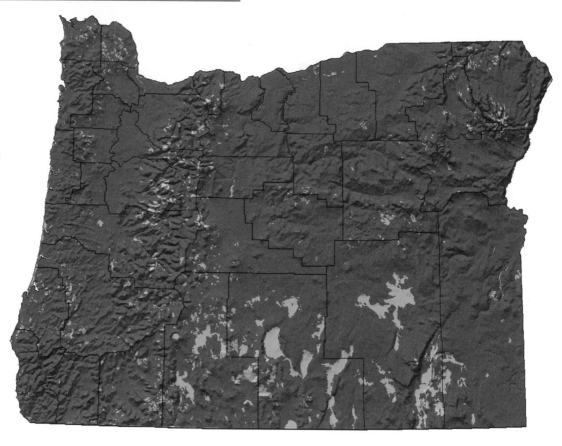

*Order: Rodentia*
*Family: Muridae*
*State Status: None*
*Federal Status: None*
*Global Rank: G5*
*State Rank: S5*
*Length: 14 in (36 cm)*

**Global Range:** Occurs as far north as the Yukon Territory and as far south as the White Mountains of eastern Arizona and ranges from the Pacific Coast east to western Nebraska and the Dakotas.

**Habitat:** These woodrats occupy a wide range of habitat types. In western Oregon, they are found in mature coniferous forests, but east of the crest of the Cascade Mountains they can be found along rimrock canyons, cliffs, and talus slopes. They also occupy deserted buildings and mineshafts, but are absent from open sagebrush-covered valleys.

**Reproduction:** The breeding season extends from January to September, but most young are born in the spring. The average litter size is 2 to 4 (range 1-6) young, and the gestation period is about 35 days. There may be 2 litters per year.

**Food Habits:** These herbivorous rodents eat forbs, twigs and shoots of trees, berries, fungi, flowers, and seeds.

**Ecology:** In forests, bushy-tailed woodrats build nests in hollow trees. They also occupy rimrock or rocky outcrops in both forests and arid habitats. Escherich (1981) suggested that, in rocky situations,

males form harems due to limited breeding habitat per female.

**Comments:** Along the southern Oregon coast, this species is displaced by the dusky-footed woodrat.

*References:* Bailey 1936, Escherich 1981, Maser et al. 1981.

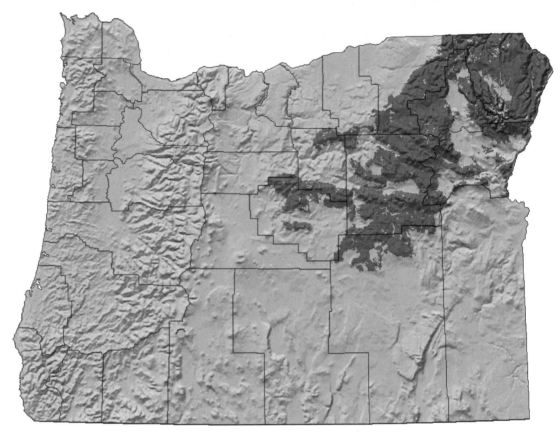

## Southern Red-backed Vole
### (Clethrionomys gapperi)

*Order: Rodentia*
*Family: Muridae*
*State Status: None*
*Federal Status: None*
*Global Rank: G5*
*State Rank: S4?*
*Length: 6 in (15 cm)*

**Global Range:** The name "southern" red-backed vole makes sense only when its range is compared with that of the northern red-backed vole (*C. rutilus*), which is found in Alaska and northern Canada. The southern species occurs in the southern half of Canada from coast to coast, is found in Washington and northeastern Oregon, and extends south along the Appalachian Mountains to northern Georgia and along the Rocky Mountains to the isolated mountain ranges of eastern Arizona and western New Mexico.

**Habitat:** In Oregon, this vole is found in open ponderosa pine or Engelmann spruce forests of the Blue Mountains, both along streams and in the drier portions of the forest. It prefers the cover of brushy areas or logs on the forest floor.

**Reproduction:** The southern red-backed vole breeds from March to October. After a 17-19 day gestation period, a litter of 4 to 6 young is born. Litter size varies with elevation and latitude.

**Food Habits:** This vole seems to be an opportunistic omnivore. During summer months, it feeds on hypogeous fungi, but turns to seeds during winter. It also will eat green parts of plants, berries, mosses, lichens, ferns, and some insects.

**Ecology:** Little is known about the populations occurring in northeastern Oregon. These voles are active year round, traveling under snow in winter. The home range varies from a fraction of a hectare to about 1.5 hectares. Mature females are territorial.

**Comments:** Unlike most voles, this species does not dig tunnels, but uses burrows of moles or other small mammals.

*References:* Bailey 1936, Maser and Maser 1988, Merritt 1981.

**399**

# Western Red-backed Vole
## (Clethrionomys californicus)

*Order: Rodentia*
*Family: Muridae*
*State Status: None*
*Federal Status: None*
*Global Rank: G5*
*State Rank: S4*
*Length: 6.5 in (17 cm)*

**Global Range:** This species reaches the northern limit of its range at the Columbia River in western Oregon. It is found in the Cascade and Coast Range mountains south into northern California.

**Habitat:** This species is a forest dweller. Along the coast and at lower elevations, it is found in Douglas-fir, western hemlock, and Sitka spruce forests. At higher elevations in the Cascade Mountains, it is found in mountain hemlock, true fir, Engelmann spruce, and whitebark pine habitats. Small clearings created by fallen trees and the presence of decaying downed logs are important microhabitat elements for this vole.

**Reproduction:** At lower elevations, this vole breeds throughout the year, but in the mountains there is no winter breeding. A litter of 2 to 4 (range 2-7) is born following a 17-21 day gestation period. There may be up to 3 or 4 litters per year.

**Food Habits:** In spring, the sporocarps of hypogeous fungi are the primary food of this vole. Some sporocarps may be stored for use into the summer. The vole will also eat seeds of coniferous trees, some green plant material, insects, and lichens.

**Ecology:** This species is found in moist microhabitats of mature coniferous forests. It does not inhabit burned or cut-over areas. It is active throughout the year. In the Cascade Mountains, it is nocturnal, but coastal populations are active day and night. The home range may be as large as 4.6 hectares for males, but is only 0.8 hectares for females.

**Comments:** Because it spreads the spores of hypogeous fungi in its scat, this species is thought to be beneficial to forest regeneration.

*References:* Alexander and Verts 1992, MacNab and Dirks 1941, Maser and Maser 1988, Maser et al. 1981, Mills 1995.

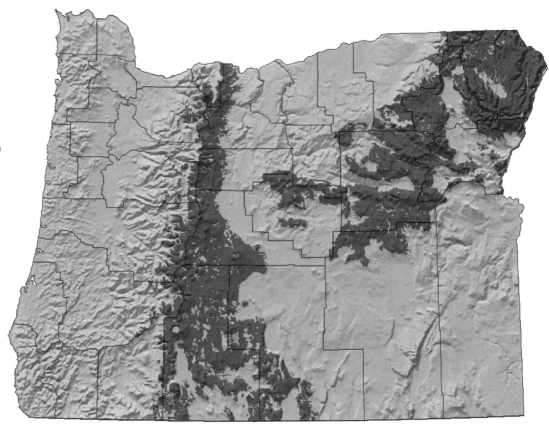

## Heather Vole
### *(Phenacomys intermedius)*

*Order: Rodentia*
*Family: Muridae*
*State Status: None*
*Federal Status: None*
*Global Rank: G5*
*State Rank: S4*
*Length: 5.5 in (14 cm)*

**Global Range:** Widespread throughout central and southern Canada, from nearly the Pacific Coast to the Atlantic Ocean. Occurs south in the Rocky, Cascade, Blue, and Sierra Nevada mountains to central California and northern New Mexico.

**Habitat:** This is a subalpine and alpine species found in the forest-meadow transition or in dry, open forests in clumps of heather or huckleberry. It is rarely found in dense forest. Trees in areas surrounding heather vole habitat include true firs, mountain hemlock, Engelmann spruce, and lodgepole pine. This vole often is found in alpine meadows above timberline.

**Reproduction:** In its high-elevation environment, the breeding season lasts from May to October. Most litters of 3 to 4 young (range 4-6) are born from mid-June to early September, after a gestation period of 19-24 days.

**Food Habits:** In summer, heather voles eat green vegetation, including leaves of willow and manzanita. Summer diets may also include seeds and berries. Heather voles cache food year round, and in winter their diet turns to twigs and bark of shrubs

and deciduous trees. Some hypogeous fungi may also be eaten.

**Ecology:** The heather vole is active throughout the year, and has a tendency to be most active at night. Densities of from 0.5 to 10 per acre have been reported.

**Comments:** While generally thought of as uncommon, the heather vole can undergo population explosions during which it is abundant. During one such event, a heather vole was found in the stomach of a trout!

*References:* Edwards 1955, Foster 1961, McAllister and Hoffmann 1988.

# White-footed Vole
*(Phenacomys albipes)*

*Order: Rodentia*
*Family: Muridae*
*State Status: Sensitive*
*Federal Status: SC*
*Global Rank: G4*
*State Rank: S3*
*Length: 6.5 in (17 cm)*

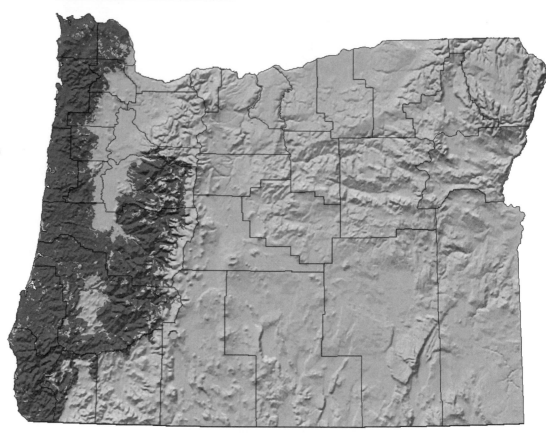

**Global Range:** Found in western Oregon and northern California, from the Columbia River south to the Cape Mendocino area. Generally, it is a species of the Coast Range, but it has been found on the west slope of the Cascade Mountains (e.g., 23 miles southeast of Vida, Lane County).

**Habitat:** The white-footed vole is most often found in riparian (especially alder) areas in coniferous forests, usually Douglas-fir, western hemlock, noble fir, or Sitka spruce. Small clearings supporting a growth of forbs may also be an important habitat component.

**Reproduction:** This vole may breed during any season. The usual litter size is 3 (range 2-4). The gestation period has not been reported.

**Food Habits:** White-footed voles are thought to eat a variety of green plant materials, especially leaves of shrubs and forbs, and also to feed on the roots of plants.

**Ecology:** Based on the combination of small eyes and large claws, this species is presumed to be a burrowing rodent. It is nocturnal and active throughout the year.

**Comments:** This species once competed with the spotted bat for the title of "rarest mammal in North America." It has been found more frequently in recent years, but still is uncommon. Some feel it may benefit from timber harvest practices that create early successional conditions. Some mammalogists place white-footed voles in the genus *Arborimus*.

*References:* Marshall 1992, Maser and Johnson 1967, Maser et al. 1981, Verts and Carraway 1995.

## Red Tree Vole
### *(Phenacomys longicaudus)*

*Order: Rodentia*
*Family: Muridae*
*State Status: None*
*Federal Status: None*
*Global Rank: G4*
*State Rank: S4*
*Length: 7 in (18 cm)*

**Global Range:** As currently classified, this species is found from the Columbia River south, in coastal mountains, to the San Francisco Bay. There is a break in distribution in the Siskiyou Mountains, and recently it has been suggested that the California populations are a different species, *Arborimus pomo* (Johnson and George 1991).

**Habitat:** This species may be found in dense, moist coniferous forests that contain sufficient numbers of large Douglas-fir trees. Other tree species in these forests include grand fir and western hemlock. It also can be found along the coast in Sitka spruce forests that include a Douglas-fir component.

**Reproduction:** The red tree vole breeds throughout the year. The litter size is usually 2 or 3 (range 1-4) young. The gestation period is 28 days, but is extended in lactating females.

**Food Habits:** This is one of the world's most specialized voles. Its diet consists almost exclusively of Douglas-fir needles. It sometimes eats needles of western hemlock, Sitka spruce, or true firs. Occasionally, the bark or interior of twigs is also eaten.

**Ecology:** Red tree voles really do live in trees, building nests of Douglas-fir needles about 50 feet (but as high as 150 feet) off the ground. They are nocturnal and are rarely seen. Females will defend an area of the tree around the nest. Their distribution is discontinuous and usually coincides with that of late-successional forests.

**Comments:** They are a major food item for the Spotted Owl. Some mammalogists place red tree voles in the genus *Arborimus*.

*References:* Corn and Bury 1986b, Hayes 1996, Johnson and George 1991, Maser et al. 1981.

**403**

# Montane Vole
## (Microtus montanus)

Order: Rodentia
Family: Muridae
State Status: None
Federal Status: None
Global Rank: G5
State Rank: S5
Length: 7 in (18 cm)

**Global Range:** Occurs from southern British Columbia south to the southern Sierra Nevada of California and the White Mountains of Arizona. It does not occur west of the Cascades, and is found as far east as the mountains of southeastern Colorado.

**Habitat:** These voles prefer moist mountain meadows. They occur in alpine meadows and grassy parts of mountain valleys, frequenting grassy areas near streams. They also can be abundant in marshes and hay meadows in otherwise arid valleys of eastern Oregon.

**Reproduction:** The montane vole breeds from about March to October. Litters of 4 to 7 (range 1-9) young are born after a 21-day gestation period. There can be 2 or 3 litters per year. Young of the year are capable of breeding.

**Food Habits:** Like most voles, the montane vole has a diet of grasses, forbs, sedges, and rushes. It will eat the roots of a wide variety of forbs. It can become a serious pest in hay meadows.

**Ecology:** The population numbers of this vole undergo cyclic fluctuations at intervals of 3 to 4 years. They are prolific breeders and can reach densities of 500 or more per hectare. They are an important part of the diet of a wide variety of carnivores.

**Comments:** In arid eastern Oregon, these voles have a discontinuous distribution.

*References:* Goertz 1964, Pinter 1986, Tamarin 1985, Zeveloff 1988.

*Order: Rodentia*
*Family: Muridae*
*State Status: None*
*Federal Status: None*
*Global Rank: G4*
*State Rank: S4*
*Length: 7 in (18 cm)*

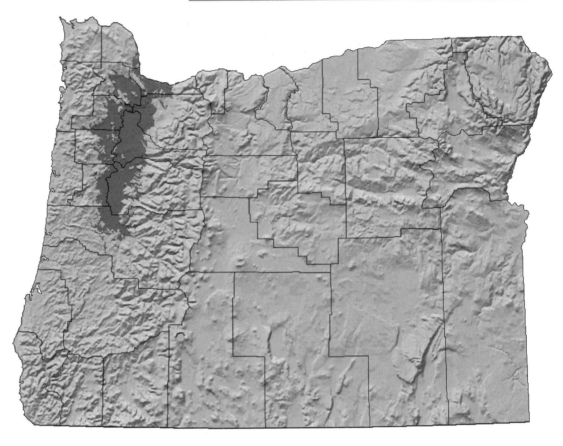

**Global Range:** Generally, this species is restricted to the lower elevations of the Willamette Valley of Oregon. However, it is known from two localities north of the Columbia River in Clark County, Washington (south of Ridgefield and 7 miles northeast of Vancouver).

**Habitat:** This vole is closely associated with pastures and other grassy areas. It has invaded agricultural areas and spread along railroad and highway right-of-ways to areas that were previously forested but which have been converted to agriculture. Originally, it was found in fire-maintained prairie that covered much of the Willamette Valley.

**Reproduction:** Most data come from captive animals. Litters of 5 (range 2-8) young are born after a 21-day gestation period. The species may breed throughout the year. These voles are capable of breeding by the time they are 1 month old.

**Food Habits:** Gray-tailed voles eat grasses and grass seeds, legumes, forbs, clover, false dandelions, and wild onions. Their original diet probably consisted of similar native species.

**Ecology:** This species can become abundant enough to require control measures (reaching densities of up to 141 voles per hectare). Gray-tailed voles construct extensive runway systems and also use tunnels of the Camas pocket gopher. They cache food and can survive in flooded fields by virtue of air trapped in their nest chambers.

**Comments:** The gray-tailed vole may have been endemic to Oregon, since the Washington populations could have been accidentally introduced by humans. The gray-tailed vole was formerly considered a subspecies of the montane vole (*Microtus montanus*).

*References:* Modi 1986, Verts and Carraway 1987b.

# California Vole
## (Microtus californicus)

*Order: Rodentia*
*Family: Muridae*
*State Status: None*
*Federal Status: None*
*Global Rank: G5*
*State Rank: S4*
*Length: 7 in (18 cm)*

**Global Range:** Found from the interior valleys of western Oregon south throughout most of California and into northern Baja California. It has a discontinuous distribution in arid transmontane California.

**Habitat:** This species occurs in dry meadows and pastures, grassy slopes, marshes, and the edges of rivers and streams. It can also invade agricultural areas and is sometimes found in the grassy understory of open oak woodlands.

**Reproduction:** In its low-elevation habitats, it breeds throughout the year. The litter size averages 4 (range 1-9) young, and the gestation period is about 21 days. Two to 5 litters may be produced each year.

**Food Habits:** California voles feed on green grass in winter and spring, and on dry grass and seeds in summer and fall. They also will consume roots, bulbs, and bark.

**Ecology:** This species is active day and night throughout the year. It builds a complex series of runways through the grass, eating as it goes. Its runway systems are used by a variety of other species of small mammals. Voles are staple prey items in the diet of a wide variety of carnivores.

**Comments:** California voles display 3- or 4-year cycles of abundance. During the peaks, densities of 160 voles per acre may be reached. Researchers have attempted to explain these cycles, attributing them to resource depletion, predation, genetics, or a combination of factors.

*References:* Batzli and Pitelka 1970, 1971, Tamarin 1985.

**406**

## Townsend's Vole
### *(Microtus townsendii)*

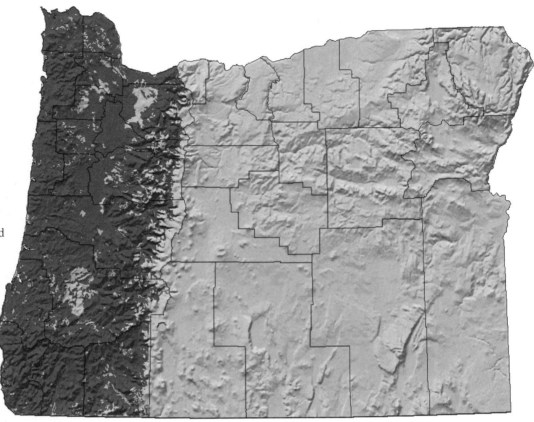

*Order: Rodentia*

*Family: Muridae*

*State Status: None*

*Federal Status: None*

*Global Rank: G5*

*State Rank: S4*

*Length: 7 in (18 cm)*

*References:* Cornely and Verts 1988, Goertz 1964, Maser et al. 1981.

**Global Range:** Found from Vancouver Island, three other small offshore islands, and the mainland of southern British Columbia south to the area of Cape Mendocino, California. It occurs west of the crest of the Cascade Mountains.

**Habitat:** Townsend's voles are found primarily in marshy areas, such as wet meadows, willow/sedge marshes, wet pastures, and riparian thickets. They freely enter the water, but also occur in boggy areas some distance from water. They avoid the forest proper.

**Reproduction:** At lower elevations, this species breeds throughout the year. Litters of 4 to 6 (range 2-10) young are born after a gestation period of about 23 days.

**Food Habits:** Townsend's voles feed on a variety of green vegetation, including rushes, tules, clovers, grasses, and the bark of trees. One report from Washington indicates they store the bulbs of the mint, *Mentha canadensis*.

**Ecology:** These voles build extensive tunnel and runway systems. During winter, when their meadow or marsh habitat is flooded, they build grass nests on high ground. They are active day and night.

**Comments:** One study in Oregon found that, during some winter months, Townsend's voles formed a major portion of the diet of Great Blue Herons.

# Long-tailed Vole
## *(Microtus longicaudus)*

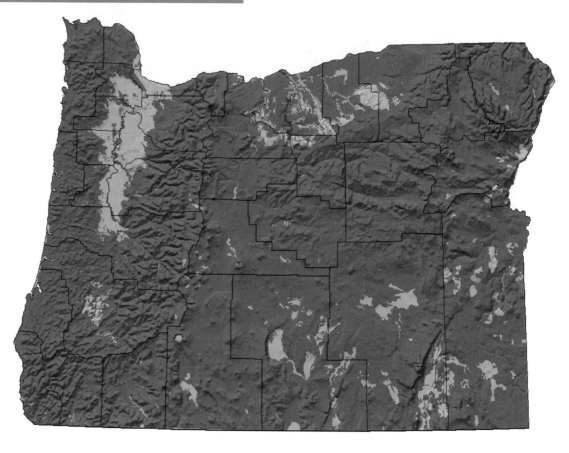

*Order: Rodentia*
*Family: Muridae*
*State Status: None*
*Federal Status: None*
*Global Rank: G5*
*State Rank: S5*
*Length: 8 in (20 cm)*

**Global Range:** Occurs from Alaska and the Yukon Territory south to southern California, Arizona, and New Mexico and as far east as western South Dakota.

**Habitat:** These voles select moist riparian vegetation and thickets within forests. In eastern Oregon, they descend along stream courses into desert valleys, but generally are animals of the mountains. At Malheur National Wildlife Refuge, they are restricted to marshes. They also can be found at the forest-meadow transition. They have been reported to take up residence in clear-cuts in forests.

**Reproduction:** The breeding season lasts from May to October. Litters usually consist of 4 or 5 (range 2-7) young. There may be 2 litters during the breeding season, but few females survive to their second year.

**Food Habits:** They eat seeds and berries as well as green plant material from grasses and forbs. In winter, they eat roots and bark.

**Ecology:** This vole does not build extensive runway systems. In habitats that receive heavy snow, it builds a nest above ground but under snow cover in winter, but lives in burrows at other times of the

year. Population densities fluctuate widely from only a few voles to densities of over 100 per acre.

**Comments:** Along the northern Oregon coast, long-tailed voles are nocturnal, but they may be active at any time of day along the southern Oregon coast.

*References:* Colvin and Colvin 1970, Maser et al. 1981, Smolen and Keller 1987.

## Creeping Vole
### (Microtus oregoni)

*Order: Rodentia*
*Family: Muridae*
*State Status: None*
*Federal Status: None*
*Global Rank: G5*
*State Rank: S4*
*Length: 5.5 in (14 cm)*

**Global Range:** Found from the Cascade Mountains to the coast from southern British Columbia south to the Mendocino Coast of northern California.

**Habitat:** The creeping vole inhabits brushy areas on the floor of moist coniferous forests. It is more abundant in clear-cuts than in mature forests. If the grass is sufficiently thick to provide good cover, this vole may be found along the Oregon coast in grassy meadows near the forest edge.

**Reproduction:** Creeping voles usually breed from February to September, but may breed throughout the year at low elevations near the coast. Four or 5 (range 1-8) young are born after a 24-day gestation period. There can be 4 or 5 litters per year, so this species has a high reproductive potential.

**Food Habits:** Creeping voles eat a variety of green vegetation (leaves of forbs and grasses), berries, roots, bulbs, and hypogeous fungi.

**Ecology:** Creeping voles are adapted for burrowing in soft soil. They construct extensive tunnel systems. They are more active at night, and have home

ranges of a fraction of a hectare. A study in Mt. Hood National Forest found that densities varied from 0.8 to 6 individuals per acre.

**Comments:** These voles consume few seeds of coniferous trees and are not considered a forest pest.

*References:* Carraway and Verts 1985, Gashwiler 1972, Maser et al. 1981.

# Water Vole
## (Microtus richardsoni)

*Order: Rodentia*
*Family: Muridae*
*State Status: None*
*Federal Status: None*
*Global Rank: G5*
*State Rank: S4*
*Length: 9 in (23 cm)*

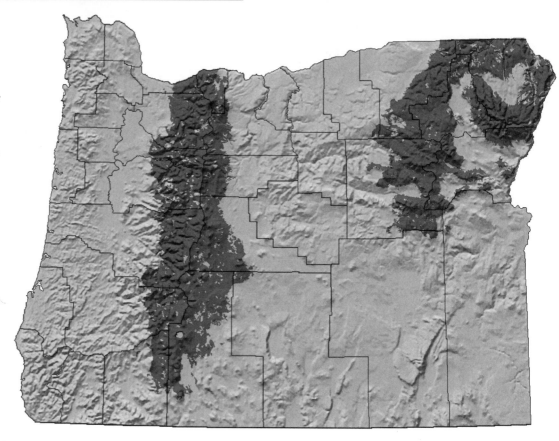

**Global Range:** Found in the Cascade Mountains from southern British Columbia to southern Oregon and in the Rocky and Blue mountains from British Columbia south to Utah.

**Habitat:** Water voles are residents of clear, cold, high-mountain streams and, occasionally, marshes or lakes. They prefer streams with dense vegetation along the banks. The alpine and subalpine meadows they use are found in a matrix of various forest types and, occasionally, they are found along streams in closed-canopy forests.

**Reproduction:** Their breeding season begins when the snow melts and green vegetation appears, often in late May or June. They breed through August or September, and may have 2 litters of 5 or 6 (range 2-9) young per year. The gestation period is about 22 days.

**Food Habits:** Forbs are an important food during summer, but water voles also eat grasses, sedges, bulbs, willow buds, and some conifer seeds. In some areas, they eat roots and buds of perennial plants. They do not store food for the winter.

**Ecology:** Water voles are unusual members of their genus. They are large (nearly a foot long, including

the tail), occur at low densities, and are closely associated with water. They dig burrows in stream banks; the burrow entrance may be under water. In winter, they disperse under snow cover, and build nests on the ground surface. One survey found densities from 0.2 to 12.2 per hectare in streamside habitat. They are most active at night.

**Comments:** These voles have a very discontinuous distribution.

*References:* Ludwig 1984, 1988, Maser and Hooven 1970.

## Sagebrush Vole
### (*Lemmiscus curtatus*)

*Order: Rodentia*
*Family: Muridae*
*State Status: None*
*Federal Status: None*
*Global Rank: G5*
*State Rank: S4*
*Length: 5 in (13 cm)*

**Global Range:** Found throughout the Great Basin, south to southwestern Arizona, and east into the northern Great Plains. It occurs from southern Canada to eastern California, and from eastern Oregon to western North Dakota.

**Habitat:** Typically, this species is found in arid shrub habitats dominated by sagebrush or rabbitbrush with a grass understory. It occurs in grasslands lacking a significant shrub component if the grass is thick enough to provide good cover.

**Reproduction:** In the more moderate parts of its range, it is capable of breeding throughout the year. The litter size is usually 4 to 6 (range 2-13) young, and the gestation period is about 25 days. There may be up to 3 litters per year. Young are sexually mature in 2 months.

**Food Habits:** This vole feeds almost exclusively on green vegetation, including grasses, leaves, flowers and stalks of herbaceous plants, and green seed pods of grasses and dicots. They do not eat ripe grass seeds, but in late fall and early winter will eat sagebrush flowers.

**Ecology:** The sagebrush vole is active day and night throughout the year. It forages along runways and

builds tunnels under the snow. Population densities fluctuate widely; in southeastern Idaho, densities have been found to range from 4 to 16 per hectare.

**Comments:** This species is known to harbor fleas that carry bubonic plague. Some mammalogists place this species in the genus *Lagurus*.

*References:* Carroll and Genoways 1980, Mullican and Keller 1986, 1987.

411

# *Muskrat*
## *(Ondatra zibethicus)*

*Order: Rodentia*
*Family: Muridae*
*State Status: None*
*Federal Status: None*
*Global Rank: G5*
*State Rank: S5*
*Length: 20 in (51 cm)*

**Global Range:** The native range of the muskrat includes most of North America, from Alaska to northernmost Mexico. However, muskrats have been widely introduced around the world, and there are established populations in much of western Europe and central Asia.

**Habitat:** The muskrat is an aquatic mammal. It occurs in or near water, including ponds, lakes, streams, rivers, marshes, and irrigation ditches. It is absent from much of arid eastern Oregon, except near permanent water.

**Reproduction:** The reproductive biology of muskrats varies with climate. In warmer areas, they breed throughout the year, but in Oregon they breed from May to October. A litter of 6 to 8 young (range 1-12) is born after a 28-30 day gestation period. In the southern United States, there can be 6 litters a year, but 2 or 3 litters per year is more typical of muskrats living in the Pacific Northwest.

**Food Habits:** Muskrats feed on tules, rushes, cattails, bulbs, grasses, and ferns. They also eat crawfish, fish, turtles, snails, and salamanders. Along the coast they eat mussels.

**Ecology:** In eastern Oregon, muskrats build large nests of vegetation and mud. The nest rests on the bottom of the lake or stream, but the top of the nest may be 4 or 5 feet above water level. In western Oregon, they more typically dig burrows in banks, tunneling upward so that the nest chambers are above water level. Population densities can be as high as 3 to 10 per hectare of surface water.

**Comments:** The muskrat is a furbearer. In the early 1970s, 25,000 to 35,000 muskrats were taken per year in Oregon.

*References:* Bailey 1936, Maser et al. 1981, Willner et al. 1980.

412

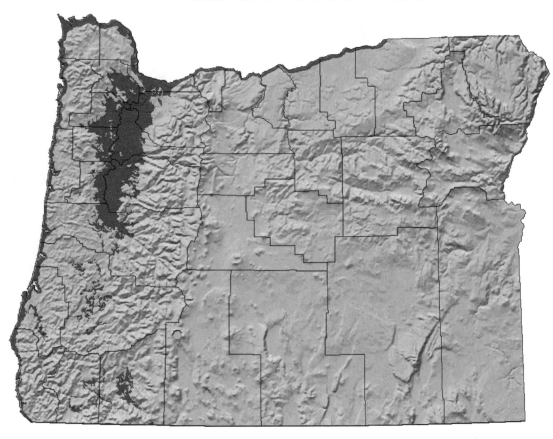

# Norway Rat
*(Rattus norvegicus)*

*Order: Rodentia*
*Family: Muridae*
*State Status: None*
*Federal Status: None*
*Global Rank: G5*
*State Rank: SE*
*Length: 15 in (38 cm)*

**Global Range:** Native to Asia, where it occupied streambanks. It appeared in Europe in the 16th century and was first noticed in North America in the 1770s. It has spread to coastal cities and towns throughout the world.

**Habitat:** This species is closely associated with human activity. It is found in sewers, garbage dumps, barns, food storage or processing facilities, buildings, and suburban gardens. Rarely, it establishes populations in agricultural areas, pastures, and riparian areas away from human development.

**Reproduction:** Norway rats breed throughout the year, but reproduction is concentrated in warmer seasons. A litter of about 9 (range 2-18) young is born after a gestation period of about 3 weeks. Young are capable of breeding at 3 months of age.

**Food Habits:** Norway rats are opportunistic omnivores. Their range of food items is remarkable, including garbage, carrion, grain, fruit, insects, snails, seeds, small mammals, birds, and eggs. They are capable of killing chickens in poultry houses and have been reported killing young of larger mammals such as pigs and sheep and attacking the feet of livestock.

**Ecology:** The Norway rat is colonial, with a dominant male defending a territory around a burrow containing several females. Juvenile rats are driven from the colony. This rat becomes nocturnal around human activity, but can be active throughout the day. It is semiaquatic and readily enters water.

**Comments:** Because it is a vector for many diseases (plague, typhus, tularemia) and ectoparasites, and because of the direct economic damage it causes by consuming and contaminating food, the Norway rat is considered the worst mammalian pest species.

*References:* Fitzgerald et al.1994, Maser et al. 1981, Strahan 1983.

413

# House Mouse
## *(Mus musculus)*

*Order: Rodentia*
*Family: Muridae*
*State Status: None*
*Federal Status: None*
*Global Rank: G5*
*State Rank: SE*
*Length: 7 in (18 cm)*

**Global Range:** Like the Norway rat, the house mouse was probably native to central Asia, but it now has a worldwide distribution. It was widespread in the towns of eastern North America by the 18th century.

**Habitat:** The house mouse lives wherever there are towns, buildings, farms, or other human developments. It is more successful than the Norway rat at becoming established in the wild, and is known to occupy riparian areas, croplands, grain fields, abandoned pastures, highway right-of-ways, and fence rows.

**Reproduction:** In temperate climates, house mice are capable of reproducing throughout the year. Litters of 4 or 5 (range 3-16) young are born after a 3-week gestation period. Males are polygamous, and females within a colony may use a communal nest. Female mice begin breeding at 2 months of age.

**Food Habits:** House mice are opportunistic omnivores; however, due to their small size, they do not prey on small vertebrates. They eat grain, insects, roots, any type of available food in dwellings, and a variety of unlikely but apparently nourishing household items (e.g., glue, leather, soap).

**Ecology:** Males defend a territory around nesting females. The home range is a few hundred square meters. Hoffmeister (1986) states that house mice are dependent on sources of drinking water, and thus are unable to exploit arid environments away from human habitation. They are primarily nocturnal and are active throughout the year.

**Comments:** Although house mice carry fewer diseases than Norway rats, they consume or foul vast quantities of foodstuff. They also do considerable damage by gnawing through walls or storage containers. They can start fires when gnawing through wiring.

*References:* Fitzgerald et al. 1994, Lidicker 1966, Maser et al. 1981, Strahan 1983.

414

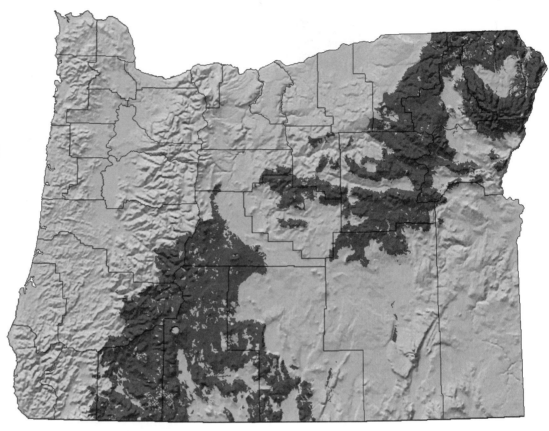

*Order: Rodentia*
*Family: Dipodidae*
*State Status: None*
*Federal Status: None*
*Global Rank: G5*
*State Rank: S4*
*Length: 9.5 in (24 cm)*

**Global Range:** Found from the Yukon Territory south to the Sierra Nevada of California and the mountains of eastern Arizona and western New Mexico. It occurs throughout the Rocky Mountains and is present in the northern Great Plains.

**Habitat:** The western jumping mouse is found in moist mountain meadows and in riparian thickets in forests. In the scattered desert mountain ranges of eastern Oregon, it can be found under willows growing along creeks.

**Reproduction:** Depending on climate, the breeding season begins in May or June. After a short 18-day gestation period, a litter of 5 (range 2-7) young is born. In areas with extended growing seasons, there may be 2 litters per year.

**Food Habits:** The diet varies with the season. In spring, western jumping mice add insects to their diet of green vegetation, seeds, and fungi. Later in the year, berries are eaten. Considerable body fat is accumulated prior to hibernation.

**Ecology:** These rodents hibernate during winter, although they have brief periods of activity throughout the winter. They are solitary and nocturnal. Nests are spherical, made of grass, and are about 6 inches in diameter. The nest may be underground or suspended in tall grass. Densities of this species have been reported between 8 and 32 mice per hectare.

**Comments:** At higher elevations, these mice may be active for less than 3 months a year. There is high winter mortality, and only a third of the juveniles may survive. Some mammalogists place jumping mice in their own family, *Zapodidae*.

*References:* Cranford 1983, Falk and Millar 1987, Jones et al. 1978, Zeveloff 1988.

415

# Pacific Jumping Mouse
## *(Zapus trinotatus)*

*Order: Rodentia*
*Family: Dipodidae*
*State Status: None*
*Federal Status: None*
*Global Rank: G5*
*State Rank: S4*
*Length: 9 in (23 cm)*

**Global Range:** Found from southern British Columbia south through western Washington and Oregon to San Francisco Bay. It is restricted to coastal mountains from central Oregon south, but is also found in the Cascade Range to the north.

**Habitat:** This jumping mouse is found in wet, grassy areas, near marsh edges, and in thickets along streams. It occurs in salal thickets in forest undergrowth and in moist areas grown to skunkcabbage.

**Reproduction:** The breeding season begins in June. Most litters are born in July, following a gestation period of 18-23 days. There is a single litter of 4 to 6 young (range 3-8) per year.

**Food Habits:** The diet of these jumping mice consists primarily of seeds of herbaceous plants and grasses. They also eat some berries, mosses, and fungi. Insects are a small part of their diet. Body fat is accumulated in preparation for hibernation.

**Ecology:** These mice hibernate from around September to May. They are primarily nocturnal, but may be active at any time of day or night. Like the western jumping mouse, they build spherical nests of grass or moss. They are capable of jumping nearly 2 meters.

**Comments:** These mice nearly double their body weight in preparation for winter hibernation.

*References:* Gannon 1988, Jones et al. 1978, Maser et al. 1981.

## Common Porcupine
### *(Erethizon dorsatum)*

*Order: Rodentia*
*Family: Erethizontidae*
*State Status: None*
*Federal Status: None*
*Global Rank: G5*
*State Rank: S5*
*Length: 28 in (71 cm)*

**Global Range:** Found throughout the United States and Canada, except for the southeastern United States and southern California. Its range reaches its southern limit in northern Mexico.

**Habitat:** Common porcupines are associated with trees. Therefore, they are restricted to forests or to riparian vegetation in non-forested areas. They use both young and older forest communities, including juniper woodland, and occur in willow and other woody vegetation around marshes. In some areas, they invade agricultural cropland.

**Reproduction:** One young is born each year after a gestation period of 205-217 days. Common porcupines mate in fall, and are sexually mature in about a year and a half.

**Food Habits:** Their diet varies seasonally. During winter, they consume the inner bark of trees and the needles of conifers. In spring and summer, they feed on developing buds, roots, grasses, leaves, berries, and fruits. They also eat acorns and some seeds.

**Ecology:** Common porcupines are active throughout the year. They wander about a home range of around 15 hectares in winter. They are solitary and active mostly at night. The fisher (*Martes pennanti*) is the main carnivore capable of avoiding the common porcupines' quills and taking the species as prey.

**Comments:** Allegedly, common porcupines damage commercial timber. However, this has not been firmly established.

*References:* Maser et al. 1981, Roze 1989, Woods 1973.

**417**

# Nutria or Coypu
## (Myocastor coypus)

Order: Rodentia
Family: Myocastoridae
State Status: None
Federal Status: None
Global Rank: G5
State Rank: SE
Length: 35 in (89 cm)

References:
Larrison 1943,
Maser et al.
1981, Walker
1964, Woods et
al. 1992.

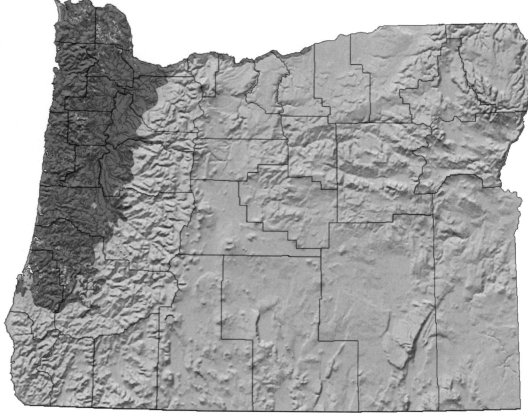

**Global Range:** Native to South America, from Bolivia south to Patagonia. Widely introduced in North America, Europe, Asia, and Africa.

**Habitat:** Nutria are aquatic mammals that are found along streams and the shores of lakes and marshes. In agricultural areas, they may occupy irrigation canals, causing some damage. They burrow into the banks of streams and find concealment in emergent vegetation growing along banks and shorelines. Rarely, burrows may be over half a mile from water.

**Reproduction:** Nutria are capable of breeding throughout the year. Nests may be in burrows from 3 to 18 feet long, usually with multiple entrances. Sometimes beaver or muskrat dens are appropriated. Litters of 3 to 6 (range 1-11) young are born after a gestation period of about 135 days. The precocial young swim shortly after birth, but remain with their mother for about 6 weeks. Nutria first breed at about 8 months of age.

**Food Habits:** Nutria eat a variety of aquatic plants (cattails, rushes, duckweed, sedges), feeding on their leaves, stems, and roots. They may also eat grass growing near water. In agricultural areas, they eat alfalfa, clover, corn, and ryegrass. In winter, they also eat tree bark, and may damage orchards and tree plantations. Occasionally, they may eat molluscs.

**Ecology:** Nutria reach densities of up to 10 per acre. Their home range may exceed 10 acres, although the male home range is about twice that of females. They may disperse long distances (over 30 miles). Nutria are mostly nocturnal and remain active throughout the year, although there may be considerable mortality during extremely cold weather.

**Comments:** Nutria are thought to have been first introduced to Oregon in 1937, with further introductions through the 1940s. While not officially a furbearer in Oregon, they have been commercially trapped. Many feral populations have resulted from the release of captive populations being raised for fur. Maser et al. (1981) suggest *coypu* is a better common name for this species, since *nutria* means "otter" in Spanish.

# Coyote
*(Canis latrans)*

*Order: Carnivora*
*Family: Canidae*
*State Status: None*
*Federal Status: None*
*Global Rank: G5*
*State Rank: S5*
*Length: 44 in (111 cm)*

**Global Range:** Occurs throughout most of North America, coast to coast, from Alaska south to Costa Rica.

**Habitat:** Coyotes prefer open habitats such as deserts and grasslands. Once they were uncommon west of the Cascade Mountains, for they do not frequent dense timber. With increased clear-cutting, they have spread into the forest matrix, using logging roads to travel between patches of open habitat. Several dens have been reported at the edge of the Portland metropolitan area.

**Reproduction:** It has 1 litter of 5 to 7 pups (range 4-19) per year. The gestation period is 58-65 days. Coyotes are sexually mature at 1 year of age.

**Food Habits:** Coyotes have a remarkably varied diet, but most of their calories come from carrion (especially in the winter) and small mammals. In addition, they eat birds, reptiles, amphibians, fish, insects, berries, and other fruit.

**Ecology:** Coyotes are quite mobile. In one study, individuals moved an average of 4 kilometers per day. Trips of over 160 kilometers have been documented. As a result, they may move through many different habitat types, selecting open areas that suit their hunting activities.

**Comments:** In some areas, coyotes are thought to be predators of livestock. However, it is not always clear whether they are actually making their own kills or feeding on carrion. Coyotes respond to control measures with increased fecundity, suggesting that other ways of reducing coyote predation (e.g., shepherds, sheepdogs) might be more cost-effective.

*References:* Bekoff 1977, Knowlton 1972, Maser et al. 1981.

419

# Red Fox
## *(Vulpes vulpes)*

*Order: Carnivora*
*Family: Canidae*
*State Status: None*
*Federal Status: None*
*Global Rank: G5*
*State Rank: S4?*
*Length: 38 in (97 cm)*

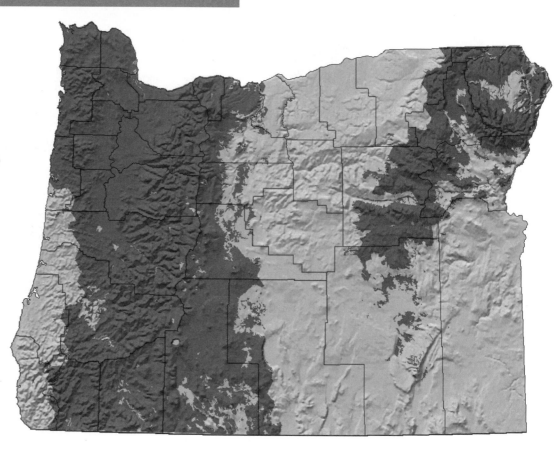

**Global Range:** The red fox is holarctic, being found throughout the cooler parts of Europe, Asia, and North America. It is also found in North Africa and is introduced in most of Australia. It is absent from hot, arid regions. The southernmost record in North America is just north of the Mexican border.

**Habitat:** This fox prefers open habitats such as meadows and grasslands that are interspersed with patches of brush or timber for cover. It is found from sea level to high elevations and frequents open woodlands and alpine or subalpine meadows.

**Reproduction:** The red fox breeds in January and February. It has 1 litter of 5 or 6 (range 1-10) young per year. Young are born in March or April, following a 49-56 day gestation period.

**Food Habits:** Like most carnivores, red foxes have a varied diet, depending on which foods are available. They eat small mammals up to the size of rabbits, birds and their eggs, reptiles, amphibians, fruit, and some insects.

**Ecology:** Red foxes are territorial and mate for life. Their territory size varies with habitat quality, but usually is several square miles. They often establish fixed foraging routes, and are active mainly from dusk to dawn.

**Comments:** This species is a furbearer in Oregon, but less than a thousand pelts are taken each year.

*References:* Ables 1975, Larivière and Pasitschniak-Arts 1996, Lloyd 1980, Maser et al. 1981.

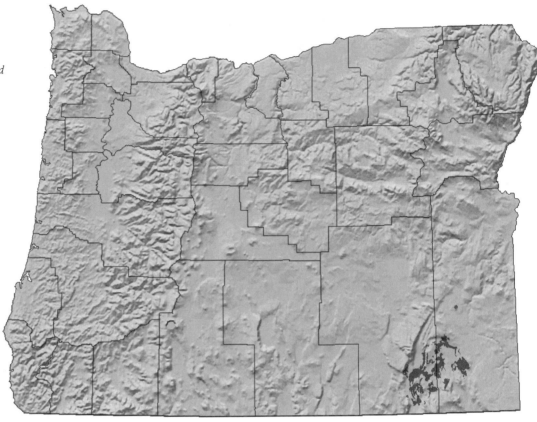

## Kit or Swift Fox
*(Vulpes macrotis)*

*Order: Carnivora*
*Family: Canidae*
*State Status: Threatened*
*Federal Status: None*
*Global Rank: G4*
*State Rank: S?*
*Length: 30 in (76 cm)*

**Global Range:** A desert species that reaches its northern limit in southeastern Oregon. It is found south through California to Baja California, and from Utah south and east to central Mexico. It occurs no farther east than west Texas.

**Habitat:** In Oregon, the kit fox is found in arid desert valleys dominated by halophytic plants such as shadscale and greasewood. These plant communities may be surrounded by or contain patches of sagebrush. It prefers areas with easily worked soil.

**Reproduction:** The breeding season begins around January. Most litters are born from February to May after a 49-56 day gestation period. Females have one litter of 4 to 5 pups per year.

**Food Habits:** Kit foxes are opportunistic carnivores, feeding primarily on the most abundant rodent or rabbit species in the area. They also take a few birds, reptiles, and insects. Because they share a preference for sandy soils with kangaroo rats, these rodents are the kit fox's chief prey in some areas.

**Ecology:** The kit fox is a nocturnal hunter with a home range of 1 to 2 square miles. It digs a shallow den with several entrances, and usually uses more than one den during the breeding season.

**Comments:** The kit fox has always been rare in Oregon. Its exact status in the southeastern deserts is undetermined and its habitat is uncommon and patchy. Recent surveys detected 4 foxes in 1990 and 7 in 1991. Some mammalogists consider the kit fox (*Vulpes macrotis*) conspecific with the swift fox (*Vulpes velox*).

*References:* Bailey 1936, Daneke et al. 1984, Hall 1946, Marshall et al. 1996, McGrew 1979.

**421**

# Common Gray Fox
## (Urocyon cinereoargenteus)

*Order: Carnivora*
*Family: Canidae*
*State Status: None*
*Federal Status: None*
*Global Rank: G5*
*State Rank: S4*
*Length: 36 in (91 cm)*

**Global Range:** Found throughout most of the United States and south into extreme northern South America. On the West Coast, the Columbia River marks the northern boundary of its range.

**Habitat:** In contrast to the red fox, the common gray fox inhabits wooded areas, preferring open woodland and riparian forests to dense coniferous forests. In the eastern United States, the species prefers deciduous forests. In Oregon, it is most often encountered in woodlands with a broadleaf component, such as Oregon white oak, tanoak, madrone, maple, alder, or California bay.

**Reproduction:** The breeding season begins in February. This fox has one litter of 3 or 4 young (range 2-7) per year. Young are born around April, after a gestation period of 63 days.

**Food Habits:** The diet of common gray foxes varies with the season. In winter, and to some extent during the remainder of the year, they feed on small mammals. However, in summer they take a surprising quantity of insects and fruit, mainly berries, including cultivated fruits and berries.

**Ecology:** The home range of the common gray fox is a few square miles. It mates for life (up to 12 years),

and the home range of each pair does not overlap with that of other pairs. The common gray fox is the only canid in North America that regularly climbs trees. It is quite at ease in trees and will leap from limb to limb.

**Comments:** The common gray fox is a furbearer in Oregon, but only a few hundred are trapped per year.

*References:* Fritzell and Haroldson 1982, Grinnell et al. 1937, Maser et al. 1981.

# Black Bear
## *(Ursus americanus)*

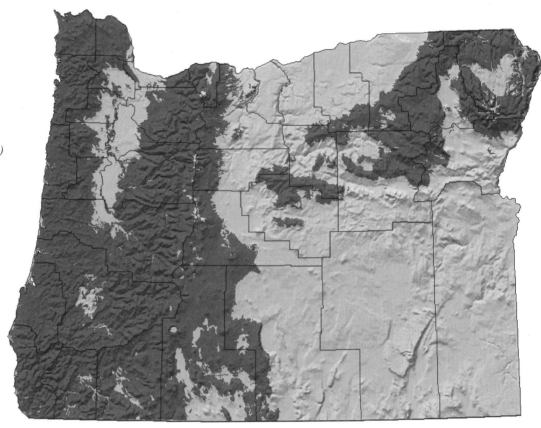

*Order: Carnivora*
*Family: Ursidae*
*State Status: None*
*Federal Status: None*
*Global Rank: G5*
*State Rank: S4*
*Length: 60 in (152 cm)*

**Global Range:** Found throughout most of forested North America south to the mountain ranges of central Mexico.

**Habitat:** Black bears prefer mixed deciduous-coniferous forests with dense understories, but also are found foraging in a variety of habitats within the forest mosaic, including clear-cuts. They were reported as very abundant in the coastal forests of western Oregon in the early 20th century.

**Reproduction:** These bears mate in early to mid-summer, but implantation is delayed until December. The young are born in late January or February, after 45-60 days of active embryonic development. There is a single litter per year, and the average litter size is 2 (range 1-5).

**Food Habits:** Black bears have an omnivorous diet that varies seasonally and by region, according to what food is available. They eat a considerable amount of green vegetable material, and are fond of fruits and berries. They also eat some insects, roots, fungi, fish, small rodents, birds and their eggs, and carrion. They sometimes strip the bark off trees to feed on the cambium layer, a practice that does not endear them to foresters.

**Ecology:** Black bears are solitary except for females with cubs. Their home range can be tens to hundreds of square miles. Because of their dependence on trees for cover, they do not occur in arid parts of southeastern Oregon. Bears overwinter in a dormant state in dens, often in caves or tree cavities.

**Comments:** The black bear is a game species in Oregon. As of this writing, it is illegal to hunt them with the aid of dogs.

*References:* Grinnell et al. 1937, Horner and Powell 1990, Maser et al. 1981.

# Ringtail
### (Bassariscus astutus)

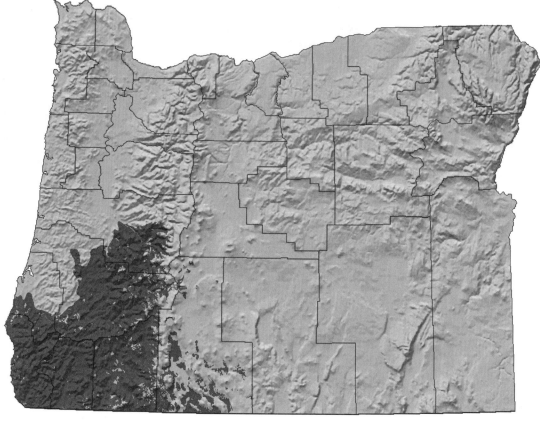

*Order: Carnivora*
*Family: Procyonidae*
*State Status: Sensitive*
*Federal Status: None*
*Global Rank: G5*
*State Rank: S3*
*Length: 28 in (71 cm)*

*References: Marshall 1992, Maser et al. 1981, Poglayen-Neuwall and Toweill 1988.*

**Global Range:** Reaches the northern limit of its range in southwestern Oregon; from there it is found southward and eastward to southern Mexico and Louisiana.

**Habitat:** Outside of Oregon, ringtails are found in a wide variety of habitats, including deserts, dry forests and woodlands, and even coniferous forests. They require some vertical structure to their habitat, such as cliff faces, steep canyon walls, or tall gallery forests along rivers and streams. In Oregon, they prefer woodlands containing tanoak which are near rocky areas and rivers. They can be found in coniferous forests, especially in riparian areas.

**Reproduction:** The ringtail breeds in the spring (usually in March and April), and a litter of 3 to 4 (range 1-5) young is born in May or June, after a 51-54 day gestation period. There is one litter per year.

**Food Habits:** Ringtails eat a variety of foods, but small mammals are especially important in winter. They also eat reptiles, some insects (in the fall), and birds and their eggs. Like the common raccoon, they eat fruits and berries.

**Ecology:** Ringtails are primarily nocturnal and are active throughout the year. They have home ranges of several hundred acres. They are adept climbers, but are secretive and seldom seen. They make their dens in rock shelters or hollow trees. In favorable habitat (gallery forests along rivers), they may reach densities of 20 per square kilometer.

**Comments:** Ringtails are rarely sighted in Oregon, and populations may be declining.

*Order: Carnivora*
*Family: Procyonidae*
*State Status: None*
*Federal Status: None*
*Global Rank: G5*
*State Rank: S5*
*Length: 32 in (81 cm)*

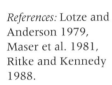

*References:* Lotze and Anderson 1979, Maser et al. 1981, Ritke and Kennedy 1988.

**Global Range:** Found from southern Canada to Panama and from coast to coast; they are absent only from high mountains and arid deserts of the interior West.

**Habitat:** Common raccoons are versatile omnivores that occur in a wide variety of habitat types. Given appropriate woods and brush for cover, they occur in agricultural and urban settings, and are found from the seashore to mid-elevation forests in the Cascade and Blue mountains. They have a discontinuous distribution in eastern Oregon, being restricted to canyons and river valleys with some permanent water supply.

**Reproduction:** The common raccoon breeds in late winter (late January to mid-March). A litter of 3 or 4 (range 1-7) young is produced after a gestation period of 63-65 days, usually in late April to early May. There is one litter per year.

**Food Habits:** Raccoons eat a broad range of food items. When food is abundant, they are selective, but when food is scarce they eat almost anything. They are not primarily carnivorous, but will eat small mammals, fish, frogs, turtles, birds and their eggs, mussels, earthworms, and insects. The vegetable part of their diet includes fruit, berries, nuts, and seeds, including acorns.

**Ecology:** Common raccoons are nocturnal, and are active throughout the year. They accumulate fat for the winter, but do not hibernate. Their home range can be as large as 1,000-2,000 acres. They commonly sleep in trees during the day.

**Comments:** The common raccoon is a furbearer in Oregon. In the early 1970s, about 4,000 pelts per year were taken in the state.

# American Marten
## (Martes americana)

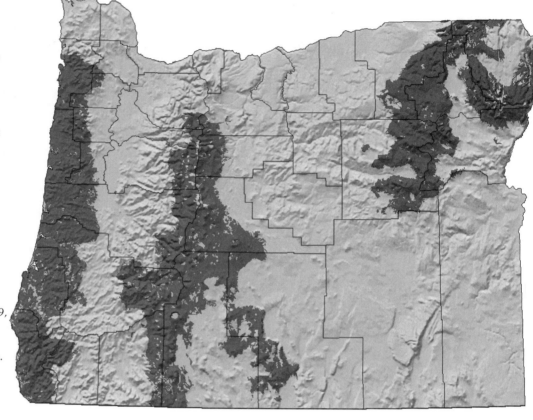

*Order: Carnivora*
*Family: Mustelidae*
*State Status: Sensitive*
*Federal Status: None*
*Global Rank: G4G5*
*State Rank: S3*
*Length: 26 in (66 cm)*

*References:* Buskirk
and McDonald 1989,
Clark et al. 1987,
Marshall 1992,
Zielinski et al. 1983.

**Global Range:** Occurs throughout Alaska and Canada, its range extending south of the Canadian border in the mountains of the West, the upper Midwest, and portions of New England. Along the northern California and southern Oregon coast, it reaches the Pacific Ocean.

**Habitat:** American martens are associated with forested habitats at any elevation, but will wander through openings and even up into alpine areas. They prefer mature forests with closed canopies, but sometimes use openings in forests if there are sufficient downed logs to provide cover. The type of forest is less important to martens than forest structure. They are not found in dry woodlands.

**Reproduction:** The American marten breeds in summer, but implantation is delayed until the following spring. After implantation, gestation lasts only another 27 days. A litter of 2 to 4 (range 1-5)

young is born in April or May. American martens are sexually mature at 1 year of age.

**Food Habits:** American martens are primarily carnivorous, feeding on small mammals such as shrews, voles, woodrats, rabbits, squirrels, and mountain beavers. They also take some birds, insects, fruit, berries, and carrion.

**Ecology:** This small carnivore is active throughout the year, sometimes traveling in tunnels under the snow during winter. The home range is variable, but usually is less than 10 square kilometers. The home range of one male may overlap with those of several females. They den under logs or in hollow tree stumps.

**Comments:** In Oregon, populations may be declining due to loss of mature forest habitat. This species is a furbearer in Oregon, but only about 100 a year were being taken in the late 1980s.

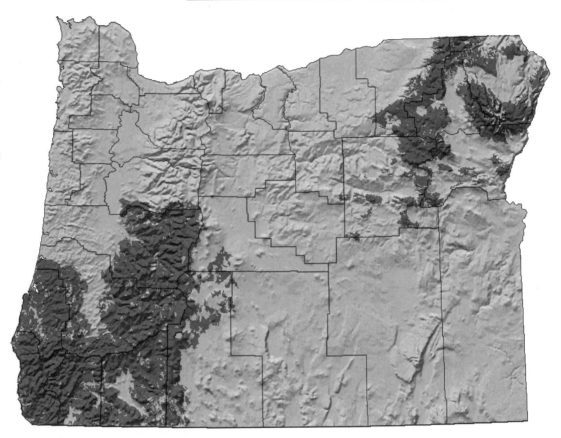

# Fisher
## *(Martes pennanti)*

*Order: Carnivora*
*Family: Mustelidae*
*State Status: Sensitive*
*Federal Status: SC*
*Global Rank: G5*
*State Rank: S2*
*Length: 36 in (91 cm)*

**Global Range:** Found in southern Canada from coast to coast. In the United States, they are restricted to the north woods of New England, the upper Midwest, the northern Rocky Mountains, and western mountains south to the Sierra Nevada of California.

**Habitat:** Fishers primarily use mature, closed-canopy coniferous forests with some deciduous component, frequently along riparian corridors. They are known to occasionally use cut-over areas, but this is not their optimal habitat.

**Reproduction:** Fishers breed from February to April. A litter of 3 (range 1-4) young is born about a year after fertilization. However, implantation is delayed for about 10-11 months.

**Food Habits:** The fisher is an opportunistic carnivore whose diet includes small rodents, rabbits, squirrels, mountain beavers, porcupines, amphibians, reptiles, and birds and their eggs. They eat some carrion, fruits, and berries.

**Ecology:** Fishers are active throughout the year, and can be active at any time of day or night. They are very mobile and may travel nearly 50 miles over a 3-day period. Their normal home range is several

hundred square kilometers. They find den sites in hollow logs, brush piles, or under rocks.

**Comments:** The fisher is very rare in Oregon. Most sightings are in the Coast and Cascade mountains. It is a furbearer with a closed season. Habitat loss and trapping have nearly extirpated this species from our state. Reintroductions have been attempted in Klamath, Union, and Wallowa counties. In recent years, fishers have also been sighted in Clackamas, Deschutes, Linn, Lincoln, and Wasco counties.

*References:* Arthur et al. 1989, Marshall 1992, Marshall et al. 1996, Maser et al. 1981, Powell 1981.

427

# *Ermine*
## (Mustela erminea)

*Order: Carnivora*
*Family: Mustelidae*
*State Status: None*
*Federal Status: None*
*Global Rank: G5*
*State Rank: S5*
*Length: 10 in (25 cm)*

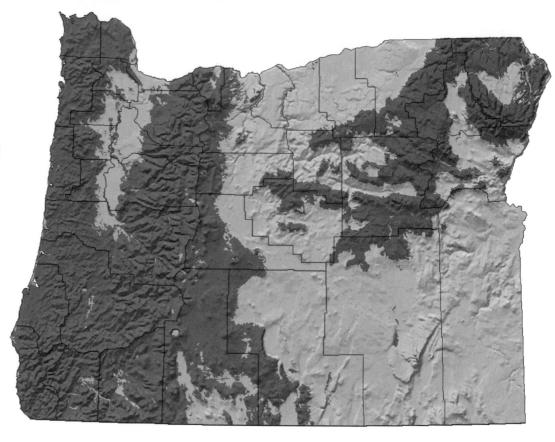

**Global Range:** Occurs throughout most of northern North America. In the eastern United States, it occurs as far south as Virginia, and in the West its southernmost record is in northern New Mexico. It is found along the Pacific Coast as far south as Mendocino County, California. It is a holarctic species that is widely distributed in Eurasia. It has been introduced into New Zealand.

**Habitat:** Ermines are found from the coast to high in the Blue Mountains in a variety of forest, riparian, and brush habitats. They do not occur in arid scrublands or open pastures, and tend to frequent brushy areas and edges of forests rather than the forest interior.

**Reproduction:** Like most weasels, the ermine has a long gestation period (280 days) with delayed implantation. It breeds in the summer, and young are born around mid-April of the following year. Litter size is usually 4 to 8 (range 4-18). Females may breed their first summer, but males are not sexually mature until the following year.

**Food Habits:** Ermine are carnivorous, feeding mostly on small mammals up to the size of rabbits.

Rarely, they eat birds, frogs, fish, earthworms, or insects.

**Ecology:** Ermines are primarily nocturnal, but, on occasion, can be seen during the day. They are active year round. At higher elevations, they hunt prey under a layer of snow. Their home range is about 12-16 hectares. They use hollow logs, spaces underneath rocks, or burrows of other species for dens.

**Comments:** The ermine (also known as the short-tailed weasel) remains brown all year in western Oregon, but in the mountains it molts into a white winter coat.

*References:* Bailey 1936, King 1983, Maser et al. 1981.

## Long-tailed Weasel
### *(Mustela frenata)*

*Order: Carnivora*
*Family: Mustelidae*
*State Status: None*
*Federal Status: None*
*Global Rank: G5*
*State Rank: S5*
*Length: 16 in (41 cm)*

**Global Range:** Found from southern Canada south, coast to coast, through the United States, Mexico, Central America and into northern South America.

**Habitat:** Long-tailed weasels occupy a wide variety of habitat types, and venture far into open areas. They occur in riparian areas and brushy areas in forests, but they also can be found in alpine communities above timberline and far out into arid desert communities of eastern Oregon.

**Reproduction:** They breed in the summer, with a peak in July. A very long (205-337 days) gestation period features delayed implantation. A litter of 6 to 8 (range 4-9) young is born from mid-April to mid-May. Embryos implant about 27 days prior to birth.

**Food Habits:** This small weasel can take mammals up to the size of rabbits. It preys mainly on burrowing rodents and pikas, but will also eat some birds and, occasionally, insects.

**Ecology:** These weasels are small enough to pursue fossorial rodents, such as pocket gophers and ground squirrels, through their burrow systems. They are active during daylight hours. Each pair has a home range of at least 1 square mile.

**Comments:** Maser et al. (1981) report that, in some cases, adult rabbits are able to fight off the attacks of long-tailed weasels. This weasel maintains nearly the same coat color throughout the year.

*References:* Bailey 1936, Maser et al. 1981, Quick 1951, Zeveloff 1988.

**429**

# Mink
## (Mustela vison)

*Order: Carnivora*
*Family: Mustelidae*
*State Status: None*
*Federal Status: None*
*Global Rank: G5*
*State Rank: S5*
*Length: 24 in (61 cm)*

**Global Range:** Found throughout North America north of Mexico, except for the desert Southwest.

**Habitat:** Of our smaller weasels, mink are most closely associated with water and riparian habitats. They are semiaquatic animals, with partly webbed feet for swimming. They are absent from forest interiors, alpine areas, and broad desert valleys, but present in rivers, streams, and marshes wherever such habitats are found in Oregon.

**Reproduction:** Mink have a much shorter gestation period than other weasels, only about 40-75 days. Implantation is delayed, but young are born the same year as mating. Breeding commences in late February. The litter size is about 4 (range 2-10), and young remain with their mothers until fall.

**Food Habits:** These aquatic weasels take prey in or near water, and, consequently, have a diet primarily of aquatic animals. Muskrats are often their most important food, but they also eat fish, frogs, crawfish, small mammals, and birds found near water.

**Ecology:** Mink are active year round and are primarily nocturnal. They will den in muskrat or beaver lodges. They are solitary except during the breeding season. Mink have linear home ranges along rivers and move considerable distances in streams and on river banks.

**Comments:** This species is a furbearer in Oregon. One to two thousand pelts a year were taken in Oregon in the early 1970s.

*References:* Bailey 1936, Maser et al. 1981, Mitchell 1961, Zeveloff 1988.

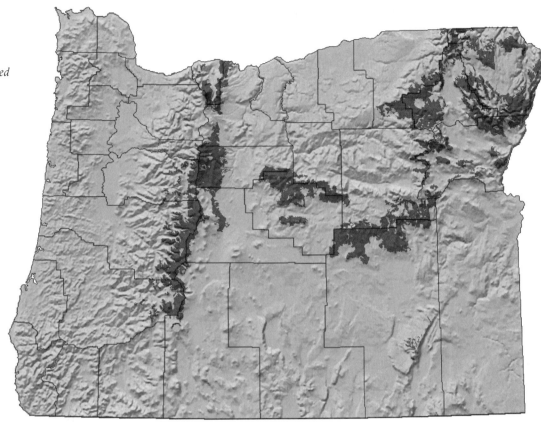

# Wolverine
### (Gulo gulo)

*Order: Carnivora*
*Family: Mustelidae*
*State Status: Threatened*
*Federal Status: SC*
*Global Rank: G4*
*State Rank: S2*
*Length: 36 in (91 cm)*

**Global Range:** A holarctic species found from Europe and northern Asia throughout most of northern North America. In the western part of North America, its range extends south through the Sierra Nevada of California and the Rocky Mountains. Historically, the range of the species extended as far south as southern Indiana in the eastern United States.

**Habitat:** In Oregon, the wolverine typically is found in open forests at higher elevations and in alpine areas. Farther north, it is a species of the taiga and tundra. It crosses clear-cuts, but avoids young, dense regenerating forests or brushy areas.

**Reproduction:** Wolverines breed in early summer. The gestation period is 7 to 9 months, but implantation is delayed. Young are born in early spring (February to March). The average litter size is 2 (range 1-4).

**Food Habits:** Wolverines primarily feed on small to medium-size rodents, marmots, and hares, and on carrion, such as ungulate carcasses. They also eat birds and their eggs, insects, fish, and a variety of roots and berries. They have been known to attack animals as large as moose that are foundering in deep snow.

**Ecology:** Wolverines are solitary animals with large home ranges (up to several hundred square miles for males). They are active throughout the year, and at any time of day or night. They have dens in caves, rock crevices, or hollow logs.

**Comments:** Prior to 1973, wolverines were considered furbearers in Oregon. They were always rare in Oregon, although recent sightings, tracks, and a road kill document their continued presence at low densities in the state. Large males can weigh over 30 pounds. Some mammalogists consider the North American wolverine a distinct species, *Gulo luscus*.

*References:* Bailey 1936, Hornocker and Hash 1981, Marshall et al. 1996, Pasitschniak-Arts and Larivière 1995, Zeveloff 1988.

# American Badger

*(Taxidea taxus)*

Order: *Carnivora*
Family: *Mustelidae*
State Status: *None*
Federal Status: *None*
Global Rank: *G5*
State Rank: *S4*
Length: *25 in (64 cm)*

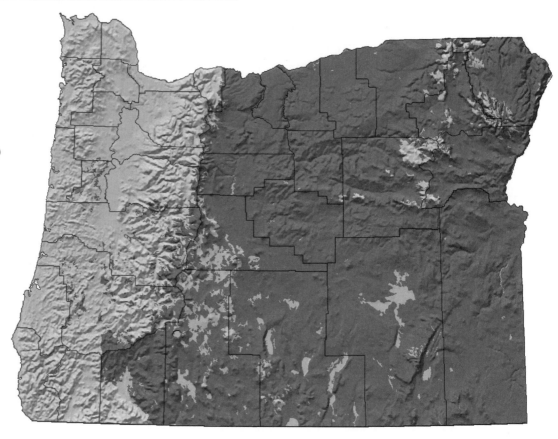

**Global Range:** Found in North America from southern Canada south to central Mexico from the Pacific Coast as far east as Ohio.

**Habitat:** American badgers are animals of open habitats of all types, from low desert valleys to alpine zones. They can be found in meadows and other sparsely vegetated areas within an open forest matrix, but avoid dense forests. In Oregon, they are most typical of the sagebrush deserts of the eastern part of the state.

**Reproduction:** Breeding season is in mid- to late summer, and implantation is delayed until December to February. A litter of 3 (range 2-5) young is born in March to early April. The young leave the family group in autumn.

**Food Habits:** Major food items include ground squirrels, gophers, kangaroo rats, and assorted small rodents. They will also eat reptiles, birds, and insects. They will eat seeds, roots, or green vegetation if animal prey is unavailable.

**Ecology:** American badgers are adapted to digging ground-dwelling animals out of their burrows. They do best where soil conditions favor rapid excavation. They are solitary and primarily nocturnal. At high elevations, they hibernate during winter, but they are active year round in the desert valleys of eastern Oregon.

**Comments:** American badgers sometimes team up with coyotes to flush rodents from cover. There is an old record and are occasional reports of American badgers from the Willamette Valley and Coast Range; however the climate and soils of the region make it unlikely badger country.

*References:* Bailey 1936, Long 1973, Messick and Hornocker 1981.

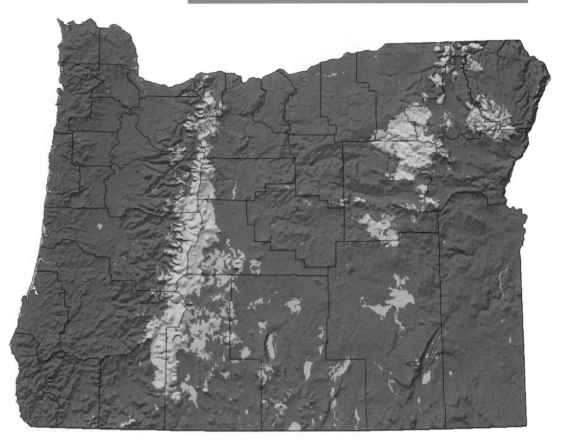

## Western Spotted Skunk
### (Spilogale gracilis)

*Order: Carnivora*
*Family: Mustelidae*
*State Status: None*
*Federal Status: None*
*Global Rank: G5*
*State Rank: S4*
*Length: 16 in (41 cm)*

**Global Range:** Found from extreme southwestern British Columbia south to central Mexico. It occurs west of the Great Plains and central Texas.

**Habitat:** In Oregon, the western spotted skunk occurs in brushy areas in a variety of habitat types. It lives in coniferous forests, riparian woodlands, thickets of canyons, and rocky hillsides. This skunk is absent from the higher mountains, and from flat desert valleys except where there is rimrock or riparian woodland.

**Reproduction:** Breeding occurs in early fall (late September to October). Implantation is delayed,

resulting in a gestation period of 230 (210-253) days, with 28-31 days of active embryonic development. A litter of about 4 (range 2-6) young is born in April or May.

**Food Habits:** This omnivore varies its diet with the season. In winter, it feeds on small mammals up to the size of cottontail rabbits, but it eats insects in summer. It also eats birds and their eggs, reptiles, carrion, fruits, and berries.

**Ecology:** Ordinarily, these skunks confine their wanderings to home ranges of 1 to 4 square miles. Adults are solitary, and are active mostly at night. They do not hibernate. They find shelter in dens under rocks, in hollow trees, or under buildings.

**Comments:** Like its relative, the striped skunk, the western spotted skunk is able to expel a musky fluid from its anal glands to discourage potential predators. Owls are about the only animal to successfully prey on these skunks. Some mammalogists consider *Spilogale gracilis* conspecific with the eastern spotted skunk (*S. putorius*).

*References:* Bailey 1936, Maser et al. 1981, Zeveloff 1988.

# Striped Skunk
## (Mephitis mephitis)

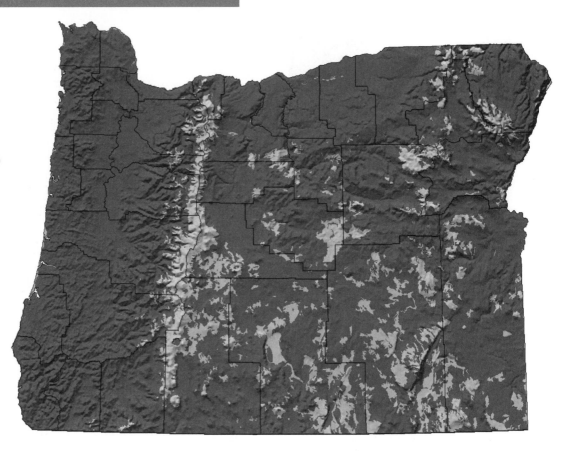

*Order: Carnivora*
*Family: Mustelidae*
*State Status: None*
*Federal Status: None*
*Global Rank: G5*
*State Rank: S5*
*Length: 24 in (61 cm)*

**Global Range:** Found coast to coast from central Canada south to northern Mexico. It is absent only from the deserts of southern Nevada, southeastern California, and Baja California.

**Habitat:** Striped skunks wander through a variety of habitat types in their search for food. They prefer brushy or rocky areas, but will forage in open pastures. They are usually found near water. Consequently, their distribution in eastern Oregon is scattered along canyons, marshes, or riparian areas within reach of water. They are not usually found at high elevations or in dense forests.

**Reproduction:** Striped skunks breed in February or March, and a litter of 6 to 8 (range 2-10) young is born in May and June, after a gestation period of 62 to 64 days. Implantation is not delayed in this mustelid.

**Food Habits:** These skunks are more insectivorous than spotted skunks, but will prey on small mammals in winter, when insects are scarce. They also eat birds and their eggs and include a considerable amount of vegetable material, especially fruits and berries, in their diets.

**Ecology:** Striped skunks are primarily nocturnal, and may be inactive for considerable periods during winter snows. During the breeding season, males defend a territory of up to 100 acres. These skunks have powerful claws, and often dig their own burrows, but they also den in hollow logs, brush piles, or under rocks.

**Comments:** While it is not officially a furbearer in Oregon, several hundred striped skunks are trapped each year for their pelts.

*References:* Maser et al. 1981, Verts 1967, Wade-Smith and Verts 1982.

# Northern River Otter
## *(Lutra canadensis)*

*Order: Carnivora*
*Family: Mustelidae*
*State Status: None*
*Federal Status: None*
*Global Rank: G5*
*State Rank: S4?*
*Length: 50 in (127 cm)*

**Global Range:** Found throughout North America north of Mexico, except for arid southwestern deserts.

**Habitat:** Northern river otters are found in and along streams, lakes, swamps, marshes, and the seashore. During the breeding season, males may travel considerable distances over land.

**Reproduction:** Northern river otters mate during the winter months. The gestation period is variable and quite long (9 months to over a year), due to delayed implantation. Active embryo development lasts about 2 months. A litter of about 3 (range 1-4) young is born in March or April.

**Food Habits:** As would be expected for an animal that spends most of its time in the water, northern river otters consume many aquatic organisms including fish, frogs, crayfish, turtles, some small mammals, birds, and carrion. They may also eat some berries and aquatic invertebrates. Along the coast, they eat mussels and nesting seabirds.

**Ecology:** Otters are active throughout the year and are primarily nocturnal. They are active swimmers, and may move as far as 50 miles up and down a stream. They often occur in pairs, both during and outside of the breeding season. They have a very scattered distribution in arid southeastern Oregon, and are present only in permanent waters.

**Comments:** The northern river otter is a furbearer in Oregon. In 1972, nearly 300 pelts were taken.

*References:* Bailey 1936, Maser et al. 1981, Melquist and Hornocker 1983, Toweill 1974.

# Mountain Lion
## (Felis concolor)

*Order: Carnivora*
*Family: Felidae*
*State Status: None*
*Federal Status: None*
*Global Rank: G5*
*State Rank: S4?*
*Length: 72 in (183 cm)*

**Global Range:** Historically, mountain lions were distributed widely in temperate and tropical parts of the Western Hemisphere. They occur now from western Canada south through the western United States into Mexico and points south. There are scattered populations east of the Rocky Mountains.

**Habitat:** Mountain lions are mobile animals that are found in a wide variety of habitat types, from dense forests to open woodlands and canyons. They are absent from broad desert valleys and flats, where neither suitable cover nor prey are found. In Oregon, they are present in the Coast Ranges and the Cascade and Blue mountains.

**Reproduction:** Mountain lions breed throughout the year, but most births occur from April to September. A litter of 2 (range 1-6) is born after a gestation period of 82-96 days. Females breed every other year, and young remain with their mother for 1 to 2 years.

**Food Habits:** Mountain lions consume both large and small mammals including deer, elk, rabbits, mice, and squirrels. Where plentiful, deer are their staple prey. At times, they eat some plant material and insects.

**Ecology:** Generally, mountain lions are solitary except for mothers with kittens and during mating. They are active day and night throughout the year but are quite secretive and are rarely seen. They have very large home ranges (15-45 square miles), and move long distances (an average of 5.5 miles per night) while searching for food.

**Comments:** Mountain lions are game animals in Oregon, but as of this writing it is illegal to hunt them using dogs. Occasionally, they kill livestock and, rarely, attack humans. Other vernacular names for the mountain lion are cougar or puma.

*References:* Beier 1993, Currier 1983, Maser et al. 1981, Robinette et al. 1961.

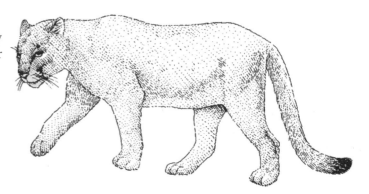

**436**

# Lynx
## *(Lynx canadensis)*

*Order: Carnivora*
*Family: Felidae*
*State Status: None*
*Federal Status: SC*
*Global Rank: G4G5*
*State Rank: S1*
*Length: 36 in (91 cm)*

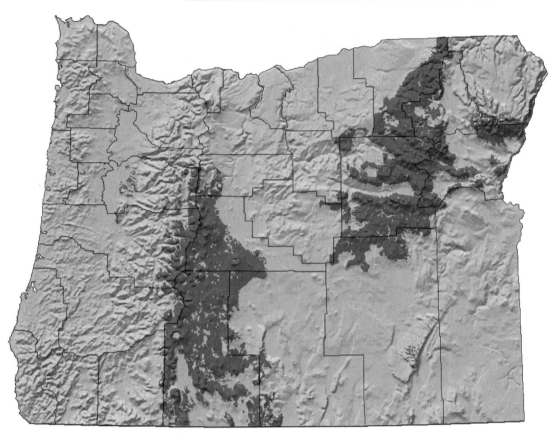

**Global Range:** Primarily a boreal species, found from Alaska and Canada south to southern Colorado in the Rocky Mountains. It occurs, or did occur, from coast to coast.

**Habitat:** Lynx typically occupy dense boreal forests that have some openings, such as meadows, bogs, or rocky outcrops.

**Reproduction:** Mating occurs from February to April. A litter of 2 or 3 (range 1-4) young is born from mid-April to mid-June, after a gestation period of 60-67 days.

**Food Habits:** In much of its range, the snowshoe hare is the primary food item of the lynx. It also takes some birds, other small mammals, and carrion and can take mammals as large as young deer.

**Ecology:** These are solitary cats that have home ranges of 14 square miles or more. They make dens in hollow trees, under logs, or in thick brush. They are primarily nocturnal.

**Comments:** Lynx have always been rare in Oregon, which is the southern limit of their range on the West Coast. The last confirmed specimen was taken in Corvallis in 1974, but most other records are from east of the Cascade Mountains. The Latin names *Felis lynx* and *Lynx lynx* have been applied to this species.

*References:* Bailey 1936, Koehler et al. 1979, Tumlison 1987.

437

# Bobcat
## (Lynx rufus)

Order: Carnivora
Family: Felidae
State Status: None
Federal Status: None
Global Rank: G5
State Rank: S4
Length: 30 in (76 cm)

**Global Range:** Occurs coast to coast, from southern Canada to southern Mexico.

**Habitat:** Three subspecies of bobcat occur in Oregon. One subspecies lives in arid desert areas where appropriate rimrock or lava formations afford it daytime shelter. Another lives in the lower mountains and open forests of the east slope of the Cascades. The third subspecies lives in the dense forests, thickets, and clear-cuts of western Oregon, including chaparral in the southwestern part of the state. Bobcats will hunt in clear-cuts.

**Reproduction:** These cats breed from mid-winter through spring. A litter of 2 or 3 (range 1-7) kittens is born after a gestation period of about 2 months. Young are weaned in another 2 months and stay with their mother until fall.

**Food Habits:** Bobcats are opportunistic carnivores. Rabbits are one of their primary food items, but they also eat a variety of small rodents, shrews, and birds. They occasionally kill deer.

**Ecology:** Bobcats are solitary and mostly nocturnal. They hunt by stealth, and usually seek heavy cover. The young are born in a den that may be located in a hollow log, under a fallen tree, or in a rock shelter.

They establish regular hunting routes that cross their home ranges, which may be as large as 60 square kilometers.

**Comments:** The bobcat is a furbearer in Oregon. In 1974, about 1,600 pelts were taken in the state. They are preyed upon by coyotes, which may limit their population in some areas.

*References:* Knick 1990, Maser et al. 1981, Young 1958.

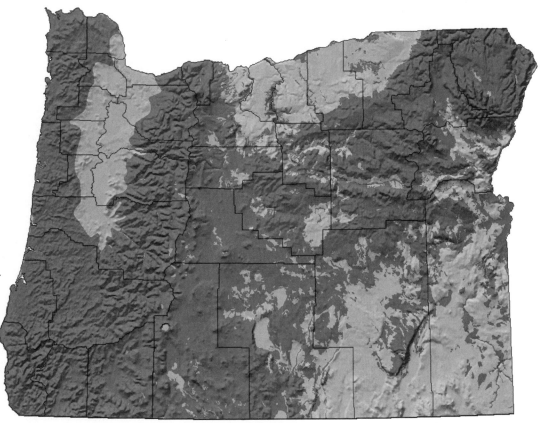

## *Wapiti or Elk*
### *(Cervus elaphus)*

*Order: Artiodactyla*
*Family: Cervidae*
*State Status: None*
*Federal Status: None*
*Global Rank: G5*
*State Rank: S5*
*Length: 84 in (213 cm)*

*References:* Maser et al. 1981, Thomas and Toweill 1982.

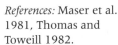

**Global Range:** In North America, elk occurred historically from western and southern Canada south to Pennsylvania, southern California, northern Mexico, and Louisiana. They are now primarily found in the western United States and Canada, although some local populations have been reintroduced in the East.

**Habitat:** Elk are now found primarily in forests, meadows, mountain valleys, and foothills. They may once have occupied shrublands of eastern Oregon. They are mobile, and move downslope out of deep snow to warmer, more open habitats in winter. They may invade agricultural areas.

**Reproduction:** Elk breed in late summer and fall. Usually, females produce only one young (sometimes there are twins). Young are born in late May or early June, after a 255-day gestation period.

**Food Habits:** Elk are primarily grazing animals, preferring a diet of grasses and some forbs. In winter, they turn to browsing the tips of twigs from willow, alder, aspen, oak, or other woody vegetation.

**Ecology:** In North America, elk are the second largest member of the deer family, and the largest deer found regularly in Oregon (a moose occasionally may stray into the Blue Mountains). They are social animals that form herds led by an older female. In breeding season, males contest for breeding rights to the herd. They may move long distances between their summer and winter ranges, but their daily movements are rarely more than 1,600 meters in any direction. They are most active at night.

**Comments:** In Oregon, elk are important game animals. In 1973, over 24,000 elk were taken by hunters. Elk were reduced in numbers by early settlers, but have been reintroduced to most forested parts of the state. The name *Cervus canadensis* was once applied to North American populations.

439

# Mule Deer and Black-tailed Deer
## (Odocoileus hemionus)

*Order: Artiodactyla*
*Family: Cervidae*
*State Status: None*
*Federal Status: None*
*Global Rank: G5*
*State Rank: S5*
*Length: 66 in (168 cm)*

*References:*
Anderson and
Wallmo 1984,
Bailey 1936, Maser
et al. 1981.

**Global Range:** Residents of western North America found from the Yukon Territory south to the tip of Baja California and central Mexico. Occur as far east as Texas and Minnesota.

**Habitat:** Two subspecies occur in Oregon. Generally, deer west of the Cascade Mountains are black-tailed deer (*O. h. columbianus*), and those east of the Cascades are mule deer (*O. h. hemionus*). In western Oregon, black-tailed deer are found in brushy areas at the edges of forests and chaparral thickets, not in dense forests. They prefer early successional stages created by clear-cuts. In eastern Oregon, mule deer once ranged into sagebrush plains in canyons or rimrock, but they now are confined mainly to open woods or isolated mountain ranges. In winter, they descend to lower valleys.

**Reproduction:** These deer breed from mid-November to mid-December. One or 2 fawns are born in May or June, following a gestation period of 203 days. Litter size is influenced by food supply.

**Food Habits:** These deer both graze and browse. They eat grasses and forbs, as well as the growing tips of woody vegetation. They also eat acorns, ferns, and some conifer shoots.

**Ecology:** These deer are frequently found in clans led by an older female. They are active throughout the year, and concentrate their daily activities around dawn and dusk. Their home range usually is less than 100 acres, but can be as large as 600 acres in less favorable habitat. Mountain lions and coyotes are their main predators, aside from humans.

**Comments:** Deer are important game animals in Oregon. They sometimes invade agricultural areas, and may do considerable damage. The standard English name for the species is *mule deer* (Jones et al. 1992).

**440**

## White-tailed Deer
### (*Odocoileus virginanus*)

*Order: Artiodactyla*

*Family: Cervidae*

*State Status: Endangered (O. v. leucurus)*

*Federal Status: Endangered*

*Global Rank: G5*

*State Rank: SU*

*Length: 60 in (152 cm)*

**Global Range:** Occurs from southern Canada to northern South America. Absent from most of the southwestern U.S., but otherwise occurs from coast to coast.

**Habitat:** In Oregon, Columbian white-tailed deer (*O. v. leucurus*) were once abundant in the wet meadows, grasslands, and riparian and oak woodlands of the Willamette Valley. Now, they are restricted to a few islands in the Columbia River and to Oregon white-oak woodlands around Roseburg. Another subspecies, the yellow-tailed deer (*O. v. ochrourus*), inhabits thickets and willow bottoms of mountain valleys and open timber. It occurs in Idaho and Montana, and has been expanding its range in the Blue Mountains in recent decades. Formerly it was more widespread in Oregon, occurring along the east slope of the Cascades.

**Reproduction:** Breeding takes place in November, and 1 or 2 fawns are born around mid-June. Gestation lasts 208-212 days.

**Food Habits:** In western Oregon, this deer is primarily a grazing animal, with about 80% of its diet consisting of grasses and forbs, the rest of branches of woody vegetation.

**Ecology:** The Columbian white-tailed deer reaches densities of about 20-30 per square kilometer. Its home range is about 100 acres. It is most active at dawn and dusk, and remains active throughout the year. On islands of the Columbia River, it grazes in pastures, and takes shelter in willow thickets.

**Comments:** The taxonomy of the Columbian white-tailed deer is muddled. It may hybridize with black-tailed deer and may not be genetically distinct from the yellow-tailed deer. This species is a game animal in Oregon, but hunting is not permitted west of the Cascades. Recent surveys estimate that the Columbia River population numbers about 1,000 individuals, while about 6,000 animals live in the Roseburg area.

*References:* Gavin and May 1988, Gavin et al. 1984, Marshall et al. 1996, Smith 1991.

# Pronghorn
## *(Antilocapra americana)*

*Order: Artiodactyla*
*Family: Bovidae*
*State Status: None*
*Federal Status: None*
*Global Rank: G5*
*State Rank: S4*
*Length: 58 in (147 cm)*

**Global Range:** This antelope-like species is restricted to western North America, from the southern prairie provinces of Canada south to northern Mexico. In Baja California, it reaches the Pacific Coast. Generally, it is a species of arid inland valleys and plains. It occurs as far east as Texas and South Dakota.

**Habitat:** In Oregon, this is a species of grasslands, sagebrush flats, and shadscale-covered valleys of the central and southeastern part of the state. Low sagebrush is an important habitat component. They were once found in the Rogue River Valley and, in summer, made their way up to flat areas along the crest of the central Cascade Mountains.

**Reproduction:** Like most ungulates, pronghorns breed in the fall, and 1 to 2 young are born in May or June, following a gestation period of about 250 days.

**Food Habits:** In spring and summer, broad-leaved herbaceous vegetation is the preferred food. They will browse on tips of sagebrush in winter, and occasionally eat some grasses. Common food plants include longleaf phlox, wallflower, and Hooker's balsamroot.

**Ecology:** Pronghorns remain active throughout the year, and have peak daily activity periods at dawn and dusk. They form small bands throughout the year. Males form bachelor herds in spring and summer. Home ranges of summer herds are about 10 to 20 square kilometers.

**Comments:** Pronghorn are game animals in Oregon. Most herds are found within 8 kilometers of a water source. Pronghorns were formerly placed in their own family, *Antilocapridae*.

*References:* Bailey 1936, O'Gara 1978, Zeveloff 1988.

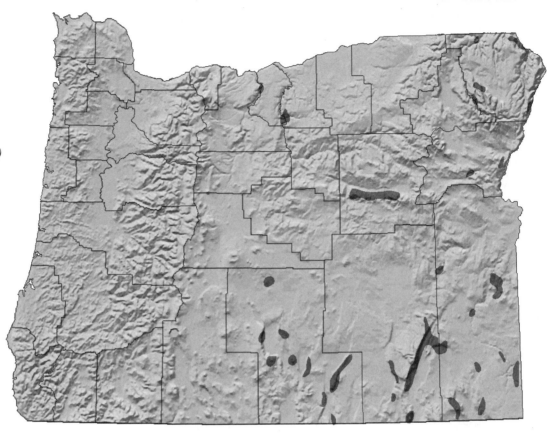

## Mountain Sheep
*(Ovis canadensis)*

*Order: Artiodactyla*
*Family: Bovidae*
*State Status: None*
*Federal Status: SC*
*Global Rank: G4G5*
*State Rank: S2*
*Length: 60 in (158 cm)*

**Global Range:** In the Rocky Mountains, populations are found from interior British Columbia and Alberta south to northern New Mexico. Interior mountain ranges, from the Sierra Nevada and desert mountains of California and Baja California east to Texas, are also occupied.

**Habitat:** Mountain sheep occur in steep, rocky mountain ranges and rarely traverse the valley floors in between. In the mountain ranges of eastern Oregon, they occur in high mountain meadows and steep canyons. Formerly, they could be found in foothills and river valleys.

**Reproduction:** The breeding season takes place from September to October. One young (rarely twins) is born in May or June, after a gestation period of about 180 days.

**Food Habits:** Grass is the primary item in the diet, especially bluebunch wheatgrass and cheatgrass. The diet changes seasonally, and mountain sheep may browse on woody plants during winter. Some forbs are eaten.

**Ecology:** Mountain sheep are social animals that occur in mixed-sex groups in spring and during breeding season, but otherwise are segregated into ram and female-young herds. They migrate considerable distances downslope in winter. Home ranges are from 20 to 40 square kilometers. Mountain sheep are active throughout the year.

**Comments:** Extirpated from Oregon by the early 1900s, this species has been reintroduced to much of its former range. The Rocky Mountain subspecies was native to the Blue Mountains, but all reintroductions are from California populations. Reintroduced herds are at risk from diseases passed on by flocks of domestic sheep. These sheep are also known as bighorn sheep.

*References:* Berger 1990, Miller and Gaud 1989, Shackleton 1985.

443

# Glossary

**Allozyme**—A protein, usually occurring in several identifiable variations, that is coded by one part of an organism's genetic material. The frequency of different allozymes may vary between populations and is thought to reflect the degree of evolutionary divergence between them.

**Altricial**—A young animal born or hatched in a relatively undeveloped state. Altricial young usually have their eyelids closed, do not have fur or feathers, and cannot move about independently for a considerable time.

**Annelid**— A member of the phylum of invertebrate animals that includes earthworms, marine worms, and leeches.

**Breeding season**—That period of the year when members of a species are engaged in activity leading to the production of offspring. Includes courtship, nest building, gestation or incubation, and care for the young prior to their ability to survive independently of their parents.

**Broad-leaved**—Trees with flat leaves usually attached to stems with a petiole or stalk. The leaves may or may not be shed during the annual cycle.

**Brood**—A group of young that are cared for together by their parents.

**Buteo**—A genus of the family Accipitridae characterized by large hawks with broad wings adapted for soaring on rising air currents.

**Cache**—A store of food or the act of storing food.

**Cambium**—The layer of plant tissue between the bark and trunk of woody plants. It produces xylem, which carries water and minerals up from the roots, and phloem, which transports sugars away from the leaves.

**Carrion**—The tissue of dead animals.

**Cismontane**—Literally, this side of the mountains, usually meaning the area west of the crest of the Cascade and Sierra Nevada mountains in California.

**Closed-canopy**—A description of a vegetation type (usually a forest) in which the foliage of neighboring trees is in contact.

**Clutch**—The eggs laid by a female in one place during a breeding episode. Some species lay multiple clutches during the breeding season.

**Cold-blooded**—Animals that lack the ability to regulate their body temperature by metabolic processes. Such animals are also called poikilothermic. Although modern reptiles are "cold-blooded," many will maintain relatively high and constant body temperatures during active periods through behavioral mechanisms.

**Colony**—A group of animals of the same species breeding in close proximity to one another.

**Composite**—A plant of the family Asteraceae (formerly Compositae) in which flower heads are made up of many smaller flowers.

**Conspecific**—Individuals or populations of the same species.

**Conterminous**—A set of areas with adjoining borders, used to describe the 48 United States on the mainland of North America south of Canada.

**Cosmopolitan**—A species whose distribution is worldwide or nearly worldwide. A cosmopolitan species occurs in both the Northern and Southern Hemispheres and in both the New and Old World.

**Cover**—Physical objects in the environment that an animal uses for shelter from the environment or predators. Cover often consists of plants or places under rocks or logs.

**Crepuscular**—Active during periods of the day when light is low, usually around dawn and dusk.

**Crustacean**—A member of the class Crustecea of the phylum Arthropoda. Crustaceans include crabs, lobsters, shrimp, crayfish and many smaller planktonic and terrestrial species.

**Dabbling duck**—A duck, such as the Mallard, that feeds on the surface of the water or on that part of the bottom that can be reached while floating on the surface.

**Deciduous**—A tree that sheds its leaves during part of the annual cycle. Most deciduous trees are broad-leaved angiosperms, but the western larch, a conifer, is also deciduous.

**Delayed fertilization**—A reproductive strategy used by many bats in which sperm are stored by the female for a period of time after mating. This allows fertilization and gestation to be timed so that young are born at a favorable season.

**Delayed implantation**—A reproductive strategy used by many mustelids in which development of the fertilized egg is arrested for a period of time. Implantation and active gestation are thus timed so that young are born at a favorable season.

**Detritus**—Particles of organic material in various stages of decomposition. Detritus often accumulates on the bottom of lakes and ponds or on the ground.

**Disjunct**—A population of a species that is isolated by a considerable distance from the area usually occupied by that species.

**Diurnal**—Active during the daytime.

**Diving duck**—A duck, such as the Canvasback, that forages by diving below the surface of the water.

**Electrophoresis**—A method of separating genetic variants of the same protein (often an enzyme) by using an electric charge to drive the proteins through a gelatinous sheet. The more strongly charged variants travel faster through the gel. The ratio, and even the presence, of different variants may vary between populations, and is thought to reflect the degree of evolutionary divergence between populations.

**Emergent vegetation**—Plants that are rooted below the surface of water but whose vegetative parts grow above water level.

**Endemic**—Native to a specified geographic region.

**Ephemeral**—Occurring or existing during a relatively short period of time.

**Euphausid**—A shrimp-like aquatic crustacean.

**Exotic**—A species that does not naturally occur in an area.

**Extirpated species**—A species that no longer occurs in a specified region, as compared to an *extinct species*, which no longer occurs anywhere.

**Fossorial**—Adapted to, living in, and burrowing in the ground.

**Fecundity**—A measure of the reproductive potential of a population.

**Fell-field**—A barren, high elevation area with a sparse cover of alpine plants growing in stony soil.

**Fledgling**—A young bird that can fly but may still be fed by its parents.

**Forb**—A low-growing, non-woody dicotyledonous plant.

**Forest**—An area dominated by trees with a total canopy cover over 60%.

**Friable**—Loose and easily worked, as soil.

**Gestation**—The period of active embryo development in mammals.

**Glean**—To pick food or other items from surfaces such as leaves, stems, bark, or the ground.

**Guild**—A group of animals that use a similar foraging strategy to capture similar food items.

**Halophytic**—Literally, "salt-loving," applied to plants adapted to growing in salty or alkaline soils.

**Hawk**—As a noun this refers to a bird in the family Accipitridae. As a verb, it refers to the activity of a bird catching prey while in flight, usually flying out from a perch to catch insects. Generally synonymous with *sally*.

**Herbs**—Non-woody plants, including both grasses and forbs.

**Hibernate**—Entering into a prolonged period of lowered metabolism and reduced activity during periods of cold weather.

**Host**—An individual or species that provides biological services to an unrelated individual or species.

**Holarctic**—Occurring throughout the Northern Hemisphere, in both the Old and New Worlds.

**Home range**—An area defined by the farthest limits of an individual animal's travels. The breeding home range encompasses only the area an animal usually visits during the breeding season, which is considerably smaller than the lifetime home range of many migratory birds.

**Hypogeous**—Literally, "below the ground," referring to the subterranean mycelia and fruiting bodies of some fungi.

**Incubate**—The act of a parent providing a favorable environment for the development of eggs.

**Interspecific**—Interactions between different species or between individuals of different species.

**Intraspecific**—Interactions between individuals of the same species.

**Juvenile**—The young of a species which is developed or old enough to be usually independent of its parents but which has not yet attained reproductive maturity.

**Kite**—The noun refers to one type of hawk. The verb describes an aerial foraging strategy of hovering in one place while searching for prey on the ground.

**Larva**—One or more morphologically distinct stages an individual passes through during development prior to assuming adult morphology.

**Lepidoptera**—The order of insects that includes butterflies and moths.

**Litter**—A collective term used to describe the young born as a result of a single reproductive episode (there may be multiple litters in a year). Also used to describe accumulated objects on a surface, as leaf litter in a forest.

**Mast**—A generic term used to describe the hard, usually protected seeds of hardwood trees. Acorns are a type of mast.

**Metamorphosis**—The ontogenetic process of transforming from larval to adult morphology. For example, aquatic tadpoles of frogs lose their gills and develop lungs when they metamorphose into the adult form.

**Microhabitat**—A small area with environmental parameters different than most of the surrounding matrix, as a pond in a forest or desert.

**Microtine**—A rodent of the subfamily Arvicolinae (formerly Microtinae) of the family Muridae. Microtine rodents have prism-shaped cusps on their teeth adapted to eating grasses and other coarse vegetation. They have a high reproductive potential, tend to display population cycles, and are important prey to many predators.

**Monotypic**—A taxonomic unit with a single representative. The mountain beaver is the only species in the monotypic family Aplodontidae. It is also the only species in the monotypic genus *Aplodontia*.

**Mysid**—A shrimp-like aquatic crustacean.

**Neotenic**—A larval individual that is capable of reproduction.

**Nest parasitism**—Also called brood parasitism. The practice of a bird laying its eggs in the nest of another species so that the host species performs the tasks of incubating the eggs and rearing the young. Most brood parasites are members of the cuckoo family, but not all cuckoos are brood parasites.

**Nestling**—A young bird between the time of hatching and fledging.

**Nocturnal**—Occurring or active during the night.

**Nymph**—One of several stages in the growth of many arthropods prior to attaining adult form.

**Open canopy**—An area dominated by trees whose individual branches do not touch. The total area covered by tree canopies determines whether the stand will be a forest or a woodland.

**Paedomorphic**—An individual that retains larval morphology for an extended period of time.

**Parthenogenetic**—Reproduction of viable and fertile female offspring by females without the assistance of males. All offspring produced this way are genetically identical.

**Passerine**—A member of the order of perching birds, Passeriformes.

**Pelage**—The hair or fur covering a mammal.

**Piscivorous**—A species with a diet consisting mainly of fish.

**Plumage**—The feathers covering a bird.

**Polyandry**—A mating system in which one female mates with several males, who usually take responsibility for rearing her clutches.

**Polygyny**—A mating system in which one male mates with several females, who nest within a territory defended by the male. The male rarely assists in rearing the young.

**Precocial**—A young animal born or hatched in a relatively mature state of development. Precocial young are born or hatched with a coat of fur or feathers, their eyes are open, and they become mobile shortly after birth or hatching.

**Primary cavity nester**—A bird that usually excavates its own nest chamber in a tree.

**Raptor**—A carnivorous bird. In North America, raptors usually belong to the orders Falconiformes or Strigiformes.

**Recruitment**—The addition of individuals to a population through reproduction or immigration.

**Roost**—A place where one or more individuals of a population will rest during periods of inactivity.

**Runway**—A trail worn in herbaceous vegetation by the activities of small mammals.

**Sally**—To fly from a perch and catch an insect in the air.

**Secondary cavity nester**—A bird that nests in a hole in a tree that was excavated by another species. Usually the original occupant has abandoned the hole, but sometimes it is evicted by the secondary cavity nester.

**Seral stage**—One definable stage in a series of plant communities that replace one another through time following disturbance to an area.

**Shorebird**—A collective term for families in the order Charadriiformes that usually occupy habitat at the edges of bodies of water. Includes plovers, sandpipers, curlews, avocets, stilts, and snipe.

**Snag**—A standing dead tree.

**Succession**—The ecological process of the replacement of plant communities through time until a stable, or climax, community is reached. As each non-climax community grows, it changes the environmental conditions in such a way as to be more beneficial to the next community.

**Succulent**—A plant whose leaves or stems have a relatively high water content.

**Sympatric**—Two or more species who live in the same area.

**Systematics**—A branch of biology that investigates the evolutionary relationships between groups of organisms.

**Talus**—An accumulation of rocks with little or no soil development.

**Taxonomy**—The classification and naming of organisms.

**Territory**—The area defended by an individual against others of the same species. Sometimes territories are also defended against other species that are competitors, predators, or merely intruders.

**Transmontane**—Literally, "the other side of the mountains." Used to refer to the area east of the crest of the Cascade and Sierra Nevada mountains in California.

**True bug**—A member of the insect order Hemiptera.

**Understory**—The plants growing beneath a tree or shrub canopy.

**Viviparous**—Giving birth to young, as opposed to oviparous (laying eggs).

**Woodland**—An area dominated by trees with a total canopy cover of 25-60%. The area between the trees is usually dominated by shrubs or grasses or both.

# References

Ables, E. D. 1975. Ecology of the red fox in North America. Pages 216-36 in M. W. Fox, editor, *The Wild Canids: Their systematics, behavioral ecology and evolution.* Van Norstrand Reinhold, New York.

Adams, M. J. 1993. Summer nests of the tailed frog (*Ascapus truei*) from the Oregon Coast Range. *Northwestern Naturalist* 74:15-18.

Aldrich, J. W., and F. C. James. 1991. Ecogeographic variation in the American Robin. *Auk* 108:230-49.

Alexander, L. F. 1996. A morphometric analysis of geographic variation within *Sorex monticolus* (Insectivora:Soricidae). University of Kansas, Museum of Natural History, Miscellaneous Publication No. 88:1-54.

Alexander, L. F., and B. J. Verts. 1992. Clethrionomys californicus. Mammalian Species No. 406:1-6, American Society of Mammalogists.

Alexander, W. C. 1983. Differential sex distribution of wintering diving ducks (*Aythyini*) in North America. *American Birds* 37:26-29.

Altig, R., and E. D. Brodie, Jr. 1971. Foods of Plethodon larselli, Plethodon dunni and Ensatina eschscholtzi in the Columbia River Gorge, Multnomah County, Oregon. *American Midland Naturalist* 85:226-28.

American Ornithologists' Union. 1983. *Check-list of North American Birds*, 6th edition. Allen Press, Lawrence, Kansas.

American Ornithologists' Union. 1985. Thirty-fifth Supplement to the American Ornithologists' Union *Check-list of North American Birds. Auk* 102:680-86.

American Ornithologists' Union. 1989. Thirty-seventh Supplement to the American Ornithologists' Union *Check-list of North American Birds. Auk* 106:532-38.

American Ornithologists' Union. 1993. Thirty-ninth supplement to the American Ornithologists' Union *Check-list of North American Birds. Auk* 110:675-82.

American Ornithologists' Union. 1995. Fortieth supplement to the American Ornithologists' Union *Check-list of North American Birds. Auk* 112:819-30.

Anderson, A. E., and O. C. Wallmo. 1984. Odocoileus hemionus. Mammalian Species No. 219:1-9, American Society of Mammalogists.

Anderson, D. A. 1988. Where do you find a Harlequin Duck in Oregon? *Oregon Birds* 14:217.

Anderson, J. D. 1967. A comparison of the life histories of coastal and montane populations of Ambystoma macrodactylum in California. *American Midland Naturalist* 77:323-55.

Anderson, J. D. 1968. A comparison of the food habits of *Ambystoma macrodactylum sigillatum, Ambystoma macrodactylum croceum,* and *Ambystoma tigrinum californiense. Herpetologica* 24:273-84.

Anderson, S. H. 1970. The avifaunal composition of Oregon white oak stands. *Condor* 72:417-23.

Anderson, S. H. 1972. Seasonal variations in forest birds of western Oregon. *Northwest Science* 46:194-206.

Anderson, S. H. 1976. Comparative food habits of Oregon nuthatches. *Northwest Science* 50:213-21.

Antonelli, A. L., R. A. Nussbaum, and S. D. Smith. 1972. Comparative food habits of four species of stream-dwelling vertebrates (*Dicamptodon ensatus, D. copei, Cottus tenuis, Salmo gairdneri*). *Northwest Science* 46:277-89.

Armitage, K. B., and J. F. Downhower. 1974. Demography of yellow-bellied marmot populations. *Ecology* 55:1233-45.

Armstrong, D. M., and J. K. Jones, Jr. 1971. Sorex merriami. Mammalian Species No. 2:1-2, American Society of Mammalogists.

Armstrong, D. P. 1987. Economics of breeding territoriality in male Calliope Hummingbirds. *Auk* 104:242-53.

Armstrong, E. A. 1956. Territory in the wren Troglodytes troglodytes. *Ibis* 98:430-37.

Armstrong, J. T. 1965. Breeding home range in the nighthawk and other birds; its evolutionary and ecological significance. *Ecology* 46:619-29.

Arthur, S. M., W. B. Krohn, and J. R. Gilbert. 1989. Home range characteristics of adult fishers. *Journal of Wildlife Management* 53:674-79.

Ashmole, N. P. 1968. Competition and interspecific territoriality in *Empidonax* flycatchers. *Systematic Zoology* 17:210-12.

Bailey, V. 1936. The mammals and life zones of Oregon. U.S. Department of Agriculture, Bureau of Biological Survey, North American Fauna No. 55:1-416.

Baird, P. H. 1991. Optimal foraging and intraspecific competition in the Tufted Puffin. *Condor* 93:503-15.

Baker, M. C., and A. E. M. Baker. 1990. Reproductive behavior of female buntings: isolating mechanisms in a hybridizing pair of species. *Evolution* 44:332-38.

Bakus, G. J. 1959. Observations on the life history of the Dipper in Montana. *Auk* 76:190-207.

Baldwin, P. H., and N. K. Zaczkowski. 1963. Breeding biology of the Vaux Swift. *Condor* 65:400-406.

Banko, W. E. 1960. The trumpeter swan: its history, habits, and population in the United States. U.S. Fish and Wildlife Service, North American Fauna No. 63:1-214.

Banks, R. C. 1988. Geographic variation in the Yellow-billed Cuckoo. *Condor* 90:473-77.

Banks, R. C., R. W. McDiarmid, and A. L. Gardner. 1987. Checklist of vertebrates of the United States, the U.S. territories, and Canada. United States Department of the Interior, Fish and Wildlife Service, Resource Publication 166. Washington, D.C.

Barash, D. P. 1989. *Marmots: Social behavior and ecology.* Stanford University Press, Stanford, California.

Barbour, R. W., and W. H. Davis. 1969. *Bats of America.* University Press of Kentucky, Lexington.

Barclay, R. M. R., P. A. Faure, and D. R. Farr. 1988. Roosting behavior and roost selection by migrating silver-haired bats (*Lasionycteris noctivagans*). *Journal of Mammalogy* 69:821-25.

Bartels, M. A., and D. P. Thompson. 1993. Spermophilus lateralis. Mammalian Species No. 440:1-8, American Society of Mammalogists.

Baskett, T. S., M. W. Sayre, and R. E. Tomlinson. 1993. *Ecology and Management of the Mourning Dove*. Stackpole Books and The Wildlife Management Institute, Harrisburg, Pennsylvania.

Batzli, G. O., and F. A. Pitelka. 1970. Influence of meadow mouse populations on a California grassland. *Ecology* 51:1027-39.

Batzli, G. O., and F. A. Pitelka. 1971. Condition and diet of cycling populations of the California vole, *Microtus californicus*. *Journal of Mammalogy* 52:141-63.

Bayer, R. D., R. W. Loe, and R. E. Loeffel. 1991. Persistent summer mortalities of Common Murres along the Oregon central coast. *Condor* 93:516-25.

Beason, R. C., and E. C. Franks. 1974. Breeding behavior of the Horned Lark. *Auk* 91:65-74.

Beaver, D. L., and P. H. Baldwin. 1975. Ecological overlap and the problem of competition and sympatry in the Western and Hammond's Flycatchers. *Condor* 77:1-13.

Bechard, M. J. 1982. Effect of vegetative cover on foraging site selection by Swainson's Hawk. *Condor* 84:153-59.

Beecher, M. D., and I. M. Beecher. 1979. Sociobiology of bank swallows: reproductive strategy of the male. *Science* 205:1282-85.

Behle, W. H. 1976. Systematic review, intergradation, and clinal variation in Cliff Swallows. *Auk* 93:66-77.

Beier, P. 1993. Determining minimum habitat areas and habitat corridors for cougars. *Conservation Biology* 7:94-108.

Bekoff, M. 1977. Canis latrans. Mammalian Species No. 79:1-9, American Society of Mammalogists.

Bekoff, M., A. C. Scott, and D. A. Conner. 1987. Nonrandom nest-site selection in Evening Grosbeaks. *Condor* 89:819-29.

Belk, M. C., and H. D. Smith. 1991. Ammospermophilus leucurus. Mammalian Species No. 368:1-8, American Society of Mammalogists.

Bellrose, F. C. 1980. *Ducks, Geese, and Swans of North America*, Third Edition. Stackpole Books, Harrisburg, Pennsylvania.

Bellrose, F. C., K. L. Johnson, and T. U. Meyers. 1964. Relative value of natural cavities and nesting houses for wood ducks. *Journal of Wildlife Management* 28:661-76.

Bendell, J. F., and P. W. Elliot. 1966. Habitat selection in Blue Grouse. *Condor* 68:431-46.

Beneski, J. T., Jr., and D. W. Stinson. 1987. Sorex palustris. Mammalian Species No. 296:1-6, American Society of Mammalogists.

Benkman, C. W. 1990. Intake rates and the timing of crossbill reproduction. *Auk* 107:376-86.

Bennett, W. A. 1990. Scale of investigation and the detection of competition: an example from the House Sparrow and House Finch introductions in North America. *American Naturalist* 135:725-47.

Bennion, R. S., and W. S. Parker. 1976. Field observations on courtship and aggressive behavior in desert striped whipsnakes, *Masticophis t. taeniatus*. *Herpetologica* 32:30-35.

Berger, J. 1990. Persistence of different-sized populations: an empirical assessment of rapid extinctions in bighorn sheep. *Conservation Biology* 4:91-98.

Bergman, R. D., P. Swain, and N. W. Weller. 1970. A comparative study of nesting Forster's and Black Terns. *Wilson Bulletin* 82:435-44.

Best, T. L. 1996. Lepus californicus. Mammalian Species No. 530:1-10, American Society of Mammalogists.

Betts, B. J. 1990. Geographic distribution and habitat preferences of Washington ground squirrels (*Spermophilus washingtoni*). *Northwestern Naturalist* 71:27-37.

Blanchard, F. N. 1942. The ring-neck snakes, genus *Diadophis*. *Bulletin of the Chicago Academy of Science* 7:1-44.

Blaney, R. M. 1977. Systematics of the common kingsnake, *Lampropeltis getulus* (Linnaeus). *Tulane Studies in Zoology and Botany* No. 19:74-103.

Blaustein, A. R., and D. B. Wake. 1990. Declining amphibian populations: A global phenomenon? *Trends in Ecology and Evolution* 5:203-4.

Block, W. M., and L. A. Brennan. 1987. Characteristics of Lewis' Woodpecker habitat on the Modoc Plateau, California. *Western Birds* 18:209-12.

Bock, C. E. 1969. Intra- vs. interspecific aggression in Pygmy Nuthatch flocks. *Ecology* 50:903-5.

Bock, C. E. 1970. The ecology and behavior of the Lewis woodpecker (Asyndesmus lewis). *University of California Publications in Zoology* 92:1-100.

Bock, C. E., and J. H. Bock. 1974. On the geographical ecology and evolution of the three-toed woodpeckers, Picoides tridactylus and Picoides arcticus. *American Midland Naturalist* 92:397-405.

Bollinger, E. K., and T. A. Gavin. 1989. The effects of site quality on breeding-site fidelity in Bobolinks. *Auk* 106:584-94.

Bragg, A. N. 1965. *Gnomes of the Night*. University of Pennsylvania Press, Philadelphia, Pennsylvania.

Brawn, J. D., and R. P. Balda. 1988. Population biology of cavity nesters in northern Arizona: do nest sites limit breeding densities. *Condor* 90:61-71.

Bray, O. E., K. H. Larsen, and D. F. Mott. 1975. Winter movements and activities of radio-equipped starlings. *Journal of Wildlife Management* 39:795-801.

Brennan, L. A., W. M. Block, and R. J. Gutierrez. 1987. Habitat use by Mountain Quail in northern California. *Condor* 89:66-74.

Brennan, L. A., and M. L. Morrison. 1991. Long-term trends of chickadee populations in western North America. *Condor* 93:130-37.

Brewer, R. 1963. Ecological and reproductive relationships of Black-capped and Carolina Chickadees. *Auk* 80:9-47.

Briggs, J. L., and R. M. Storm. 1970. Growth and population structure of the Cascade frog, *Rana cascadae* Slater. *Herpetologica* 26:283-300.

Briggs, J. L., Sr. 1987. Breeding biology of the Cascade frog, *Rana cascadae*, with comparisons to *R. aurora* and *R. pretiosa. Copeia* 1987:241-45.

Brigham, R. M. 1991. Flexibility in foraging and roosting behaviour by the big brown bat (*Eptesicus fuscus*). *Canadian Journal of Zoology* 69:117-21.

Broadbrooks, H. E. 1961. Ring-billed Gulls nesting on Columbia River islands. *Murrelet* 42:7-8.

Broadbrooks, H. E. 1970. Home ranges and territorial behavior of the yellow-pine chipmunk, *Eutamias amoenus. Journal of Mammalogy* 51:310-26.

Brodie, E. D., Jr. 1968. Investigations of the skin toxin of the adult rough-skinned newt, *Taricha granulosa. Copeia* 1968:307-13.

Brodie, E. D., Jr. 1970. Western salamanders of the genus *Plethodon*: systematics and geographic variation. *Herpetologica* 26:468-516.

Brodie, E. D., Jr., R. A. Nussbaum, and R. M. Storm. 1969. An egg-laying aggregation of five species of Oregon reptiles. *Herpetologica* 25:223-27.

Brooks, R. P., and W. J. Davis. 1987. Habitat selection by breeding belted kingfishers (Ceryle alcyon). *American Midland Naturalist* 117:63-70.

Brown, C. W. 1974. Hybridization among the subspecies of the plethodontid salamander Ensatina eschscholtzi. *University of California Publications in Zoology* 98:1-57.

Brown, C. W., and M. B. Brown. 1988. The costs and benefits of egg destruction by conspecifics in colonial Cliff Swallows. *Auk* 105:737-48.

Brown, H. A. 1975. Temperature and development of the tailed frog, *Ascaphus truei. Comparative Biochemistry and Physiology* 50A:397-405.

Brown, J. L. 1964. The integration of agonistic behavior in the Steller's jay, Cyanocitta stelleri (Gmelin). *University of California Publications in Zoology* 60:223-328.

Brown, J. L. 1974. Alternate routes to sociality in jays—with a theory for the evolution of altruism and communal breeding. *American Zoologist* 14:63-80.

Brown, W. S., and W. S. Parker. 1976. Movement ecology of *Coluber constrictor* near communal hibernacula. *Copeia* 1976:225-42.

Browning, M. R., and W. W. English. 1972. Breeding birds of selected Oregon coastal islands. *Murrelet* 53:1-7.

Buchanan, J. B., R. W. Lundquist, and K. B. Aubry. 1990. Winter populations of Douglas' squirrels in different-aged Douglas-fir forests. *Journal of Wildlife Management* 54:577-81.

Bull, E. L. 1987. Ecology of the pileated woodpecker in northeastern Oregon. *Journal of Wildlife Management* 51:472-81.

Bull, E., and R. G. Anderson. 1978. Notes on Flammulated Owls in northeastern Oregon. *Murrelet* 59:26-28.

Bull, E. L., and M. G. Henjum. 1990. Ecology of the Great Gray Owl. U.S.D.A. Forest Service, Pacific Northwest Research Station, General Technical Report PNW-GTR-265.

Bull, E. L., G. E. Hohmann, and M. G. Henjum. 1987. Northern Pygmy-Owl nests in northeastern Oregon. *Journal of Raptor Research* 21:77-78.

Bull, E. L., and J. A. Jackson. 1995. Pileated Woodpecker (*Dryocopus pileatus*). *In The Birds of North America*, No. 148 (A. Poole and F. Gill, eds.). The Academy of Natural Sciences, Philadelphia, PA, and The American Ornithologists' Union, Washington, D.C.

Bull, E. L., A. L. Wright, and M. G. Henjum. 1989. Nesting and diet of Long-eared Owls in conifer forests, Oregon. *Condor* 91:908-12.

Bull, E. L., A. L. Wright, and M. G. Henjum. 1990. Nesting habitat of Flammulated Owls in Oregon. *Journal of Raptor Research* 24:52-55.

Bump, S. R. 1986. Yellow-headed Blackbird nest defense: aggressive responses to Marsh Wrens. *Condor* 88:328-35.

Burns, K. J., and R. M. Zink. 1990. Temporal and geographic homogeneity of gene frequencies in the Fox Sparrow (*Passerella iliaca*). *Auk* 107:421-25.

Burton, D. H., and H. C. Black. 1978. Feeding habits of Mazama pocket gophers in south-central Oregon. *Journal of Wildlife Management* 42:383-90.

Bury, R. B. 1972. Small mammals and other prey in the diet of the Pacific giant salamander (Dicamptodon ensatus). *American Midland Naturalist* 87:524-26.

Bury, R. B. 1986. Feeding ecology of the turtle, *Clemmys marmorata. Journal of Herpetology* 20:515-21.

Bury, R. B., and C. R. Johnson. 1965. Note on the food of *Plethodon elongatus* in California. *Herpetologica* 21:67-68.

Buskirk, S. W., and L. L. McDonald. 1989. Analysis of variability in home-range size of the American marten. *Journal of Wildlife Management* 53:997-1004.

Butterfield, B. R., B. Csuti, and J. M. Scott. 1994. Modeling vertebrate distributions for Gap Analysis. Pages 53-68 *in Mapping the Diversity of Nature* (R. I. Miller, editor). Chapman & Hall, London.

Caccamise, D. F. 1974. Competitive relationships of the Common and Lesser Nighthawks. *Condor* 56:1-20.

Cairns, D. K., K. A. Bredin, and W. A. Montevecchi. 1987. Activity budgets and foraging ranges of breeding Common Murres. *Auk* 104:218-24.

Calder, W. A. 1971. Temperature relationships and nesting of the Calliope Hummingbird. *Condor* 73:314-21.

Carey, A. B., S. P. Horton, and B. L. Biswell. 1992. Northern Spotted Owls: influence of prey base and landscape character. *Ecological Monographs* 62:223-50.

Carey, A. B., J. A. Reid, and S. P. Horton. 1990. Spotted owl home range and habitat use in southern Oregon Coast Ranges. *Journal of Wildlife Management* 54:11-17.

Carraway, L. N. 1985. Sorex pacificus. Mammalian Species No. 231:1-5, American Society of Mammalogists.

Carraway, L. N. 1990. A morphologic and morphometric analysis of the "Sorex vagrans species complex" in the Pacific Coast region. Special Publications, The Museum, Texas Tech University, No. 32:1-76.

Carraway, L. N., L. F. Alexander, and B. J. Verts. 1993. Scapanus townsendii. Mammalian Species No. 434:1-7, American Society of Mammalogists.

Carraway, L. N., and P. K. Kennedy. 1993. Genetic variation in *Thomomys bulbivorous*, an endemic to the Willamette Valley, Oregon. *Journal of Mammalogy* 74:952-62.

Carraway, L. N., E. Yensen, B. J. Verts, and L. F. Alexander. 1993. Range extension and habitat of *Peromyscus truei* in eastern Oregon. *Northwestern Naturalist* 74:81-84.

Carraway, L. N., and B. J. Verts. 1985. Microtus oregoni. Mammalian Species No. 233:1-6, American Society of Mammalogists.

Carraway, L. N., and B. J. Verts. 1991a. Neurotrichus gibbsii. Mammalian Species No. 387:1-7, American Society of Mammalogists.

Carraway, L. N., and B. J. Verts. 1991b. Neotoma fuscipes. Mammalian Species No. 386:1-10, American Society of Mammalogists.

Carraway, L. N., and B. J. Verts. 1993. Aplodontia rufa. Mammalian Species No. 431:1-10, American Society of Mammalogists.

Carraway, L. N., and B. J. Verts. 1994. Sciurus griseus. Mammalian Species No. 474:1-7, American Society of Mammalogists.

Carroll, L. E., and H. H. Genoways. 1980. Lagurus curtatus. Mammalian Species No. 124:1-6, American Society of Mammalogists.

Cassirer, E. F., C. R. Groves, and R. L. Wallen. 1991. Distribution and population status of Harlequin Ducks in Idaho. *Wilson Bulletin* 103:723-25.

Chamberlain-Auger, J. A., P. J. Auger, and E. G. Strauss. 1990. Breeding biology of American Crows. *Wilson Bulletin* 102:615-22.

Chapman, J. A. 1971. Orientation and homing in the brush rabbit (*Sylvilagus bachmani*). *Journal of Mammalogy* 52:686-99.

Chapman, J. A. 1974. Sylvilagus bachmani. Mammalian Species No. 34:1-4, American Society of Mammalogists.

Chapman, J. A. 1975. Sylvilagus nuttallii. Mammalian Species No. 56:1-3, American Society of Mammalogists.

Chapman, J. A., J.G. Hockman, and M.M. Ojeda. 1980. Sylvilagus floridanus. Mammalian Species No. 136:1-8, American Society of Mammalogists.

Christensen, G. C. 1970. The Chukar Partridge: its introduction, life history, and management. Biological Bulletin 4, Nevada Department of Fish and Game. State Printing Office, Reno, Nevada.

Clark, R. J. 1975. A field study of the Short-eared Owl in North America. Wildlife Monographs No. 47:1-67.

Clark, T. W., E. Anderson, C. Douglas, and M. Strickland. 1987. Martes americana. Mammalian Species No. 289:1-8, American Society of Mammalogists.

Cochran, J. F., and S. H. Anderson. 1987. Comparison of habitat attributes at sites of stable and declining long-billed curlew populations. *Great Basin Naturalist* 47:459-66.

Cody, M. L., and C. B. J. Cody. 1972. Territory size, clutch size, and food in populations of wrens. *Condor* 74:473-77.

Coleman, J. S., and J. D. Fraser. 1989. Habitat use and home ranges of black and turkey vultures. *Journal of Wildlife Management* 53:782-92.

Collazo, J. A. 1981. Some aspects of the breeding ecology of the Great Blue Heron at Heyburn State Park. *Northwest Science* 55:293-97.

Collins, J. T. 1990. Standard common and current scientific names for North American amphibians and reptiles, Third Edition. Society for the Study of Amphibians and Reptiles, Herpetological Circular No. 19:i-iii, 1-41.

Collopy, M. W. 1977. Food caching by female American Kestrels in winter. *Condor* 79:63-68.

Collopy, M. W. 1984. Parental care and feeding ecology of Golden Eagle nestlings. *Auk* 101:753-60.

Colvin, M. A., and D. V. Colvin. 1970. Breeding and fecundity of six species of voles (*Microtus*). *Journal of Mammalogy* 51:417-19.

Colwell, M. A. 1986. The first documented case of polyandry for Wilson's Phalarope (*Phalaropus tricolor*). *Auk* 103:611-12.

Conaway, C. H. 1952. Life history of the water shrew (Sorex palustris navigator). *American Midland Naturalist* 48:219-48.

Congdon, J. D., and R. E. Gatten, Jr. 1989. Movements and energetics of nesting *Chrysemys picta. Herpetologica* 45:94-100.

Connelly, J. W., H. W. Browers, and R. J. Gates. 1988. Seasonal movements of sage grouse in southeastern Idaho. *Journal of Wildlife Management* 52:116-22.

Connelly, J. W., W. L. Wakkinen, A. D. Apa, and K. P. Reese. 1991. Sage grouse use of nest sites in southeastern Idaho. *Journal of Wildlife Management* 55:521-24.

Contreras, A. 1988. Northern Waterthrush summer range in Oregon. *Western Birds* 19:41-42.

Contreras, A. 1992. Winter status of the Sora in the Pacific Northwest. *Western Birds* 23:137-42.

Cook, S.F., Jr. 1960. On the occurrence and life history of *Contia tenius. Herpetologica* 16: 163-73.

Corn, P. S., and R. B. Bury. 1986a. Morphological variation and zoogeography of racers (*Coluber constrictor*) in the central Rocky Mountains. *Herpetologica* 42:258-64.

Corn, P. S., and R. B. Bury. 1986b. Habitat use and terrestrial activity by red tree voles (*Arborimus longicaudus*) in Oregon. *Journal of Mammalogy* 67:404-6.

Corn, P. S., and R. B. Bury. 1989. Logging in western Oregon: responses of headwater habitats and stream amphibians. *Forest Ecology and Management* 29:39-57.

Cornely, J. E., L. N. Carraway, and B. J. Verts. 1992. Sorex preblei. Mammalian Species No. 416:1-3, American Society of Mammalogists.

Cornely, J. E., and B. J. Verts. 1988. Microtus townsendii. Mammalian Species No. 325:1-9, American Society of Mammalogists.

Coulter, M. C. 1975. Post-breeding movements and mortality in the Western Gull (*Larus occidentalis*). *Condor* 77:243-49.

Coutlee, E. L. 1968. Comparative breeding behavior of Lesser and Lawrence's Goldfinches. *Condor* 70:228-42.

Cranford, J. A. 1983. Ecological strategies of a small hibernator, the western jumping mouse *Zapus princeps*. *Canadian Journal of Zoology* 61:232-40.

Crockett, A. B., and H. H. Hadow. 1975. Nest site selection by Williamson's and Red-naped Sapsuckers. *Condor* 77:365-68.

Cross, S. P., J. M. Barss, and J. Peterson. 1990. Winter records of bats in Oregon and Washington. *Northwestern Naturalist* 71:59-62.

Crouch, G. L. 1968. Clipping of woody plants by mountain beaver. *Journal of Mammalogy* 49:151-52.

Cruz, A. 1976. Food and foraging ecology of the American Kestrel in Jamaica. *Condor* 78:409-12.

Csuti, B. A. 1971. Distribution of some southern California kangaroo rats. *Bulletin of the Southern California Academy of Sciences* 70:50-51.

Csuti, B. A. 1979. Patterns of adaptation and variation in the Great Basin kangaroo rat (*Dipodomys microps*). *University of California Publications in Zoology* 111:1-69.

Csuti, B. A. 1996. Mapping animal distribution areas for Gap Analysis. Pages 135-145 *in Gap Analysis: A landscape approach to biodiversity planning* (J. M. Scott, T. H. Tear, and F. W. Davis, editors). American Society for Photogrammetry and Remote Sensing, Bethesda, Maryland.

Currier, M. J. P. 1983. Felis concolor. Mammalian Species No. 200:1-7, American Society of Mammalogists.

Cuthbert, F. J. 1985. Mate retention in Caspian Terns. *Condor* 87:74-78.

Cuthbert, F. J. 1988. Reproductive success and colony-site tenacity in Caspian Terns. *Auk* 105:339-44.

Dalquest, W. W. 1947. Notes on the natural history of the bat, Myotis yumanensis, in California, with a description of a new race. *American Midland Naturalist* 38:224-47.

Dalquest, W. W., and D. R. Orcutt. 1942. The biology of the least shrew-mole, Neurotrichus gibbsii minor. *American Midland Naturalist* 27:387-401.

Daly, J. C., and J. L. Patton. 1986. Growth, reproduction, and sexual dimorphism in *Thomomys bottae* pocket gophers. *Journal of Mammalogy* 67:256-65.

Daly, J. C., and J. L. Patton. 1990. Dispersal, gene flow, and allelic diversity between local populations of

*Thomomys bottae* pocket gophers in the coastal ranges of California. *Evolution* 44:1283-94.

Daneke, D., M. Sunquist, and S. Berwick. 1984. Notes on kit fox biology in Utah. *Southwestern Naturalist* 29:361-62.

Daugherty, C. H., and A. L. Sheldon. 1982. Age-specific movement patterns of the frog *Ascaphus truei*. *Herpetologica* 38:468-74.

Davis, C. M. 1978. A nesting study of the Brown Creeper. *Living Bird* 17:237-63.

Davis, D. E. 1955. Observations on the breeding biology of kingbirds. *Condor* 57:208-12.

Davis, J. 1951. Distribution and variation of the brown towhees. *University of California Publications in Zoology* 52:1-120.

Davis, J. 1957. Comparative foraging behavior of the Spotted and Brown towhees. *Auk* 74:129-66.

Davis, J. 1960. Nesting behavior of the Rufous-sided Towhee in coastal California. *Condor* 62:434-56.

Davis, J., G. F. Fisler, and B. S. Davis. 1963. The breeding biology of the Western Flycatcher. *Condor* 65:337-82.

Davis, J., and R. G. Ford. 1983. Home range in the western fence lizard (*Sceloporus occidentalis occidentalis*). *Copeia* 1983:933-40.

Davis, J., and N. A. M. Verbeek. 1972. Habitat preferences and the distribution of *Uta stansburiana* and *Sceloporus occidentalis* in coastal California. *Copeia* 1972:643-49.

Davis, W. C., and V. C. Twitty. 1964. Courtship behavior and reproductive isolation in the species of *Taricha* (Amphibia, Caudata). *Copeia* 1964:601-10.

DeGraaf, R. M., and D. D. Rudis. 1983. New England Wildlife: Habitat, Natural History, and Distribution. Northeastern Forest Experiment Station, General Technical Report NE-108.

DeGraaf, R. M., V. E. Scott, R. H. Hamre, L. Ernst, and S. H. Anderson. 1991. Forest and rangeland birds of the United States: Natural history and habitat use. U.S. Department of Agriculture, Forest Service, Agriculture Handbook 688.

Derrickson, S. R. 1978. The mobility of breeding Pintails. Auk 95:104-114.

Desrochers, A., S. J. Hannon, and K. E. Nordin. 1988. Winter survival and territory acquisition in a northern population of Black-capped Chickadees. *Auk* 105:727-36.

Dick, J. A., and J. D. Rising. 1965. A comparison of foods eaten by eastern kingbirds and western kingbirds in Kansas. *Kansas Ornithological Society Bulletin* 16:23-24.

Dickson, J. G. 1992. *The Wild Turkey: Biology and management*. Stackpole Books, Harrisburg, Pennsylvania.

Diller, L. V., and D. R. Johnson. 1988. Food habits, consumption rates, and predation rates of western rattlesnakes and gopher snakes in southwestern Idaho. *Herpetologica* 44:228-33.

Diller, L. V., and R. L. Wallace. 1984. Reproductive biology of the northern Pacific rattlesnake (*Crotalus viridis oreganus*) in northern Idaho. *Herpetologica* 40:182-93.

Diller, L. V., and R. L. Wallace. 1986. Aspects of the life history and ecology of the desert night snake, *Hypsiglena torquata deserticola*: Colubridae, in southwestern Idaho. *Southwestern Naturalist* 31:55-64.

Dinsmore, J. J. 1973. Foraging success of cattle egrets, Bubulcus ibis. *American Midland Naturalist* 89:242-46.

Dixon, K. L. 1949. Behavior of the Plain Titmouse. *Condor* 51:110-36.

Dixon, K. L. 1954. Some ecological relations of chickadees and titmice in central California. *Condor* 56:113-24.

Dixon, K. L. 1956. Territoriality and survival in the Plain Titmouse. *Condor* 58:169-82.

Doak, D. 1989. Spotted Owls and old growth logging in the Pacific Northwest. *Conservation Biology* 3:389-96.

Dobkin, D. S., A. C. Rich, J. A. Pretare, and W. H. Pyle. 1995. Nest-site relationships among cavity-nesting birds of riparian and snow-pocket aspen woodlands in the northwestern Great Basin. *Condor* 97:694-707.

Dobson, F. S., and D. E. Davis. 1986. Hibernation and sociality in the California ground squirrel. *Journal of Mammalogy* 67:416-21.

Douglas, C. L. 1969. Comparative ecology of pinyon mice and deer mice in Mesa Verde National Park, Colorado. University of Kansas Publications, Museum of Natural History 18:421-504.

Douglas, D. C., J. T. Ratti, R. A. Black, and J. R. Alldredge. 1992. Avian habitat associations in riparian zones of Idaho's Centennial Mountains. *Wilson Bulletin* 104:485-500.

Drilling, N. E., and C. F. Thompson. 1988. Natal and breeding dispersal in House Wrens (*Troglodytes aedon*). *Auk* 105:480-91.

Dumas, P. C. 1955. Eggs of the salamander *Plethodon dunni* in nature. *Copeia* 1955:65.

Dumas, P. C. 1956. The ecological relations of sympatry in *Plethodon dunni* and *Plethodon vehiculum*. *Ecology* 37:484-95.

Dumas, P. C. 1964. Species-pair allopatry in the genera *Rana* and *Phrynosoma*. *Ecology* 45:178-81.

Dumas, P. C. 1966. Studies of the *Rana* species complex in the Pacific Northwest. *Copeia* 1966:60-74.

Eckhardt, R. C. 1977. Effects of a late spring storm on a local Dusky Flycatcher population. *Auk* 94:362.

Edwards, R. Y. 1955. The habitat preferences of the boreal Phenacomys. *Murrelet* 36:35-38.

Eisenberg, J. F. 1964. Studies on the behavior of Sorex vagrans. *American Midland Naturalist* 72:417-25.

Eisenman, E. 1971. Range expansion and population increase in North and Middle America of the White-tailed Kite. *American Birds* 25:529-36.

Eklund, C. R. 1942. Ecological and mortality factors affecting the nesting of the Chinese Pheasant in the Willamette Valley. *Journal of Wildlife Management* 6:225-30.

Elliott, C. L., and J. T. Flinders. 1991. Spermophilus columbianus. Mammalian Species No. 372:1-9, American Society of Mammalogists.

Emlen, S. T., J. D. Rising, and W. L. Thompson. 1975. A behavioral and morphological study of sympatry in the Indigo and Lazuli Buntings of the Great Plains. *Wilson Bulletin* 87:145-77.

Erickson, M. M. 1938. Territory, annual cycle, and numbers in a population of wren-tits (Chamea fasciata). *University of California Publications in Zoology* 42:247-334.

Erpino, M. J. 1968. Nest-related activities of Black-billed Magpies. *Condor* 70:154-65.

Escherich, P. C. 1981. Social biology of the bushy-tailed woodrat, *Neotoma cinerea*. *University of California Publications in Zoology* 110:1-132.

Ettinger, A. O., and J. R. King. 1980. Time and energy budgets of the Willow Flycatcher (*Empidonax trallii*) during the breeding season. *Auk* 97:533.

Falk, J. W., and J. S. Millar. 1987. Reproduction by female *Zapus princeps* in relation to age, size, and body fat. *Canadian Journal of Zoology* 65:568-71.

Farner, D. S. 1947. Notes on the food habits of the salamanders of Crater Lake, Oregon. *Copeia* 1947:259-61.

Feare, C. 1984. *The Starling*. Oxford University Press, Oxford, United Kingdom.

Fears, O. T., III. 1975. Observations on the aerial drinking performance of a Poor-will. *Wilson Bulletin* 87:284.

Fenton, M. B., and R. M. R. Barclay. 1980. Myotis lucifugus. Mammalian Species No. 142:1-8, American Society of Mammalogists.

Fenton, M. B., and G. P. Bell. 1979. Echolocation and feeding behaviour in four species of *Myotis* (Chiroptera). *Canadian Journal of Zoology* 57:1271-77.

Fenton, M. B., D. C. Tennant, and J. Wyszecki. 1983. A survey of the distribution of *Euderma maculatum* (Chiroptera: Vespertilionidae) throughout its known range in the United States and Canada by monitoring its audible echolocation calls. U.S. Fish and Wildlife Service, Albuquerque, New Mexico. 44pp.

Fenton, M. B., D. C. Tennant, and J. Wyszecki. 1987. Using echolocation calls to measure distribution of bats: the case of *Euderma maculatum*. *Journal of Mammalogy* 68:142-44.

Ferguson, D. E. 1961. The geographic variation of Ambystoma macrodactylum Baird, with the description of two new subspecies. *American Midland Naturalist* 65:311-38.

Ficken, M. S., and R. W. Ficken. 1965. Territorial display as a population-regulating mechanism in the Yellow Warbler. *Auk* 82:274-75.

Ficken, M. S., and R. W. Ficken. 1966. Notes on mate and habitat selection in the Yellow Warbler. *Wilson Bulletin* 78:232-33.

Findley, J. S. 1955. Speciation of the wandering shrew. University of Kansas Publications, Museum of Natural History 9:1-68.

Findley, J. S., and C. Jones. 1964. Seasonal distribution of the hoary bat. *Journal of Mammalogy* 45:461-70.

**453**

Fitch, H. S. 1965. An ecological study of the garter snake, *Thamnophis sirtalis*. University of Kansas Publications, Museum of Natural History 15:493-564.

Fitch, H. S. 1975. A demographic study of the ringneck snake (*Diadophis punctatus*) in Kansas. University of Kansas, Museum of Natural History, Miscellaneous Publication 62:1-53.

Fitch, H. S. 1980. *Thamnophis sirtalis. Catalog of American Amphibians and Reptiles* 270:1-4.

Fitch, H. S. 1984. *Thamnophis couchii. Catalog of American Amphibians and Reptiles* 351:1-3.

Fitch, H. S., and W. S. Brown. 1981. *Coluber mormon*, a species distinct from *Coluber constrictor*. Transactions of the Kansas Academy of Science.

Fitzgerald, J. P., C. A. Meaney, and D. M. Armstrong. 1994. *Mammals of Colorado*. Denver Museum of Natural History and University Press of Colorado, Niwot, Colorado.

Forsman, E., and C. Maser. 1970. Saw-whet Owl preys on red tree mice. *Murrelet* 51:10.

Foster, J. B. 1961. Life history of the Phenacomys vole. *Journal of Mammalogy* 42:181-98.

Franklin, A. B. 1988. Breeding biology of the Great Gray Owl in southeastern Idaho and northwestern Wyoming. *Condor* 90:689-96.

Franklin, J. F., and C. T. Dyrness. 1973. Natural vegetation of Oregon and Washington. USDA Forest Service, General Technical Report PNW-8. Pacific Northwest Forest and Range Experiment Station, Portland, Oregon. Reprinted Oregon State University Press, 1988.

Franzreb, K. E. 1976. Nest site competition between Mountain Chickadees and Violet-green Swallows. *Auk* 93:836-37.

Franzreb, K. E., and R. D. Ohmart. 1978. The effects of timber harvesting on breeding birds in a mixed-coniferous forest. *Condor* 80:431-41.

Frase, B. A., and R. S. Hoffmann. 1980. Marmota flaviventris. Mammalian Species No. 135:1-8, American Society of Mammalogists.

French, N. R. 1954. Notes on breeding activities and on gular sacs in the Pine Grosbeak. *Condor* 56:83-85.

French, N. R. 1959. Life history of the Black Rosy Finch. *Auk* 76:159-80.

Friedmann, H., L. F. Kiff, and S. I. Rothstein. 1977. A further contribution to knowledge of the host relations of the parasitic cowbirds. *Smithsonian Contributions in Zoology* No. 235:1-75.

Friesen, V.L., J.F. Piatt, and A.J. Baker. 1996. Evidence from cytochrome *B* sequences and allozymes for a 'new' species of alcid: the Long-billed Murrelet (*Brachyramphus perdix*). *Condor* 98: 681-90.

Fritzell, E. K., and K. J. Haroldson. 1982. Urocyon cinereoargenteus. Mammalian Species No. 189:1-8, American Society of Mammalogists.

Frost, D. R. 1983. *Sonora semiannulata. Catalog of American Amphibians and Reptiles* 333:1-4.

Frost, D. R., and R. M. Timm. 1992. Phylogeny of plecotine bats (Chiroptera:"Vespertilionidae"):

proposal of a logically consistent taxonomy. *American Museum Novitates* 3034:1-16.

Gabrielson, I. N., and S. G. Jewett. 1940. *Birds of Oregon*. Oregon State College, Corvallis.

Gaines, D., and S. A. Laymon. 1984. Decline, status and preservation of the Yellow-billed Cuckoo in California. *Western Birds* 15:49-80.

Galati, R. 1991. *Golden-crowned Kinglets: Treetop nesters of the north woods*. Iowa State University Press.

Gannon, W. L. 1988. Zapus trinotatus. Mammalian Species No. 315:1-5, American Society of Mammalogists.

Gannon, W. L., and R. B. Forbes. 1995. Tamias senex. Mammalian Species No. 502:1-6, American Society of Mammalogists.

Gannon, W. L., and T. E. Lawlor. 1989. Variation of the chip vocalizations of three species of Townsend chipmunks (genus *Eutamias*). *Journal of Mammalogy* 70:740-53.

Garrison, T. E., and T. L. Best. 1990. Dipodomys ordii. Mammalian Species No. 353:1-10, American Society of Mammalogists.

Gashwiler, J. S. 1972. Life history notes on the Oregon vole, *Microtus oregoni. Journal of Mammalogy* 53:558-69.

Gashwiler, J. S. 1976a. Notes on reproduction of the Trowbridge shrews in western Oregon. *Murrelet* 57:58-62.

Gashwiler, J. S. 1976b. Biology of Townsend's chipmunks in western Oregon. *Murrelet* 57:26-31.

Gashwiler, J. S. 1977. Bird populations in four vegetational types in central Oregon. Special Scientific Report, Wildlife, No. 205:1-20. United States Department of the Interior, Fish and Wildlife Service, Washington, D.C.

Gashwiler, J. S., and A. L. Ward. 1966. Western redcedar seed, a food of pine siskins. *Murrelet* 47:73-75.

Gass, C. L., G. Angehr, and J. Centa. 1976. Regulation of food supply by feeding territoriality in the rufous hummingbird. *Canadian Journal of Zoology* 54:2046-54.

Gates, J. M. 1962. Breeding biology of the Gadwall in northern Utah. *Wilson Bulletin* 74:43-67.

Gavin, T. A., and B. May. 1988. Taxonomic status and genetic purity of Columbian white-tailed deer. *Journal of Wildlife Management* 52:1-10.

Gavin, T. A., L. H. Suring, P.A. Vohs, Jr., and E. C. Meslow. 1984. Population characteristics, spatial organization, and natural mortality in the Columbian white-tailed deer. *Wildlife Monograph* No. 91:1-41.

Genoways, H. H., and J. H. Brown. 1993. Biology of the Heteromyidae. Special Publication No. 10, American Society of Mammalogists.

George, S. B. 1988. Systematics, historical biogeography, and evolution of the genus *Sorex. Journal of Mammalogy* 69:443-61.

George, S. B. Sorex trowbridgii. 1989. Mammalian Species No. 337:1-5, American Society of Mammalogists.

Gettinger, R. D. 1984. A field study of activity patterns of *Thomomys bottae*. *Journal of Mammalogy* 65:76-84.

Geupel, G. R., and D. F. DeSante. 1990. Incidence and determinants of double brooding in Wrentits. *Condor* 92:67-75.

Gibb, J. 1956. Food, feeding habits and territory of the Rock Pipit *Anthus spinoletta*. *Ibis* 98:506-30.

Gibbons, J. W. 1968. Population structure and survivorship in the painted turtle, *Chrysemys picta*. *Copeia* 1968:260-68.

Gibson, F. 1971. The breeding biology of the American Avocet (*Recurvirostra americana*) in central Oregon. *Condor* 73:444-54.

Giger, R. D. 1973. Movements and homing in Townsend's mole near Tillamook, Oregon. *Journal of Mammalogy* 54:648-59.

Gill, F. B., and B. Slikas. 1992. Patterns of mitochondrial DNA divergence in North American crested titmice. *Condor* 94:20-28.

Gill, R. E. 1976. Notes on the foraging of nesting Caspian terns, *Hydroprogne caspia* (Pallas). *California Fish and Game* 62:155.

Gill, R. E., Jr., and L. R. Mewaldt. 1983. Pacific Coast Caspian Terns: dynamics of an expanding population. *Auk* 100:369-81.

Gilligan, J., M. Smith, D. Rogers, and A. Contreras. 1994. *Birds of Oregon: Status and distribution.* Cinclus Publications, McMinnville, Oregon.

Gladson, J. 1981. Green Heron. *Oregon Wildlife* 36(6):11.

Glover, F. A. 1956. Nesting and production of the blue-winged teal (*Anas discors* Linnaeus) in northwest Iowa. *Journal of Wildlife Management* 20:28-46.

Godfrey, W. E. 1966. *The birds of Canada.* National Museums of Canada, Bulletin No. 203, Biological Series No. 73.

Goertz, J. W. 1964. Habitats of three Oregon voles. *Ecology* 45:846-48.

Goldberg, S. R. 1972. Reproduction in the southern alligator lizard *Gerrhonotus multicarinatus*. *Herpetologica* 28:267-73.

Good, D. A. 1988. Allozyme variation and phylogenetic relationships among the species of *Elgaria* (Squamata: Anguidae). *Herpetologica* 44:154-62.

Good, D. A. 1989. Hybridization and cryptic species in *Dicamptodon* (Caudata:Dicamptodontidae). *Evolution* 43:728-44.

Good, D. A., and D. B. Wake. 1992. Geographic variation and speciation in the torrent salamanders of the genus *Rhyacotriton* (Caudata:Rhyacotritionidae). *University of California Publications in Zoology* 126:1-91.

Good, D. A., G. Z. Wurst, and D. B. Wake. 1987. Patterns of geographic variation in allozymes of the Olympic salamander, *Rhyacotriton olympicus* (Caudata:Dicamptodontidae). *Fieldiana:Zoology, New Series*, No. 32:1-15.

Goodwin, D. 1986. *Crows of the World*, Second Edition. University of Washington Press, Seattle.

Graf, W. 1955. Cottontail rabbit introductions and distribution in western Oregon. *Journal of Wildlife Management* 19:184-88.

Green, D. M. 1986. Systematics and evolution of western North American frogs allied to *Rana aurora* and *Rana boylii*: electrophoretic evidence. *Systematic Zoology* 35:283-96.

Green, D.M., H. Kaiser, T.F. Sharbel, J. Kearsley, and K.R. McAllister. 1997. Cryptic species of spotted frogs, *Rana pretiosa* complex, in western North America. *Copeia* 1997: 1-8.

Green, D. M., T. F. Sharbel, J. Kearsley, and H. Kaiser. 1996. Postglacial range fluctuation, genetic subdivision and speciation in the western North American spotted frog complex, *Rana pretiosa*. *Evolution* 50:374-90.

Green, G. A., and R. G. Anthony. 1989. Nesting success and habitat relationships of Burrowing Owls in the Columbia Basin, Oregon. *Condor* 91:347-54.

Green, J. S., and J. T. Flinders. 1980. Brachylagus idahoensis. Mammalian Species No. 125:1-4, American Society of Mammalogists.

Greene, H. W. 1984. Taxonomic status of the western racer, *Coluber constrictor mormon*. *Journal of Herpetology* 18:210-11.

Gregory, P. T. 1978. Feeding habits and diet overlap of three garter snakes (*Thamnophis*) on Vancouver Island. *Canadian Journal of Zoology* 56:1967-74.

Grim, J. N. 1958. Feeding habits of the southern California mole. *Journal of Mammalogy* 39:265-68.

Grinnell, J., J. Dixon, and J. M. Linsdale. 1937. *Fur-bearing Mammals of California.* University of California Press, Berkeley.

Grinnell, J., and A. H. Miller. 1944. The distribution of the birds of California. *Pacific Coast Avifauna* 27:1-608.

Gross, J. E., C. L. Stoddart, and F. H. Wagner. 1974. Demographic analysis of a northern Utah jackrabbit population. *Wildlife Monograph* No. 40:1-68.

Groth, J. G. 1993. Evolutionary differentiation in morphology, vocalizations, and allozymes among nomadic sibling species in the North American Red Crossbill (*Loxia curvirostra*) complex. *University of California Publications in Zoology* 127:1-143.

Gutierrez, R. J., and W. D. Koenig. 1978. Characteristics of storage trees used by acorn woodpeckers in two California woodlands. *Journal of Forestry* 76:162-64.

Ha, J. C., and P. N. Lehner. 1990. Notes on Gray Jay demographics in Colorado. *Wilson Bulletin* 102:698-702.

Hafner, J. C. 1985. New kangaroo mice, genus *Microdipodops* (Rodentia: Heteromyidae), from Idaho and Nevada. *Proceedings of the Biological Society of Washington* 98:1-9.

Hagan, J. M., III, and D. W. Johnson. 1992. *Ecology and Conservation of Neotropical Migrant Landbirds.* Smithsonian Institution Press, Washington, D.C.

Hall, E. R. 1946. *Mammals of Nevada.* University of California Press, Berkeley.

Hall, E. R. 1981. *The Mammals of North America*, Second Edition. John Wiley & Sons, New York.

Hamilton, R. B. 1975. Comparative behavior of the American Avocet and the Black-necked Stilt (Recurvirostridae). *Ornithological Monographs* No. 17:1-98.

**455**

Hamilton, T. H. 1962. Species relationships and adaptations for sympatry in the avian genus Vireo. *Condor* 64:40-68.

Hammerson, G. A. 1982. Bullfrog eliminating leopard frogs in Colorado? *Herpetological Review* 13:115-16.

Hanlin, H. G., J. J. Beatty, and S. W. Hanlin. 1979. A nest site of the western red-backed salamander *Plethodon vehiculum* (Cooper). *Journal of Herpetology* 13:214-16.

Hansen, R. M., and E. E. Remmenga. 1961. Nearest neighbor concept applied to pocket gopher populations. *Ecology* 42:812-14.

Harrison, C. 1978. *A Field Guide to the Nests, Eggs, and Nestlings of North American Birds.* William Collins Sons & Company Ltd., Glasgow.

Harte, J., and E. Hoffman. 1989. Possible effects of acidic deposition on a Rocky Mountain population of the tiger salamander *Ambystoma tigrinum. Conservation Biology* 3:149-58.

Hartman, G. D., and T. L. Yates. 1985. Scapanus orarius. Mammalian Species No. 253:1-5, American Society of Mammalogists.

Haskell, H. S., and H. G. Reynolds. 1947. Growth, developmental food requirements, and breeding activity of the California jack rabbit. *Journal of Mammalogy* 28:129-36.

Hawbecker, A. C. 1942. A life history study of the White-tailed Kite. *Condor* 44:267-76.

Hawes, M. L. 1977. Home range, territoriality, and ecological separation in sympatric shrews, *Sorex vagrans* and *Sorex obscurus. Journal of Mammalogy* 58:354-67.

Hayes, J. P. 1996. Arborimus longicaudus. Mammalian Species No. 532:1-5, American Society of Mammalogists.

Hayes, M. P., and M. R. Jennings. 1986. Decline of ranid frog species in western North America: are bullfrogs (*Rana catesbiana*) responsible? *Journal of Herpetology* 20:490-509.

Hayssen, V. 1991. Dipodomys microps. Mammalian Species No. 389:1-9, American Society of Mammalogists.

Hayward, C. L., C. Cottam, A. M. Woodbury, and H. H. Frost. 1976. Birds of Utah. *Great Basin Naturalist Memoirs* No. 1:1-229.

Hayward, G. D., and E. O. Garton. 1988. Resource partitioning among forest owls in the River of No Return Wilderness, Idaho. *Oecologia* 75:253-65.

Hedges, S. B. 1986. An electrophoretic analysis of holarctic hylid frog evolution. *Systematic Zoology* 35:1-21.

Hendricks, F. S., and J. R. Dixon. 1984. Population structure of *Cnemidophorus tigris* (Reptilia: Teiidae) east of the continental divide. *Southwestern Naturalist* 29:137-40.

Hendrickson, J. R. 1954. Ecology and systematics of salamanders of the genus Batrachoseps. *University of California Publications in Zoology* 54:1-46.

Hennings, D., and R. S. Hoffmann. 1977. A review of the taxonomy of the *Sorex vagrans* species complex from western North America. University of Kansas, Occasional Papers, Museum of Natural History 68:1-35.

Herd, R. M. 1987. Electrophoretic divergence of *Myotis leibii* and *Myotis ciliolabrum* (Chiroptera: Vespertilionidae). *Canadian Journal of Zoology* 65:1857-60.

Hering, L. 1948. Nesting birds of the Black Forest, Colorado. *Condor* 50:49-56.

Herman, S. G., J. B. Bulger, and J. B. Buchanan. 1988. The Snowy Plover in southeastern Oregon and western Nevada. *Journal of Field Ornithology* 59:13-21.

Hermanson, J. W., and T. J. O'Shea. 1983. Antrozous pallidus. Mammalian Species No. 213:1-8, American Society of Mammalogists.

Herrington, R. E., and J. H. Larsen. 1985. Current status, habitat requirements and management of the Larch Mountain salamander *Plethodon larselli* Burns. *Biological Conservation* 34:169-79.

Herrington, R. E., and J. H. Larsen, Jr. 1987. Reproductive biology of the Larch Mountain salamander (*Plethodon larselli*). *Journal of Herpetology* 21:48-56.

Hertz, P. E., J. V. Remsen, Jr., and S. I. Zones. 1976. Ecological complementarity of three sympatric parids in a California oak woodland. *Condor* 78:307-16.

Hespenheide, H. A. 1964. Competition and the genus *Tyrannus. Wilson Bulletin* 76:265-81.

Hill, B. G., and M. R. Lein. 1989. Territory overlap and habitat use of sympatric chickadees. *Auk* 106:259-68.

Hines, J. E. 1977. Nesting and brood ecology of Lesser Scaup at Waterhen Marsh, Saskatchewan. *Canadian Field Naturalist* 91:248-55.

Hoffman, W., J. A. Wiens, and J. M. Scott. 1978. Hybridization between gulls (*Larus glaucescens* and *L. occidentalis*) in the Pacific Northwest. *Auk* 95:441-58.

Hoffmeister, D. F. 1981. Peromyscus truei. Mammalian Species No. 161:1-5, American Society of Mammalogists.

Hoffmeister, D. F. 1986. *Mammals of Arizona.* University of Arizona Press, Tucson, and the Arizona Department of Game and Fish, Phoenix.

Holland, D. C. 1993. A synopsis of the distribution and current status of the western pond turtle (*Clemmys marmorata*) in Oregon. Report to the Nongame Program, Wildlife Division, Oregon Department of Fish and Wildlife, Portland, Oregon.

Holm, C. H. 1973. Breeding sex ratios, territoriality and reproductive success in the Red-winged Blackbird (*Agelaius phoeniceus*). *Ecology* 54:356-65.

Holmes, R. T., T. W. Sherry, and L. Reitsma. 1989. Population structure, territoriality and overwinter survival of two migrant warbler species in Jamaica. *Condor* 91:545-61.

Hooven, E. F., R. F. Hoyer, and R. M. Storm. 1975. Notes on the vagrant shrew, *Sorex vagrans*, in the Willamette Valley of western Oregon. *Northwest Science* 49:163-73.

Hopkins, D. D., and R. B. Forbes. 1979. Size and reproductive patterns of the Virginia opossum in northwestern Oregon. *Murrelet* 60:95-98.

Hopkins, D. D., and R. B. Forbes. 1980. Dietary patterns of the Virginia opossum in an urban environment. *Murrelet* 61:20-30.

Horn, H. S. 1968. The adaptive significance of colonial nesting in the Brewer's blackbird. *Ecology* 49:682-94.

Horn, H. S. 1970. Social behavior of nesting Brewer's Blackbird's. *Condor* 72:15-23.

Horn, K. M., and D. B. Marshall. 1975. Status of poor-will in Oregon and possible extension due to clearcut timber harvest methods. *Murrelet* 56:4-5.

Horner, M. A., and R. A. Powell. 1990. Internal structure of home ranges of black bears and analyses of home-range overlap. *Journal of Mammalogy* 71:402-10.

Hornocker, M. G., and H. S. Hash. 1981. Ecology of the wolverine in northwestern Montana. *Canadian Journal of Zoology* 59:1286-1301.

Horvath, O. 1964. Seasonal differences in rufous hummingbird nest height in relation to nest climate. *Ecology* 45:235-41.

Howard, J. H., R. L. Wallace, and J. H. Larsen, Jr. 1983. Genetic variation and population divergence in the Larch Mountain salamander (*Plethodon larselli*). *Herpetologica* 39:41-47.

Howe, M. A. 1974. Observations on the terrestrial wing displays of breeding Willets. *Wilson Bulletin* 86:286-88.

Howe, M. A. 1975. Behavioral aspects of the pair bond in Wilson's Phalarope. *Wilson Bulletin* 87:248-70.

Howell, T. R. 1952. Natural history and differentiation in the Yellow-bellied Sapsucker. *Condor* 54:237-82.

Hoyer, R. F. 1974. Description of a rubber boa (*Charina bottae*) population from western Oregon. *Herpetologica* 30:275-83.

Hubbard, J. P. 1969. The relationships and evolution of the *Dendroica coronata* complex. *Auk* 86:393-432.

Hunter, J. E., R. J. Gutierrez, and A. B. Franklin. 1995. Habitat configuration around Spotted Owl sites in northwestern California. *Condor* 97:684-93.

Hunter, W. C., R. D. Ohmart, and B. W. Anderson. 1988. Use of exotic saltcedar (*Tamarix chinensis*) by birds in arid riparian systems. *Condor* 90:113-23.

Huntly, N. J., A. T. Smith, and B. L. Ivins. 1986. Foraging behavior of the pika (*Ochotona princeps*), with comparisons of grazing versus haying. *Journal of Mammalogy* 67:139-48.

Hupp, J. W., J. T. Ratti, and L. M. Smith. 1988. Gray Partridge foraging ecology in eastern South Dakota. *Great Basin Naturalist* 48:202-5.

Ingold, D. J. 1989. Nesting phenology and competition for nest sites among Red-headed and Red-bellied Woodpeckers and European Starlings. *Auk* 106:209-17.

Ivey, G., M. A. Stern, and C. G. Carey. 1988. An increasing White-faced Ibis population in Oregon. *Western Birds* 19:105-8.

Jackson, J. A. 1970. A quantitative study of the foraging ecology of Downy Woodpeckers. *Ecology* 51:318-23.

James, F. C., and H. H. Shugart, Jr. 1974. The phenology of the nesting season of the American Robin (*Turdus migratorius*) in the United States. *Condor* 76:159-68.

Janes, S. W. 1983. Status, distribution, and habitat selection of the Grasshopper Sparrow in Morrow County, Oregon. *Murrelet* 64:51-54.

Janes, S. W. 1987. Status and decline of Swainson's Hawks in Oregon: the role of interspecific competition. *Oregon Birds* 13:165-79.

Janzen, D. H. 1983. *Costa Rican Natural History.* University of Chicago Press, Chicago.

Jehl, J. R. 1988. Biology of the Eared Grebe and Wilson's Phalarope in the nonbreeding season: A study of adaptations to saline lakes. *Studies in Avian Biology* No. 12:1-74, Cooper Ornithological Society.

Jenkins, S. H., and P. E. Busher. 1979. Castor canadensis. Mammalian Species No. 120:1-8, American Society of Mammalogists.

Jenkins, S. H., and B. D. Eshelman. 1984. Spermophilus beldingi. Mammalian Species No. 221:1-8, American Society of Mammalogists.

Jenni, D. A. 1969. A study of the ecology of four species of herons during the breeding season at Lake Alice, Alachua County, Florida. *Ecological Monographs* 39:245-70.

Jewett, S. G. 1955. Free-tailed bats and melanistic mice in Oregon. *Journal of Mammalogy* 36:458-59.

Jewett, S. G., W. P. Taylor, W. T. Shaw, and J. W. Aldrich. 1953. *Birds of Washington State.* University of Washington Press, Seattle.

Johnsgard, P. A. 1987. *Diving Birds of North America.* University of Nebraska Press, Lincoln.

Johnson, D. W., and D. M. Armstrong. 1987. Peromyscus crinitus. Mammalian Species No. 287:1-8, American Society of Mammalogists.

Johnson, M. L., and C. W. Clanton. 1954. Natural history of *Sorex merriami* in Washington State. *Murrelet* 35:1-4.

Johnson, M. L., and S. B. George. 1991. Species limits within the *Arborimus longicaudus* species-complex (Mammalia: Rodentia) with a description of a new species from California. Natural History Museum of Los Angeles County, *Contributions in Science* No. 429:1-16.

Johnson, N. K. 1963. Biosystematics of sibling species of flycatchers in the Empidonax hammondii-oberholseri-wrightii complex. *University of California Publications in Zoology* 66:79-238.

Johnson, N. K. 1966. Bill size and the question of competition in allopatric and sympatric populations of Dusky and Gray Flycatchers. *Systematic Zoology* 15:70-87.

Johnson, N. K. 1976. Breeding distribution of Nashville and Virginia's Warblers. *Auk* 93:219-30.

Johnson, N. K. 1980. Character variation and evolution of sibling species in the *Empidonax difficilis-flavescens* complex (Aves: Tyrannidae). *University of California Publications in Zoology* 112:1-151.

**457**

Johnson, N. K. 1995. Speciation in vireos. I. Macrogeographic patterns of allozymic variation in the *Vireo solitarius* complex in the contiguous United States. *Condor* 97:903-19

Johnson, N. K., and C. B. Johnson. 1985. Speciation in sapsuckers (*Sphyrapicus*): II. Sympatry, hybridization, and mate preference in *S. ruber daggetti* and *S. nuchalis. Auk* 102:1-15.

Johnson, N. K., and J. A. Marten. 1988. Evolutionary genetics of flycatchers. II. Differentiation in the *Empidonax difficilis* complex. *Auk* 105:177-91.

Johnson, N. K., and J. A. Marten. 1992. Macrogeographic patterns of morphometric and genetic variation in the Sage Sparrow complex. *Condor* 94:1-19.

Johnson, N. K., and R. M. Zink. 1983. Speciation in sapsuckers (*Sphyrapicus*): I. Genetic differentiation. *Auk* 100:871-84.

Johnson, N. K., and R. M. Zink. 1985. Genetic evidence for relationships among the Red-eyed, Yellow-green, and Chivi vireos. *Wilson Bulletin* 97:421-35.

Johnson, N. K., R. M. Zink, and J. A. Marten. 1988. Genetic evidence for relationships in the avian family Vireonidae. *Condor* 90:428-45.

Johnson, R. E. 1975. New breeding localities for *Leucosticte* in the contiguous western United States. *Auk* 92:586-89.

Johnson, R. G., and S. A. Temple. 1990. Nest predation and brood parasitism of tallgrass prairie birds. *Journal of Wildlife Management* 54:106-11.

Johnston, D. W. 1949. Populations and distribution of summer birds of Latah County, Idaho. *Condor* 51:140-49.

Jones, G. S., J. O. Whitaker, Jr., and C. Maser. 1978. Food habits of jumping mice (*Zapus trinotatus* and *Zapus princeps*) in western North America. *Northwest Science* 52:57-60.

Jones, J. K., Jr., D. M. Armstrong, and J. R. Choate. 1985. *Guide to Mammals of the Plains States.* University of Nebraska Press, Lincoln.

Jones, J. K., R. S. Hoffmann, D. W. Rice, C. Jones, R. J. Baker, and M. D. Engstrom. 1992. Revised checklist of North American mammals north of Mexico, 1991. Occasional Papers, The Museum, Texas Tech University No. 146:1-23.

Jones, L. L. C., R. B. Bury, and P. S. Corn. 1990. Field observation of the development of a clutch of Pacific giant salamander (*Dicamptodon tenebrosus*) eggs. *Northwestern Naturalist* 71:93-94.

Jones, L. L. C., and P. S. Corn. 1989. Third specimen of a metamorphosed Cope's giant salamander (*Dicamptodon copei*). *Northwestern Naturalist* 70:37-38.

Kaufman, G. A., D. W. Kaufman, and E. J. Finck. 1988. Influence of fire and topography on habitat selection by *Peromyscus maniculatus* and *Reithrodontomys megalotis* in ungrazed tallgrass prairie. *Journal of Mammalogy* 69:342-52.

Keast, A., and S. Saunders. 1991. Ecomorphology of the North American Ruby-crowned (*Regulus calendula*) and Golden-crowned (*R. satrapa*) Kinglets. *Auk* 108:880-88.

Kebbe, C. E. 1958. Nesting records of the red-necked grebe in Oregon. *Murrelet* 39:14.

Keith, L. B., and L. A. Windberg. 1978. A demographic analysis of the snowshoe hare cycle. *Wildlife Monograph* No. 58:6-70.

Keller, B. L. 1987. Analysis of the bat species present in Idaho, with special attention to the spotted bat, *Euderma maculatum.* Department of Biological Sciences, Idaho State University, Pocatello. 25 pp.

Kelt, D. A. 1988. Dipodomys californicus. Mammalian Species No. 324, American Society of Mammalogists.

Kenagy, G. J. 1973. Adaptations for leaf eating in the Great Basin Kangaroo Rat, *Dipodomys microps. Oecologia* 12:383-412.

Kenagy, G. J., and B. M. Barnes. 1988. Seasonal reproductive patterns in four coexisting rodent species from the Cascade Mountains, Washington. *Journal of Mammalogy* 69:274-92.

Kenagy, G. J., and G. W. Bartholomew. 1985. Seasonal reproductive patterns in five coexisting California desert rodent species. *Ecological Monographs* 55:371-97.

Kerpez, T. A., and N. S. Smith. 1990. Nest-site selection and nest-cavity characteristics of Gila Woodpeckers and Northern Flickers. *Condor* 92:193-98.

Kilham, L. 1965. Differences in feeding behavior of male and female Hairy Woodpeckers. *Wilson Bulletin* 77:134-45.

Kilham, L. 1968. Reproductive behavior of White-breasted Nuthatches. I. Distraction display, bill-sweeping, and nest hole defense. *Auk* 85:477-92.

Kilham, L. 1971. Roosting habits of White-breasted Nuthatches. *Condor* 73:113-14.

Kilham, L. 1972. Reproductive behavior of White-breasted Nuthatches. II. Courtship. *Auk* 89:115-29.

Kilham, L. 1973. Reproductive behavior of the Red-breasted Nuthatch. I. Courtship. *Auk* 90:597-609.

Kilham, L. 1974a. Biology of young belted kingfishers. *American Midland Naturalist* 92:245-47.

Kilham, L. 1974b. Covering of stores by White-breasted and Red-breasted Nuthatches. *Condor* 76:108-9.

Kilham, L. 1975. Associations of Red-breasted Nuthatches with chickadees in a hemlock cone year. *Auk* 92:160-62.

Kindschy, R. R., Jr. 1964. Ecological studies on the Rock Dove in southeastern Oregon. *Northwest Science* 38:138-40.

King, C. M. 1983. Mustela erminea. Mammalian Species No. 195:1-8, American Society of Mammalogists.

King, J. A. 1968. Biology of *Peromyscus* (Rodentia). Special Publication No. 2, American Society of Mammalogists.

Kirk, J. J. 1979. *Thamnophis ordinoides. Catalog of American Amphibians and Reptiles* 233:1-2.

Klauber, L. M. 1972. *Rattlesnakes,* Second Edition. University of California Press, Berkeley.

Kluyver, H. N. 1961. Food consumption in relation to habitat in breeding chickadees. *Auk* 78:532-50.

Knick, S. T. 1990. Ecology of bobcats relative to exploitation and a prey decline in southeastern Idaho. *Wildlife Monograph* No 108:1-42.

Knight, R. L., and A. W. Erickson. 1977. Ecological notes on Long-eared and Great Horned Owls along the Columbia River. *Murrelet* 58:2-6.

Knorr, O. A. 1957. Communal roosting of the Pygmy Nuthatch. *Condor* 59:398.

Knorr, O. A. 1961. The geographical and ecological distribution of the Black Swift in Colorado. *Wilson Bulletin* 73:155-70.

Knowlton, F. F. 1972. Preliminary interpretations of coyote population mechanics with some management implications. *Journal of Wildlife Management* 36:369-82.

Knox, A. G. 1990. The sympatric breeding of Common and Scottish crossbills *Loxia curvirostra* and *L. scotia* and the evolution of crossbills. *Ibis* 132:454-66.

Koehler, G. M., M. G. Hornocker, and H. S. Hash. 1979. Lynx movements and habitat use in Montana. *Canadian Field Naturalist* 93:441-42.

Koenig, W. D., and R. L. Mumme. 1987. *Population Ecology of the Cooperatively Breeding Acorn Woodpecker.* Princeton University Press, Princeton, New Jersey.

Koford, C. B., and M. R. Koford. 1948. Breeding colonies of bats, *Pipistrellus hesperus* and *Myotis subulatus melanorhinus. Journal of Mammalogy* 29:417-18.

Koprowski, J. L. 1994a. Sciurus carolinensis. Mammalian Species No. 480:1-9, American Society of Mammalogists.

Koprowski, J. L. 1994b. Sciurus niger. Mammalian Species No. 479:1-9, American Society of Mammalogists.

Krebs, J. R. 1974. Colonial nesting and social feeding as strategies for exploiting food resources in the Great Blue Heron (*Ardea herodias*). *Behaviour* 51:99-134.

Kritzman, E. B. 1974. Ecological relationships of *Peromyscus maniculatus* and *Perognathus parvus* in eastern Washington. *Journal of Mammalogy* 55:172-88.

Kroodsma, D. E. 1973. Coexistence of Bewick's Wrens and House Wrens in Oregon. *Auk* 90:341-52.

Kroodsma, R. L. 1974. Species-recognition behavior of territorial male Rose-breasted and Black-headed Grosbeaks (*Pheucticus*). *Auk* 91:54-64.

Kuhn, L. W., W. Q. Wick, and R. J. Pedersen. 1966. Breeding nests of Townsend's mole in Oregon. *Journal of Mammalogy* 47:239-49.

Kunz, T. H. 1982. Lasionycteris noctivagans. Mammalian Species No. 172:1-5, American Society of Mammalogists.

Kunz, T. H., and R. A. Martin. 1982. Plecotus townsendii. Mammalian Species No. 175:1-6, American Society of Mammalogists.

Kurta, A., and R. H. Baker. 1990. Eptesicus fuscus. Mammalian Species No. 356:1-10, American Society of Mammalogists.

Kushlan, J. A. 1976. Foraging behavior of North American herons. *Auk* 93:86-94.

Lair, H. 1985. Length of gestation in the red squirrel, *Tamiasciurus hudsonicus. Journal of Mammalogy* 66:809-10.

Lamberson, R. H., R. McKelvey, B. R. Noon, and C. Voss. 1992. A dynamic analysis of Northern Spotted Owl viability in a fragmented forest landscape. *Conservation Biology* 6:505-12.

Lamberson, R. H., B. R. Noon, C. Voss, and K. S. McKelvey. 1994. Reserve design for territorial species: the effects of patch size and spacing on the viability of the Northern Spotted Owl. *Conservation Biology* 8:185-95.

Larivière, S., and M. Pasitschniak-Arts. 1996. Vulpes vulpes. Mammalian Species No. 537:1-11, American Society of Mammalogists.

Larrison, E. J. 1940. The Anthony green heron in the state of Washington. *Murrelet* 21:1-3.

Larrison, E. J. 1943. Feral coypus in the Pacific Northwest. *Murrelet* 24:3-9.

Laudenslayer, W. F., Jr., and R. P. Balda. 1976. Breeding bird use of a pinyon-juniper-ponderosa pine ecotone. *Auk* 93:571-86.

Lawrence, G. E. 1950. The diving and feeding activity of the Western Grebe on the breeding grounds. *Condor* 52:3-16.

Lawrence, L. de K. 1948. Comparative study of the nesting behavior of Chestnut-sided and Nashville Warblers. *Auk* 65:204-19.

Lawrence, L. de K. 1967. A comparative life-history study of four species of woodpeckers. *Ornithological Monograph* No. 5:1-156.

Lawson, R. 1987. Molecular studies of thamnophiine snakes: 1. The phylogeny of the genus *Nerodia. Journal of Herpetology* 21:140-57.

Laymon, S. A., and M. D. Halterman. 1987. Can the western subspecies of the Yellow-billed Cuckoo be saved from extinction? *Western Birds* 18:19-25.

Lederer, R. J. 1977. Winter territoriality and foraging behavior of the Townsend's solitaire. *American Midland Naturalist* 97:101-9.

Legg, K., and F. A. Pitelka. 1956. Ecological overlap of Allen and Anna Hummingbirds nesting at Santa Cruz, California. *Condor* 58:393-405.

Lenington, S., and T. Mace. 1975. Mate fidelity and nesting site tenacity in the Killdeer. *Auk* 92:149-51.

Leonard, M. L., and J. Picman. 1986. Why are nesting Marsh Wrens and Yellow-headed Blackbirds spatially segregated? *Auk* 103:135-40.

Leonard, W. P., H. A. Brown, L. L. C. Jones, K. R. McAllister, R. M. Storm. 1993. *Amphibians of Washington and Oregon.* Seattle Audubon Society, Seattle, Washington.

Leopold, A. S. 1959. *Wildlife of Mexico.* University of California Press, Berkeley.

Leopold, A. S. 1977. *The California Quail.* University of California Press, Berkeley.

Levenson, H., and R. S. Hoffmann. 1984. Systematic relationships among taxa in the Townsend chipmunk group. *Southwestern Naturalist* 29:157-68.

Lewis, R. A. 1975. Reproductive biology of the White-crowned Sparrow. II. Environmental control of reproductive and associated cycles. *Condor* 77:111-24.

Licht, L. E. 1971. Breeding habits and embryonic thermal requirements of the frogs, *Rana aurora aurora* and *Rana pretiosa pretiosa*, in the Pacific Northwest. *Ecology* 52:116-24.

Lidicker, W. Z., Jr. 1966. Ecological observations on a feral house mouse population declining to extinction. *Ecological Monographs* 36:27-50.

Lidicker, W. Z., Jr. 1971. Corrections and additions to our knowledge of the pocket gopher *Thomomys mazama pugetensis*. *Murrelet* 52:12-13.

Ligon, J. D. 1973. Foraging behavior of the White-headed Woodpecker in Idaho. *Auk* 90:862-69.

Lim, B. K. 1987. Lepus townsendii. Mammalian Species No. 288:1-6, American Society of Mammalogists.

Linsdale, J. M. 1936. The birds of Nevada. *Pacific Coast Avifauna* No. 23:1-145.

Linsdale, J. M. 1946. *The California Ground Squirrel.* University of California Press, Berkeley.

Linsdale, J. M. 1957. Goldfinches on the Hastings Natural History Reservation. *American Midland Naturalist* 57:1-119.

Linsdale, J. M., and L. P. Tevis, Jr. 1951. *The Dusky-footed Woodrat.* University of California Press, Berkeley.

Littlefield, C. D. 1990. *Birds of Malheur National Wildlife Refuge, Oregon.* Oregon State University Press, Corvallis.

Littlefield, C. D., and D. G. Paullin. 1990. Effects of land management on nesting success of sandhill cranes in Oregon. *Wildlife Society Bulletin* 18:63-65.

Littlefield, C. D., and S. P. Thompson. 1981. History and status of the Franklin's Gull on Malheur National Wildlife Refuge, Oregon. *Great Basin Naturalist* 41:440-44.

Livezey, R. L. 1959. The egg mass and larvae of *Plethodon elongatus* Van Denburgh. *Hepertologica* 15:41-42.

Lloyd, H. G. 1980. *The Red Fox.* B. T. Batsford, Ltd., London.

Long, C. A. 1973. Taxidea taxus. Mammalian Species No. 26:1-4, American Society of Mammalogists.

Lotze, J-H., and S. Anderson. 1979. Procyon lotor. Mammalian Species No. 119:1-8, American Society of Mammalogists.

Ludwig, D. R. 1984. Microtus richardsoni. Mammalian Species No. 223:1-6, American Society of Mammalogists.

Ludwig, D. R. 1988. Reproduction and population dynamics of the water vole, *Microtus richardsoni*. *Journal of Mammalogy* 69:532-41.

Lundquist, R. W., and J. M. Mariani. 1991. Nesting habitat and abundance of snag-dependent birds in the southern Washington Cascade Range. Pages 221-240 *in* (L. F. Ruggiero, K. B. Aubry, A. B. Carey, and M. H. Huff, editors) *Wildlife and vegetation of unmanaged Douglas-fir forests.* U.S.D.A. Forest Service, Pacific Northwest Research Station, General Technical Report PNW-GTR-285, Portland, Oregon.

Lunk, W. A. 1962. The Rough-winged Swallow (*Stelgidopteryx ruficollis* (Vieillot)), a study based on its breeding biology in Michigan. Nuttall Ornithological Club, Publication No. 4:1-155.

Lynch, J. F. 1981. Patterns of ontogenetic and geographic variation in the black salamander, *Aneides flavipunctatus* (Caudata: Plethodontidae). *Smithsonian Contributions in Zoology* 324:1-53.

Lynch, J. F. 1985. Feeding ecology of *Aneides flavipunctatus* and sympatric plethodontid salamanders in northwestern California. *Journal of Herpetology* 19:328-52.

Macartney, J. M. 1989. Diet of the northern Pacific rattlesnake, *Crotalus viridis oreganus*, in British Columbia. *Herpetologica* 45:299-304.

Macartney, J. M., and P. T. Gregory. 1988. Reproductive biology of female rattlesnakes (*Crotalus virdis*) in British Columbia. *Copeia* 1988:47-57.

MacNab, J. A., and J. C. Dirks. 1941. The California red-backed mouse in the Oregon Coast Range. *Journal of Mammalogy* 22:174-80.

MacRoberts, M. H., and B. R. MacRoberts. 1976. Social organization and behavior of the Acorn Woodpecker in central coastal California. *Ornithological Monograph* No. 21:1-115, American Ornithologists Union.

Maiorana, V. C. 1976. Size and environmental predictability for salamanders. *Evolution* 30:599-613.

Mannan, R. W. 1984. Habitat use by Hammond's Flycatchers in old-growth forests, northeastern Oregon. *Murrelet* 65:84-86.

Manning, R. W., and J. K. Jones, Jr. 1988. A new subspecies of fringed myotis, *Myotis thysanodes*, from the northwestern coast of the United States. Occasional Paper No. 123:1-6, The Museum, Texas Tech University.

Manning, R. W., and J. K. Jones, Jr. 1989. Myotis evotis. Mammalian Species No. 329:1-5, American Society of Mammalogists.

Manolis, T. 1977. Foraging relationships of Mountain Chickadees and Pygmy Nuthatches. *Western Birds* 8:13-20.

Manuwal, D. A. 1970. Notes on the territoriality of Hammond's Flycatcher (*Empidonax hammondii*) in western Montana. *Condor* 72:364-65.

Marcellini, D., and J. P. Mackey. 1970. Habitat preferences of the lizards, *Sceloporus occidentalis* and *Sceloporus graciosus* (Lacertilia, Iguanidae). *Herpetologica* 26:51-56.

Marks, J. S. 1984. Feeding ecology of breeding Long-eared Owls in southwestern Idaho. *Canadian Journal of Zoology* 62:1528-33.

Marks, J. S., J. H. Doremus, and A. R. Bammann. 1980. Black-throated Sparrows breeding in Idaho. *Murrelet* 61:112-13.

Marks, J. S., J. H. Doremus, and R. J. Cannings. 1989. Polygyny in the Northern Saw-whet Owl. *Auk* 106:732-34.

Marshall, D. B. 1988. Status of the Marbled Murrelet in North America: with special emphasis on populations in California, Oregon, and Washington. U.S. Fish and Wildlife Service, Biological Report No. 88.

Marshall, D. B. 1992. Sensitive vertebrates of Oregon. Oregon Department of Fish and Wildlife, Portland.

Marshall, D. B., M. Chilcote, and H. Weeks. 1996. Species at risk: sensitive, threatened and endangered vertebrates of Oregon. 2nd edition. Oregon Department of Fish and Wildlife, Portland.

Marshall, D. B., and H. F. Deubbert. 1965. Nesting of the ring-necked duck in Oregon in 1963 and 1964. *Murrelet* 46:43.

Marti, C. D. 1974. Feeding ecology of four sympatric owls. *Condor* 76:45-61.

Marti, C. D., and P. W. Wagner. 1985. Winter mortality in Common Barn-Owls and its effect on population density and reproduction. *Condor* 87:111-15.

Martin, S. G. 1970. The agonistic behavior of Varied Thrushes (*Ixoreus naevius*) in winter assemblages. *Condor* 72:452-59.

Marvil, R. E., and A. Cruz. 1989. Impact of Brown-headed Cowbird parsitism on the reproductive success of the Solitary Vireo. *Auk* 106:476-80.

Marzluff, J. M., and R. P. Balda. 1988. The advantages of, and constraints forcing, mate fidelity in Pinyon Jays. *Auk* 105:286-95.

Marzluff, J. M., and R. P. Balda. 1992. *The Pinyon Jay: Behavioral ecology of a colonial and cooperative breeder.* Academic Press, New York.

Mascarello, J. T. 1978. Chromosomal, biochemical, mensural, penile, and cranial variation in desert woodrats (*Neotoma lepida*). *Journal of Mammalogy* 59:477-95.

Maser, C., and E. F. Hooven. 1970. Low altitude records of *Microtus richardsoni* in the Oregon Cascades. *Murrelet* 51:12.

Maser, C., E. W. Hammer, and S. H. Anderson. 1971. Food habits of the burrowing owl in central Oregon. *Northwest Science* 45:19-26.

Maser, C., and E. F. Hooven. 1974. Notes on the behavior and food habits of captive Pacific shrews, *Sorex pacificus pacificus*. *Northwest Science* 48:81-95.

Maser, C., and M. L. Johnson. 1967. Notes on the white-footed vole (*Phenacomys albipes*). *Murrelet* 48:24-27.

Maser, C., and Z. Maser. 1988. Interactions among squirrels, mycorrhizal fungi, and coniferous forests in Oregon. *Great Basin Naturalist* 48:358-69.

Maser, C., B. R. Mate, J. F. Franklin, and C. T. Dyrness. 1981. Natural history of Oregon coast mammals. USDA Forest Service General Technical Report PNW-133. Pacific Northwest Forest and Range Experiment Station, Portland, Oregon.

Mason, C. F., and S. M. MacDonald. 1976. Aspects of the breeding biology of the snipe. *Bird Study* 23:33-38.

Matson, J. O. 1976. The distribution of rodents in Owens Lake region, Inyo County, California. Natural History Museum of Los Angeles County, *Contributions in Science* No. 276:1-27.

Maya, J. E., and P. Malone. 1989. Feeding habits and behavior of the whiptail lizard, *Cnemidophorus tigris tigris*. *Journal of Herpetology* 23:309-11.

Mayhew, W. W. 1958. The biology of the Cliff Swallow in California. *Condor* 60:7-37.

McAllister, J. A., and R. S. Hoffmann. 1988. Phenacomys intermedius. Mammalian Species No. 305:1-8, American Society of Mammalogists.

McAllister, K. R. 1995. Distribution of amphibians and reptiles in Washington state. *Northwest Fauna* 3:81-112.

McAllister, N. M. 1958. Courtship, hostile behavior, nest-establishment and egg-laying in the Eared Grebe (*Podiceps caspicus*). *Auk* 75:290-311.

McCarty, R. 1978. Onychomys leucogaster. Mammalian Species No. 87:1-6, American Society of Mammalogists.

McGillivray, W. B. 1989. Geographic variation in size and reverse size dimorphism of the Great Horned Owl in North America. *Condor* 91:777-86.

McGrew, J. C. 1979. Vulpes macrotis. Mammalian Species No. 123:1-6, American Society of Mammalogists.

McKay, D. O., and B. J. Verts. 1978. Estimates of some attributes of a population of Nuttall's cottontails. *Journal of Wildlife Management* 42:159-68.

McKeever, S. 1964. The biology of the golden-mantled ground squirrel, *Citellus lateralis*. *Ecological Monographs* 34:383-401.

McKenzie, D. S. 1970. Aspects of the autecology of the plethodontid salamander *Aneides ferreus* (Cope). Unpublished Ph.D. Dissertation, Oregon State University, Corvallis.

McKenzie, D. S., and R. M. Storm. 1970. Patterns of habitat selection in the clouded salamander, *Aneides ferreus* (Cope). *Herpetologica* 26:450-54.

McKnight, D. E. 1974. Dry-land nesting by redheads and ruddy ducks. *Journal of Wildlife Management* 38:112-19.

McManus, J. J. 1974. Didelphis virginiana. Mammalian Species No. 40:1-6, American Society of Mammalogists.

Medin, D. E. 1990. Birds of a shadscale (*Atriplex confertifolia*) habitat in east central Nevada. *Great Basin Naturalist* 50:295-98.

Mellen, T. K., E. C. Meslow, and R. W. Mannan. 1992. Summertime home range and habitat use of pileated woodpeckers in western Oregon. *Journal of Wildlife Management* 56:96-103.

Melquist, W. E., and M. G. Hornocker. 1983. Ecology of river otters in west central Idaho. *Wildlife Monograph* No. 83:1-60.

Merritt, J. F. 1981. Clethrionomys gapperi. Mammalian Species No. 146:1-9, American Society of Mammalogists.

Messick, J. P., and M. G. Hornocker. 1981. Ecology of the badger in southwestern Idaho. *Wildlife Monograph* No. 76:6-53.

Metter, D. E. 1964. A morphological and ecological comparison of two populations of the tailed frog, *Ascaphus truei* Stejneger. *Copeia* 1964:181-95.

Mewaldt, L. R. 1956. Nesting behavior of the Clark Nutcracker. *Condor* 58:3-23.

Mewaldt, L. R., and J. R. King. 1985. Breeding site faithfulness, reproductive biology, and adult survivorship in an isolated population of Cassin's Finches. *Condor* 87:494-510.

Millar, J. S., and D. G. L. Innes. 1985. Breeding by *Peromyscus maniculatus* over an elevational gradient. *Canadian Journal of Zoology* 63:124-29.

Miller, A. H. 1939. Status of the breeding Lincoln's Sparrows of Oregon. *Auk* 56:342-43.

Miller, A. H., and R. C. Stebbins. 1964. *The Lives of Desert Animals in Joshua Tree National Monument.* University of California Press, Berkeley.

Miller, G. D., and W. S. Gaud. 1989. Composition and variability of desert bighorn sheep diets. *Journal of Wildlife Management* 53:597-606.

Mills, L. S. 1995. Edge effects and isolation: red-backed voles on forest remnants. *Conservation Biology* 9:395-403.

Milne, K. A., and S. J. Hejl. 1989. Nest-site characteristics of white-headed woodpeckers. *Journal of Wildlife Management* 53:50-55.

Mitchell, J. C. 1984. Observations on the ecology and reproduction of the leopard lizard, *Gambelia wislizenii* (Iguanidae), in southeastern Arizona. *Southwestern Naturalist* 29:509-11.

Mitchell, J. L. 1961. Mink movements and populations on a Montana river. *Journal of Wildlife Management* 25:48-54.

Modi, W. S. 1986. Karyotypic differentiation among two sibling species pairs of New World microtine rodents. *Journal of Mammalogy* 67:159-65.

Moll, E. O. 1973. Latitudinal and intersubspecific variation in reproduction of the painted turtle, *Chrysemys picta*. *Herpetologica* 29:307-18.

Montanucci, R. R. 1983. Natural hybridization between two species of collared lizards (*Crotaphytus*). *Copeia* 1983:1-11.

Morrison, M. L. 1982. The structure of western warbler assemblages: Ecomorphological analysis of the Black-throated Gray and Hermit Warblers. *Auk* 99:503-13.

Morse, D. H. 1970. Ecological aspects of some mixed-species foraging flocks of birds. *Ecological Monographs* 40:119-68.

Morse, D. H. 1972. Habitat differences of Swainson's and Hermit Thrushes. *Wilson Bulletin* 84:206-8.

Morse, T. E., J. L. Jakabosky, and V. P. McCrow. 1969. Some aspects of the breeding biology of the hooded merganser. *Journal of Wildlife Management* 33:596-604.

Morton, E. S., and K. C. Derrickson. 1990. The biological significance of age-specific return schedules in breeding Purple Martins. *Condor* 92:1040-50.

Mullican, T. R., and B. L. Keller. 1986. Ecology of the sagebrush vole (*Lemmiscus curtatus*) in southeastern Idaho. *Canadian Journal of Zoology* 64:1218-23.

Mullican, T. R., and B. L. Keller. 1987. Burrows of the sagebrush vole (*Lemmiscus curtatus*) in southeastern Idaho. *Great Basin Naturalist* 47:276-79.

Mullins, W. H., and E. G. Bizeau. 1978. Summer foods of Sandhill Cranes in Idaho. *Auk* 95:175-78.

Murray, G. A. 1976. Geographic variation in the clutch sizes of seven owl species. *Auk* 93:602-13.

Nagorsen, D. W. 1985. A morphometric study of geographic variation in the snowshoe hare (*Lepus americanus*). *Canadian Journal of Zoology* 63:567-79.

Nicholls, T. H., and D. W. Warner. 1972. Barred owl habitat use as determined by radiotelemetry. *Journal of Wildlife Management* 36:213-24.

Noon, B., and C. M. Biles. 1990. Mathematical demography of spotted owls in the Pacific Northwest. *Journal of Wildlife Management* 54:18-27.

Norris, R. A. 1958. Comparative biosystematics and life history of the nuthatches Sitta pygmaea and Sitta pusilla. *University of California Publications in Zoology* 56:119-300.

Nuechterlein, G. L., and D. P. Buitron. 1989. Diving differences between Western and Clark's Grebes. *Auk* 106:467-70.

Nussbaum, R. A. 1969. Nests and eggs of the Pacific giant salamander, *Dicamptodon ensatus* (Eschscholtz). *Herpetologica* 25:257-62.

Nussbaum, R. A. 1970. *Dicamptodon copei*, n. sp., from the Pacific Northwest, U.S.A. (Amphibia:Caudata: Ambystomatidae). *Copeia* 1970:506-14.

Nussbaum, R. A. 1974. A report on the distributional ecology and life history of the Siskiyou Mountain salamander, *Plethodon stormi*, in relation to the potential impact of the proposed Applegate Reservoir on this species. U. S. Army Corps of Engineers, Portland, Oregon.

Nussbaum, R. A., E. D. Brodie, Jr., and R. M. Storm. 1983. *Amphibians and Reptiles of the Pacific Northwest.* University Press of Idaho, Moscow, Idaho.

Nussbaum, R. A., and G. R. Clothier. 1973. Population structure, growth, and size of larval *Dicamptodon ensatus* (Eschscholtz). *Northwest Science* 47:218-27.

Nussbaum, R. A., and L. V. Diller. 1976. The life history of the side-blotched lizard, *Uta stansburiana* Baird and Girard, in north-central Oregon. *Northwest Science* 50:243-60.

Nussbaum, R. A., and R. F. Hoyer. 1974. Geographic variation and the validity of subspecies in the rubber boa, *Charina bottae* (Blainville). *Northwest Science* 48:219-29.

Nussbaum, R. A., and C. K. Tait. 1977. Aspects of the life history and ecology of the Olympic salamander, Rhyachotriton olympicus (Gaige). *American Midland Naturalist* 98:176-99.

O'Farrell, M. J., and A. R. Blaustein. 1974. Microdipodops megacephalus. Mammalian Species No. 46:1-3, American Society of Mammalogists.

O'Farrell, M. J., and E. H. Studier. 1975. Population structure and emergence activity patterns in *Myotis thysanodes* and *M. lucifugus* (Chiroptera: Vespertilionidae) in northeastern New Mexico. *American Midland Naturalist* 93:368-76.

O'Farrell, M. J., and E. H. Studier. 1980. Myotis thysanodes. Mammalian Species No. 137:1-5, American Society of Mammalogists.

O'Gara, B. W. 1978. Antilocapra americana. Mammalian Species No. 90:1-7, American Society of Mammalogists.

O'Neil, T. A., R. J. Steidl, W. D. Edge, and B. Csuti. 1995. Using wildlife communities to classify vegetation for assessing biodiversity. *Conservation Biology* 9:1482-91.

O'Shea, T. J., and T. A. Vaughan. 1977. Nocturnal and seasonal activities of the pallid bat, *Antrozous pallidus*. *Journal of Mammalogy* 58:269-84.

Olendorff, R. R. 1976. The food habits of North American golden eagles. *American Midland Naturalist* 95:231-36.

Olson, D. H. 1989. Predation on breeding western toads (*Bufo boreas*). *Copeia* 1989:391-97.

Oregon Natural Heritage Program. 1995. Rare, threatened and endangered plants and animals of Oregon. Oregon Natural Heritage Program, Portland, Oregon.

Orians, G. H. 1961. The ecology of blackbird (*Agelaius*) social systems. *Ecological Monographs* 31:285-312.

Orians, G. H., and H. S. Horn. 1969. Overlap in foods and foraging of four species of blackbirds in the Potholes of central Washington. *Ecology* 50:930-38.

Oring, L. W. 1964. Behavior and ecology of certain ducks during the postbreeding period. *Journal of Wildlife Management* 28:223-33.

Oring, L. W., D. B. Lank, and S. J. Maxson. 1983. Population studies of the polyandrous Spotted Sandpiper. *Auk* 100:272-85.

Ouellet, H. 1970. Further observations on the food and predatory habits of the gray jay. *Canadian Journal of Zoology* 48:327-30.

Pace, A. E. 1974. Systematic and biological studies of the leopard frogs (*Rana pipiens* complex) of the United States. Miscellaneous Publications No. 148:1-140, Museum of Zoology, University of Michigan.

Paczolt, M. 1987. Winter nesting of Anna's Hummingbird in Medford, Oregon. *Oregon Birds* 13:31-33.

Page, G. W., L. E. Stenzel, and C. A. Ribic. 1985. Nest site selection and clutch predation in the Snowy Plover. *Auk* 102:347-53.

Pampush, G. J., and R. G. Anthony. 1993. Nest success, habitat utilization and nest-site selection of Long-billed Curlews in the Columbia Basin, Oregon. *Condor* 95:957-67.

Parker, W. S., and W. S. Brown. 1974. Notes on the ecology of regal ringneck snakes (*Diadophis punctatus regalis*) in northern Utah. *Journal of Herpetology* 8:262-63.

Parker, W. S. and W. S. Brown. 1980. Comparative ecology of two colubrid snakes, *Masticophis t. taeniatus* (Hallowell) and *Pituophis melanoleucus deserticola* Stejneger in northern Utah. Milwaukee Public Museum, *Publication in Biology and Geology* No. 7:1-104.

Parker, W. S., and E. R. Pianka. 1975. Comparative ecology of populations of the lizard *Uta stansburiana*. *Copeia* 1975:615-32.

Pasitschniak-Arts, M., and S. Larivière. 1995. Gulo gulo. Mammalian Species No. 499:1-10, American Society of Mammalogists.

Pattie, D. 1973. Sorex bendirii. Mammalian Species No. 27:1-2, American Society of Mammalogists.

Patton, J. L., H. MacArthur, and S. Y. Yang. 1976. Systematic relationships of the four-toed populations of *Dipodomys heermanni*. *Journal of Mammalogy* 57:159-63.

Paullin, D. G., G. L. Ivey, and C. D. Littlefield. 1988. The re-establishment of American White Pelican nesting in the Malheur-Harney Lakes Basin, Oregon. *Murrelet* 69:61-64.

Payne, R. B. 1969. Breeding season and reproductive physiology of Tricolored and Red-winged Blackbirds. *University of California Publications in Zoology* 90:1-115.

Peacock, R. L., and R. A. Nussbaum. 1973. Reproductive biology and population structure of the western red-backed salamander, *Plethodon vehiculum* (Cooper). *Journal of Herpetology* 7:215-24.

Peakall, D. B. 1990. Prospects for the Peregrine Falcon, *Falco peregrinus*, in the Nineties. *Canadian Field-Naturalist* 104:168-73.

Pearson, O. P., M. R. Koford, and A. K. Pearson. 1952. Reproduction of the lump-nosed bat (*Corynorhinus rafinesquii*) in California. *Journal of Mammalogy* 33:273-320.

Pekins, P. J., F. G. Lindzey, and J. A. Gessaman. 1991. Physical characteristics of Blue Grouse winter-use trees and roost sites. *Great Basin Naturalist* 51:244-48.

Perrill, S. A., and R. E. Daniel. 1983. Multiple egg clutches in *Hyla regilla, H. cinerea,* and *H. gratiosa. Copeia* 1983:513-16.

Petersen, K. L., and L. B. Best. 1985. Nest-site selection by Sage Sparrows. *Condor* 87:217-21.

Petersen, K. L., and L. B. Best. 1991. Nest-site selection by Sage Thrashers in southeastern Idaho. *Great Basin Naturalist* 51:261-66.

Peterson, A. J. 1955. The breeding cycle in the Bank Swallow. *Wilson Bulletin* 67:235-86.

Peterson, R. T. 1990. *A Field Guide to Western Birds,* Third edition. Houghton Mifflin Company, Boston.

Petrinovich, L., and T. L. Patterson. 1983. The White-Crowned Sparrow: reproductive success (1975-1980). *Auk* 100:811-25.

Philips, G. L. 1966. Ecology of the big brown bat (Chiroptera: Vespertilionidae) in northeastern Kansas. *American Midland Naturalist* 75:168-98.

Pianka, E. R. 1970. Comparative autecology of the lizard *Cnemidophorus tigris* in different parts of its geographic range. *Ecology* 51:703-20.

Pianka, E. R., and W. S. Parker. 1975. Ecology of horned lizards: a review with special reference to *Phrynosoma platyrhinos. Copeia* 1975:141-62.

Picman, J. 1977. Destruction of eggs by the long-billed marsh wren (*Telmatodytes palustris palustris*). *Canadian Journal of Zoology* 55:1914-20.

Pimentel, R. A. 1960. Inter- and intrahabitat movements of the rough-skinned newt, Taricha torosa granulosa (Skilton). *American Midland Naturalist* 63:470-96.

Pinter, A. J. 1986. Population dynamics and litter size of the montane vole, *Microtus montanus. Canadian Journal of Zoology* 64:1487-90.

Pitelka, F. A. 1951a. Ecological overlap and interspecific strife in breeding populations of Anna and Allen hummingbirds. *Ecology* 32:641-61.

Pitelka, F. A. 1951b. Speciation and ecologic distribution in the American jays of the genus Aphelocoma. *University of California Publications in Zoology* 50:195-464.

Pitocchelli, J. 1990. Plumage, morphometric, and song variation in Mourning (*Oporornis philadelphia*) and MacGillivray's (*O. tolmiei*) Warblers. *Auk* 107:161-71.

Poche, R. M. 1981. Ecology of the spotted bat (*Euderma maculatum*) in southwest Utah. Utah Department of Natural Resources, Publication No. 81-1. 63pp.

Poche, R. M., and G. A. Ruffner. 1975. Roosting behavior of male *Euderma maculatum* from Utah. *Great Basin Naturalist* 35:121-22.

Podolsky, R. H., and S. W. Kress. 1989. Factors affecting colony formation in Leach's Storm-Petrel. *Auk* 106:332-36.

Poglayen-Neuwall, I., and D. E. Toweill. 1988. Bassariscus astutus. Mammalian Species No. 327:1-8, American Society of Mammalogists.

Ports, M. A., and S. B. George. 1990. *Sorex preblei* in the northern Great Basin. *Great Basin Naturalist* 50:93-95.

Powell, R. A. 1981. Martes pennanti. Mammalian Species No. 156:1-6, American Society of Mammalogists.

Power, H. W., III. 1966. Biology of the Mountain Bluebird in Montana. *Condor* 68:351-71.

Powers, D. R. 1987. Effects of variation in food quality on the breeding territoriality of the male Anna's Hummingbird. *Condor* 89:103-11.

Powers, R. A., and B. J. Verts. 1971. Reproduction in the mountain cottontail rabbit in Oregon. *Journal of Wildlife Management* 35:605-13.

Prescott, D. R. C. 1991. Winter distribution of age and sex classes in an irruptive migrant, the Evening Grosbeak (*Coccothraustes vespertinus*). *Condor* 93:694-700.

Prescott, D. R. C., and A. L. A. Middleton. 1988. Feeding-time minimization and the territorial behavior of the Willow Flycatcher (*Empidonax traillii*). *Auk* 105:17-28.

Price, D. J., and T. R. Simons. 1986. The influence of human disturbance on Tufted Puffin breeding success. *Auk* 103:214-16.

Price, F. E., and C. E. Bock. 1973. Polygyny in the Dipper. *Condor* 75:457-59.

Puchy, C. A., and D. B. Marshall. 1993. Oregon Wildlife Diversity Plan. Oregon Department of Fish and Wildlife, Portland, Oregon.

Pulliam, H. R., and F. Enders. 1971. The feeding ecology of five sympatric finch species. *Ecology* 52:557-66.

Putnam, L. S. 1949. The life history of the Cedar Waxwing. *Wilson Bulletin* 61:141-82.

Pylypec, B. 1991. Impacts of fire on bird populations in a fescue prairie. *Canadian Field-Naturalist* 105:346-49.

Quick, H. F. 1951. Notes on the ecology of weasels in Gunnison County, Colorado. *Journal of Mammalogy* 32:281-90.

Quinney, T. E., and P. C. Smith. 1980. Comparative foraging behaviour and efficiency of adult and juvenile great blue herons. *Canadian Journal of Zoology* 58:1168-73.

Ralph, C. J., G. L. Hunt, Jr., M. G. Raphael, and J. F. Piatt, Technical Editors. 1995. Ecology and conservation of the Marbled Murrelet. General Technical Report PSW-GTR-152, Pacific Southwest Research Station, Forest Service, U.S. Department of Agriculture. Albany, California.

Ralph, C. J., and C. A. Pearson. 1971. Correlation of age, size of territory, plumage, and breeding success in White-crowned Sparrows. *Condor* 73:77-80.

Raphael, M. G. 1985. Orientation of American Kestrel nest cavities and nest trees. *Condor* 87:437-38.

Rea, A. M. 1970. Winter territoriality in a Ruby-crowned Kinglet. *Western Bird-Bander* 1970:4-7.

Reese, K. P., and J. A. Kadlec. 1985. Influence of high density and parental age on the habitat selection and reproduction of Black-billed Magpies. *Condor* 87:96-105.

Reid, W. V. 1988. Population dynamics of the glaucous-winged gull. *Journal of Wildlife Management* 52:763-70.

Reynolds, R. T., E. C. Meslow, and H. M. Wight. 1982. Nesting habitat of coexisting *Accipiter* in Oregon. *Journal of Wildlife Management* 46:124-38.

Reynolds, R. T., and H. M. Wight. 1978. Distribution, density, and productivity of accipiter hawks breeding in Oregon. *Wilson Bulletin* 90:182-96.

Reynolds, T. D. 1979. The impact of Loggerhead Shrikes on nesting birds in a sagebrush environment. *Auk* 96:798-800.

Reynolds, T. D., and T. D. Rich. 1978. Reproductive ecology of the Sage Thrasher (*Oreoscoptes montanus*) on the Snake River Plain in South-central Idaho. *Auk* 95:580-82.

Richmond, S. M. 1953. The attraction of Purple Martins to an urban location in western Oregon. *Condor* 55:225-49.

Rickart, E. A. 1987. Spermophilus townsendii. Mammalian Species No. 268:1-6, American Society of Mammalogists.

Rickart, E. A. 1988. Population structure of the Piute ground squirrel (*Spermophilus mollis*). *Southwestern Naturalist* 33:91-96.

Rickart, E. A., and E. Yensen. 1991. Spermophilus washingtoni. Mammalian Species No. 371:1-5. American Society of Mammalogists.

Riddle, B. R., and J. R. Choate. 1986. Systematics and biogeography of *Onychomys leucogaster* in western North America. *Journal of Mammalogy* 67:233-55.

Ripley, S. D. 1977. *Rails of the World*. David R. Godine, Publisher, Boston, Massachusetts.

Ritke, M. E., and M. L. Kennedy. 1988. Intraspecific morphologic variation in the raccoon (*Procyon lotor*) and its relationship to selected environmental variables. *Southwestern Naturalist* 33:295-314.

Roberts, R. C. 1979. Habitat and resource relationships in Acorn Woodpeckers. *Condor* 81:1-8.

Robinette, W. L., J. S. Gashwiler, and O. W. Morris. 1961. Notes on cougar productivity and life history. *Journal of Mammalogy* 42:204-17.

Rodenhouse, N. L., and L. B. Best. 1983. Breeding ecology of vesper sparrows in corn and soybean fields. *American Midland Naturalist* 110:265-75.

Rogers, L. E., and J. D. Hedlund. 1980. A comparison of small mammal populations occupying three distinct shrub-steppe communities in eastern Oregon. *Northwest Science* 54:183-86.

Rohwer, S., and J. Manning. 1990. Differences in timing and number of molts for Baltimore and Bullock's Orioles: implications to hybrid fitness and theories of delayed plumage maturation. *Condor* 92:125-40.

Root, R. B. 1967. The niche exploitation pattern of the Blue-gray Gnatcatcher. *Ecological Monographs* 37:317-50.

Root, R. B. 1969a. Interspecific territoriality between Bewick's and House Wrens. *Auk* 86:125-27.

Root, R. B. 1969b. The behavior and reproductive success of the Blue-gray Gnatcatcher. *Condor* 71:16-31.

Root, T. 1988. *Atlas of Wintering North American Birds.* University of Chicago Press, Chicago.

Rotenberry, J. T., and J. A. Wiens. 1989. Reproductive biology of shrubsteppe passerine birds: geographical and temporal variation in clutch size, brood size, and fledging success. *Condor* 91:1-14.

Rotenberry, J. T., and J. A. Wiens. 1991. Weather and reproductive variation in shrubsteppe sparrows: a hierarchical analysis. *Ecology* 72:1325-35.

Rothfels, M., and M. R. Lein. 1983. Territoriality in sympatric populations of red-tailed and Swainson's hawks. *Canadian Journal of Zoology* 61:60-64.

Rothstein, S. I. 1971. High nest density and non-random nest placement in the Cedar Waxwing. *Condor* 73:483-85.

Roze, U. 1989. *The North American Porcupine.* Smithsonian Institution Press, Washington, D.C.

Rumble, M. A., and S. H. Anderson. 1992. Stratification of habitats for identifying habitat selection by Merriam's Turkeys. *Great Basin Naturalist* 52:139-44.

Rutter, R. J. 1969. A contribution to the biology of the Gray Jay (*Perisoreus canadensis*). *Canadian Field-Naturalist* 83:300-16.

Ryser, F. A., Jr. 1985. *Birds of the Great Basin: A natural history.* University of Nevada Press, Reno.

Sakai, H. F., and B. R. Noon. 1991. Nest-site characteristics of Hammond's and Pacific-slope Flycatchers in northwestern California. *Condor* 93:563-74.

Salomonson, M. G., and R. P. Balda. 1977. Winter territoriality of Townsend's Solitaires (*Myadestes townsendi*) in a pinon-juniper-ponderosa pine ecotone. *Condor* 79:148-61.

Salt, G. W. 1952. The relation of metabolism to climate and distribution in three finches of the genus *Carpodacus*. *Ecological Monographs* 22:121-52.

Salt, G. W., and D. C. Willard. 1971. The hunting behavior and success of Forster's Tern. *Ecology* 52:989-98.

Samallow, P. B. 1980. Selective mortality and reproduction in a natural population of *Bufo boreas*. *Evolution* 34:18-39.

Samson, F. B. 1976. Territory, breeding density, and fall departure in Cassin's Finch. *Auk* 93:477-97.

Sanger, G. A. 1972. Checklist of bird observations from the eastern North Pacific Ocean, 1955-1967. *Murrelet* 53:16-21.

Schaub, D. L., and J. H. Larsen, Jr. 1978. The reproductive ecology of the Pacific treefrog (*Hyla regilla*). *Herpetologica* 34:409-16.

Schmutz, J. K. 1989. Hawk occupancy of disturbed grasslands in relation to models of habitat selection. *Condor* 91:362-71.

Schonberger, C. F. 1945. Food of some amphibians and reptiles of Oregon and Washington. *Copeia* 1945:120-21.

Scott, A. C., and M. Bekoff. 1991. Breeding behavior of Evening Grosbeaks. *Condor* 93:71-81.

Scott, D. M., and C. D. Ankney. 1983. The laying cycle of Brown-headed Cowbirds: passerine chickens? *Auk* 100:583-92.

Scott, J. M. 1971. Interbreeding of the Glaucous-winged Gull and Western Gull in the Pacific Northwest. *California Birds* 2:129-33.

Scott, J. M., and B. Csuti. 1997. Gap analysis for biodiversity survey and maintenance. Pages 321-340 *in Biodiversity II: Understanding and protecting our biological resources* (M. L. Reaka-Kudla, D. E. Wilson, and E. O. Wilson, editors). Joseph Henry Press, Washington, D.C.

Scott, J. M., F. Davis, B. Csuti, R. Noss, B. Butterfield, C. Groves, H. Anderson, S. Caicco, F. D'Erchia, T. C. Edwards, Jr., J. Ulliman, and R. G. Wright. 1993. Gap analysis: A geographic approach to protection of biological diversity. *Wildlife Monographs* No. 123:1-41.

Sealy, S. G. 1974. Ecological segregation of Swainson's and Hermit Thrushes in Langara Island, British Columbia. *Condor* 76:350-51.

Sealy, S. G. 1990. Auks at sea. Western Foundation of Vertebrate Biology, *Studies in Avian Biology* No. 14:1-180.

Serie, J. R., and G. A. Swanson. 1976. Feeding ecology of breeding gadwalls on saline wetlands. *Journal of Wildlife Management* 40:69-81.

Servello, F. A., and R. L. Kirkpatrick. 1987. Regional variation in the nutritional ecology of ruffed grouse. *Journal of Wildlife Management* 51:749-70.

Shackleton, D. M. 1985. Ovis canadensis. Mammalian Species No. 230:1-9, American Society of Mammalogists.

Sherry, T. W., and R. T. Holmes. 1988. Habitat selection by breeding American Redstarts in response to a dominant competitor, the Least Flycatcher. *Auk* 105:350-64.

Short, L. L. 1974. Habits and interactions of North American three-toed woodpeckers (*Picoides arcticus* and *Picoides tridactylus*). *American Museum Novitates* No. 2547:1-42.

Shump, K. A., and A. U. Shump. 1982. Lasiurus cinereus. Mammalian Species No. 185:1-5, American Society of Mammalogists.

Siegfried, W. R. 1976. Breeding biology and parasitism in the Ruddy Duck. *Wilson Bulletin* 88:566-74.

Simpson, M. R. 1993. Myotis californicus. Mammalian Species No. 428:1-4, American Society of Mammalogists.

Singer, S. W., N. L. Naslund, S. A. Singer, and C. J. Ralph. 1991. Discovery and observations of two tree nests of the Marbled Murrelet. *Condor* 93:330-39.

Small, A. 1994. *California Birds: Their status and distribution.* Ibis Publishing Company, Vista, California.

Small, R. J., and D. H. Rusch. 1989. The natal dispersal of Ruffed Grouse. *Auk* 106:72-79.

Smith, A. T. 1974. The distribution and dispersal of pikas: influences of behavior and climate. *Ecology* 55:1368-76.

Smith, A. T., and M. L. Weston. 1990. Ochotona princeps. Mammalian Species No. 352:1-8, American Society of Mammalogists.

Smith, G. W. 1990. Home range and activity patterns of black-tailed jackrabbits. *Great Basin Naturalist* 50:249-56.

Smith, K. G., and D. C. Andersen. 1982. Food, predation, and reproductive ecology of the Dark-eyed Junco in northern Utah. *Auk* 99:650-61.

Smith, M. E., and M. C. Belk. 1996. Sorex monticolus. Mammalian Species No. 528:1-5, American Society of Mammalogists.

Smith, R. I. 1968. The social aspects of reproductive behavior in the Pintail. *Auk* 85:381-96.

Smith, S. M. 1973a. A study of prey-attack behaviour in young Loggerhead Shrikes, *Lanius ludovicianus* L. *Behaviour* 44:113-41.

Smith, S. M. 1973b. An aggressive display and related behavior in the Loggerhead Shrike. *Auk* 90:287-98.

Smith, S. M. 1991. *The Black-capped Chickadee: Behavioral ecology and natural history.* Cornell University Press, Ithica, New York.

Smith, W. P. 1991. Odocoileus virginianus. Mammalian Species No. 388:1-13, American Society of Mammalogists.

Smits, A. W. 1984. Activity patterns and thermal biology of the toad *Bufo boreas halophilus. Copeia* 1984:689-96.

Smolen, M. J., and B. L. Keller. 1987. Microtus longicaudus. Mammalian Species No. 271:1-7, American Society of Mammalogists.

Smyth, M., and G. A. Bartholomew. 1966. The water economy of the Black-throated Sparrow and the Rock Wren. *Condor* 68:447-58.

Snapp, B. D. 1976. Colonial breeding in the Barn Swallow (*Hirundo rustica*) and its adaptive significance. *Condor* 78:471-80.

Snyder, R. C. 1963. *Ambystoma gracile. Catalog of American Amphibians and Reptiles* 6.1-6.2.

Sorenson, M. W. 1962. Some aspects of water shrew behavior. *American Midland Naturalist* 68:445-62.

Spear, L. B., D. G. Ainley, and R. P. Henderson. 1986. Post-fledging parental care in the Western Gull. *Condor* 88:194-99.

St. John, A. D. 1982. The herpetology of Curry County, Oregon. Nongame Wildlife Program Technical Report No. 82-2-04. Oregon Department of Fish and Wildlife, Portland, Oregon.

Stebbins, R. C. 1949a. Speciation in salamanders of the plethodontid genus Ensatina. *University of California Publications in Zoology* 48:377-526.

Stebbins, R. C. 1949b. Courtship of the plethodontid salamander *Ensatina eschscholtzii. Copeia* 1949:274-81.

Stebbins, R. C. 1949c. Observations on laying, development, and hatching of the eggs of *Batrachoseps wrighti. Copeia* 1949:161-68.

Stebbins, R. C. 1954. Natural history of the salamanders of the plethodontid genus Ensatina. *University of California Publications in Zoology* 54:47-124.

Stebbins, R. C. 1985. *A Field Guide to Western Reptiles and Amphibians,* Second Edition. Houghton Mifflin Company, Boston, Massachusetts.

Steidl, R. J., C. R. Griffin, L. J. Niles, and K. E. Clark. 1991. Reproductive success and eggshell thinning of a reestablished peregrine falcon population. *Journal of Wildlife Management* 55:294-99.

Stelmock, J. J., and A. S. Harestad. 1979. Food habits and life history of the clouded salamander (*Aneides ferreus*) on northern Vancouver Island, British Columbia. *Syesis* 12:71-75.

Stenzel, L. E., H. R. Huber, and G. W. Page. 1976. Feeding behavior and diet of the Long-billed Curlew and Willet. *Wilson Bulletin* 88:314-32.

Stern, M. A., J. F. Morawski, and G. A. Rosenberg. 1993. Rediscovery and status of a disjunct population of breeding Yellow Rails in southern Oregon. *Condor* 95:1024-27.

Stern, M. A., and G. A. Rosenberg. 1985. Occurrence of a breeding Upland Sandpiper at Sycan Marsh, Oregon. *Murrelet* 66:34-35.

Stern, M. A., T. G. Wise, and K. L. Theodore. 1987. Use of natural cavity by Bufflehead nesting in Oregon. *Murrelet* 68:50.

Stewart, J. R. 1985. Growth and survivorship in a California population of Gerrhonotus coeruleus with comments on intraspecific variation in adult female size. *American Midland Naturalist* 113:30-44.

Stewart, R. M. 1973. Breeding behavior and the life history of the Wilson's Warbler. *Wilson Bulletin* 85:21-30.

Stewart, R. M., R. P. Henderson, and K. Darling. 1978. Breeding ecology of the Wilson's Warbler in the high Sierra Nevada, California. *Living Bird* 16:83-102.

Stiles, F. G. 1973. Food supply and the annual cycle of the Anna Hummingbird. *University of California Publications in Zoology* 97:1-109.

Stiles, N. W. 1980. Bird community structure in alder forests in Washington. *Condor* 82:20-30.

Stokes, D. W., and L. Q. Stokes. 1996. *Stokes Field Guide to Birds: Western region.* Little, Brown & Company, Boston.

Storm, R. M. 1960. Notes on the breeding biology of the red-legged frog (*Rana aurora aurora*). *Herpetologica* 16:251-59.

Storm, R. M., and A. R. Aller. 1947. Food habits of *Aneides ferreus. Herpetologica* 4:59-60.

Storm, R. M., and W. P. Leonard. 1995. *Reptiles of Washington and Oregon.* Seattle Audubon Society, Seattle, Washington.

Strahan, R. 1983. *The Complete Book of Australian Mammals.* Cornstalk Publishing, North Ryde, Australia.

Sullivan, R. M. 1985. Phyletic, biogeographic, and ecologic relationships among montane populations of least chipmunks (*Eutamias minimus*) in the Southwest. *Systematic Zoology* 34:419-48.

Sullivan, T. P. 1990. Responses of the red squirrel (*Tamiasciurus hudsonicus*) populations to supplemental food. *Journal of Mammalogy* 71:579-90.

Sutton, D. A. 1987. Analysis of Pacific coast Townsend chipmunks (Rodentia: Sciuridae). *Southwestern Naturalist* 32:371-76.

Sutton, D. A. 1992. Tamias amoenus. Mammalian Species No. 390:1-8, American Society of Mammalogists.

Sutton, D. A. 1993. Tamias townsendii. Mammalian Species No. 435:1-6, American Society of Mammalogists.

Sutton, D. A. 1995. Problems of taxonomy and distribution in four species of chipmnuks. *Journal of Mammalogy* 76:843-50.

Svihla, A. 1953. Diurnal retreats of the spadefoot toad *Scaphiopus hammondi. Copeia* 1953:186.

Sweet, S. S., and W. S. Parker. 1990. *Pituophis melanoleucus. Catalog of American Amphibians and Reptiles* 474:1-8.

Sydeman, W. J., M. Guntert, and R. P. Balda. 1988. Annual reproductive yield in the cooperative Pygmy Nuthatch (*Sitta pygmaea*). *Auk* 105:70-77.

Tamarin, R. H. 1985. Biology of New World *Microtus.* American Society of Mammalogists, Special Publication No. 8.

Tanner, V. M. 1939. A study of the genus Scaphiopus. *Great Basin Naturalist* 1:3-26.

Tanner, W. W. 1953. Notes on the life history of *Plethopsis wrighti* Bishop. *Herpetologica* 9:139-40.

Tanner, W. W. 1957. A taxonomic and ecological study of the western skink (*Eumeces skiltonianus*). *Great Basin Naturalist* 17:59-94.

Tanner, W. W. 1988. *Eumeces skiltonianus. Catalog of American Amphibians and Reptiles,* 447:1-4.

Tanner, W. W., and J. E. Krogh. 1973. Ecology of *Phrynosoma platyrhinos* at the Nevada Test Site, Nye County, Nevada. *Herpetologica* 29:327-42.

Tanner, W. W., and J. E. Krogh. 1974. Ecology of the leopard lizard, *Crotaphytus wislizeni* at the Nevada Test Site, Nye County, Nevada. *Herpetologica* 30:63-72.

Tanner, W. W., and C. H. Lowe. 1989. Variations in *Thamnophis elegans* with descriptions of new subspecies. *Great Basin Naturalist* 49:511-16.

Taylor, A. L., and E. D. Forsman. 1976. Recent range extensions of the Barred Owl in western North America, including the first records for Oregon. *Condor* 78:560-61.

Taylor, J. T. 1977. The behavioral ecology of larval and neotonic northwestern salamanders (*Ambystoma gracile*). Unpublished Ph.D. Thesis, Oregon State University, Corvallis.

Thomas, J. W. 1979. Wildlife habitats in managed forests: The Blue Mountains of Oregon and Washington. USDA Forest Service Agricultural Handbook No. 553. U. S. Government Printing Office, Washington, D.C.

Thomas, J. W., and D. E. Toweill. 1982. *The Elk of North America: Ecology and management.* Stackpole Books, Harrisburg, Pennsylvania.

Thompson, C. F., and V. Nolan, Jr. 1973. Population biology of the Yellow-breasted Chat (*Icteria virens* L.) in southern Indiana. *Ecological Monographs* 43:145-71.

Thompson, F. R., III, and E. K. Fritzell. 1989. Habitat use, home range, and survival of territorial male ruffed grouse. *Journal of Wildlife Management* 53:15-21.

Thut, R. N. 1970. Feeding habits of the Dipper in southwestern Washington. *Condor* 72:234-35.

Tinkle, D. W. 1967. The life and demography of the side-blotched lizard, *Uta stansburiana.* Miscellaneous Publications, Museum of Zoology, University of Michigan No. 132:1-182.

Tinkle, D. W. 1973. A population analysis of the sagebrush lizard, *Sceloporus graciosus* in southern Utah. *Copeia* 1973:284-96.

Titus, T. A. 1990. Genetic variation in two subspecies of *Ambystoma gracile* (Caudata:Ambystomatidae). *Journal of Herpetology* 24:107-11.

Tomback, D. F. 1977. Foraging strategies of Clark's Nutcrackers. *Living Bird* 16:123-61.

Tomback, D. F. 1980. How nutcrackers find their seed stores. *Condor* 82:10-19.

Tompa, F. S. 1962. Territorial behavior: the main controlling factor of a local Song Sparrow population. *Auk* 79:687-97.

Toweill, D. E. 1974. Winter food habits of river otters in western Oregon. *Journal of Wildlife Management* 38:107-11.

Trethewey, D. E. C., and B. J. Verts. 1971. Reproduction in eastern cottontail rabbits in western Oregon. *American Midland Naturalist* 86:463-76.

Trost, C. H. 1972. Adaptations of Horned Larks (*Eremophila alpestris*) to hot environments. *Auk* 89:506-27.

Tuck, L. M. 1972. The snipes: a study of the genus *Capella.* Canadian Wildlife Service, Ottawa, *Monograph Series* No. 5:1-428.

Tumlison, R. 1987. Felis lynx. Mammalian Species No. 269:1-8, American Society of Mammalogists.

Tvrdik, G. M. 1971. Pendulum display by Olive-sided Flycatcher. Auk 88:174.

Udvardy, M.D.F. 1977. *The Audubon Society Field Guide to North American Birds: Western Region.* Alfred A. Knopf, Inc., New York.

U.S. Fish and Wildlife Service. 1995. Draft Marbled Murrelet (*Brachyramphus marmoratus*)(Washington, Oregon and California Population) Recovery Plan. Portland, Oregon. 171 pp.

Van Camp, L. F., and C. J. Henny. 1975. The screech owl: its life history and population ecology in northern Ohio. North American Fauna No. 71:1-65, U.S. Fish and Wildlife Service.

van Zyll de Jong, C. G. 1984. Taxonomic relationships of Nearctic small-footed bats of the *Myotis leibii* group (Chiroptera: Vespertilionidae). *Canadian Journal of Zoology* 62:2519-26.

Vander Wall, S. B., and R. P. Balda. 1977. Coadaptations of the Clark's Nutcracker and the piñon pine for efficient seed harvest and dispersal. *Ecological Monographs* 47:89-111.

Verbeek, N. A. M. 1967. Breeding biology and ecology of the Horned Lark in alpine tundra. *Wilson Bulletin* 79:208-18.

Verbeek, N. A. M. 1970. Breeding ecology of the Water Pipit. *Auk* 87:425-51.

Verbeek, N. A. M. 1973. The exploitation system of the Yellow-billed Magpie. *University of California Publications in Zoology* 99:1-58.

Verbeek, N. A. M. 1975. Comparative feeding behavior of three coexisting tyrannid flycatchers. *Wilson Bulletin* 87:231-40.

Vermeer, K. 1970. Breeding biology of California and ring-billed gulls: a study of ecological adaptation to the inland habitat. *Canadian Wildlife Service Report Series* No. 12:1-52.

Verner, J. 1975. Interspecific aggression between Yellow-headed Blackbirds and Long-billed Marsh Wrens. *Condor* 77:328-31.

Verner, J., and G. H. Engelsen. 1970. Territories, multiple nest building, and polygyny in the Long-billed Marsh Wren. *Auk* 87:557-67.

Verts, B. J. 1967. *The Biology of the Striped Skunk.* University of Illinois Press, Urbana.

Verts, B. J. 1975. New records for three uncommon mammals in Oregon. *Murrelet* 56:22-23.

Verts, B. J., and L. N. Carraway. 1981. Dispersal and dispersion of an introduced population of *Sylvilagus floridanus. Great Basin Naturalist* 41:167-75.

Verts, B. J., and L. N. Carraway. 1987a. Thomomys bulbivorus. Mammalian Species No. 273:1-4, American Society of Mammalogists.

Verts, B. J., and L. N. Carraway. 1987b. Microtus canicaudus. Mammalian Species No. 267:1-4, American Society of Mammalogists.

Verts, B. J., and L. N. Carraway. 1995. Phenacomys albipes. Mammalian Species No. 494:1-5, American Society of Mammalogists.

Verts, B. J., S. D. Gehman, and K. J. Hundertmark. 1984. *Sylvilagus nuttallii*: a semiarboreal lagomorph. *Journal of Mammalogy* 65:131-35.

Verts, B. J., and G. L. Kirkland, Jr. 1988. Perognathus parvus. Mammalian Species No. 318:1-8, American Society of Mammalogists.

Vitt, L. J. 1973. Reproductive biology of the anguid lizard, *Gerrhonotus coeruleus principis. Herpetologica* 29:176-84.

Wade-Smith, J., and B. J. Verts. 1982. Mephitis mephitis. Mammalian Species No. 173:1-7, American Society of Mammalogists.

Wai-Ping, V., and M. B. Fenton. 1989. Ecology of spotted bat (*Euderma maculatum*) roosting and foraging behavior. *Journal of Mammalogy* 70:617-22.

Wakeley, J. S. 1978. Factors affecting the use of hunting sites by Ferruginous Hawks. *Condor* 80:316-26.

Walker, E. P. 1964. *Mammals of the World*, Volume II. The Johns Hopkins Press, Baltimore, Maryland.

Warham, J. 1991. *The Petrels: Their ecology and breeding systems.* Academic Press.

Warner, R. M., and N. J. Czaplewski. 1984. Myotis volans. Mammalian Species No. 224:1-4, American Society of Mammalogists.

Watkins, L. C. 1977. Euderma maculatum. Mammalian Species No. 77:1-4, American Society of Mammalogists.

Watt, D. J., and A. M. Dimberio. 1990. Structure of successful nests of the American Goldfinch (*Carduelis tristis*). *Journal of Field Ornithology* 61:413-18.

Wauer, R. H. 1964. Ecological distribution of the birds of the Panamint Mountains, California. *Condor* 66:287-301.

Weatherhead, P. J. 1989. Sex ratios, host-specific reproductive success, and impact of Brown-headed Cowbirds. *Auk* 106:358-66.

Webster, W. D., and J. K. Jones, Jr. 1982. Reithrodontomys megalotis. Mammalian Species No. 167:1-5, American Society of Mammalogists.

Weigand, J. P. 1980. Ecology of the Hungarian Partridge in north-central Montana. *Wildlife Monographs* 74:1-106.

Weise, J. H. 1976. Courtship and pair formation in the Great Egret. *Auk* 93:709-24.

Weiss, N. T., and B. J. Verts. 1984. Habitat and distribution of pygmy rabbits (*Sylvilagus idahoensis*) in Oregon. *Great Basin Naturalist* 44:563-69.

Weitzel, N. H. 1988. Nest-site competition between the European Starling and native breeding birds in northwestern Nevada. *Condor* 90:515-17.

Wells-Gosling, N., and L. R. Heaney. 1984. Glaucomys sabrinus. Mammalian Species No. 229:1-8, American Society of Mammalogists.

Welsh, H. H., Jr. 1990. Relictual amphibians and old-growth forests. *Conservation Biology* 4:309-19.

Wernz, J. G. 1969. Spring mating of *Ascaphus. Journal of Herpetology* 3:167-69.

Weston, H. G., Jr. 1947. Breeding behavior of the Black-headed Grosbeak. *Condor* 49:54-73.

Wheeler, R. J. 1965. Pioneering of blue-winged teal in California, Oregon, Washington, and British Columbia. *Murrelet* 46:40-42.

Wheelwright, N. T. 1986. The diet of American Robins: an analysis of U.S. Biological Survey records. *Auk* 103:710-25.

Whelton, B. D. 1989. Distribution of the Boreal Owl in eastern Washington and Oregon. *Condor* 91:712-16.

Whitaker, J. O., Jr., and C. Maser. 1976. Food habits of five western Oregon shrews. *Northwest Science* 50:102-7.

Whitaker, J. O., Jr., and C. Maser. 1981. Food habits of seven species of lizards from Malheur County, southeastern Oregon. *Northwest Science* 55:202-8.

Whitaker, J. O., Jr., C. Maser, and S. P. Cross. 1981. Food habits of eastern Oregon bats, based on stomach and scat analysis. *Northwest Science* 55:281-92.

Whitaker, J.O., C. Maser, and L.E. Keller. 1977. Food habits of bats of western Oregon. *Northwest Science* 51:46-55.

Whitaker, J. O., Jr., C. Maser, and R. J. Pedersen. 1979. Food and ectoparasitic mites of Oregon moles. *Northwest Science* 53:268-73.

White, C. M., and M. Tanner-White. 1988. Use of interstate highway overpasses and billboards for nesting by the Common Raven (*Corvus corax*). *Great Basin Naturalist* 48:64-67.

White, C. M., and T. L. Thurow. 1985. Reproduction of Ferruginous Hawks exposed to controlled disturbance. *Condor* 87:14-22.

White, D., A. J. Kimerling, and W. S. Overton. 1992. Cartographic and geometric components of a global sampling design for environmental monitoring. *Cartography and Geographic Information Systems* 19:5-22.

White, R. L., II. 1977. Prey selection by the rough skinned newt (*Taricha granulosa*) in two pond types. *Northwest Science* 51:114-18.

White, S. M., J. T. Flinders, and B. L. Welch. 1982. Preference of pygmy rabbits (*Brachylagus idahoensis*) for various populations of big sagebrush (*Artemisia tridentata*). *Journal of Range Management* 35:724-26.

Whitmore, R. C. 1975. Indigo buntings in Utah with special reference to interspecific competition with Lazuli Buntings. *Condor* 77:509-10.

Whitney, C. L., and J. R. Krebs. 1975. Spacing and calling in Pacific tree frogs, *Hyla regilla*. *Canadian Journal of Zoology* 53:1519-27.

Wiens, J. A., and R. A. Nussbaum. 1975. Model estimation of energy flow in Northwestern coniferous bird communities. *Ecology* 56:547-61.

Wiens, J. A., J. T. Rotenberry, and B. Van Horne. 1985. Territory size variations in shrubsteppe birds. *Auk* 102:500-5.

Wiens, J. A., and B. Van Horne. 1990. Comparisons of the behavior of Sage and Brewer's Sparrows in shrubsteppe habitats. *Condor* 92:264-66.

Wiese, J.H. 1976. Courtship and pair formation in Great Egrets. *Auk* 93: 704-24.

Wiggett, D. R., and D. A. Boag. 1989. Intercolony natal dispersal in the Columbian ground squirrel. *Canadian Journal of Zoology* 67:42-50.

Wight, H. M., R. U. Mace, and W. M. Batterson. 1967. Mortality estimates of an adult band-tailed pigeon population in Oregon. *Journal of Wildlife Management* 31:519-25.

Wilbur, S. R., and J. A. Jackson. 1983. *Vulture Biology and Management.* University of California Press, Berkeley.

Wilkins, K. T. 1989. Tadarida brasiliensis. Mammalian Species No. 331:1-10, American Society of Mammalogists.

Williams, D. F. 1984. Habitat associations of some rare shrews (*Sorex*) from California. *Journal of Mammalogy* 65:325-28.

Williams, J. B. 1987. Field metabolism and food consumption of Savannah Sparrows during the breeding season. *Auk* 104:277-89.

Williams, L. 1952. Breeding behavior of the Brewer Blackbird. *Condor* 54:3-47.

Williams, P. L., and W. D. Koenig. 1980. Water dependence of birds in a temperate oak woodland. *Auk* 97:339-50.

Williamson, P. 1971. Feeding ecology of the Red-eyed Vireo (*Vireo olivaceus*) and associated foliage-gleaning birds. *Ecological Monographs* 41:129-52.

Willner, G. R., G. A. Feldhamer, E. E. Zucker, and J. A. Chapman. 1980. Ondatra zibethicus. Mammalian Species No. 141:1-8, American Society of Mammalogists.

Willson, M. F. 1966. Breeding ecology of the Yellow-headed Blackbird. *Ecological Monographs* 36:51-77.

Wilson, D. E., and D. M. Reeder. 1993. *Mammal Species of the World: A taxonomic and geographic reference,* Second Edition. Smithsonian Institution Press, Washington, DC.

Wilson, U. W., and D. A. Manuwal. 1986. Breeding biology of the Rhinoceros Auklet in Washington. *Condor* 88:143-55.

Wiseman, A. J. 1975. Changes in body weight of American Goldfinches. *Wilson Bulletin* 87:390-411.

Wittenberger, J. F. 1978. The breeding biology of an isolated Bobolink population in Oregon. *Condor* 80:355-71.

Woodbury, M., and A. M. Woodbury. 1945. Life-history studies of the sagebrush lizard *Sceloporus g. graciosus* with special reference to cycles in reproduction. *Herpetologica* 2:175-96.

Woods, C. A. 1973. Erethizon dorsatum. Mammalian Species No. 29:1-6, American Society of Mammalogists.

Woods, C. A., L. Contreras, G. Willner-Chapman, and H. P. Whidden. 1992. Myocastor coypus. Mammalian Species No. 398:1-8, American Society of Mammalogists.

Woods, C. P., and T. J. Cade. 1996. Nesting habits of the Loggerhead Shrike in sagebrush. *Condor* 98:75-81.

Woodward, B. 1982. Male persistence and mating success in Woodhouse's toad (*Bufo woodhousei*). *Ecology* 63:583-85.

Woodward, P. W. 1983. Behavioral ecology of fledgling Brown-headed Cowbirds and their hosts. *Condor* 85:151-63.

Young, S. P. 1958. *The Bobcat of North America: Its history, life habits, economic status and control, with a list of currently recognized subspecies.* Stackpole Co., Harrisburg, Pennsylvania.

Zammuto, R. M., and J. S. Millar. 1985. Environmental predictability, variability, and *Spermophilus columbianus* life history over an environmental gradient. *Ecology* 66:1784-94.

Zeiner, D. C., W. F. Laudenslayer, Jr., and K. E. Mayer. 1988. California's Wildlife, Volume I: Amphibians and Reptiles. California Department of Fish and Game, Sacramento.

Zeiner, D. C., W. F. Laudenslayer, Jr., K. E. Mayer, and M. White. 1990a. California's Wildlife, Volume II: Birds. California Department of Fish and Game, Sacramento.

Zeiner, D. C., W. F. Laudenslayer, Jr., K. E. Mayer, and M. White. 1990b. California's Wildlife, Volume III: Mammals. California Department of Fish and Game, Sacramento.

Zeveloff, S. I. 1988. *Mammals of the Intermountain West.* University of Utah Press, Salt Lake City.

Zielinski, W. J., W. D. Spencer, and R. H. Barrett. 1983. Relationship between food habits and activity patterns of pine martens. *Journal of Mammalogy* 64:387-96.

Zink, R. M. 1988. Evolution of Brown Towhees: allozymes, morphometrics and species limits. *Condor* 90:72-82.

Zink, R. M., and J. C. Avise. 1990. Patterns of mitochondrial DNA and allozyme evolution in the avian genus *Ammodramus*. *Systematic Zoology* 39:148-61.

Zink, R. M., and D. L. Dittmann. 1991. Evolution of Brown Towhees: mitochondrial DNA evidence. *Condor* 93:98-105.

Zink, R. M., D. L. Dittmann, and W. L. Rootes. 1991. Mitochrondrial DNA variation and the phylogeny of *Zonotrichia*. *Auk* 108:578-84.

Zweifel, R. G. 1955. Ecology, distribution and systematics of frogs of the Rana boylei group. *University of California Publications in Zoology* 54:207-92.

Zweifel, R. G. 1974. *xmpropeltis zonata. Catalog of American Amphibians and Reptiles* 174:1-4.

Zwickel, F. C., D. A. Boag, and J. H. Brigham. 1974. The autumn diet of Spruce Grouse: a regional comparison. *Condor* 76:212-14.

With a few exceptions (explained in "Comments" for individual species), nomenclature and taxonomy for amphibians and reptiles follows Collins (1990) and the sequence follows Nussbaum et al. (1983); taxonomy, nomenclature, and sequence for birds follows American Ornithologists' Union (1983, 1985, 1989, 1993, 1995); taxonomy and nomenclature for mammals follows Jones et al. (1992) and sequence follows Hall (1981); species status in Oregon usually follows Puchy and Marshall (1993) as updated from information found in Gilligan et al. (1994). Total species = 441 (amphibians = 30; reptiles = 28; birds = 261; mammals = 122).

## Amphibians

*Ambystoma gracile* Northwestern Salamander
*Ambystoma macrodactylum* Long-toed Salamander
*Ambystoma tigrinum* Tiger Salamander
*Dicamptodon copei* Cope's Giant Salamander
*Dicamptodon tenebrosus* Pacific Giant Salamander
*Rhyacotriton variegatus* Southern Torrent Salamander
*Rhyacotriton cascadae* Cascade Torrent Salamander
*Rhyacotriton kezeri* Columbia Torrent Salamander
*Aneides ferreus* Clouded Salamander
*Aneides flavipunctatus* Black Salamander
*Batrachoseps attenuatus* California Slender Salamander
*Batrachoseps wrighti* Oregon Slender Salamander
*Ensatina eschscholtzii* Ensatina
*Plethodon dunni* Dunn's Salamander
*Plethodon elongatus* Del Norte Salamander
*Plethodon larselli* Larch Mountain Salamander
*Plethodon stormi* Siskiyou Mountains Salamander
*Plethodon vehiculum* Western Redback Salamander
*Taricha granulosa* Roughskin Newt
*Bufo boreas* Western Toad
*Bufo woodhouseii* Woodhouse's Toad
*Pseudacris regilla* Pacific Chorus Frog
*Ascaphus truei* Tailed Frog
*Scaphiopus intermontanus* Great Basin Spadefoot
*Rana aurora* Red-legged Frog
*Rana boylii* Foothill Yellow-legged Frog
*Rana cascadae* Cascades Frog
*Rana catesbeiana* Bullfrog
*Rana pipiens* Northern Leopard Frog
*Rana pretiosa* Spotted Frog

## Reptiles

*Chrysemys picta* Painted Turtle
*Clemmys marmorata* Western Pond Turtle
*Elgaria coerulea* Northern Alligator Lizard
*Elgaria multicarinata* Southern Alligator Lizard
*Crotaphytus bicinctores* Mojave Black-Collared Lizard
*Gambelia wislizenii* Longnose Leopard Lizard
*Phrynosoma douglassii* Short-horned Lizard
*Phrynosoma platyrhinos* Desert Horned Lizard
*Sceloporus graciosus* Sagebrush Lizard
*Sceloporus occidentalis* Western Fence Lizard
*Uta stansburiana* Side-blotched Lizard
*Eumeces skiltonianus* Western Skink
*Cnemidophorus tigris* Western Whiptail
*Charina bottae* Rubber Boa
*Coluber constrictor* Racer
*Contia tenuis* Sharptail Snake
*Diadophis punctatus* Ringneck Snake
*Hypsiglena torquata* Night Snake
*Lampropeltis getula* Common Kingsnake
*Lampropeltis zonata* California Mountain Kingsnake
*Masticophis taeniatus* Striped Whipsnake
*Pituophis melanoleucus* Gopher Snake
*Sonora semiannulata* Ground Snake
*Thamnophis couchii* Western Aquatic Garter Snake
*Thamnophis elegans* Western Terrestrial Garter Snake
*Thamnophis ordinoides* Northwestern Garter Snake
*Thamnophis sirtalis* Common Garter Snake
*Crotalus viridis* Western Rattlesnake

## Birds

*Podilymbus podiceps* Pied-billed Grebe
*Podiceps auritus* Horned Grebe
*Podiceps grisegena* Red-necked Grebe
*Podiceps nigricollis* Eared Grebe
*Aechmophorus occidentalis* Western Grebe
*Aechmophorus clarkii* Clark's Grebe
*Oceanodroma furcata* Fork-tailed Storm-Petrel
*Oceanodroma leucorhoa* Leach's Storm-Petrel
*Pelecanus erythrorhynchos* American White Pelican
*Phalacrocorax auritus* Double-crested Cormorant
*Phalacrocorax penicillatus* Brandt's Cormorant
*Phalacrocorax pelagicus* Pelagic Cormorant
*Botaurus lentiginosus* American Bittern
*Ixobrychus exilis* Least Bittern
*Ardea herodias* Great Blue Heron
*Ardea alba* Great Egret
*Egretta thula* Snowy Egret
*Bubulcus ibis* Cattle Egret
*Butorides virescens* Green Heron
*Nycticorax nycticorax* Black-crowned Night-Heron

*Plegadis chihi* White-faced Ibis
*Cygnus buccinator* Trumpeter Swan
*Branta canadensis* Canada Goose
*Aix sponsa* Wood Duck
*Anas crecca* Green-winged Teal
*Anas platyrhynchos* Mallard
*Anas acuta* Northern Pintail
*Anas discors* Blue-winged Teal
*Anas cyanoptera* Cinnamon Teal
*Anas clypeata* Northern Shoveler
*Anas strepera* Gadwall
*Anas americana* American Wigeon
*Aythya valisineria* Canvasback
*Aythya americana* Redhead
*Aythya collaris* Ring-necked Duck
*Aythya affinis* Lesser Scaup
*Histrionicus histrionicus* Harlequin Duck
*Bucephala islandica* Barrow's Goldeneye
*Bucephala albeola* Bufflehead
*Lophodytes cucullatus* Hooded Merganser
*Mergus merganser* Common Merganser
*Oxyura jamaicensis* Ruddy Duck
*Cathartes aura* Turkey Vulture
*Pandion haliaetus* Osprey
*Elanus leucurus* White-tailed Kite
*Haliaeetus leucocephalus* Bald Eagle
*Circus cyaneus* Northern Harrier
*Accipiter striatus* Sharp-shinned Hawk
*Accipiter cooperii* Cooper's Hawk
*Accipiter gentilis* Northern Goshawk
*Buteo swainsoni* Swainson's Hawk
*Buteo jamaicensis* Red-tailed Hawk
*Buteo regalis* Ferruginous hawk
*Aquila chrysaetos* Golden Eagle
*Falco sparverius* American Kestrel
*Falco peregrinus* Peregrine Falcon
*Falco mexicanus* Prairie Falcon
*Perdix perdix* Gray Partridge
*Alectoris chukar* Chukar
*Phasianus colchicus* Ring-necked Pheasant
*Dendragapus canadensis* Spruce Grouse
*Dendragapus obscurus* Blue Grouse
*Bonasa umbellus* Ruffed Grouse
*Centrocercus urophasianus* Sage Grouse
*Meleagris gallopavo* Wild Turkey
*Callipepla californica* California Quail
*Oreortyx pictus* Mountain Quail
*Coturnicops noveboracensis* Yellow Rail
*Rallus limicola* Virginia Rail
*Porzana carolina* Sora
*Fulica americana* American Coot
*Grus canadensis* Sandhill Crane
*Charadrius alexandrinus* Snowy Plover
*Charadrius semipalmatus* Semipalmated Plover
*Charadrius vociferus* Killdeer
*Haematopus bachmani* Black Oystercatcher
*Himantopus mexicanus* Black-necked Stilt
*Recurvirostra americana* American Avocet
*Tringa melanoleuca* Greater Yellowlegs
*Tringa solitaria* Solitary Sandpiper
*Catoptrophorus semipalmatus* Willet
*Actitis macularia* Spotted Sandpiper
*Bartramia longicauda* Upland Sandpiper

*Numenius americanus* Long-billed Curlew
*Gallinago gallinago* Common Snipe
*Phalaropus tricolor* Wilson's Phalarope
*Larus pipixcan* Franklin's Gull
*Larus delawarensis* Ring-billed Gull
*Larus californicus* California Gull
*Larus occidentalis* Western Gull
*Larus glaucescens* Glaucous-winged Gull
*Sterna caspia* Caspian Tern
*Sterna forsteri* Forster's Tern
*Chlidonias niger* Black Tern
*Uria aalge* Common Murre
*Cepphus columba* Pigeon Guillemot
*Brachyramphus marmoratus* Marbled Murrelet
*Ptychoramphus aleuticus* Cassin's Auklet
*Cerorhinca monocerata* Rhinoceros Auklet
*Fratercula cirrhata* Tufted Puffin
*Columba livia* Rock Dove
*Columba fasciata* Band-tailed Pigeon
*Zenaida macroura* Mourning Dove
*Coccyzus americanus* Yellow-billed Cuckoo
*Tyto alba* Barn Owl
*Otus flammeolus* Flammulated Owl
*Otus kennicottii* Western Screech-Owl
*Bubo virginianus* Great Horned Owl
*Glaucidium gnoma* Northern Pygmy-Owl
*Athene cunicularia* Burrowing Owl
*Strix occidentalis* Spotted Owl
*Strix varia* Barred Owl
*Strix nebulosa* Great Gray Owl
*Asio otus* Long-eared Owl
*Asio flammeus* Short-eared Owl
*Aegolius funereus* Boreal Owl
*Aegolius acadicus* Northern Saw-whet Owl
*Chordeiles minor* Common Nighthawk
*Phalaenoptilus nuttallii* Common (=Dusky) Poorwill
*Cypseloides niger* Black Swift
*Chaetura vauxi* Vaux's Swift
*Aeronautes saxatalis* White-throated Swift
*Archilochus alexandri* Black-chinned Hummingbird
*Calypte anna* Anna's Hummingbird
*Stellula calliope* Calliope Hummingbird
*Selasphorus rufus* Rufous Hummingbird
*Selasphorus sasin* Allen's Hummingbird
*Ceryle alcyon* Belted Kingfisher
*Melanerpes lewis* Lewis' Woodpecker
*Melanerpes formicivorus* Acorn Woodpecker
*Sphyrapicus nuchalis* Red-naped Sapsucker
*Sphyrapicus ruber* Red-breasted Sapsucker
*Sphyrapicus thyroideus* Williamson's Sapsucker
*Picoides pubescens* Downy Woodpecker
*Picoides villosus* Hairy Woodpecker
*Picoides albolarvatus* White-headed Woodpecker
*Picoides tridactylus* Three-toed Woodpecker
*Picoides arcticus* Black-backed Woodpecker
*Colaptes auratus* Northern Flicker
*Dryocopus pileatus* Pileated Woodpecker
*Contopus borealis* Olive-sided Flycatcher
*Contopus sordidulus* Western Wood-Pewee
*Empidonax traillii* Willow Flycatcher
*Empidonax minimus* Least Flycatcher
*Empidonax hammondii* Hammond's Flycatcher
*Empidonax oberholseri* Dusky Flycatcher

*Empidonax wrightii* Gray Flycatcher
*Empidonax difficilis* Pacific-slope Flycatcher
*Empidonax occidentalis* Cordilleran Flycatcher
*Sayornis nigricans* Black Phoebe
*Sayornis saya* Say's Phoebe
*Myiarchus cinerascens* Ash-throated Flycatcher
*Tyrannus verticalis* Western Kingbird
*Tyrannus tyrannus* Eastern Kingbird
*Eremophila alpestris* Horned Lark
*Progne subis* Purple Martin
*Tachycineta bicolor* Tree Swallow
*Tachycineta thalassina* Violet-green Swallow
*Stelgidopteryx serripennis* Northern Rough-winged Swallow
*Riparia riparia* Bank Swallow
*Hirundo pyrrhonota* Cliff Swallow
*Hirundo rustica* Barn Swallow
*Perisoreus canadensis* Gray Jay
*Cyanocitta stelleri* Steller's Jay
*Aphelocoma californica* Western Scrub-Jay
*Gymnorhinus cyanocephalus* Pinyon Jay
*Nucifraga columbiana* Clark's Nutcracker
*Pica pica* Black-billed Magpie
*Corvus brachyrhynchos* American Crow
*Corvus corax* Common Raven
*Parus atricapillus* Black-capped Chickadee
*Parus gambeli* Mountain Chickadee
*Parus rufescens* Chestnut-backed Chickadee
*Parus inornatus* Plain Titmouse
*Psaltriparus minimus* Bushtit
*Sitta canadensis* Red-breasted Nuthatch
*Sitta carolinensis* White-breasted Nuthatch
*Sitta pygmaea* Pygmy Nuthatch
*Certhia americana* Brown Creeper
*Salpinctes obsoletus* Rock Wren
*Catherpes mexicanus* Canyon Wren
*Thryomanes bewickii* Bewick's Wren
*Troglodytes aedon* House Wren
*Troglodytes troglodytes* Winter Wren
*Cistothorus palustris* Marsh Wren
*Cinclus mexicanus* American Dipper
*Regulus satrapa* Golden-crowned Kinglet
*Regulus calendula* Ruby-crowned Kinglet
*Polioptila caerulea* Blue-gray Gnatcatcher
*Sialia mexicana* Western Bluebird
*Sialia currucoides* Mountain Bluebird
*Myadestes townsendi* Townsend's Solitare
*Catharus fuscescens* Veery
*Catharus ustulatus* Swainson's Thrush
*Catharus guttatus* Hermit Thrush
*Turdus migratorius* American Robin
*Ixoreus naevius* Varied Thrush
*Chamaea fasciata* Wrentit
*Dumetella carolinensis* Gray Catbird
*Oreoscoptes montanus* Sage Thrasher
*Anthus rubescens* American Pipit
*Bombycilla cedrorum* Cedar Waxwing
*Lanius ludovicianus* Loggerhead Shrike
*Sturnus vulgaris* European Starling
*Vireo solitarius* Solitary Vireo
*Vireo huttoni* Hutton's Vireo
*Vireo gilvus* Warbling Vireo
*Vireo olivaceus* Red-eyed Vireo
*Vermivora celata* Orange-crowned Warbler

*Vermivora ruficapilla* Nashville Warbler
*Dendroica petechia* Yellow Warbler
*Dendroica coronata* Yellow-rumped Warbler
*Dendroica nigrescens* Black-throated Gray Warbler
*Dendroica townsendi* Townsend's Warbler
*Dendroica occidentalis* Hermit Warbler
*Setophaga ruticilla* American Redstart
*Seiurus noveboracensis* Northern Waterthrush
*Oporornis tolmiei* MacGillivray's Warbler
*Geothlypis trichas* Common Yellowthroat
*Wilsonia pusilla* Wilson's Warbler
*Icteria virens* Yellow-breasted Chat
*Piranga ludoviciana* Western Tanager
*Pheucticus melanocephalus* Black-headed Grosbeak
*Passerina amoena* Lazuli Bunting
*Pipilo chlorurus* Green-tailed Towhee
*Pipilo maculatus* Spotted Towhee
*Pipilo crissalis* California Towhee
*Spizella passerina* Chipping Sparrow
*Spizella breweri* Brewer's Sparrow
*Spizella atrogularis* Black-chinned Sparrow
*Pooecetes gramineus* Vesper Sparrow
*Chondestes grammacus* Lark Sparrow
*Amphispiza bilineata* Black-throated Sparrow
*Amphispiza belli* Sage Sparrow
*Passerculus sandwichensis* Savannah Sparrow
*Ammodramus savannarum* Grasshopper Sparrow
*Passerella iliaca* Fox Sparrow
*Melospiza melodia* Song Sparrow
*Melospiza lincolnii* Lincoln's Sparrow
*Zonotrichia leucophrys* White-crowned Sparrow
*Junco hyemalis* Dark-eyed Junco
*Dolichonyx oryzivorus* Bobolink
*Agelaius phoeniceus* Red-winged Blackbird
*Agelaius tricolor* Tricolored Blackbird
*Sturnella neglecta* Western Meadowlark
*Xanthocephalus xanthocephalus* Yellow-headed Blackbird
*Euphagus cyanocephalus* Brewer's Blackbird
*Molothrus ater* Brown-headed Cowbird
*Icterus bullockii* Bullock's Oriole
*Leucosticte atrata* Black Rosy-Finch
*Leucosticte tephrocotis* Gray-crowned Rosy-Finch
*Pinicola enucleator* Pine Grosbeak
*Carpodacus purpureus* Purple Finch
*Carpodacus cassinii* Cassin's Finch
*Carpodacus mexicanus* House Finch
*Loxia curvirostra* Red Crossbill
*Carduelis pinus* Pine Siskin
*Carduelis psaltria* Lesser Goldfinch
*Carduelis tristis* American Goldfinch
*Coccothraustes vespertinus* Evening Grosbeak
*Passer domesticus* House Sparrow

## Mammals

*Didelphis virginiana* Virginia Opossum
*Sorex preblei* Preble's Shrew
*Sorex vagrans* Vagrant Shrew
*Sorex monticolus* Montane Shrew
*Sorex bairdi* Baird's Shrew
*Sorex sonomae* Fog Shrew

*Sorex pacificus* Pacific Shrew
*Sorex palustris* Water Shrew
*Sorex bendirii* Pacific Marsh Shrew
*Sorex trowbridgii* Trowbridge's Shrew
*Sorex merriami* Merriam's Shrew
*Neurotrichus gibbsii* Shrew-mole
*Scapanus townsendii* Townsend's Mole
*Scapanus orarius* Coast Mole
*Scapanus latimanus* Broad-footed Mole
*Myotis californicus* California Myotis
*Myotis ciliolabrum* Western Small-footed Myotis
*Myotis yumanensis* Yuma Myotis
*Myotis lucifugus* Little Brown Myotis
*Myotis volans* Long-legged Myotis
*Myotis thysanodes* Fringed Myotis
*Myotis evotis* Long-eared Myotis
*Lasionycteris noctivagans* Silver-haired Bat
*Pipistrellus hesperus* Western Pipistrelle
*Eptesicus fuscus* Big Brown Bat
*Lasiurus cinereus* Hoary Bat
*Euderma maculatum* Spotted Bat
*Plecotus townsendii* Townsend's Big-eared Bat
*Antrozous pallidus* Pallid Bat
*Tadarida brasiliensis* Brazilian Free-tailed Bat
*Ochotona princeps* American Pika
*Brachylagus idahoensis* Pygmy Rabbit
*Sylvilagus bachmani* Brush Rabbit
*Sylvilagus floridanus* Eastern Cottontail
*Sylvilagus nuttallii* Mountain Cottontail
*Lepus americanus* Snowshoe Hare
*Lepus townsendii* White-tailed Jackrabbit
*Lepus californicus* Black-tailed Jackrabbit
*Aplodontia rufa* Mountain Beaver
*Tamias minimus* Least Chipmunk
*Tamias amoenus* Yellow-pine Chipmunk
*Tamias townsendii* Townsend's Chipmunk
*Tamias senex* (formerly part of *townsendii*) Allen's Chipmunk
*Tamias siskiyou* (formerly part of *townsendii*) Siskiyou
  Chipmunk
*Marmota flaviventris* Yellow-bellied Marmot
*Ammospermophilus leucurus* White-tailed Antelope Squirrel
*Spermophilus townsendii* Townsend's Ground Squirrel
*Spermophilus washingtoni* Washington Ground Squirrel
*Spermophilus beldingi* Belding's Ground Squirrel
*Spermophilus columbianus* Columbian Ground Squirrel
*Spermophilus beecheyi* California Ground Squirrel
*Spermophilus lateralis* Golden-mantled Ground Squirrel
*Sciurus niger* Eastern Fox Squirrel
*Sciurus griseus* Western Gray Squirrel
*Tamiasciurus hudsonicus* Red Squirrel
*Tamiasciurus douglasii* Douglas' Squirrel
*Glaucomys sabrinus* Northern Flying Squirrel
*Thomomys talpoides* Northern Pocket Gopher
*Thomomys mazama* Western Pocket Gopher
*Thomomys bulbivorus* Camas Pocket Gopher
*Thomomys bottae* Botta's Pocket Gopher
*Thomomys townsendii* Townsend's Pocket Gopher
*Perognathus parvus* Great Basin Pocket Mouse
*Perognathus longimembris* Little Pocket Mouse
*Microdipodops megacephalus* Dark Kangaroo Mouse
*Dipodomys ordii* Ord's Kangaroo Rat
*Dipodomys microps* Chisel-toothed Kangaroo Rat

*Dipodomys californicus* California Kangaroo Rat
*Castor canadensis* American Beaver
*Reithrodontomys megalotis* Western Harvest Mouse
*Peromyscus maniculatus* Deer Mouse
*Peromyscus crinitus* Canyon Mouse
*Peromyscus truei* Piñon Mouse
*Onychomys leucogaster* Northern Grasshopper Mouse
*Neotoma lepida* Desert Woodrat
*Neotoma fuscipes* Dusky-footed Woodrat
*Neotoma cinerea* Bushy-tailed Woodrat
*Clethrionomys gapperi* Southern Red-backed Vole
*Clethrionomys californicus* Western Red-backed Vole
*Phenacomys intermedius* Heather Vole
*Phenacomys albipes* White-footed Vole
*Phenacomys longicaudus* Red Tree Vole
*Microtus montanus* Montane Vole
*Microtus canicaudus* Gray-tailed Vole
*Microtus californicus* California Vole
*Microtus townsendii* Townsend's Vole
*Microtus longicaudus* Long-tailed Vole
*Microtus oregoni* Creeping Vole
*Microtus richardsoni* Water Vole
*Lemmiscus curtatus* Sagebrush Vole
*Ondatra zibethicus* Muskrat
*Rattus norvegicus* Norway Rat
*Mus musculus* House Mouse
*Zapus princeps* Western Jumping Mouse
*Zapus trinotatus* Pacific Jumping Mouse
*Erethizon dorsatum* Common Porcupine
*Myocastor coypus* Nutria or Coypu
*Canis latrans* Coyote
*Vulpes vulpes* Red Fox
*Vulpes macrotis* Kit or Swift Fox
*Urocyon cinereoargenteus* Common Gray Fox
*Ursus americanus* Black Bear
*Bassariscus astutus* Ringtail
*Procyon lotor* Common Raccoon
*Martes americana* American Marten
*Martes pennanti* Fisher
*Mustela erminea* Ermine
*Mustela frenata* Long-tailed Weasel
*Mustela vison* Mink
*Gulo gulo* Wolverine
*Taxidea taxus* American Badger
*Spilogale gracilis* Western Spotted Skunk
*Mephitis mephitis* Striped Skunk
*Lutra canadensis* Northern River Otter
*Felis concolor* Mountain Lion
*Lynx canadensis* Lynx
*Lynx rufus* Bobcat
*Cervus elaphus* Wapiti or Elk
*Odocoileus hemionus* Mule Deer (includes Black-tailed Deer)
*Odocoileus virginianus* White-tailed Deer
*Antilocapra americana* Pronghorn
*Ovis canadensis* Mountain Sheep

*Gavia stellata* Red-throated Loon
*Gavia immer* Common Loon
*Gavia adamsii* Yellow-billed Loon
*Gavia pacifica* Pacific Loon
*Podilymbus podiceps* Pied-billed Grebe
*Podiceps auritus* Horned Grebe
*Podiceps grisegena* Red-necked Grebe
*Podiceps nigricollis* Eared Grebe
*Aechmophorus occidentalis* Western Grebe
*Aechmophorus clarkii* Clark's Grebe
*Phalacrocorax auritus* Double-crested Cormorant
*Phalacrocorax penicillatus* Brandt's Cormorant
*Phalacrocorax pelagicus* Pelagic Cormorant
*Botaurus lentiginosus* American Bittern
*Ardea herodias* Great Blue Heron
*Ardea alba* Great Egret
*Egretta thula* Snowy Egret
*Bubulcus ibis* Cattle Egret
*Butorides virescens* Green Heron
*Nycticorax nycticorax* Black-crowned Night-Heron
*Cygnus columbianus* Tundra Swan
*Cygnus buccinator* Trumpeter Swan
*Anser albifrons* Greater White-fronted Goose
*Chen caerulescens* Snow Goose
*Chen rossii* Ross' Goose
*Chen canagica* Emperor Goose
*Branta bernicla* Brant
*Branta canadensis* Canada Goose
*Aix sponsa* Wood Duck
*Anas crecca* Green-winged Teal
*Anas platyrhynchos* Mallard
*Anas acuta* Northern Pintail
*Anas clypeata* Northern Shoveler
*Anas strepera* Gadwall
*Anas penelope* Eurasian Wigeon
*Anas americana* American Wigeon
*Aythya valisineria* Canvasback
*Aythya americana* Redhead
*Aythya collaris* Ring-necked Duck
*Aythya marila* Greater Scaup
*Aythya affinis* Lesser Scaup
*Histrionicus histrionicus* Harlequin Duck
*Clangula hyemalis* Oldsquaw
*Melanitta nigra* Black Scoter
*Melanitta perspicillata* Surf Scoter
*Melanitta fusca* White-winged Scoter
*Bucephala clangula* Common Goldeneye
*Bucephala islandica* Barrow's Goldeneye
*Bucephala albeola* Bufflehead
*Lophodytes cucullatus* Hooded Merganser
*Mergus merganser* Common Merganser
*Mergus serrator* Red-breasted Merganser
*Oxyura jamaicensis* Ruddy Duck
*Elanus leucurus* White-tailed Kite
*Haliaeetus leucocephalus* Bald Eagle
*Circus cyaneus* Northern Harrier

*Accipiter striatus* Sharp-shinned Hawk
*Accipiter cooperii* Cooper's Hawk
*Accipiter gentilis* Northern Goshawk
*Buteo lineatus* Red-shouldered Hawk
*Buteo jamaicensis* Red-tailed Hawk
*Buteo regalis* Ferruginous Hawk
*Buteo lagopus* Rough-legged Hawk
*Aquila chrysaetos* Golden Eagle
*Falco sparverius* American Kestrel
*Falco columbarius* Merlin
*Falco peregrinus* Peregrine Falcon
*Falco rusticolus* Gyrfalcon
*Falco mexicanus* Prairie Falcon
*Perdix perdix* Gray Partridge
*Alectoris chukar* Chukar
*Phasianus colchicus* Ring-necked Pheasant
*Dendragapus canadensis* Spruce Grouse
*Dendragapus obscurus* Blue Grouse
*Bonasa umbellus* Ruffed Grouse
*Centrocercus urophasianus* Sage Grouse
*Meleagris gallopavo* Wild Turkey
*Colinus virginianus* Northern Bobwhite
*Callipepla californica* California Quail
*Oreotryx pictus* Mountain Quail
*Rallus limicola* Virginia Rail
*Porzana carolina* Sora
*Fulica americana* American Coot
*Pluvialis squatarola* Black-bellied Plover
*Charadrius alexandrinus* Snowy Plover
*Charadrius vociferus* Killdeer
*Haematopus bachmani* Black Oystercatcher
*Tringa melanoleuca* Greater Yellowlegs
*Catoptrophorus semipalmatus* Willet
*Actitis macularia* Spotted Sandpiper
*Numenius phaeopus* Whimbrel
*Arenaria interpres* Ruddy Turnstone
*Arenaria melanocephala* Black Turnstone
*Aphriza virgata* Surfbird
*Calidris alba* Sanderling
*Calidris mauri* Western Sandpiper
*Calidris minutilla* Least Sandpiper
*Calidris ptilocnemis* Rock Sandpiper
*Calidris alpina* Dunlin
*Limnodromus scolopaceus* Long-billed Dowitcher
*Gallinago gallinago* Common Snipe
*Larus canus* Mew Gull
*Larus delawarensis* Ring-billed Gull
*Larus californicus* California Gull
*Larus argentatus* Herring Gull
*Larus thayeri* Thayer's Gull
*Larus occidentalis* Western Gull
*Larus glaucescens* Glaucous-winged Gull
*Larus hyperboreus* Glaucous Gull
*Rissa tridactyla* Black-legged Kittiwake
*Uria aalge* Common Murre
*Cepphus columba* Pigeon Guillemot
*Brachyramphus marmoratus* Marbled Murrelet

**475**

*Synthliboramphus antiquus* Ancient Murrelet
*Ptychoramphus aleuticus* Cassin's Auklet
*Cerorhinca monocerata* Rhinoceros Auklet
*Columba livia* Rock Dove
*Columba fasciata* Band-tailed Pigeon
*Zenaida macroura* Mourning Dove
*Tyto alba* Barn Owl
*Otus kennicottii* Western Screech-Owl
*Bubo virginianus* Great Horned Owl
*Nyctea scandiaca* Snowy Owl
*Glaucidium gnoma* Northern Pygmy-Owl
*Athene cunicularia* Burrowing Owl
*Strix occidentalis* Spotted Owl
*Strix varia* Barred Owl
*Strix nebulosa* Great Gray Owl
*Asio otus* Long-eared Owl
*Asio flammeus* Short-eared Owl
*Aegolius funereus* Boreal Owl
*Aegolius acadicus* Northern Saw-whet Owl
*Calypte anna* Anna's Hummingbird
*Ceryle alcyon* Belted Kingfisher
*Melanerpes lewis* Lewis' Woodpecker
*Melanerpes formicivorus* Acorn Woodpecker
*Sphyrapicus ruber* Red-breasted Sapsucker
*Sphyrapicus thyroideus* Williamson's Sapsucker
*Picoides pubsecens* Downy Woodpecker
*Picoides villosus* Hairy Woodpecker
*Picoides albolarvatus* White-headed Woodpecker
*Picoides tridactylus* Three-toed Woodpecker
*Picoides arcticus* Black-backed Woodpecker
*Colaptes auratus* Northern Flicker
*Dryocopus pileatus* Pileated Woodpecker
*Sayornis nigricans* Black Phoebe
*Eremophila alpestris* Horned Lark
*Perisoreus canadensis* Gray Jay
*Cyanocitta stelleri* Steller's Jay
*Aphelocoma californica* Western Scrub-Jay
*Gymnorhinus cyanocephalus* Pinyon Jay
*Nucifraga columbiana* Clark's Nutcracker
*Pica pica* Black-billed Magpie
*Corvus brachyrhynchos* American Crow
*Corvus corax* Common Raven
*Parus atricapillus* Black-capped Chickadee
*Parus gambeli* Mountain Chickadee
*Parus rufescens* Chestnut-backed Chickadee
*Parus inornatus* Plain Titmouse
*Psaltriparus minimus* Bushtit
*Sitta canadensis* Red-breasted Nuthatch
*Sitta carolinensis* White-breasted Nuthatch
*Sitta pygmaea* Pygmy Nuthatch
*Certhia americana* Brown Creeper
*Salpinctes obsoletus* Rock Wren
*Catherpes mexicanus* Canyon Wren
*Thryomanes bewickii* Bewick's Wren
*Troglodytes troglodytes* Winter Wren
*Cistothorus palustris* Marsh Wren
*Cinclus mexicanus* American Dipper
*Regulus satrapa* Golden-crowned Kinglet
*Regulus calendula* Ruby-crowned Kinglet
*Sialia mexicana* Western Bluebird
*Sialia currucoides* Mountain Bluebird
*Myadestes townsendi* Townsend's Solitaire

*Catharus guttatus* Hermit Thrush
*Turdus migratorius* American Robin
*Ixoreus naevius* Varied Thrush
*Chamaea fasciata* Wrentit
*Oreoscoptes montanus* Sage Thrasher
*Anthus rubescens* American Pipit
*Bombycilla garrulus* Bohemian Waxwing
*Bombycilla cedrorum* Cedar Waxwing
*Lanius excubitor* Northern Shrike
*Lanius ludovicianus* Loggerhead Shrike
*Sturnus vulgaris* European Starling
*Vireo huttoni* Hutton's Vireo
*Vermivora celata* Orange-crowned Warbler
*Dendroica coronata* Yellow-rumped Warbler
*Dendroica townsendi* Townsend's Warbler
*Pipilo maculatus* Spotted Towhee
*Pipilo crissalis* California Towhee
*Spizella arborea* American Tree Sparrow
*Spizella pallida* Clay-colored Sparrow
*Passerculus sandwichensis* Savannah Sparrow
*Passerella iliaca* Fox Sparrow
*Melospiza melodia* Song Sparrow
*Melospiza lincolnii* Lincoln's Sparrow
*Melospiza georgiana* Swamp Sparrow
*Zonotrichia albicollis* White-throated Sparrow
*Zonotrichia atricapilla* Golden-crowned Sparrow
*Zonotrichia leucophrys* White-crowned Sparrow
*Zonotrichia querula* Harris' Sparrow
*Junco hyemalis* Dark-eyed Junco
*Calcarius lapponicus* Lapland Longspur
*Plectrophenax nivalis* Snow Bunting
*Agelaius phoeniceus* Red-winged Blackbird
*Agelaius tricolor* Tricolored Blackbird
*Xanthocephalus xanthocephalus* Yellow-headed Blackbird
*Sturnella neglecta* Western Meadowlark
*Euphagus cyanocephalus* Brewer's Blackbird
*Molothrus ater* Brown-headed Cowbird
*Leucosticte atrata* Black Rosy-Finch
*Leuctosticte tephrocotis* Gray-crowned Rosy-Finch
*Pinicola enucleator* Pine Grosbeak
*Carpodacus purpureus* Purple Finch
*Carpodacus cassinii* Cassin's Finch
*Carpodacus mexicanus* House Finch
*Loxia curvirostra* Red Crossbill
*Loxia leucoptera* White-winged Crossbill
*Carduelis flammea* Common Redpoll
*Carduelis pinus* Pine Siskin
*Carduelis psaltria* Lesser Goldfinch
*Carduelis tristis* American Goldfinch
*Coccothraustes vespertinus* Evening Grosbeak
*Passer domesticus* House Sparrow

Hexagons are shaded to reflect the degree of confidence with which we predict species presence:
black = highly confident; dark gray = probable; light gray = possible

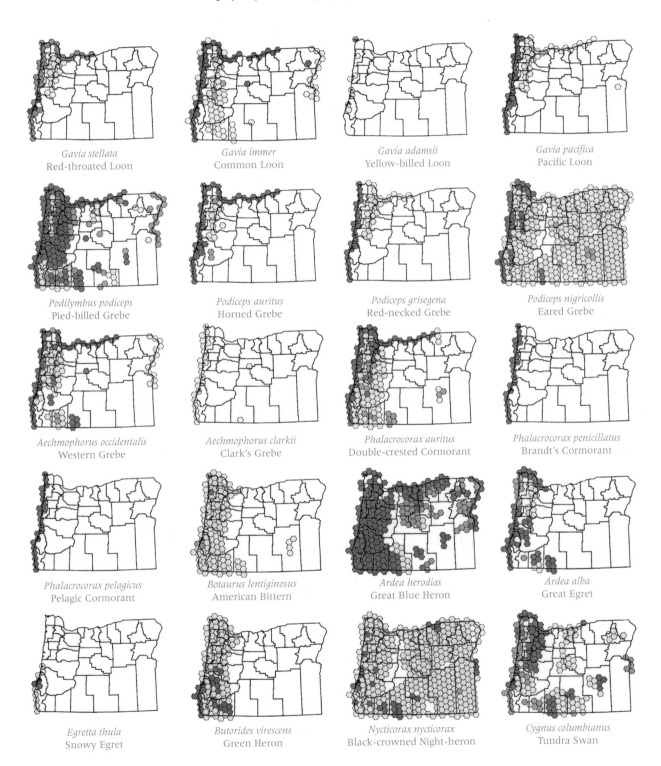

*Gavia stellata*
Red-throated Loon

*Gavia immer*
Common Loon

*Gavia adamsii*
Yellow-billed Loon

*Gavia pacifica*
Pacific Loon

*Podilymbus podiceps*
Pied-billed Grebe

*Podiceps auritus*
Horned Grebe

*Podiceps grisegena*
Red-necked Grebe

*Podiceps nigricollis*
Eared Grebe

*Aechmophorus occidentalis*
Western Grebe

*Aechmophorus clarkii*
Clark's Grebe

*Phalacrocorax auritus*
Double-crested Cormorant

*Phalacrocorax penicillatus*
Brandt's Cormorant

*Phalacrocorax pelagicus*
Pelagic Cormorant

*Botaurus lentiginosus*
American Bittern

*Ardea herodias*
Great Blue Heron

*Ardea alba*
Great Egret

*Egretta thula*
Snowy Egret

*Butorides virescens*
Green Heron

*Nycticorax nycticorax*
Black-crowned Night-heron

*Cygnus columbianus*
Tundra Swan

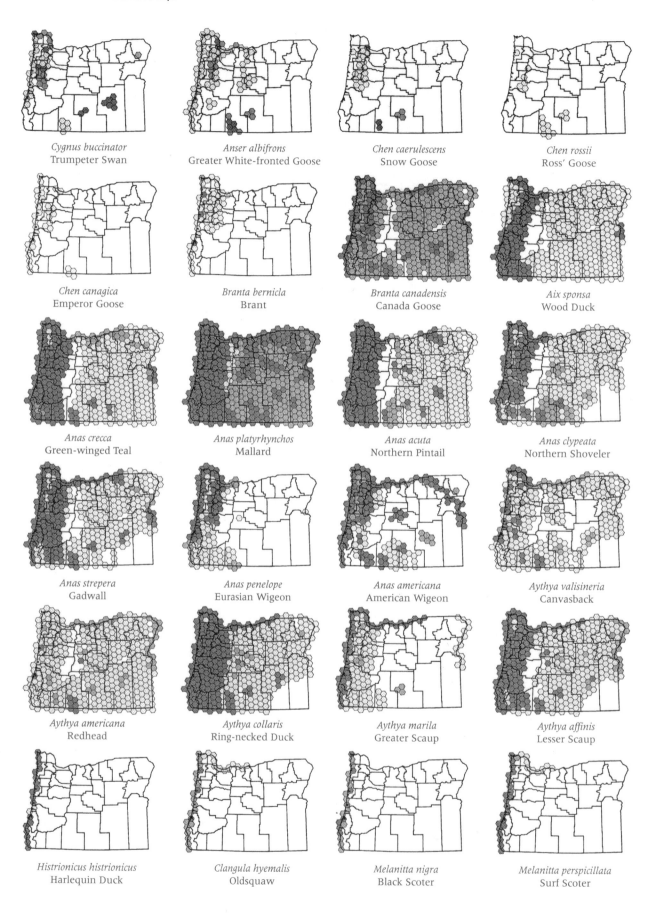

*Cygnus buccinator*
Trumpeter Swan

*Anser albifrons*
Greater White-fronted Goose

*Chen caerulescens*
Snow Goose

*Chen rossii*
Ross' Goose

*Chen canagica*
Emperor Goose

*Branta bernicla*
Brant

*Branta canadensis*
Canada Goose

*Aix sponsa*
Wood Duck

*Anas crecca*
Green-winged Teal

*Anas platyrhynchos*
Mallard

*Anas acuta*
Northern Pintail

*Anas clypeata*
Northern Shoveler

*Anas strepera*
Gadwall

*Anas penelope*
Eurasian Wigeon

*Anas americana*
American Wigeon

*Aythya valisineria*
Canvasback

*Aythya americana*
Redhead

*Aythya collaris*
Ring-necked Duck

*Aythya marila*
Greater Scaup

*Aythya affinis*
Lesser Scaup

*Histrionicus histrionicus*
Harlequin Duck

*Clangula hyemalis*
Oldsquaw

*Melanitta nigra*
Black Scoter

*Melanitta perspicillata*
Surf Scoter

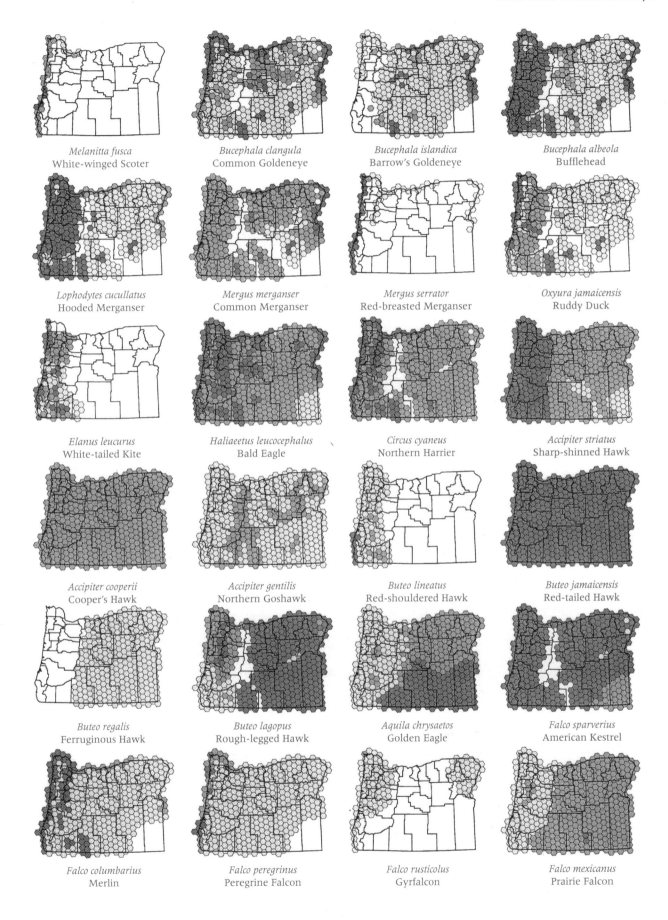

*Melanitta fusca*
White-winged Scoter

*Bucephala clangula*
Common Goldeneye

*Bucephala islandica*
Barrow's Goldeneye

*Bucephala albeola*
Bufflehead

*Lophodytes cucullatus*
Hooded Merganser

*Mergus merganser*
Common Merganser

*Mergus serrator*
Red-breasted Merganser

*Oxyura jamaicensis*
Ruddy Duck

*Elanus leucurus*
White-tailed Kite

*Haliaeetus leucocephalus*
Bald Eagle

*Circus cyaneus*
Northern Harrier

*Accipiter striatus*
Sharp-shinned Hawk

*Accipiter cooperii*
Cooper's Hawk

*Accipiter gentilis*
Northern Goshawk

*Buteo lineatus*
Red-shouldered Hawk

*Buteo jamaicensis*
Red-tailed Hawk

*Buteo regalis*
Ferruginous Hawk

*Buteo lagopus*
Rough-legged Hawk

*Aquila chrysaetos*
Golden Eagle

*Falco sparverius*
American Kestrel

*Falco columbarius*
Merlin

*Falco peregrinus*
Peregrine Falcon

*Falco rusticolus*
Gyrfalcon

*Falco mexicanus*
Prairie Falcon

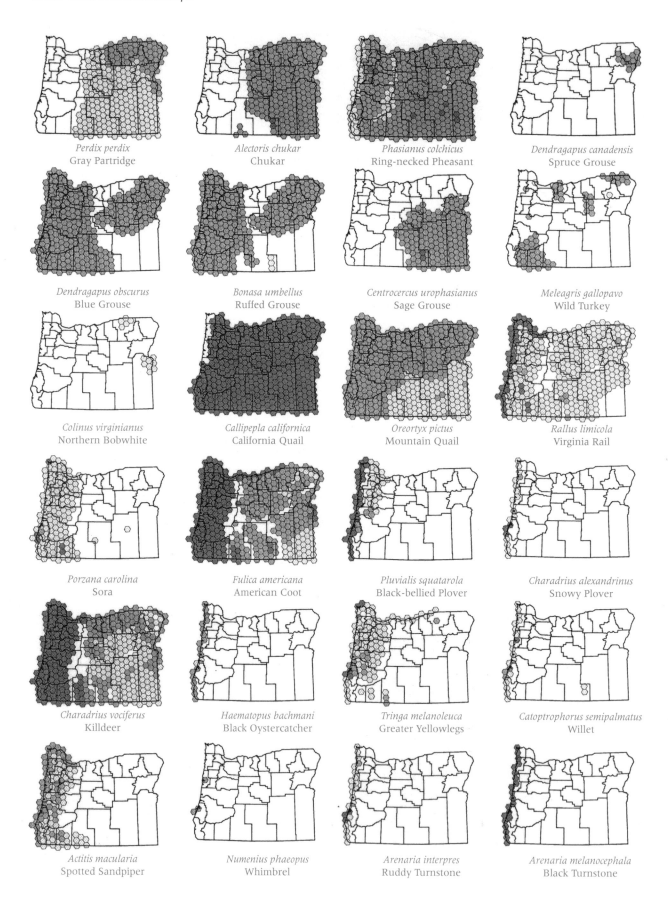

*Perdix perdix*
Gray Partridge

*Alectoris chukar*
Chukar

*Phasianus colchicus*
Ring-necked Pheasant

*Dendragapus canadensis*
Spruce Grouse

*Dendragapus obscurus*
Blue Grouse

*Bonasa umbellus*
Ruffed Grouse

*Centrocercus urophasianus*
Sage Grouse

*Meleagris gallopavo*
Wild Turkey

*Colinus virginianus*
Northern Bobwhite

*Callipepla californica*
California Quail

*Oreortyx pictus*
Mountain Quail

*Rallus limicola*
Virginia Rail

*Porzana carolina*
Sora

*Fulica americana*
American Coot

*Pluvialis squatarola*
Black-bellied Plover

*Charadrius alexandrinus*
Snowy Plover

*Charadrius vociferus*
Killdeer

*Haematopus bachmani*
Black Oystercatcher

*Tringa melanoleuca*
Greater Yellowlegs

*Catoptrophorus semipalmatus*
Willet

*Actitis macularia*
Spotted Sandpiper

*Numenius phaeopus*
Whimbrel

*Arenaria interpres*
Ruddy Turnstone

*Arenaria melanocephala*
Black Turnstone

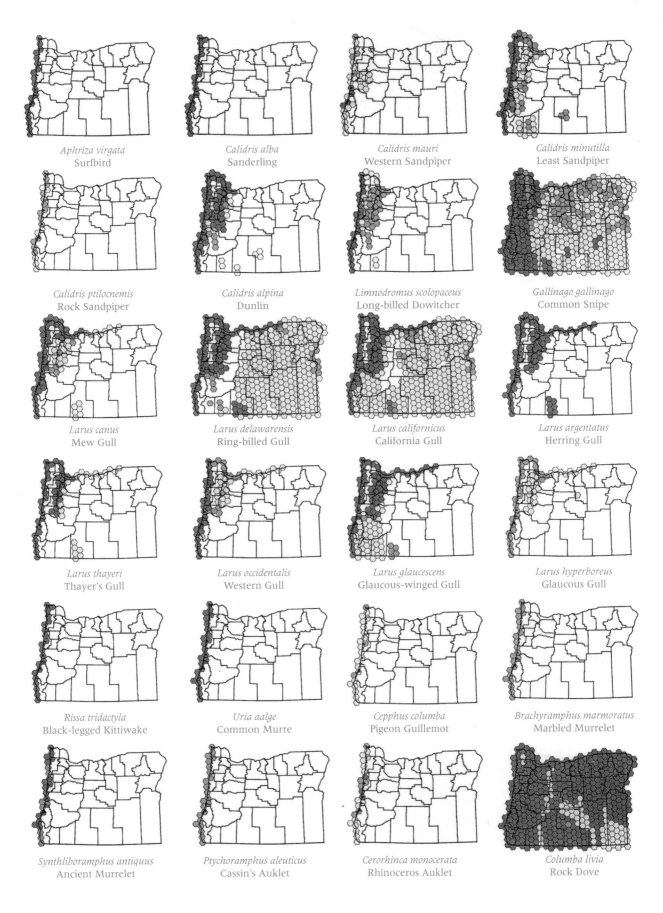

*Aphriza virgata*
Surfbird

*Calidris alba*
Sanderling

*Calidris mauri*
Western Sandpiper

*Calidris minutilla*
Least Sandpiper

*Calidris ptilocnemis*
Rock Sandpiper

*Calidris alpina*
Dunlin

*Limnodromus scolopaceus*
Long-billed Dowitcher

*Gallinago gallinago*
Common Snipe

*Larus canus*
Mew Gull

*Larus delawarensis*
Ring-billed Gull

*Larus californicus*
California Gull

*Larus argentatus*
Herring Gull

*Larus thayeri*
Thayer's Gull

*Larus occidentalis*
Western Gull

*Larus glaucescens*
Glaucous-winged Gull

*Larus hyperboreus*
Glaucous Gull

*Rissa tridactyla*
Black-legged Kittiwake

*Uria aalge*
Common Murre

*Cepphus columba*
Pigeon Guillemot

*Brachyramphus marmoratus*
Marbled Murrelet

*Synthliboramphus antiquus*
Ancient Murrelet

*Ptychoramphus aleuticus*
Cassin's Auklet

*Cerorhinca monocerata*
Rhinoceros Auklet

*Columba livia*
Rock Dove

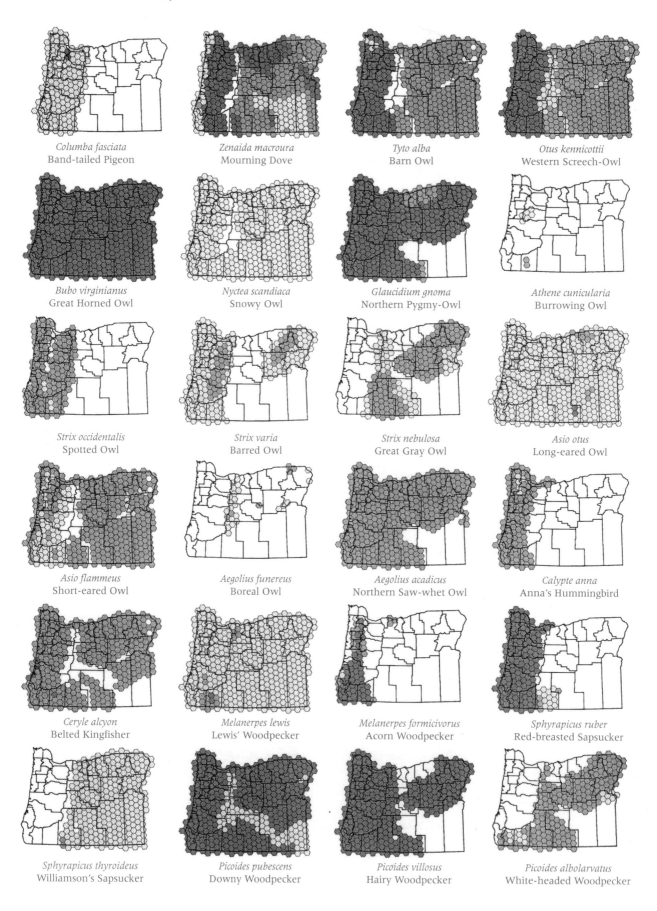

*Columba fasciata*
Band-tailed Pigeon

*Zenaida macroura*
Mourning Dove

*Tyto alba*
Barn Owl

*Otus kennicottii*
Western Screech-Owl

*Bubo virginianus*
Great Horned Owl

*Nyctea scandiaca*
Snowy Owl

*Glaucidium gnoma*
Northern Pygmy-Owl

*Athene cunicularia*
Burrowing Owl

*Strix occidentalis*
Spotted Owl

*Strix varia*
Barred Owl

*Strix nebulosa*
Great Gray Owl

*Asio otus*
Long-eared Owl

*Asio flammeus*
Short-eared Owl

*Aegolius funereus*
Boreal Owl

*Aegolius acadicus*
Northern Saw-whet Owl

*Calypte anna*
Anna's Hummingbird

*Ceryle alcyon*
Belted Kingfisher

*Melanerpes lewis*
Lewis' Woodpecker

*Melanerpes formicivorus*
Acorn Woodpecker

*Sphyrapicus ruber*
Red-breasted Sapsucker

*Sphyrapicus thyroideus*
Williamson's Sapsucker

*Picoides pubescens*
Downy Woodpecker

*Picoides villosus*
Hairy Woodpecker

*Picoides albolarvatus*
White-headed Woodpecker

*Picoides tridactylus*
Three-toed Woodpecker

*Picoides arcticus*
Black-backed Woodpecker

*Colaptes auratus*
Northern Flicker

*Dryocopus pileatus*
Pileated Woodpecker

*Sayornis nigricans*
Black Phoebe

*Eremophila alpestris*
Horned Lark

*Perisoreus canadensis*
Gray Jay

*Cyanocitta stelleri*
Steller's Jay

*Aphelocoma californica*
Western Scrub-Jay

*Gymnorhinus cyanocephalus*
Pinyon Jay

*Nucifraga columbiana*
Clark's Nutcracker

*Pica pica*
Black-billed Magpie

*Corvus brachyrhynchos*
American Crow

*Corvus corax*
Common Raven

*Parus atricapillus*
Black-capped Chickadee

*Parus gambeli*
Mountain Chickadee

*Parus rufescens*
Chestnut-backed Chickadee

*Parus inornatus*
Plain Titmouse

*Psaltriparus minimus*
Bushtit

*Sitta canadensis*
Red-breasted Nuthatch

*Sitta carolinensis*
White-breasted Nuthatch

*Sitta pygmaea*
Pygmy Nuthatch

*Certhia americana*
Brown Creeper

*Salpinctes obsoletus*
Rock Wren

*Catherpes mexicanus*
Canyon Wren

*Thryomanes bewickii*
Bewick's Wren

*Troglodytes troglodytes*
Winter Wren

*Cistothorus palustris*
Marsh Wren

*Cinclus mexicanus*
American Dipper

*Regulus satrapa*
Golden-crowned Kinglet

*Regulus calendula*
Ruby-crowned Kinglet

*Sialia mexicana*
Western Bluebird

*Sialia currucoides*
Mountain Bluebird

*Myadestes townsendi*
Townsend's Solitaire

*Catharus guttatus*
Hermit Thrush

*Turdus migratorius*
American Robin

*Ixoreus naevius*
Varied Thrush

*Chamaea fasciata*
Wrentit

*Oreoscoptes montanus*
Sage Thrasher

*Anthus rubescens*
American Pipit

*Bombycilla garrulus*
Bohemian Waxwing

*Bombycilla cedrorum*
Cedar Waxwing (263)

*Lanius excubitor*
Northern Shrike

*Lanius ludovicianus*
Loggerhead Shrike

*Sturnus vulgaris*
European Starling

*Vireo huttoni*
Hutton's Vireo

*Vermivora celata*
Orange-crowned Warbler

*Dendroica coronata*
Yellow-rumped Warbler

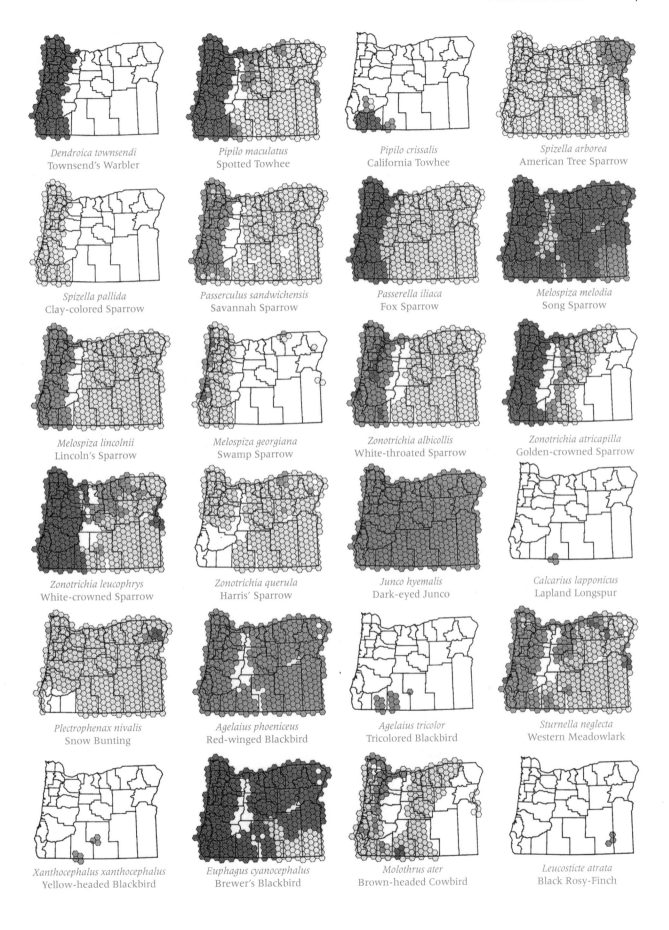

*Dendroica townsendi*
Townsend's Warbler

*Pipilo maculatus*
Spotted Towhee

*Pipilo crissalis*
California Towhee

*Spizella arborea*
American Tree Sparrow

*Spizella pallida*
Clay-colored Sparrow

*Passerculus sandwichensis*
Savannah Sparrow

*Passerella iliaca*
Fox Sparrow

*Melospiza melodia*
Song Sparrow

*Melospiza lincolnii*
Lincoln's Sparrow

*Melospiza georgiana*
Swamp Sparrow

*Zonotrichia albicollis*
White-throated Sparrow

*Zonotrichia atricapilla*
Golden-crowned Sparrow

*Zonotrichia leucophrys*
White-crowned Sparrow

*Zonotrichia querula*
Harris' Sparrow

*Junco hyemalis*
Dark-eyed Junco

*Calcarius lapponicus*
Lapland Longspur

*Plectrophenax nivalis*
Snow Bunting

*Agelaius phoeniceus*
Red-winged Blackbird

*Agelaius tricolor*
Tricolored Blackbird

*Sturnella neglecta*
Western Meadowlark

*Xanthocephalus xanthocephalus*
Yellow-headed Blackbird

*Euphagus cyanocephalus*
Brewer's Blackbird

*Molothrus ater*
Brown-headed Cowbird

*Leucosticte atrata*
Black Rosy-Finch

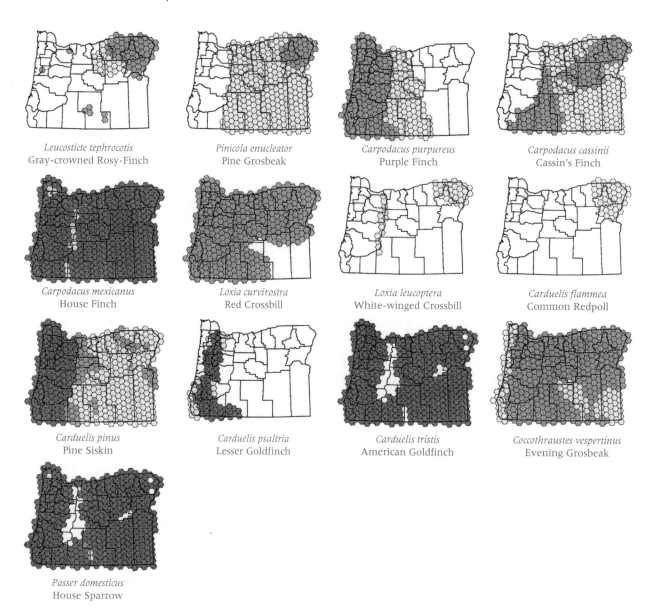

*Leucosticte tephrocotis*
Gray-crowned Rosy-Finch

*Pinicola enucleator*
Pine Grosbeak

*Carpodacus purpureus*
Purple Finch

*Carpodacus cassinii*
Cassin's Finch

*Carpodacus mexicanus*
House Finch

*Loxia curvirostra*
Red Crossbill

*Loxia leucoptera*
White-winged Crossbill

*Carduelis flammea*
Common Redpoll

*Carduelis pinus*
Pine Siskin

*Carduelis psaltria*
Lesser Goldfinch

*Carduelis tristis*
American Goldfinch

*Coccothraustes vespertinus*
Evening Grosbeak

*Passer domesticus*
House Sparrow

Academy of Natural Sciences, Philadephia
Monte L. Bean Museum, Brigham Young University
Burke Museum, University of Washington
California Academy of Sciences
Carnegie Museum of Natural History
Charles R. Connor Museum, Washington State University
Cowan Vertebrate Museum, University of British Columbia
Department of Biology, Portland State University
Department of Biology, University of California, Los
 Angeles
Department of Biology, University of Idaho
Department of Biology, Willamette University
Department of Fisheries and Wildlife, Oregon State
 University
Douglas County Museum of Natural History
Eastern Oregon State College
Linfield College, McMinnville, Oregon
Museum of Comparative Zoology, Harvard University
Museum of Natural History, University of Kansas
Museum of Southwestern Biology, University of New
 Mexico
Museum of Vertebrate Zoology, University of California,
 Berkeley
Museum of Zoology, Iowa State University
Museum of Zoology, University of Michigan
National Museum of Natural History, Smithsonian
 Institution
Natural History Museum of Los Angeles County
North Carolina State Museum of Natural Sciences
Oklahoma Museum of Natural History
Peabody Museum of Natural History, Yale University
Royal Ontario Museum, Toronto
San Diego Museum of Natural History
Southern Oregon State College
James R. Slater Museum of Natural History, University of
 Puget Sound
Tillamook County Museum
University of Arkansas Museum
University of Nebraska State Museum
Utah Museum of Natural History, University of Utah
Western Foundation of Vertebrate Zoology, Camarillo,
 California

# Index